Problems Manual

for

Grob's Basic Electronics

Twelfth Edition

Mitchel E. Schultz
Western Technical College

PROBLEMS MANUAL FOR GROB'S BASIC ELECTRONICS, TWELFTH EDITION

Published by McGraw-Hill Education, 2 Penn Plaza, New York, NY 10121.

This book is printed on acid-free paper.

3 4 5 6 7 8 QVS 22 21 20 19 18

ISBN 978-1-259-19044-5
MHID 1-259-19044-7

Senior Vice President, Products & Markets: *Kurt L. Strand*
Vice President, General Manager, Products & Markets: *Marty Lange*
Vice President, Content Design & Delivery: *Kimberly Meriwether David*
Managing Director: *Thomas Timp*
Global Publisher: *Raghu Srinivasan*
Director, Product Development: *Rose Koos*
Director, Digital Content: *Thomas Scaife, Ph.D*
Product Developer: *Vincent Bradshaw*
Marketing Manager: *Nick McFadden*
Director, Content Design & Delivery: *Linda Avenarius*
Program Manager: *Faye M. Herrig*
Content Project Managers: *Kelly Hart, Tammy Juran, Sandra Schnee*
Buyer: *Michael F. McCormick*
Design: *Studio Montage, St. Louis, MO*
Content Licensing Specialist: *DeAnna Dausener*
Cover Image: © *Getty Images/RF*
Compositor: *MPS Limited*
Typeface: *10/12 Times Roman*
Printer: *Quad/Graphics, Inc.*

www.mhhe.com

Contents

Preface

The *Problems Manual for Grob's Basic Electronics* is designed to provide students and instructors with a source of hundreds of practical problems for self-study, homework assignments, tests, and review. Every major topic of a course in basic electronics has been covered. Although the book was written to accompany *Grob's Basic Electronics,* its contents are universal and applicable to any introductory text in electricity and electronics.

The chapters in this book exactly parallel the first 26 chapters of *Grob's Basic Electronics,* twelfth edition. However, some chapters in this book provide expanded coverage of the topics presented in the textbook.

Each chapter contains a section of solved illustrative examples demonstrating, step-by-step, how representative problems on a particular topic are solved. Following these examples are sets of problems for the students to solve. Troubleshooting problems are included in appropriate areas throughout the book. Such problems fill a long-standing need in the traditional DC/AC coverage. Practical experience in electricity and electronics is emphasized by using standard component values in most problems. Not only does the student gain knowledge about these standard values, but their use also allows many of the circuits to be constructed so that problem solutions can be verified and further circuit behavior studied. New to the twelfth edition is a short true/false test at the end of every chapter in the book. Each end-of-chapter test quizzes students on the general content presented in the chapter. The answers to all of the end-of-chapter tests are at the back of the book.

The abundance of graded problem material provides a wide choice for student assignments; the huge selection also means that short-term repetition of assignments can be successfully avoided.

I would like to thank Susan Bye Bernau for her many long hours of hard work in the preparation of the original manuscript for this book. Without her help, the first edition of this book may have never happened. I would also like to thank Cathy Welch for her contributions in typing manuscript for this book. Your help has been greatly appreciated. And finally, I would like to thank my lovely wife, Sheryl, for her tremendous support and patience during the long period of manuscript preparation.

Mitchel E. Schultz

BIBLIOGRAPHY

SCHULTZ, MITCHEL E.: *Grob's Basic Electronics,* 12th Edition, McGraw-Hill, Dubuque, Iowa.

Introduction to Powers of 10

In the field of electronics, the magnitudes of the various units are often extremely small or extremely large. For example, in electronics it is not at all uncommon to work with extremely small decimal fractions such as 0.000000000047 or extremely large numbers such as 100,000,000. To enable us to work conveniently with both very small and very large numbers, powers of 10 notation is used. *Powers of ten notation* allows us to express any number, no matter how large or how small, as a decimal number multiplied by a power of 10. A power of 10 is an exponent written above and to the right of 10, which is called the base. The power or exponent indicates how many times the base is to be multiplied by itself. For example, 10^2 means 10×10 and 10^4 means $10 \times 10 \times 10 \times 10$. In electronics, the base 10 is common because multiples of 10 are used in the metric system of units. As you will see, powers of 10 allow us to keep track of the decimal point when working with extremely small and extremely large numbers.

Positive powers of 10 are used to indicate numbers greater than 1, and negative powers of 10 are used to indicate numbers less than 1. Table I-1 shows the powers of 10 ranging from 10^{-12} to 10^9. As you will discover in your study of electronics, seldom will you encounter powers of 10 which fall outside this range. From Table I-1 notice that $10^0 = 1$ and $10^1 = 10$. In the case of $10^0 = 1$, it is important to realize that any number raised to the zero power equals 1. In the case of $10^1 = 10$, it is important to realize that any number written without a power is assumed to have a power of 1.

TABLE I-1 POWERS OF 10

$1{,}000{,}000{,}000 = 10^9$	$10 = 10^1$	$0.000001 = 10^{-6}$
$100{,}000{,}000 = 10^8$	$1 = 10^0$	$0.0000001 = 10^{-7}$
$10{,}000{,}000 = 10^7$	$0.1 = 10^{-1}$	$0.00000001 = 10^{-8}$
$1{,}000{,}000 = 10^6$	$0.01 = 10^{-2}$	$0.000000001 = 10^{-9}$
$100{,}000 = 10^5$	$0.001 = 10^{-3}$	$0.0000000001 = 10^{-10}$
$10{,}000 = 10^4$	$0.0001 = 10^{-4}$	$0.00000000001 = 10^{-11}$
$1{,}000 = 10^3$	$0.00001 = 10^{-5}$	$0.000000000001 = 10^{-12}$
$100 = 10^2$		

SEC. I-1 SCIENTIFIC NOTATION

The procedure for using powers of 10 is to write the original number as two separate factors. Scientific notation is a special form of powers of 10 notation. Using scientific notation, any number can be expressed as a number between 1 and 10 times a power of 10. The power of 10 is used to place the decimal point correctly. In fact, the power of 10 indicates the number of places the decimal point has been moved to the left or right in the original number. If the decimal point is moved to the left in the original number, then the power of 10 will increase (become more positive). Conversely, if the decimal point is moved to the right in the original number then the power of 10 will decrease (become more negative).

Solved Problem

Express the numbers 27,000 and 0.000068 in scientific notation.

Answer

To express 27,000 in scientific notation the number must be expressed as a number between 1 and 10, which is 2.7 in this case, times a power of 10. To do this the decimal point must be shifted four places to the left. The number of places the decimal point has been moved to the left indicates the positive power of 10. Therefore, $27{,}000 = 2.7 \times 10^4$ in scientific notation.

To express 0.000068 in scientific notation the number must be expressed as a number between 1 and 10, which is 6.8 in this case, times a power of 10. This means the decimal point must be shifted five places to the right. The number of places the decimal point moves to the right indicates the negative power of 10. Therefore, $0.000068 = 6.8 \times 10^{-5}$ in scientific notation.

When expressing numbers in scientific notation, remember the following rules.

Rule 1 Express the number as a number between 1 and 10 times a power of 10.

Rule 2 When moving the decimal point to the left in the original number, make the power of 10 positive. When moving the decimal point to the right in the original number, make the power of 10 negative.

Rule 3 The power of 10 always equals the number of places the decimal point has been shifted to the left or right in the original number.

PRACTICE PROBLEMS

Sec. I-1 Express the following numbers in scientific notation.

1. 56,000
2. 1,200,000
3. 0.05
4. 0.000472
5. 0.000000000056
6. 120
7. 330,000
8. 8,200
9. 0.001
10. 0.000020
11. 0.00015
12. 4.7
13. 4,700,000,000
14. 0.246
15. 0.0055
16. 0.00096
17. 100,000
18. 150
19. 0.000066
20. 750,000
21. 5,000,000,000
22. 0.00000000000004
23. 215
24. 5,020
25. 0.658

Decimal Notation

Numbers that are written without powers of 10 are said to be written in decimal notation (sometimes referred to as floating decimal notation). In some cases it may be necessary to change a number written in scientific notation back into decimal notation. To convert from scientific notation to decimal notation use Rules 4 and 5.

Rule 4 If the exponent or power of 10 is positive, move the decimal point to the right, the same number of places as the exponent.

Rule 5 If the exponent or power of 10 is negative, move the decimal point to the left, the same number of places as the exponent.

Solved Problem

Convert the number 3.56×10^4 back into decimal notation.

Answer

Since the power of 10 is 4, the decimal point must be shifted four places to the right: 3.5600. Therefore, $3.56 \times 10^4 = 35{,}600$.

MORE PRACTICE PROBLEMS

Sec. I-1 Convert the following numbers expressed in scientific notation back into decimal notation.

1. 2.66×10^7
2. 3.75×10^{-8}
3. 5.51×10^{-2}
4. 1.67×10^4
5. 7.21×10^1
6. 2.75×10^{-3}
7. 1.36×10^{-11}
8. 4.4×10^9
9. 3.6×10^{-1}
10. 3.65×10^{-4}
11. 1.36×10^{-5}
12. 2.25×10^5
13. 7.56×10^8
14. 1.8×10^{-3}
15. 6.8×10^1
16. 5.5×10^2
17. 3.0×10^{-1}
18. 2.7×10^{-2}
19. 3.3×10^{-8}
20. 5.6×10^{-12}
21. 4.7×10^3
22. 1.27×10^3
23. 3.3×10^0
24. 2.33×10^{-4}
25. 4.7×10^6

SEC. I-2 ENGINEERING NOTATION AND METRIC PREFIXES

Another popular way of expressing very small and very large numbers is with engineering notation. Engineering notation is very much like scientific notation except that with engineering notation the powers of 10 are always multiples of 3 such as 10^{-12}, 10^{-9}, 10^{-6}, 10^{-3}, 10^3, 10^6, 10^9, 10^{12}, etc. More specifically, a number written in engineering notation is always written as a number between 1 and 1,000 times a power of 10 that is a multiple of 3.

Solved Problem

Express the number 330,000 in engineering notation.

Answer

To express the number 330,000 in engineering notation, it must be written as a number between 1 and 1,000 times a power of 10 that is a multiple of 3. It is often helpful to begin by expressing the number in scientific notation: $330{,}000 = 3.3 \times 10^5$. Next, examine the power of 10 to see if it should be increased to 10^6 or decreased to 10^3. If the power of 10 is increased to 10^6, then the decimal point in the number 3.3 would have to be moved one place to the left. Since 0.33 is not a number between 1 and 1,000, the answer of 0.33×10^6 is not representative of engineering notation. If the power of 10 were decreased to 10^3, however, then the decimal point in the number 3.3 would have to be moved two places to the right and the answer would be 330×10^3, which is representative of engineering notation. In summary: $330{,}000 = 3.3 \times 10^5 = 330 \times 10^3$.

Solved Problem

Express the number 0.000015 in engineering notation.

Answer

To express the number 0.000015 in engineering notation, it must be written as a number between 1 and 1,000 times a power of 10 that is a multiple of 3. Begin by expressing the number in scientific notation: $0.000015 = 1.5 \times 10^{-5}$. Next, examine the power of 10 to see if it should be increased to 10^{-3} or decreased to 10^{-6}. If the power of 10 were increased to 10^{-3}, then the decimal point in the number 1.5 would have to be moved two places to the left. Since 0.015 is not a number between 1 and 1,000, the answer of 0.015×10^{-3} is not representative of engineering notation. If the power of 10 were decreased to 10^{-6}, however, then the decimal point in the number 1.5 would have to be moved one place to the right and the answer would be 15×10^{-6}, which is representative of engineering notation. In summary: $0.000015 = 1.5 \times 10^{-5} = 15 \times 10^{-6}$.

PRACTICE PROBLEMS

Sec. I-2 Express the following numbers in engineering notation.

1. 47,200
2. 0.00047
3. 0.65
4. 22,000,000
5. 1,875
6. 39,000
7. 0.075
8. 0.00000055
9. 0.000000000082
10. 910,000
11. 1,680,000
12. 0.0072
13. 0.00065
14. 0.350
15. 12,500
16. 15,000,000,000,000
17. 0.000000000470
18. 0.0005
19. 0.00000033
20. 156,000
21. 68,000
22. 25,030,000
23. 0.000000000068
24. 0.057
25. 0.0088

Metric Prefixes

The metric prefixes represent the powers of 10 that are multiples of 3. In electronics, engineering notation is preferred over scientific notation because most values of voltage, current, resistance, power, etc. are specified in terms of the metric prefixes. Table I-2 lists the most common metric prefixes and their corresponding powers of 10. Notice that uppercase letters are used for the abbreviations for the prefixes involving positive powers of 10, and lowercase letters are used for negative powers of 10. There is one exception to the rule, however; the lowercase letter "k" is used for kilo, corresponding to 10^3. Because metric prefixes are used so often in electronics, it is common practice to express the value of a given quantity in engineering notation so that the power of 10 (that is a multiple of 3) can be replaced with its corresponding metric prefix. For example, a resistor whose value is 2,700 Ω can be expressed in engineering notation as 2.7×10^3 Ω. In Table I-2, we see that the metric prefix kilo (k) corresponds to 10^3. Therefore, 2,700 Ω or 2.7×10^3 Ω can be expressed as 2.7 kΩ. As another example, a current of 0.025 A can be expressed in engineering notation as 25×10^{-3} A. In Table I-2, we see that the metric prefix milli (m) corresponds to 10^{-3}. Therefore, 0.025 A or 25×10^{-3} A can be expressed as 25 mA. In general, when using the metric prefixes to express the value of a quantity, write the original number in engineering notation and then substitute the appropriate metric prefix corresponding to the power of 10 involved. As this procedure shows, the metric prefixes are just a substitute for the powers of 10 used in engineering notation.

TABLE I-2 METRIC PREFIXES

Power of 10	Prefix	Abbreviation
10^{12}	tera	T
10^{9}	giga	G
10^{6}	mega	M
10^{3}	kilo	k
10^{-3}	milli	m
10^{-6}	micro	μ
10^{-9}	nano	n
10^{-12}	pico	p

Table I-3 lists many of the electrical quantities that you will encounter in your study of electronics. For each electrical quantity listed in Table I-3, take special note of the units and symbols shown. In the practice problems that follow, we will use several numerical values with the various units from this table.

TABLE I-3 ELECTRICAL QUANTITIES WITH THEIR UNITS AND SYMBOLS

Quantity	Unit	Symbol
Current	Ampere (A)	I
Electromotive force (voltage)	Volt (V)	V
Resistance	Ohm (Ω)	R
Frequency	Hertz (Hz)	f
Capacitance	Farad (F)	C
Inductance	Henry (H)	L
Power	Watt (W)	P

Solved Problem

Express the resistance value of 2,200,000 Ω using the appropriate metric prefix from Table I-2.

Answer

First, express 2,200,000 Ω in engineering notation: 2,200,000 Ω = 2.2×10^6 Ω. Next, replace 10^6 with its corresponding metric prefix. Since the uppercase letter "M" (abbreviation for mega) represents 10^6, the value 2,200,000 Ω can be expressed as 2.2 MΩ. In summary: 2,200,000 Ω = 2.2×10^6 Ω = 2.2 MΩ.

Solved Problem

Express the current value of 0.0005 A using the appropriate metric prefix from Table I-2.

Answer

First, express 0.0005 A in engineering notation: 0.0005 A = 500×10^{-6} A. Next, replace 10^{-6} with its corresponding metric prefix. Since the metric prefix micro (μ) corresponds to 10^{-6}, the value of 0.0005 A can be expressed as 500 μA. In summary: 0.0005 A = 500×10^{-6} A = 500 μA.

Solved Problem

Express the power value of 560 W using the appropriate metric prefix from Table I-2.

Answer

In this case, it is not necessary or desirable to use any of the metric prefixes listed in Table I-2. The reason is that 560 W cannot be expressed as a number between 1 and 1,000 times a power of 10 that is a multiple of 3. In other words, 560 W cannot be expressed in engineering notation. The closest we can come is 0.56×10^3 W, which is not representative of engineering notation. Although 10^3 can be replaced with the metric prefix kilo (k), it is usually preferable to express the power as 560 W and not 0.56 kW.

In summary, whenever the value of a quantity lies between 1 and 1,000, only the basic unit of measure should be used for the answer. As examples, 150 Ω should be expressed as 150 Ω and not 0.15 kΩ. Also, 39 V should be expressed as 39 V and not 0.039 kV, etc.

MORE PRACTICE PROBLEMS

Sec. I-2 Express the following values using the appropriate metric prefixes from Table I-2.

1. 39,000 Ω
2. 0.004 A
3. 0.000000042 V
4. 0.018 W
5. 270,000 Ω
6. 1,000,000 Ω
7. 0.00000000033 F
8. 0.000000000100 F
9. 0.000000068 F
10. 16,000 V
11. 1,500,000,000 Hz
12. 10,000,000,000 Hz
13. 0.15 W
14. 0.033 H
15. 0.000075 W
16. 4,700 Ω
17. 56,000 Ω
18. 180,000 Ω
19. 12 Ω
20. 28 W
21. 1,800,000 Hz
22. 0.00015 F
23. 0.000033 H
24. 0.39 A
25. 1,490,000 Hz

SEC. I-3 CONVERTING BETWEEN METRIC PREFIXES

As you have learned in the previous section, metric prefixes can be substituted for the powers of 10 that are a multiple of 3. This is true even when the value of a quantity may not be expressed in proper engineering notation. It is sometimes necessary to convert from one metric prefix to another. Converting from one metric prefix to another is, in fact, a change in the power of 10. Care must be taken to make sure that the numerical part of the expression is also changed so that the value of the original number remains unchanged. To convert from one metric prefix to another, observe Rule 6.

Rule 6 When converting from a larger metric prefix to a smaller one, increase the numerical coefficient by the same factor that the metric prefix has been decreased. Conversely, when converting from a smaller metric prefix to a larger one, decrease the numerical coefficient by the same factor that the metric prefix has been increased.

Solved Problem

Convert 15 mV to microvolts (μV) and 1,490 kHz to megahertz (MHz).

Answer

To convert 15 mV to μV, recall that the prefix milli (m) corresponds to 10^{-3} and that the prefix micro (μ) corresponds to 10^{-6}. Since 10^{-6} is less than 10^{-3} by a factor of 1,000 (10^3), the numerical coefficient will have to be increased by a factor of 1,000 (10^3). Therefore, 15 mV = 15×10^{-3} V = $15{,}000 \times 10^{-6}$ V or 15 mV = 15,000 μV. To convert 1,490 kHz to MHz, recall that the prefix kilo (k) corresponds to 10^3 and that the prefix mega (M) corresponds to 10^6. Since 10^6 is larger than 10^3 by a factor of 1,000 (10^3), the numerical coefficient will have to be decreased by a factor of 1,000 (10^3). Therefore, 1,490 kHz = $1{,}490 \times 10^3$ Hz = 1.49×10^6 Hz or 1,490 kHz = 1.49 MHz.

PRACTICE PROBLEMS

Sec. I-3 Make the following conversions.

1. 22 mA = _______μA
2. 3,500 μA = _______mA
3. 4.7 MΩ = _______kΩ
4. 0.047 MΩ = _______kΩ
5. 1,510 kHz = _______MHz
6. 1.13 MHz = _______kHz
7. 330 μF = _______nF
8. 40 mH = _______μH
9. 500 μH = _______mH
10. 1,500 MHz = _______GHz
11. 100 nF = _______pF
12. 5,000 nF = _______pF
13. 30,000 nF = _______μF
14. 150 μH = _______mH
15. 2,500 nF = _______μF
16. 2,200 kΩ = _______MΩ
17. 5.6 MΩ = _______kΩ
18. 1,000 μA = _______mA
19. 47,000 kΩ = _______GΩ
20. 100,000 MΩ = _______GΩ
21. 5,500 pF = _______μF
22. 0.22 H = _______μH
23. 5,600 kΩ = _______Ω
24. 40 mA = _______μA
25. 1 pF = _______nF

SEC. I-4 ADDITION AND SUBTRACTION USING POWERS OF 10

When adding or subtracting numbers expressed as a power of 10, all the terms must have the same power of 10 before the numbers can be added or subtracted. When both terms have the same power of 10, just add or subtract the numerical parts of each term and then multiply the sum by the power of 10 common to both terms. Finally, express the answer in scientific notation.

Solved Problem

Add 25×10^4 and 75×10^3.

Answer

First, express both terms using either 10^3 or 10^4 as the common power of 10. Either one can be used. For this example, we will use 10^3 as the common power of 10 for each term. Rewriting 25×10^4 gives us 250×10^3. The numerical value of this term remains the same even though the power of 10 is different. Next, add the two terms as follows: $(250 \times 10^3) + (75 \times 10^3) = 325 \times 10^3$. Expressed in scientific notation, the answer is 3.25×10^5.

Solved Problem

Subtract 400×10^3 from 2.0×10^6.

Answer

First, express both terms using either 10^3 or 10^6 as the common power of 10. For this example we will use 10^6 as the common power of 10. Rewriting 400×10^3 gives us 0.4×10^6. The numerical value of this term remains the same even though the power of 10 is different. Next, subtract the two terms: $(2.0 \times 10^6) - (0.4 \times 10^6) = 1.6 \times 10^6$. Notice that the answer is already expressed in scientific notation.

PRACTICE PROBLEMS

Sec. I-4 Add the following numbers. Express all answers in scientific notation.

1. $(250 \times 10^3) + (1.5 \times 10^6)$
2. $(100 \times 10^{-3}) + (5{,}000 \times 10^{-6})$
3. $(250 \times 10^{-6}) + (1 \times 10^{-3})$
4. $(330 \times 10^3) + (1.17 \times 10^6)$
5. $(56 \times 10^{-3}) + (14{,}000 \times 10^{-6})$
6. $(100 \times 10^{-4}) + (20 \times 10^{-3})$
7. $(2{,}000 \times 10^{-9}) + (40{,}000 \times 10^{-12})$
8. $(100 \times 10^{-12}) + (0.2 \times 10^{-9})$
9. $(1{,}500 \times 10^3) + (2.0 \times 10^6)$
10. $(220 \times 10^0) + (0.080 \times 10^3)$

Subtract the following numbers. Express all answers in scientific notation.

11. $(54 \times 10^2) - (2.7 \times 10^3)$
12. $(560 \times 10^{-6}) - (20 \times 10^{-5})$
13. $(500 \times 10^5) - (2.5 \times 10^6)$
14. $(3{,}300 \times 10^{-5}) - (250 \times 10^{-6})$
15. $(1{,}250 \times 10^1) - (7.5 \times 10^3)$
16. $(40 \times 10^{-3}) - (1{,}000 \times 10^{-6})$
17. $(1.5 \times 10^6) - (75 \times 10^4)$
18. $(205 \times 10^{-4}) - (6.5 \times 10^{-3})$
19. $(2.5 \times 10^2) - (0.15 \times 10^3)$
20. $(150 \times 10^8) - (0.5 \times 10^6)$

SEC. I-5 MULTIPLICATION AND DIVISION USING POWERS OF 10

When multiplying or dividing numbers expressed as a power of 10, observe the following rules.

Rule 7 When multiplying numbers expressed as a power of 10, multiply the numerical coefficients and powers of 10 separately. When multiplying powers of 10, simply add the exponents to obtain a new power of 10.

Rule 8 When dividing numbers expressed as a power of 10, divide the numerical coefficients and powers of 10 separately. When dividing powers of 10, subtract the power of 10 in the denominator from the power of 10 in the numerator to obtain a new power of 10.

Solved Problem

Multiply (2.0×10^4) by (30×10^2).

Answer

First, multiply 2.0 × 30 to obtain 60. Next, multiply 10^4 by 10^2 to obtain $10^4 \times 10^2 = 10^{4+2} = 10^6$. To review: $(2.0 \times 10^4) \times (30 \times 10^2) = 60 \times 10^6$ or 6.0×10^7.

Solved Problem

Divide (500×10^5) by (25×10^2).

Answer

First, divide 500 by 25 to obtain 20. Next, divide 10^5 by 10^2 to obtain $10^{5-2} = 10^3$. Finally, as an answer we have 20×10^3 or 2.0×10^4. To review:

$$\frac{500 \times 10^5}{25 \times 10^2} = \frac{500}{25} \times \frac{10^5}{10^2} = 20 \times 10^3$$

or

$$2.0 \times 10^4$$

PRACTICE PROBLEMS

Sec. I-5 Multiply the following numbers. Express all answers in scientific notation.

1. $(25 \times 10^0) \times (2 \times 10^3)$
2. $(100 \times 10^{-3}) \times (4 \times 10^{-4})$
3. $(22 \times 10^4) \times (3.3 \times 10^2)$
4. $(4.7 \times 10^{-6}) \times (2.0 \times 10^{-3})$
5. $(15 \times 10^2) \times (300 \times 10^{-5})$
6. $(4.5 \times 10^3) \times (2.0 \times 10^{-2})$
7. $(2.5 \times 10^1) \times (3.0 \times 10^0)$
8. $(3{,}300 \times 10^{-3}) \times (3.0 \times 10^6)$
9. $(2{,}000 \times 10^5) \times (5.0 \times 10^{-5})$
10. $(250 \times 10^{-2}) \times (2.0 \times 10^0)$

Divide the following numbers. Express all answers in scientific notation.

11. $(100 \times 10^{-6}) \div (2.5 \times 10^{-4})$
12. $(5.0 \times 10^4) \div (2.5 \times 10^3)$
13. $(600 \times 10^8) \div (3.0 \times 10^6)$
14. $(12 \times 10^{-12}) \div (6.0 \times 10^{-10})$
15. $(1{,}000 \times 10^0) \div (1{,}000 \times 10^0)$
16. $(480 \times 10^{-7}) \div (2.4 \times 10^{-5})$
17. $(800 \times 10^3) \div (125 \times 10^1)$
18. $(550 \times 10^{-4}) \div (1.10 \times 10^{-2})$
19. $(2{,}000 \times 10^5) \div (50{,}000 \times 10^2)$
20. $(1{,}250 \times 10^{-6}) \div (62.5 \times 10^{-4})$

SEC. I-6 RECIPROCALS WITH POWERS OF 10

Taking the reciprocal of a power of 10 is really a special case of division using powers of 10. This is because a numerator of 1 can be written as 10^0 since $10^0 = 1$. Taking the reciprocal of a power of 10 results in a sign change for the power of 10.

Solved Problem

Find the reciprocals for 10^6 and 10^{-4}.

Answer

$$\frac{1}{10^6} = \frac{10^0}{10^6} = 10^{0-6} = 10^{-6}, \text{ therefore } \frac{1}{10^6} = 10^{-6}$$

$$\frac{1}{10^{-4}} = \frac{10^0}{10^{-4}} = 10^{0-(-4)} = 10^4, \text{ therefore } \frac{1}{10^{-4}} = 10^4$$

Sec. I-6 Find the reciprocal for each power of 10 listed.

1. 10^8
2. 10^{-3}
3. 10^1
4. 10^{-2}
5. 10^{-3}
6. 10^5
7. 10^{-12}
8. 10^9
9. 10^{-6}
10. 10^3
11. 10^{15}
12. 10^{-8}
13. 10^{14}
14. 10^0
15. 10^{-25}

SEC. I-7 SQUARING NUMBERS EXPRESSED AS A POWER OF 10

Many calculations involve squaring a number expressed as a power of 10. When squaring a number expressed as a power of 10, square the numerical coefficient and double the power of 10.

Solved Problem

Square 2.0×10^2.

Answer

First, square 2.0 to obtain 4.0. Next, square 10^2 to obtain $(10^2)^2 = 10^4$. Therefore, $(2.0 \times 10^2)^2 = 4.0 \times 10^4$.

PRACTICE PROBLEMS

Sec. I-7 Express the following answers in scientific notation.

1. $(30 \times 10^4)^2$
2. $(6.0 \times 10^{-3})^2$
3. $(10 \times 10^0)^2$
4. $(15 \times 10^{-3})^2$
5. $(40 \times 10^0)^2$
6. $(5 \times 10^{-3})^2$
7. $(11 \times 10^{-4})^2$
8. $(1.2 \times 10^{-6})^2$
9. $(7.0 \times 10^2)^2$
10. $(8.0 \times 10^{-6})^2$
11. $(20 \times 10^{-6})^2$
12. $(90 \times 10^{-4})^2$
13. $(500 \times 10^{-3})^2$
14. $(1.0 \times 10^{-9})^2$
15. $(3{,}000 \times 10^2)^2$

SEC. I-8 SQUARE ROOTS OF NUMBERS EXPRESSED AS A POWER OF 10

Many calculations involve taking the square root of a number expressed as a power of 10. When taking the square root of a number expressed as a power of 10, take the square root of the numerical coefficient and divide the power of 10 by 2.

Solved Problem

Find the square root of 9.0×10^4.

Answer

$$\sqrt{9.0 \times 10^4} = \sqrt{9.0} \times \sqrt{10^4}$$
$$= 3.0 \times 10^2$$

Solved Problem

Find the square root of 40×10^3.

Answer

The problem can be simplified if we increase the power of 10 to 10^4 and decrease the numerical coefficient to 4.0. This gives us:

$$\sqrt{40 \times 10^3} = \sqrt{4.0 \times 10^4} = \sqrt{4.0} \times \sqrt{10^4}$$
$$= 2.0 \times 10^2$$

PRACTICE PROBLEMS

Sec. 1-8 Express the following answers in scientific notation.

1. $\sqrt{4{,}000 \times 10^2}$
2. $\sqrt{16 \times 10^{-2}}$
3. $\sqrt{36 \times 10^{-12}}$
4. $\sqrt{100 \times 10^0}$
5. $\sqrt{25 \times 10^{-3}}$
6. $\sqrt{15 \times 10^{-6}}$
7. $\sqrt{40 \times 10^{-5}}$
8. $\sqrt{16 \times 10^{-3}}$
9. $\sqrt{900 \times 10^{-5}}$
10. $\sqrt{2.0 \times 10^5}$
11. $\sqrt{35 \times 10^{-6}}$
12. $\sqrt{1.4 \times 10^4}$
13. $\sqrt{50 \times 10^7}$
14. $\sqrt{600 \times 10^9}$
15. $\sqrt{5.0 \times 10^4}$

END OF CHAPTER TEST

Chapter 1: Introduction to Powers of 10 Answer True or False.

1. $10^0 = 1$.
2. In scientific notation, 0.000039 equals 3.9×10^{-5}.
3. In engineering notation, 0.0001 equals 100×10^{-6}.
4. 8.235×10^4 equals 8235 in decimal notation.
5. In engineering notation, the powers of 10 are always multiples of 3.
6. The metric prefix micro (μ) corresponds to 10^{-6}.
7. 3300 Ω is the same amount of resistance as 3.3 kΩ.
8. When multiplying powers of 10, add the exponents.
9. When dividing powers of 10, subtract the exponents.
10. 75,000 mA = 75 μA.
11. 10^6 can be replaced with the metric prefix milli (m).
12. When adding or subtracting numbers expressed in powers of 10 notation, all terms must have the same power of 10 before they can be added or subtracted.
13. $\dfrac{1}{10^4} = 10^{-4}$.
14. $(6 \times 10^3)^2 = 3.6 \times 10^7$.
15. Positive powers of 10 are used for numbers greater than 1, whereas negative powers of 10 are used for numbers less than 1.

CHAPTER

1

Electricity

Electricity is a form of energy, where energy refers to the ability to do work. More specifically, electrical energy refers to the energy associated with electric charges. In our homes and industries, we depend on electricity for lighting, heating, air conditioning, and the operation of appliances, computers, and home entertainment systems, as a few examples. Although the applications of electricity are extensive and almost limitless, electricity itself can be explained in terms of electric charge, voltage, and current.

SEC. 1-1 THE COULOMB UNIT OF ELECTRIC CHARGE

An atom is the smallest particle of an element. Therefore, all the substances, which include solids, liquids, and gases, are made up of individual atoms. All atoms are made up of electrons and protons, which are the most basic particles of electric charge. The electron is the smallest particle of negative (−) charge, whereas the proton is the smallest particle of positive (+) charge. The charge of an electron or proton has an invisible field of force that extends outward in all directions. This invisible field of force can do the work of moving another charge.

In common applications of electricity, the charge of a single electron or proton is not noticeable or useful. Instead, the charge of many billions of electrons or protons is needed to provide any practical effect of electricity. Therefore, it is convenient to define electric charge in quantities much larger than the charge of a single electron or proton. A practical unit of electric charge is the coulomb (C). One coulomb of electric charge equals 6.25×10^{18} electrons or protons or $1\ \text{C} = 6.25 \times 10^{18}$ electrons or protons. The symbol for electric charge is Q, where Q stands for quantity. The charge of a single electron or proton is 0.16×10^{-18} C, which is the reciprocal of 6.25×10^{18}. It is important to note that the charge of a single electron or proton is identical but their polarities are opposite. It is also important to note that charges of opposite polarity attract each other, but charges of the same polarity repel each other. This means that electrons and protons will be attracted to each other by the force of attraction between opposite charges. Since the weight of a proton is 1,840 times more than that of an electron, the force of attraction tends to make electrons move toward protons. When charges are of the same polarity, such as two negative charges or two positive charges, they repel each other.

Storing Electric Charge

A material with atoms in which electrons tend to stay in their own orbits is an insulator because it cannot conduct electricity very easily. Insulators are able to hold or store electric charge because of the fact that electrons cannot easily move from one atom to the next in the material. Insulators such as glass, plastic, rubber, air, paper, or mica are also called dielectrics, meaning they can store an electric charge.

Figure 1-1*a* shows a glass dielectric having a net charge, Q, of zero. The neutral condition ($Q = 0$ C) exists when there is a balance of electrons and protons in the material. If electrons are somehow added to the neutral dielectric, as in Fig. 1-1*b* then the dielectric will possess a negative charge. The reason is that the electrons now outnumber the protons. As shown in Fig. 1-1*b*, the negative charge of the glass dielectric is labeled $-Q$. Now imagine that electrons are somehow removed from the neutral dielectric, as in Fig. 1-1*c*. This condition results in a positively charged dielectric, labeled $+Q$ because the protons now outnumber the electrons.

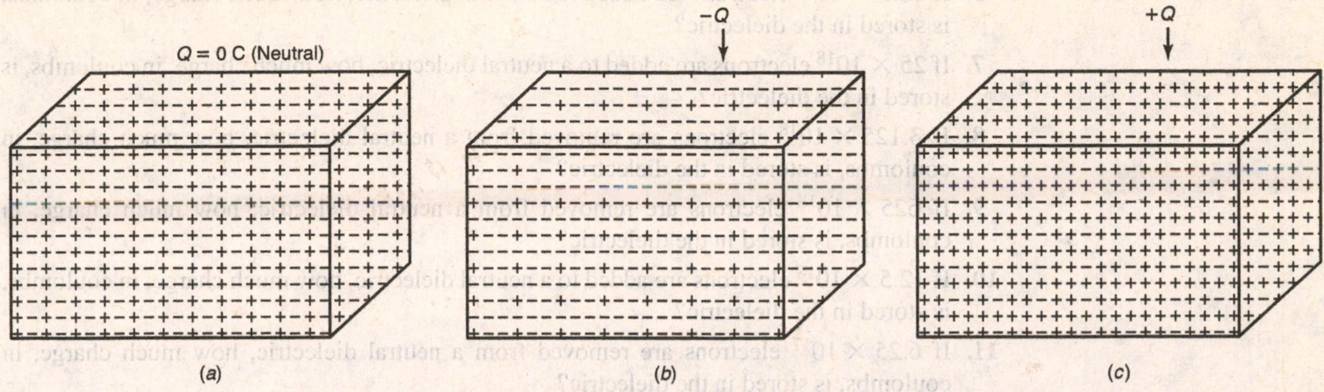

Fig. 1-1 Storing an electric charge in a glass dielectric. (*a*) $Q = 0$ C. The number of electrons equals the number of protons. This is called the neutral condition. (*b*) Dielectric with more electrons than protons. This results in a negative charge labeled $-Q$. (*c*) Dielectric with more protons than electrons. This results in a positive charge labeled $+Q$.

Solved Problem

Suppose 18.75×10^{18} electrons are added to a neutral dielectric. How much charge in coulombs is stored in the dielectric?

Answer

To determine the charge, Q, in coulombs, divide 18.75×10^{18} electrons by 6.25×10^{18} electrons, which is the quantity of electrons in 1 coulomb.

$$-Q = \frac{18.75 \times 10^{18}\ \textit{Electrons}}{6.25 \times 10^{18}\ \textit{Electrons/C}} = 3\ \text{C}$$

Since electrons were added to the neutral dielectric, the charge, Q, is negative. Therefore, the answer is expressed as $-Q = 3$ C.

Solved Problem

Suppose 31.25×10^{18} electrons are removed from a neutral dielectric. How much charge in coulombs is stored in the dielectric?

Answer

To determine the charge, Q, in coulombs, divide 31.25×10^{18} electrons by 6.25×10^{18} electrons, which is the quantity of electrons in 1 coulomb.

$$+Q = \frac{31.25 \times 10^{18}\ \textit{Electrons}}{6.25 \times 10^{18}\ \textit{Electrons/C}} = 5\ \text{C}$$

Since electrons were removed from the neutral dielectric, the charge, Q, is positive. Therefore, the answer is expressed as $+Q = 5$ C.

PRACTICE PROBLEMS

Sec. 1-1 Answer the following questions.

1. What is the smallest particle of (a) negative charge; (b) positive charge?
2. Define 1 coulomb of electric charge. What is the symbol for charge?
3. How do like and unlike charges react to each other?
4. What is the charge, in coulombs, of a single electron or proton?
5. What is another name for an insulator?

6. If 62.5×10^{18} electrons are added to a neutral dielectric, how much charge, in coulombs, is stored in the dielectric?
7. If 25×10^{18} electrons are added to a neutral dielectric, how much charge, in coulombs, is stored in the dielectric?
8. If 3.125×10^{18} electrons are removed from a neutral dielectric, how much charge, in coulombs, is stored in the dielectric?
9. If 625×10^{18} electrons are removed from a neutral dielectric, how much charge, in coulombs, is stored in the dielectric?
10. If 12.5×10^{18} electrons are added to a neutral dielectric, how much charge, in coulombs, is stored in the dielectric?
11. If 6.25×10^{17} electrons are removed from a neutral dielectric, how much charge, in coulombs, is stored in the dielectric?
12. A dielectric has a charge of $-Q = 2$ C. If 31.25×10^{18} electrons are removed from the dielectric, what is the charge, in coulombs, stored by dielectric?
13. A dielectric has a charge of $+Q = 12$ C. If 56.25×10^{18} electrons are added to the dielectric, how much charge, in coulombs, is stored by the dielectric?
14. If 10,000 electrons are removed from a neutral dielectric, how much charge, in coulombs, is stored by the dielectric?
15. If 500 million electrons are added to a neutral dielectric, how much charge, in coulombs, is stored by the dielectric?

SEC. 1-2 POTENTIAL DIFFERENCE BETWEEN ELECTRIC CHARGES

An electric charge has the potential to do the work of moving another charge, by either attraction or repulsion. When we consider two unlike charges in close proximity to each other, there is a difference of potential between them. The term *potential* refers to the ability to do work, such as moving electrons. See Fig. 1-2 where two dissimilarly charged objects are close to each other. The object on the left has a negative charge, whereas the object on the right has a positive charge. Between the two dissimilarly charged objects is an invisible field of force, called an electric field. The electric field is made up of electric field lines, as shown. These lines show the path an electron would move in the field. In Fig. 1-2, the excess electrons in the object on the left are straining toward the excess number of protons in the object on the right. This is a result of unlike charges being attracted to each other. If an electron were placed between the two dissimilarly charged objects in Fig. 1-2, the negatively charged object on the left would repel the electron toward the positively charged object on the right. Not only would the electron be repelled to the right, but it would also be attracted to the right by the positively charged object. Therefore, an electron would move from left to right in Fig. 1-2 as a result of both repulsion and attraction. The work done in moving one electron between two dissimilarly charged objects is made possible by the potential difference between the two points.

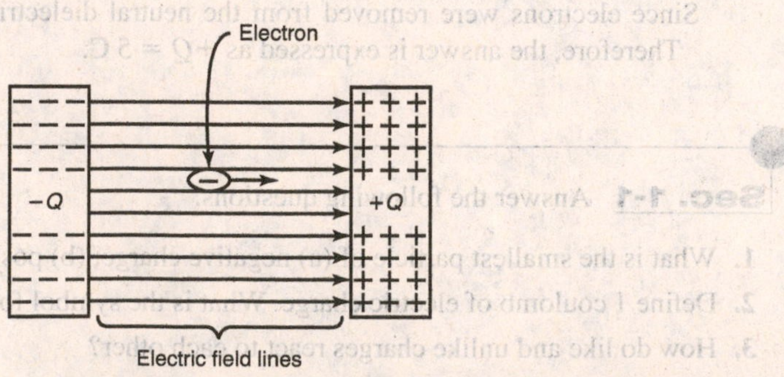

Fig. 1-2 Two dissimilarly charged objects in close proximity to each other have a potential difference between them.

The Volt Unit of Potential Difference

The basic unit of potential difference is the volt, with the symbol V. In most cases, potential difference is referred to as voltage. In general, voltage is a measure of the amount of work or energy needed to move a specific quantity of electric charge between two points. The basic unit of work or energy is the joule (J). The symbol for work or energy is W. When one joule of energy is expended in moving one coulomb (6.25×10^{18} electrons) of charge between two points, the potential difference or voltage between those two points is one volt. This can be expressed as

$$1\text{V} = \frac{1\text{ J}}{1\text{ C}}$$

In general: $V = \frac{W}{Q}$, where W represents the work or energy in joules, Q represents the electric charge in coulombs, and V represents the potential difference in volts.

Solved Problem

How much is the voltage between two points if 9 J of energy is expended in moving 1 C of charge between those two points?

Answer

Use the equation $V = \frac{W}{Q}$.

$$V = \frac{W}{Q}$$

$$V = \frac{9\text{ J}}{1\text{ C}}$$

$$= 9\text{ V}$$

Solved Problem

How much energy is expended when 3 C of charge is moved between two points having a potential difference of 6 V?

Answer

Since $V = \frac{W}{Q}$, then $W = V \times Q$. Therefore,

$$W = V \times Q$$

$$= 6\text{ V} \times 3\text{ C}$$

$$= 18\text{ J}$$

PRACTICE PROBLEMS

Sec. 1-2 Answer the following questions.

1. What is meant by the term *potential difference*?
2. What is another name for potential difference?
3. What is the basic unit of potential difference?
4. What is the basic unit of work or energy?
5. In general, define what voltage is.
6. How much is the voltage between two points if 12 J of energy is expended in moving 2 C of charge between those two points?

7. How much is the voltage between two points if 6 J of energy is expended in moving 0.5 C of charge between those points?
8. How much is the voltage between two points if 2.4 J of energy is expended in moving 1.6 C of charge between those two points?
9. How much energy, in joules, is expended when 5 C of charge is moved between two points having a potential difference of 24 V?
10. How much energy, in joules, is expended when 0.25 C of charge is moved between two points having a potential difference of 12 V?

SEC. 1-3 CHARGE IN MOTION IS CURRENT

When a potential difference forces particles of charge, such as electrons, to move between two points, the result is an electric current. In a metal conductor, such as a copper wire, there are billions of free electrons that can be forced to move with relative ease by a potential difference. This is illustrated in Fig. 1-3. The negative potential of the battery repels free electrons to the right through the copper wire. At the same time, the positive potential of the battery attracts free electrons to the right. The result is a flow of electrons, to the right, through the copper wire, made possible by the potential difference of the applied voltage. It is important to note that only the free electrons move, not the potential difference. If the potential difference in Fig. 1-3 were increased, the forces of attraction and repulsion would cause more electrons to move to the right thus producing more charge in motion. A larger amount of charge moving during a given time means more electric current. If the potential difference were decreased in Fig. 1-3, fewer electrons would be forced through the copper wire, which means less electric current.

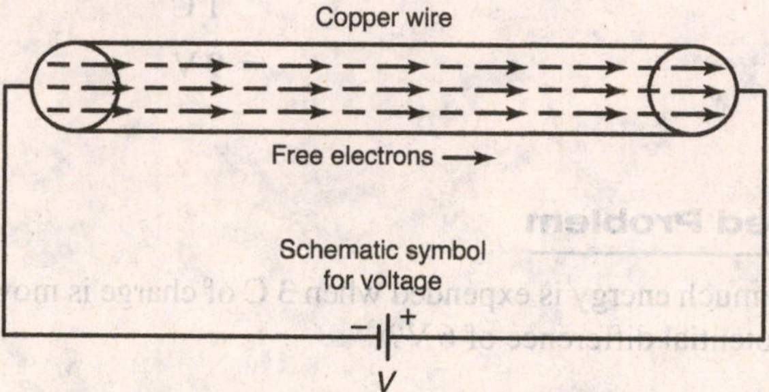

Fig. 1-3 Flow of free electrons in a copper wire.

The Ampere Unit of Current

Current can be defined as the rate of flow of electric charge. The symbol for electric current is *I* for intensity, since the current is a measure of how intense or concentrated the charge flow is. The basic unit of electric current is the ampere (A). When one coulomb (6.25×10^{18} electrons) of charge moves past a given point each second, the current, *I*, equals one ampere. In general, the formula for the current, *I* can be stated as

$$I = \frac{Q}{T}$$

where *I* is in amperes, *Q* is in coulombs, and *T* is in seconds.

Based on the formula for *I*, $1\text{A} = \frac{1\text{ C}}{1\text{ s}}$. Therefore, one ampere of current results when one coulomb of charge moves past a given point in 1 s.

It is important to understand the difference between static electric charge and current. Static electric charge is a specific quantity of charge accumulated in a dielectric, which is an insulator. The charge is considered static or at rest, without any motion. When the charge moves, usually in a conductor, the current, *I*, indicates the intensity of the charge in motion. In summary, electric charge is an amount, current is a rate, similar to miles versus miles per hour.

The fundamental definition of current can also be used to consider the charge as equal to the product of current multiplied by time. In general, $Q = I \times T$. The charge is generally accumulated in the dielectric of a capacitor or at the terminals of a battery.

Solved Problem

A charge of 2 C moves past a given point every second. How much is the current, I, in amperes?

Answer

Recall that $I = \frac{Q}{T}$. Therefore,

$$I = \frac{Q}{T}$$

$$= \frac{2\ \text{C}}{1\ \text{s}}$$

$$= 2\ \text{A}$$

Solved Problem

A current of 5 A charges a dielectric for 2 s. How much charge is accumulated in the dielectric?

Answer

Recall that $Q = I \times T$. Therefore,

$$Q = I \times T$$

$$= 5\ \text{A} \times 2\ \text{s}$$

$$= 10\ \text{C}$$

PRACTICE PROBLEMS

Sec. 1-3 Answer the following questions.

1. In your own words, define what electric current is.
2. Explain the difference between charge and current.
3. What is the symbol for electric current?
4. What is the basic unit of electric current?
5. Is electric current possible without a potential difference?
6. Define one ampere of electric current.
7. A charge, Q, of 1 C moves past a given point every $\frac{1}{10}$ second. How much is the current, I, in amperes?
8. A charge, Q, of 0.25 C moves past a given point every 0.5 s. How much is the current, I, in amperes?
9. A charge, Q, of 100 C moves past a given point every 0.01 s. How much is the current, I, in amperes?
10. A charge, Q, of 0.01 C moves past a given point every $\frac{1}{10}$ second. How much is the current, I, in amperes?
11. A current of 0.1 A charges a dielectric for 10 s. How much charge is accumulated in the dielectric?

12. A current of 2.5 A charges a dielectric for 4 s. How much charge is accumulated in the dielectric?
13. A current of 20 mA charges a dielectric for 50 ms. How much charge is accumulated in the dielectric?
14. A current of 400 mA charges a dielectric for 300 ms. How much charge is accumulated in the dielectric?
15. How long will it take a neutral dielectric to obtain a charge of 0.5 C if the charging current is 25 mA?

SEC. 1-4 RESISTANCE IS OPPOSITION TO CURRENT

The opposition to the flow of current is called resistance. The symbol for resistance is *R* and the basic unit is the ohm. The abbreviation for the ohm unit is the Greek letter omega, written as Ω. The atoms in a copper wire have a large number of free electrons, which are easily moved by a potential difference. Therefore, a copper wire offers very little opposition to the flow of free electrons, which corresponds to a very low value of resistance. All the metals, such as silver, copper, gold, and aluminum, have many free electrons and are therefore good conductors with low resistance. Materials with atoms having very few free electrons offer more resistance to the flow of current. Carbon, for example, has fewer free electrons than copper. When the same amount of potential difference is applied to the carbon as was applied to the copper wire, fewer free electrons will flow. It should be noted that just as much current can be made to flow in the carbon as the copper wire by increasing the voltage. For the same voltage, however, carbon has less current and therefore a higher resistance.

Conductance

The opposite of resistance is conductance. The symbol for conductance is *G* and the basic unit of measure is the siemens (S). In fact, resistance and conductance are reciprocals of each other; $R = \frac{1}{G}$ and $G = \frac{1}{R}$. The higher the resistance, the lower the conductance and vice versa.

Solved Problem

Calculate the conductance, *G*, in siemens, for each of the following resistance values: $R = 10\ \Omega$, $R = 50\ \Omega$, and $R = 100\ \Omega$.

Answer

Use the formula $G = \frac{1}{R}$.

For $R = 10\ \Omega$; $G = \frac{1}{10\ \Omega} = 0.1\text{ S}$

For $R = 50\ \Omega$; $G = \frac{1}{50\ \Omega} = 0.02\text{ S}$

For $R = 100\ \Omega$; $G = \frac{1}{100\ \Omega} = 0.01\text{ S}$

Notice that as the resistance, *R*, increases, the conductance, *G*, decreases.

Solved Problem

Calculate the resistance, R, in ohms, for each of the following conductance values: $G = 0.1$ S, $G = 1$ S, and $G = 10$ S.

Answer

Use the formula $R = \frac{1}{G}$

For $G = 0.1$ S $\quad R = \frac{1}{0.1\text{ S}}$

$= 10\ \Omega$

For $G = 1$ S $\quad R = \frac{1}{1\text{ S}}$

$= 1\ \Omega$

For $G = 10$ S $\quad R = \frac{1}{10\text{ S}}$

$= 0.1\ \Omega$

Notice that a higher value of conductance, G, corresponds to a lower value of resistance.

PRACTICE PROBLEMS

Sec. 1-4 Answer the following questions.

1. Define the term *resistance* as it relates to electricity.
2. What is the letter symbol used for resistance?
3. What is the basic unit of measure for resistance?
4. What has more resistance, a material with very few free electrons or one with many free electrons? (The physical dimensions of the materials are the same.)
5. What is conductance and how is it related to resistance?
6. What is the letter symbol used for conductance?
7. What is the basic unit of measure for conductance?
8. Calculate the conductance, G, in siemens, for the following resistance values: (a) $R = 1{,}000\ \Omega$, (b) $R = 2{,}000\ \Omega$, (c) $R = 10{,}000\ \Omega$.
9. Calculate the conductance, G, in siemens, for the following resistance values: (a) $R = 0.01\ \Omega$, (b) $R = 0.0025\ \Omega$, (c) $R = 0.0004\ \Omega$.
10. Calculate the resistance, R, in ohms, for the following conductance values: (a) $G = 0.04$ S, (b) $G = 0.00003333$ S, (c) $G = 0.0333$ S.
11. Calculate the resistance, R, in ohms, for the following conductance values: (a) $G = 0.05$ S, (b) $G = 0.00667$ S, (c) $G = 0.1333$ S.

SEC. 1-5 CHARACTERISTICS OF AN ELECTRIC CIRCUIT

Figure 1-4 shows a simple electric circuit, which consists of a 12 V DC voltage source and a load resistor, R_L. The voltage source and load resistance are connected to each other using low-resistance wire conductors. Ideally, the resistance of the wire conductors is insignificant in comparison to the load resistance, R_L. The DC voltage source provides the electrical pressure or force to move free electrons around the circuit. The unified movement of free electrons is the current flow in the circuit. Since like charges repel each other and unlike charges attract, the negative terminal of the DC voltage source repels free electrons in the connecting wires and load resistance, whereas the positive terminal of the voltage source attracts free electrons.

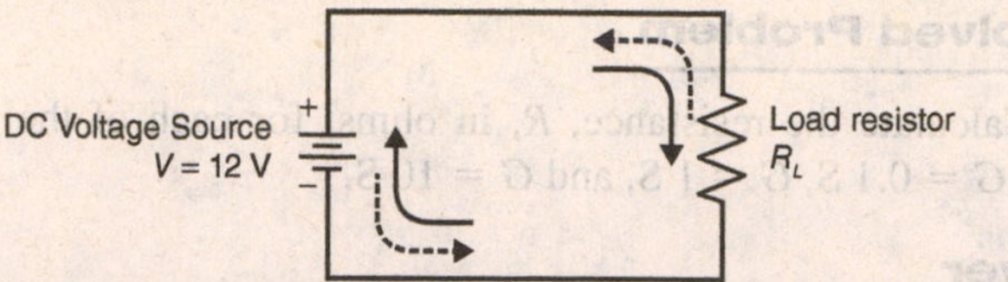

Fig. 1-4 A simple electric circuit consisting of a DC voltage source and load resistor, R_L. The dashed arrows indicate the direction of electron flow, whereas the solid arrows indicate the direction of conventional current flow.

The movement and/or direction of free electrons in the circuit is considered "electron flow" and is always from the negative (−) side of the voltage source to the positive (+) side. In Fig. 1-4, electron flow is in a counterclockwise direction. Sometimes, it is assumed that the movement of electric charges is a movement of positive charges. In this case, it is assumed that positive charges flow from the positive (+) side of the voltage source through the external circuit returning to the negative (−) side. The movement and/or direction of positive charges in the circuit is considered "conventional current flow" and is always from the positive (+) side of the voltage source to the negative (−) side. In Fig. 1-4, the direction of conventional current flow is in a clockwise direction.

There are three important characteristics of any electric circuit:

1. There must be a source of potential difference or voltage. Without voltage, free electric charges are not forced to move through the circuit to produce current flow.
2. There must be a complete path for current flow. Without a complete path, electrons cannot flow from the negative side of the potential difference, through the external circuit, and back to the positive side.
3. Every circuit has some resistance which limits the amount of current flow. The resistance serves as a load on the voltage source and may be a light bulb, motor, or other useful device.

Open Circuit

In Fig. 1-4, a break or open in any part of the circuit interrupts the path for current flow. The open or break has infinitely high resistance. This is shown in Fig. 1-5. As a result of the open circuit's infinitely high resistance, the current, I, is zero. It is important to note however, that even though the current, I, is zero, the voltage of 12 V is still present in the circuit. This is because the voltage source generates its potential difference internally and is not dependent on whether or not there is current flowing in the circuit. Always remember, an open circuit has infinitely high resistance and zero current.

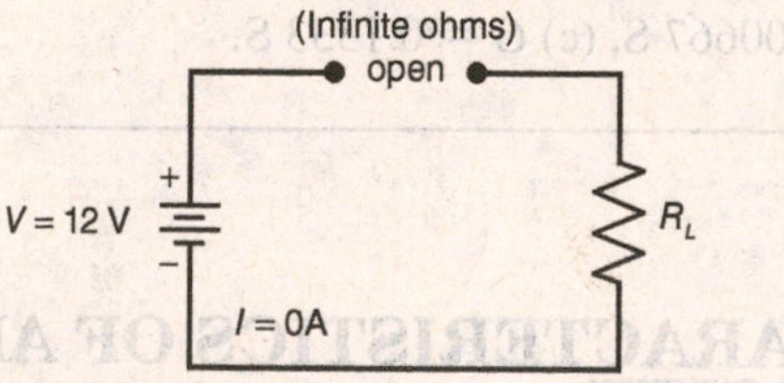

Fig. 1-5 An open circuit has infinitely high resistance and zero current.

Short Circuit

The opposite of an open circuit is a short circuit. Although an open circuit has a resistance that approaches infinity, a short circuit has practically zero resistance. An example of a short circuit is shown in Fig. 1-6. In this example, a low-resistance wire serves as a bypass or short circuit around the load resistor, R_L. Although the short circuit provides a complete path for

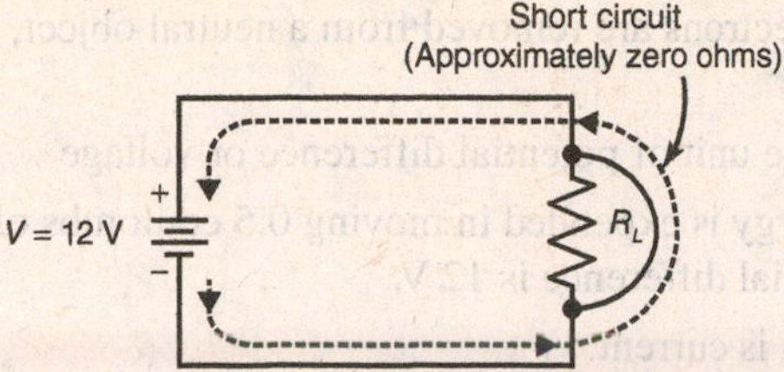

Fig. 1-6 A short circuit across R_L results in excessively high current in the connecting wires and in the short circuit itself. However, there is no current in the load resistor, R_L.

current flow, its extremely low resistance means the voltage source sees little or no opposition to the flow of current. The result is excessively high current in the connecting wires and short circuit. However, there is no current in the load, R_L because the short circuit path around R_L has a much lower resistance than R_L itself. It is important to note that the high current caused by the short circuit could cause the wire conductors to overheat. This in turn, could cause the wire's insulation to melt or burn. Therefore, a short circuit can be a safety hazard because it may cause a fire. In most cases, a fuse is placed in series with the voltage source so that the path for current flow is interrupted when the current exceeds the rating of the fuse.

One more point: In Fig. 1-6, it is unlikely that the voltage source could maintain its voltage of 12 V while supplying an extremely high amount of current to the short circuit. If the current drawn from a voltage source becomes excessively high, the output voltage drops to a much lower value and can even drop to near zero volts. The amount of current a voltage source can supply to a circuit, while still maintaining its rated output voltage, depends on several different factors. This topic is beyond the scope of this chapter and is therefore not covered here.

PRACTICE PROBLEMS

Sec. 1-5 Answer the following questions.

1. List the three characteristics of an electric circuit.
2. Can current exist without a source of potential difference or voltage?
3. Can potential difference or voltage exist when there is no current flow?
4. Define the difference between electron flow and conventional current flow.
5. Describe what an open circuit is.
6. Is there current flow in an open circuit?
7. Is there voltage present in an open circuit?
8. Describe what a short circuit is.
9. What is the danger of a short circuit?
10. If a component is short-circuited, is there any current flow in the component itself?

END OF CHAPTER TEST

Chapter 1: Electricity Answer True or False.

1. An electron has negative polarity.
2. A proton has positive polarity.
3. A neutral object has an equal number of protons and electrons.
4. Unlike charges repel each other and like charges attract.
5. 1 coulomb of electric charge equals 6.25×10^{18} electrons or protons.
6. The term dielectric is another name for insulator.

7. If 25.0×10^{18} electrons are removed from a neutral object, it has a negative charge, $-Q$, of 4 C.
8. The volt (V) is the unit of potential difference or voltage.
9. If 6 joules of energy is expended in moving 0.5 coulombs of electric charge between two points, the potential difference is 12 V.
10. Charge in motion is current.
11. $1\text{ A} = \frac{1\,C}{1\,s}$.
12. Resistance is the opposition to the flow of current.
13. Resistance and conductance are reciprocals of each other.
14. The electric charge stored by an insulator is considered static, or at rest, without any motion.
15. A 100 Ω resistor has a higher conductance than a 10 Ω resistor.
16. The resistance of an open circuit is approximately zero ohms.
17. In order for current to flow in a circuit, there must be a source of potential difference or voltage.
18. The direction of electron flow is from the negative side of the voltage source, through the external circuit, and back to the positive side of the voltage source.
19. A short circuit has very low resistance and excessively high current.
20. A requirement of any circuit is that there must be a complete path for current flow.

CHAPTER

Resistor Color Coding

Resistors are perhaps the most common of all electronic components. Their main purpose is to limit current flow or to provide a desired drop in voltage. Resistors come in a variety of sizes, shapes, and values. The physical size of a resistor determines its power-handling capabilities. However, there is no direct correlation between its physical size and its resistance value.

SEC. 2-1 FOUR-BAND RESISTOR COLOR CODE

Carbon resistors use colored bands or stripes to indicate their resistance value in ohms. A colored band or stripe is also used to indicate the tolerance rating of the resistor. The color stripes represent numerical values. The colors, and the numerical values they represent, are shown in Table 2-1.

TABLE 2-1 RESISTOR COLOR CODE

Color	Value
Black	0
Brown	1
Red	2
Orange	3
Yellow	4
Green	5
Blue	6
Violet	7
Gray	8
White	9

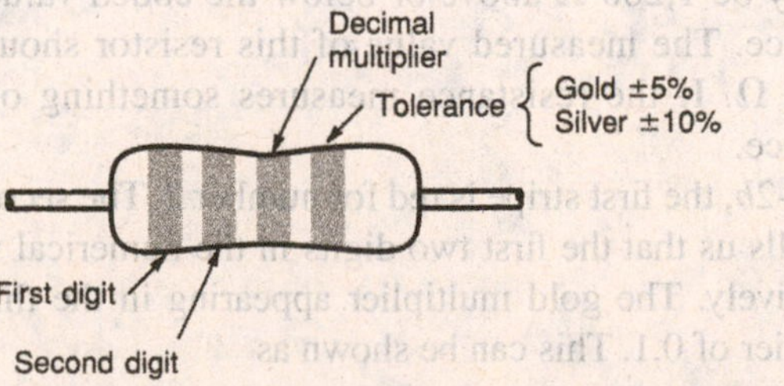

Fig. 2-1 How to read color stripes on carbon resistors.

Figure 2-1 shows a carbon resistor employing the use of the color-code system.

The first two color stripes in Fig. 2-1 indicate the first two digits in the numerical value of the resistance. The third color stripe tells how many zeros must be added to the first two digits, or that this multiplier is 10^n, where n is the number represented by the color stripe. The fourth band or stripe tells how far the measured resistance can be from the coded value, which is called the *tolerance* rating.

Although the colors gold and silver are not shown in Table 2-1, they are also used when color-coding resistors. When either color appears in the fourth band, it is being used to indicate resistor tolerance. Gold in the fourth band indicates a tolerance of ±5 percent. Silver in the fourth band indicates a tolerance of ±10 percent. (No fourth band indicates a tolerance of ±20 percent.) When either color appears in the third band, it is being used to indicate a fractional multiplier. Gold in the third band indicates a fractional multiplier of 0.1. Silver in the third band indicates a fractional multiplier of 0.01. It should be noted that the colors gold and silver will never appear in the first and second bands.

Solved Problem

For each resistor shown in Fig. 2-2, find the color-coded value of resistance along with its tolerance rating.

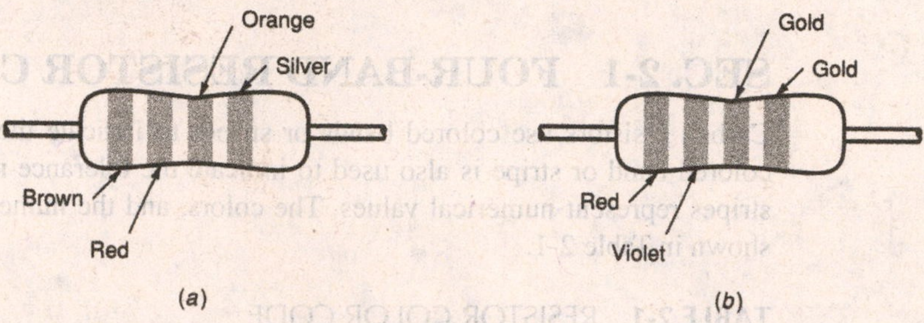

Fig. 2-2 Examples of color-coded *R* values.

Answer

In Fig. 2-2*a*, the first stripe is brown for number 1. The second stripe is red for number 2. This tells us that the first two digits in the numerical value of resistance are 1 and 2, respectively. The orange multiplier appearing in the third stripe means that three zeros are to be added to the first two digits, or that the first two digits are to be multiplied by 10^3. This can be shown as

$$\begin{array}{cccc} \text{Brown} & \text{Red} & & \text{Orange} \\ 1 & 2 & \times & 1{,}000 \end{array} = 12{,}000\ \Omega \textit{ or } 12\ \text{k}\Omega$$

The fourth stripe is silver for a tolerance of ±10 percent; and 10 percent of 12,000 Ω is 12,000 Ω × 0.1 = 1,200 Ω. This tells us that the measured resistance value can actually be 1,200 Ω above or below the coded value of 12 kΩ and still be within tolerance. The measured value of this resistor should fall between 10,800 Ω and 13,200 Ω. If the resistance measures something outside this range, it is out of tolerance.

In Fig. 2-2*b*, the first stripe is red for number 2. The second stripe is violet for number 7. This tells us that the first two digits in the numerical value of resistance are 2 and 7, respectively. The gold multiplier appearing in the third stripe indicates a fractional multiplier of 0.1. This can be shown as

$$\begin{array}{cccc} \text{Red} & \text{Violet} & & \text{Gold} \\ 2 & 7 & \times & 0.1 \end{array} = 2.7\ \Omega$$

The fourth stripe is gold for a tolerance of ±5 percent; 5 percent of 2.7 Ω is 2.7 Ω × 0.05 = 0.135 Ω. This indicates that the measured resistance value can actually be 0.135 Ω above or below the coded value of 2.7 Ω and still be within tolerance. The measured value of this resistor should fall between 2.565 Ω and 2.835 Ω. If the resistance measures something outside this range, it is out of tolerance.

Sec. 2-1 For the resistors shown in Figs. 2-3 through 2-32, indicate the resistance and tolerance values. For each resistor, indicate the ohmic range permissible for the specified tolerance.

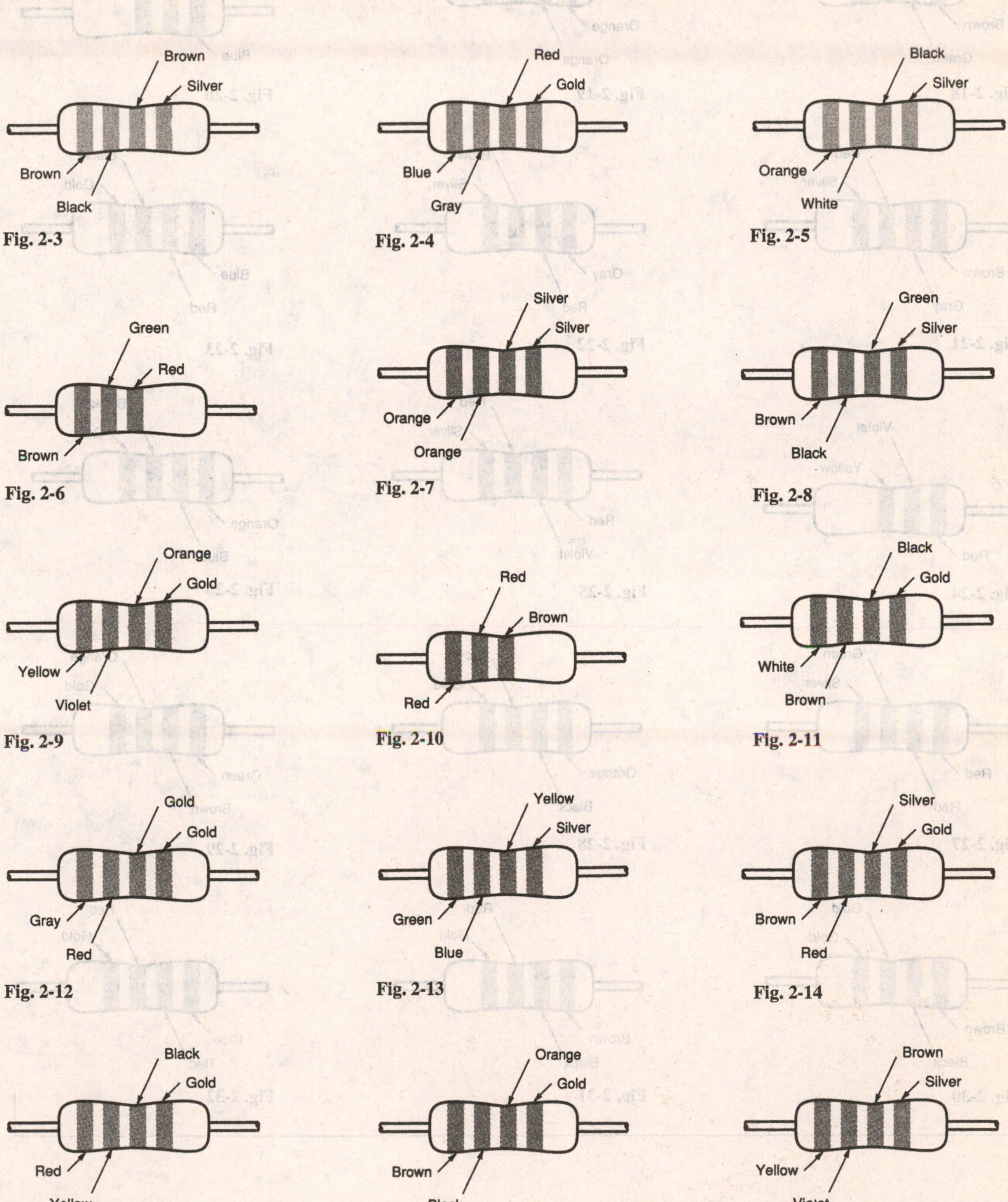

Fig. 2-3

Fig. 2-4

Fig. 2-5

Fig. 2-6

Fig. 2-7

Fig. 2-8

Fig. 2-9

Fig. 2-10

Fig. 2-11

Fig. 2-12

Fig. 2-13

Fig. 2-14

Fig. 2-15

Fig. 2-16

Fig. 2-17

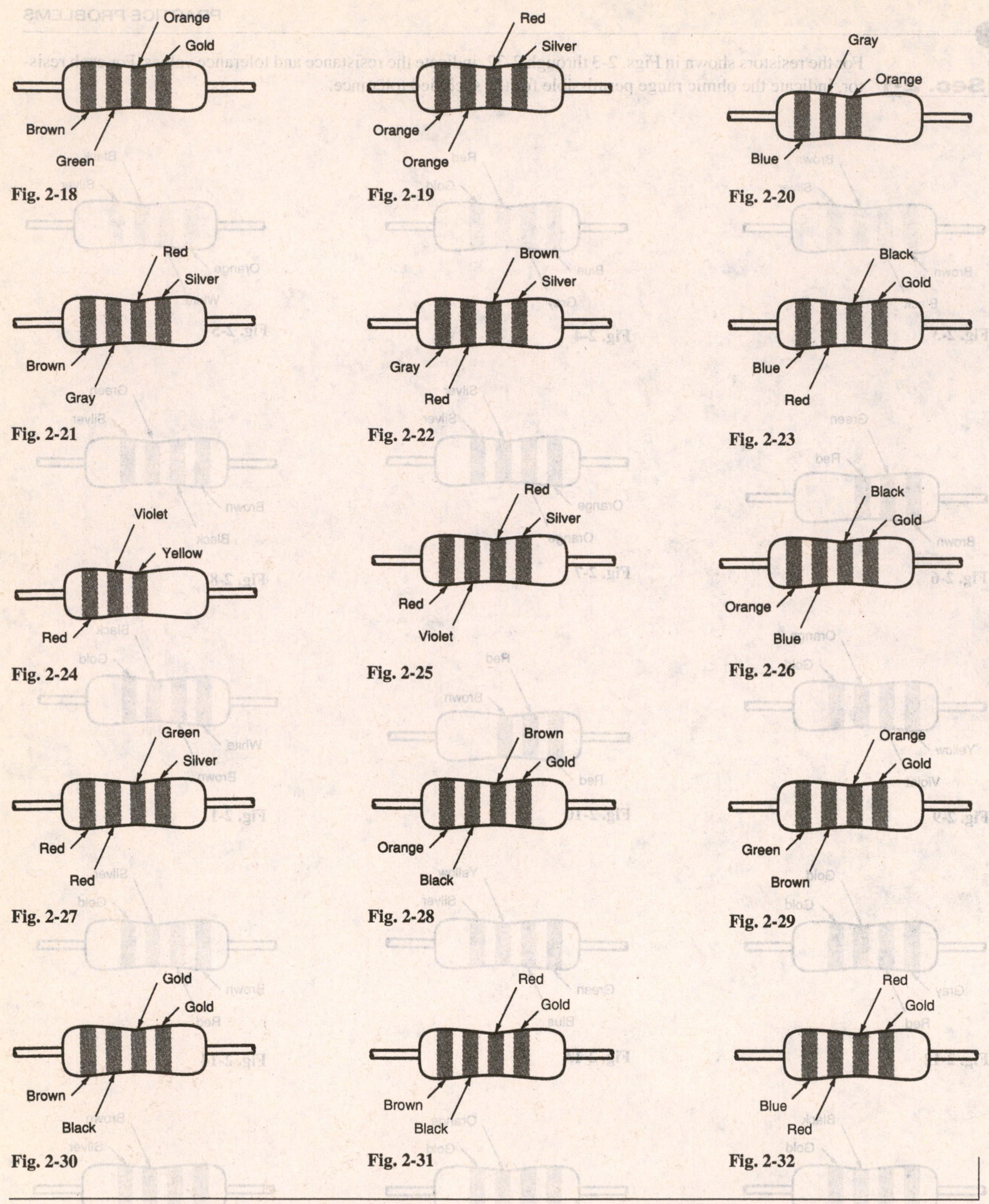

Orange
Gold
Brown
Green
Fig. 2-18
Red
Silver
Orange
Orange
Fig. 2-19
Gray
Orange
Blue
Fig. 2-20
Red
Silver
Brown
Gray
Fig. 2-21
Brown
Silver
Gray
Red
Fig. 2-22
Black
Gold
Blue
Red
Fig. 2-23
Violet
Yellow
Red
Fig. 2-24
Red
Silver
Red
Violet
Fig. 2-25
Black
Gold
Orange
Blue
Fig. 2-26
Green
Silver
Red
Red
Fig. 2-27
Brown
Gold
Orange
Black
Fig. 2-28
Orange
Gold
Green
Brown
Fig. 2-29
Gold
Gold
Brown
Black
Fig. 2-30
Red
Gold
Brown
Black
Fig. 2-31
Red
Gold
Blue
Red
Fig. 2-32

Sec. 2-1 For Probs. 1–25, identify the color of each band for the resistor values listed.

	First Band	Second Band	Third Band	Fourth Band	Resistor Value, %
1.	______	______	______	______	680 kΩ±5
2.	______	______	______	______	27 Ω±10
3.	______	______	______	______	1 kΩ±5
4.	______	______	______	______	100 kΩ±10
5.	______	______	______	______	56 Ω±10
6.	______	______	______	______	1.5 Ω±5
7.	______	______	______	______	910 kΩ±10
8.	______	______	______	______	18 Ω±5
9.	______	______	______	______	2.2 kΩ±5
10.	______	______	______	______	510 Ω±5
11.	______	______	______	______	33 kΩ±20
12.	______	______	______	______	240 kΩ±5
13.	______	______	______	______	12 kΩ±5
14.	______	______	______	______	5.6 kΩ±10
15.	______	______	______	______	12 Ω±10
16.	______	______	______	______	4.7 Ω±20
17.	______	______	______	______	0.82 Ω±5
18.	______	______	______	______	390 Ω±5
19.	______	______	______	______	75 kΩ±5
20.	______	______	______	______	360 kΩ±5
21.	______	______	______	______	1.5 MΩ±10
22.	______	______	______	______	330 Ω±20
23.	______	______	______	______	75 Ω±5
24.	______	______	______	______	1.8 kΩ±5
25.	______	______	______	______	0.47 Ω±10

SEC. 2-2 FIVE-BAND RESISTOR COLOR CODE

Some resistors use five color bands rather than four to indicate their resistance value. Five color bands allow more precise resistance values. Furthermore, resistors having five color bands are available with tolerances of ±0.1 percent, ±0.25 percent, ±0.5 percent, ±1 percent, and ±2 percent. This is a considerable improvement over the 5 and 10 percent tolerance values available with the four-band color code.

Figure 2-33 shows a precision resistor employing the use of the five-band color code. The first three color bands indicate the first three digits in the numerical value of the resistance; the fourth band indicates the multiplier; and the fifth band indicates tolerance. In the fifth band, the colors brown, red, green, blue, and violet represent the following tolerances:

Brown	±1 percent
Red	±2 percent
Green	±0.5 percent
Blue	±0.25 percent
Violet	±0.1 percent

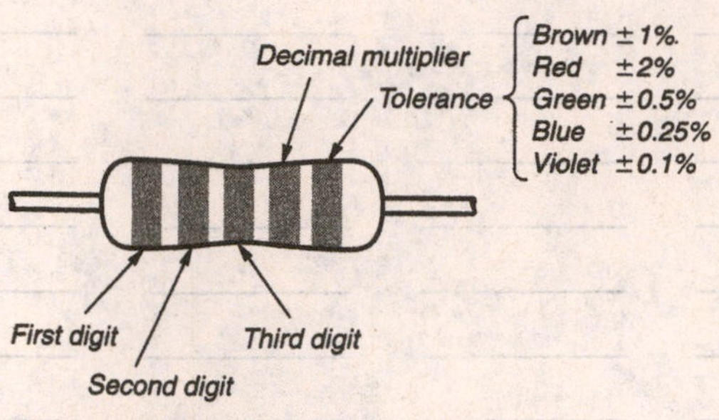

Fig. 2-33 Five-band resistor color code.

Solved Problem

For each resistor in Fig. 2-34, indicate the resistance, tolerance, and permissible ohmic range.

Answer

In Fig. 2-34*a*, the first band is green for 5, the second band is white for 9, and the third band is violet for 7. Therefore, the first three digits in the resistance value are 5, 9, and 7, respectively. The fourth band, which is gold, indicates a multiplier of 0.1. Therefore, $R = 597 \times 0.1 = 59.7\ \Omega$. The fifth band, which is blue, indicates a tolerance of ±0.25 percent. The permissible ohmic range is calculated as $59.7\ \Omega \times 0.0025 = \pm 0.14925\ \Omega$, which is 59.55 Ω to 59.85 Ω, approximately.

In Fig. 2-34*b*, the first color band is red for 2, the second band is green for 5, and the third band is red for 2. Therefore, the first three digits in the resistance value are 2, 5, and 2, respectively. The fourth band, which is brown, indicates a multiplier of 10^1 or 10. Therefore, $R = 252 \times 10^1\ \Omega = 2{,}520\ \Omega$ or 2.52 kΩ. The fifth band is green, indicating a tolerance of ±0.5 percent. The permissible ohmic range is calculated as $2{,}520 \times 0.005 = \pm 12.6\ \Omega$, which is 2,507.4 Ω to 2,532.6 Ω.

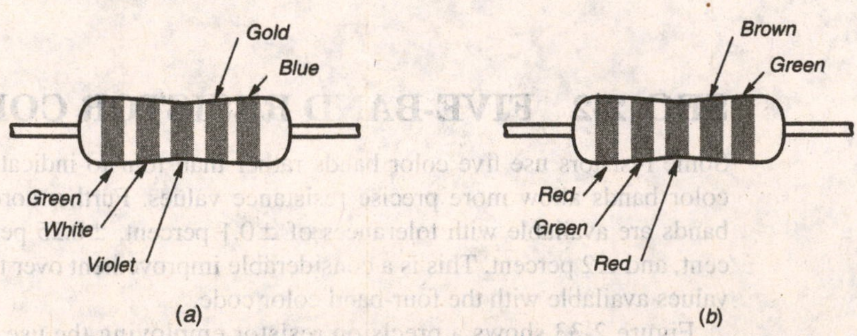

Fig. 2-34 Examples of color-coded R values using the five-band system.

Sec. 2-2 For the resistors shown in Figs. 2-35 through 2-58, indicate the resistance, tolerance, and permissible ohmic range. Indicate the permissible ohmic range as $\pm\Omega$.

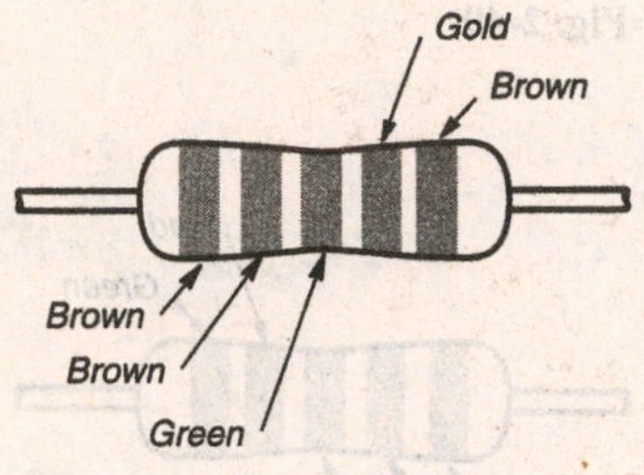

Fig. 2-35

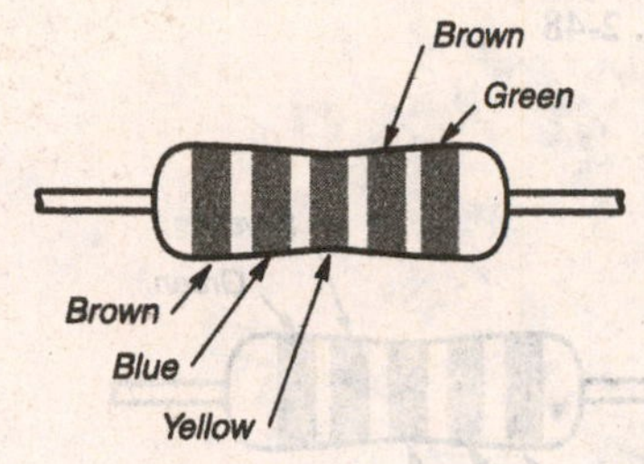

Fig. 2-36

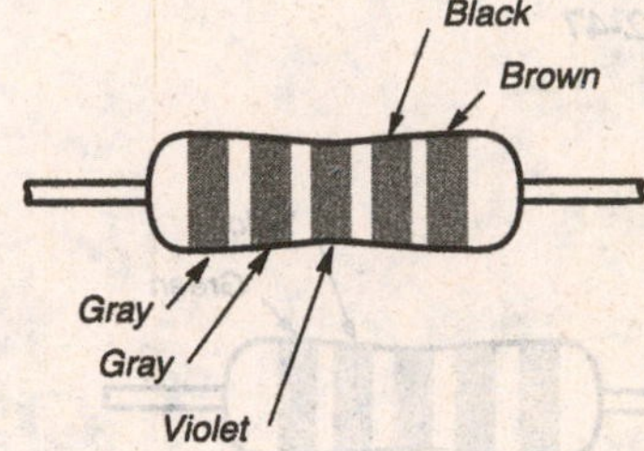

Fig. 2-37

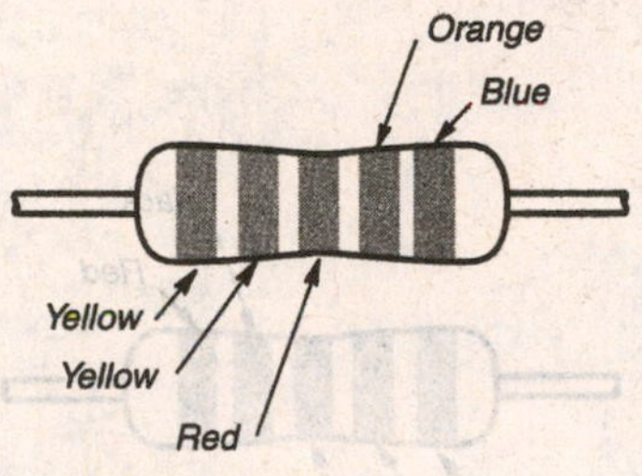

Fig. 2-38

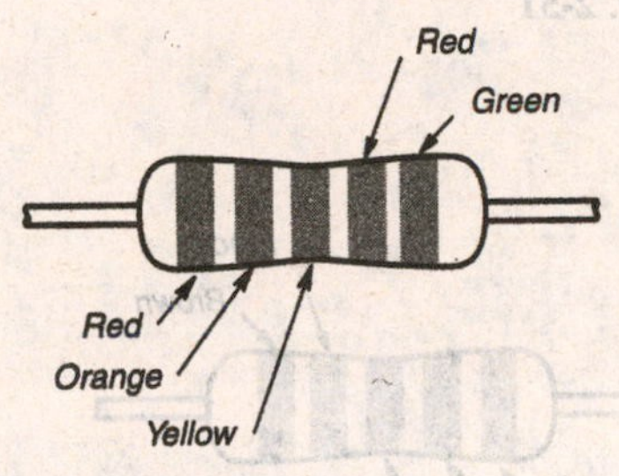

Fig. 2-39

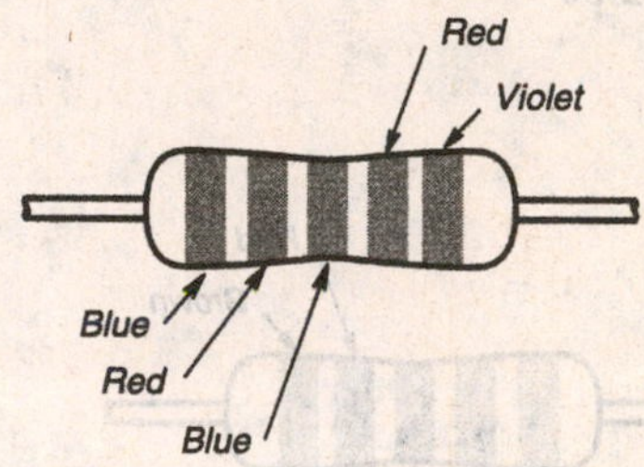

Fig. 2-40

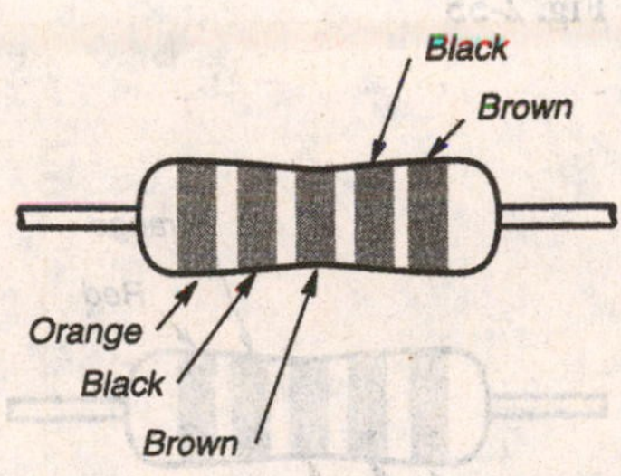

Fig. 2-41

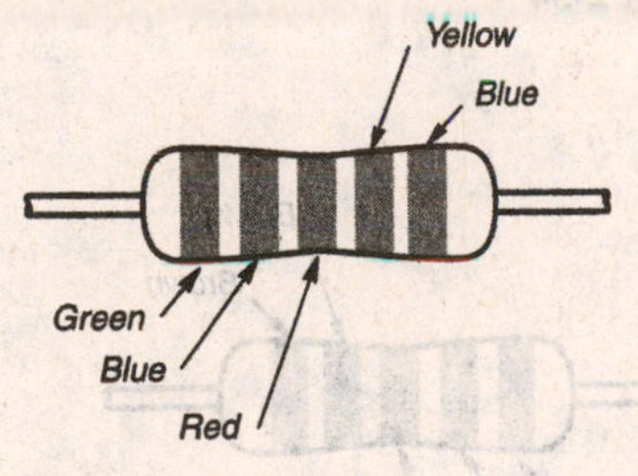

Fig. 2-42

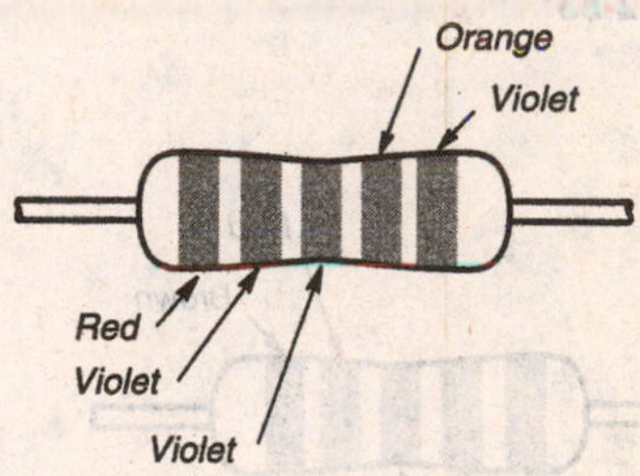

Fig. 2-43

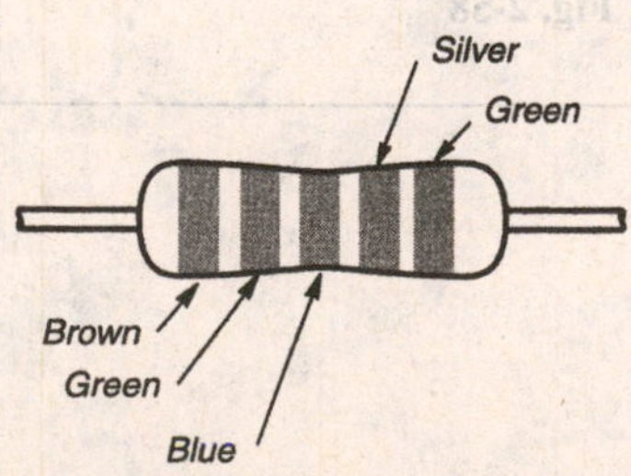

Fig. 2-44

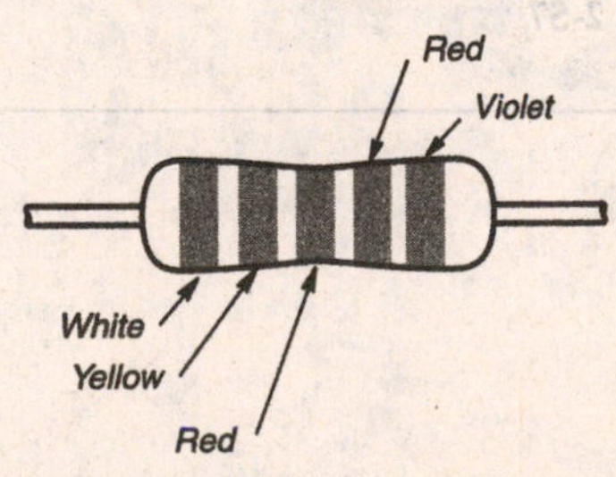

Fig. 2-45

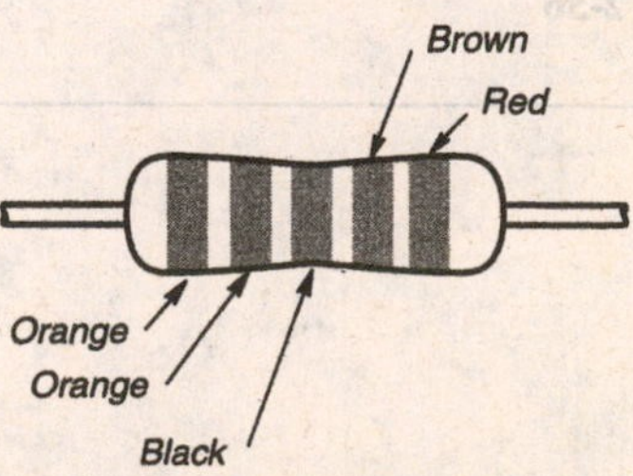

Fig. 2-46

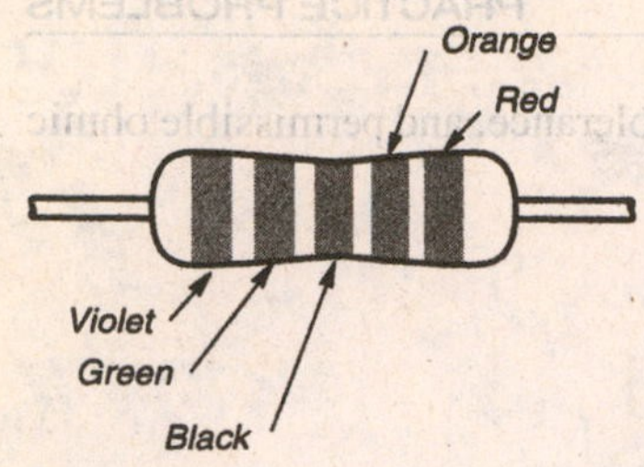

Fig. 2-47

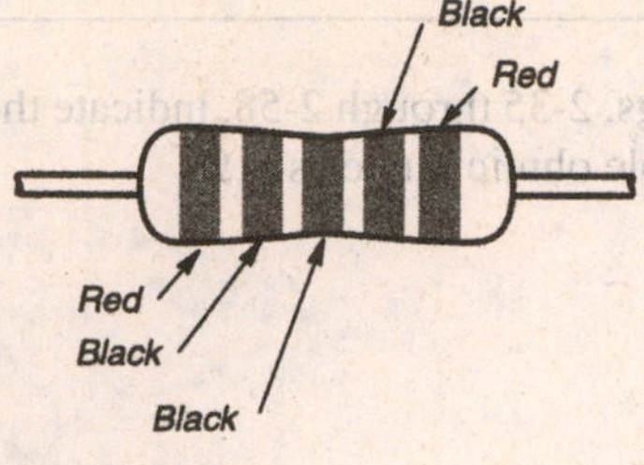

Fig. 2-48

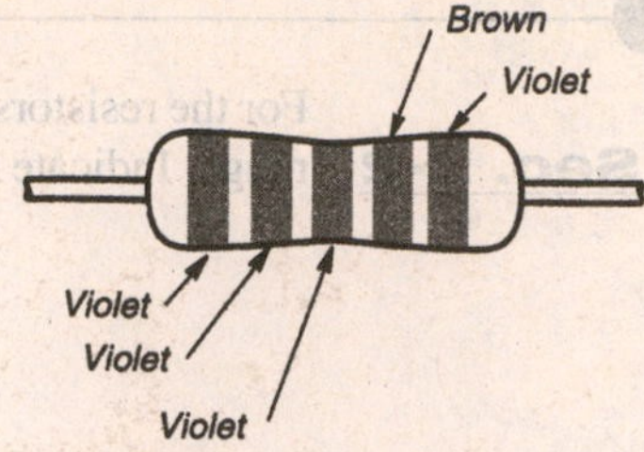

Fig. 2-49

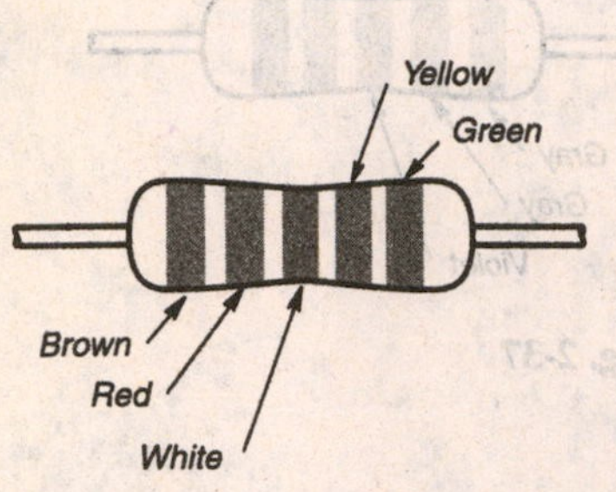

Fig. 2-50

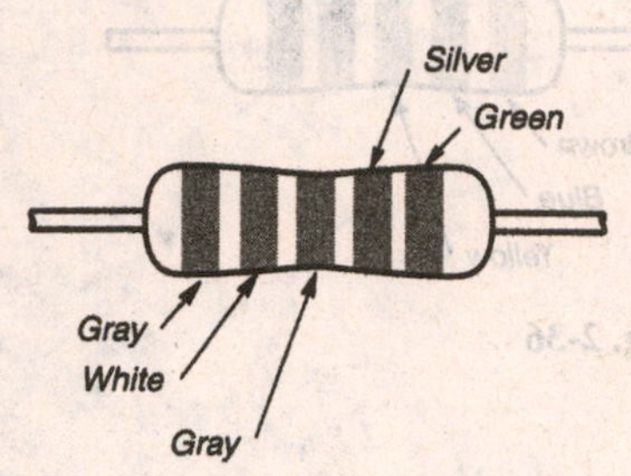

Fig. 2-51

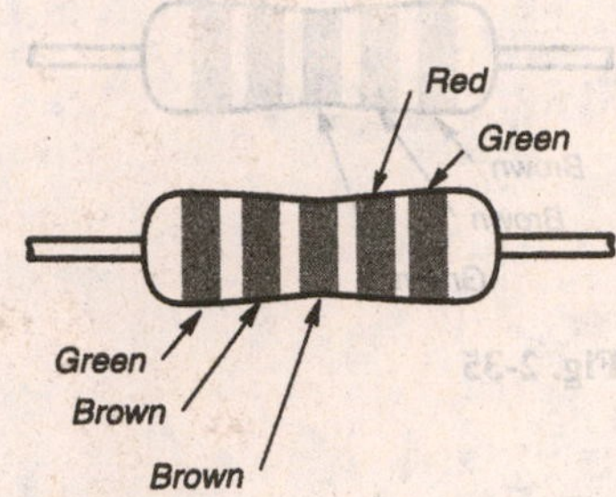

Fig. 2-52

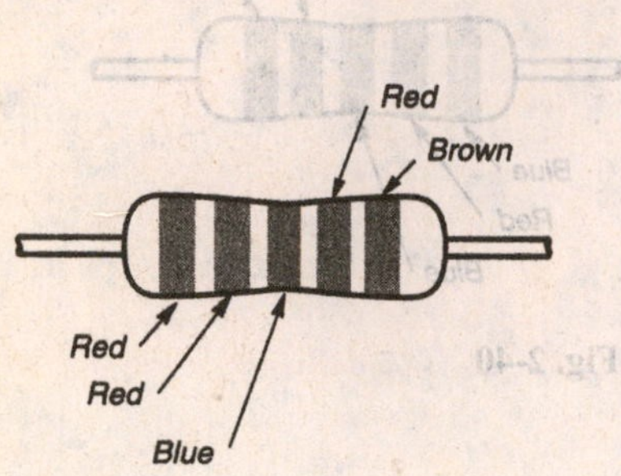

Fig. 2-53

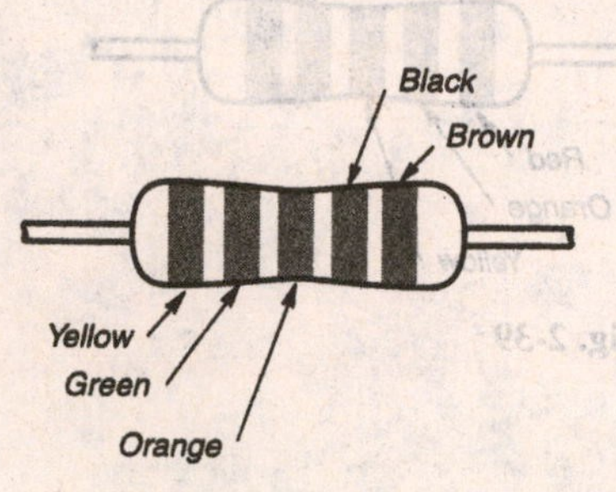

Fig. 2-54

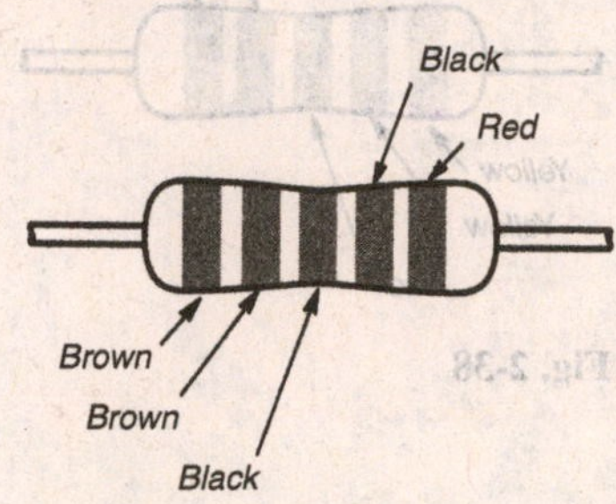

Fig. 2-55

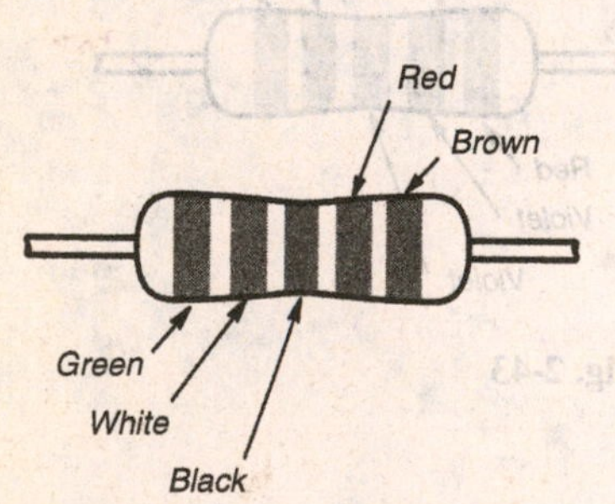

Fig. 2-56

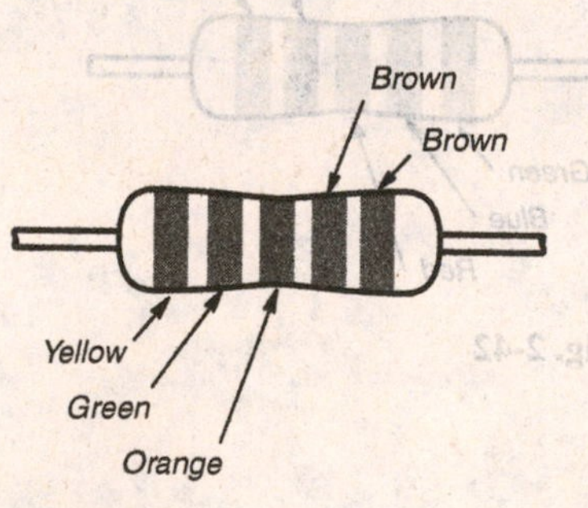

Fig. 2-57

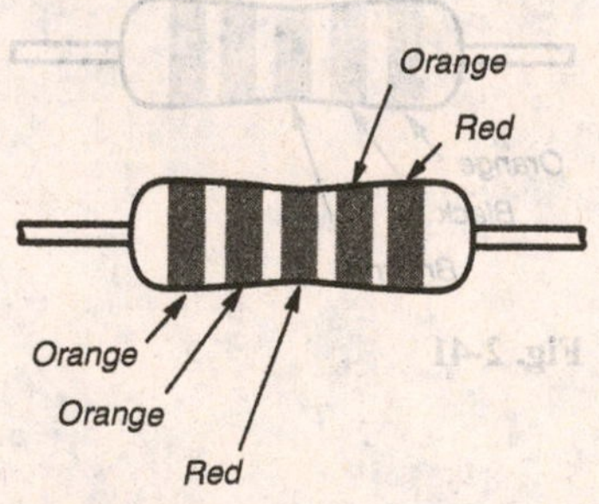

Fig. 2-58

Sec. 2-2 For Probs. 1–15, identify the color of each band for the resistor values listed.

	First Band	Second Band	Third Band	Fourth Band	Fifth Band	Resistor Value, %
1.	____	____	____	____	____	56.2 Ω±1
2.	____	____	____	____	____	25.2 Ω±0.5
3.	____	____	____	____	____	98.8 kΩ±0.1
4.	____	____	____	____	____	336 kΩ±0.25
5.	____	____	____	____	____	4.42 kΩ±0.25
6.	____	____	____	____	____	12.0 Ω±2
7.	____	____	____	____	____	665 Ω±0.1
8.	____	____	____	____	____	2.84 MΩ±0.1
9.	____	____	____	____	____	8.56 Ω±0.5
10.	____	____	____	____	____	36.0 kΩ±2
11.	____	____	____	____	____	18.2 kΩ±0.5
12.	____	____	____	____	____	583 kΩ±0.1
13.	____	____	____	____	____	7.41 Ω±0.1
14.	____	____	____	____	____	2.80 Ω±0.1
15.	____	____	____	____	____	357 Ω±0.5

SEC. 2-3 SURFACE-MOUNT RESISTOR CODING

Unlike axial-lead resistors, surface-mount resistors do not use colored bands or stripes to indicate their resistance value in ohms. Instead, surface-mount resistors use a three- or four-digit number printed on the film or body side of the component. This is shown in Fig. 2-59. When a three-digit number is shown, the first two digits indicate the first and second digits of the resistance value. The third digit indicates how many zeros must be added after the first two digits. If a four-digit number is shown, the first, second, and third digits indicate the first three digits of the resistance value, and the fourth digit indicates how many zeros must be added.

For values between 1 and 10 Ω, the letter *R* is used to signify a decimal point. For example, a surface-mount resistor coded 3R9 indicates a resistance value of 3.9 Ω. It is important to note that surface-mount resistors do not include a tolerance rating in their coding system.

Solved Problem

For each resistor shown in Fig. 2-59, indicate the resistance value in ohms.

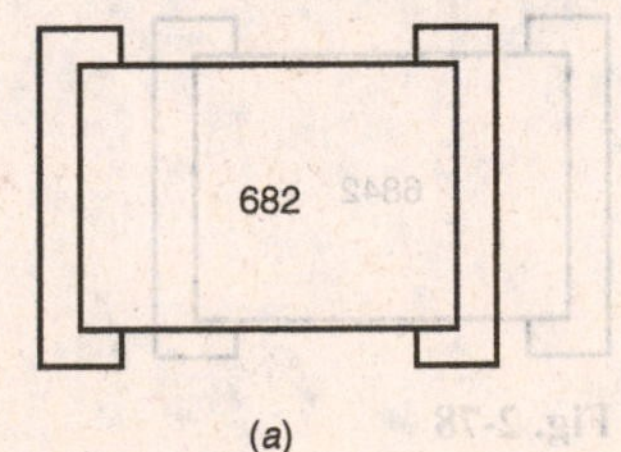

(*a*)

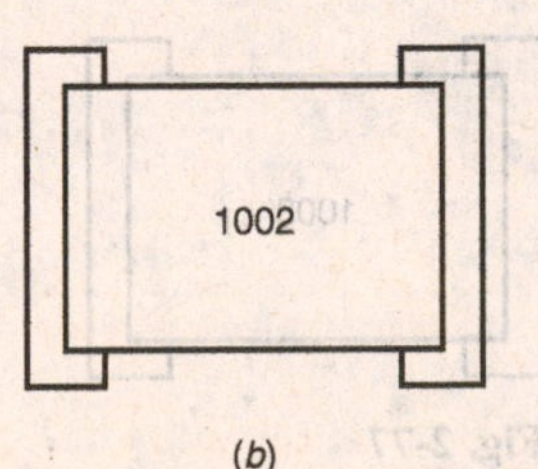

(*b*)

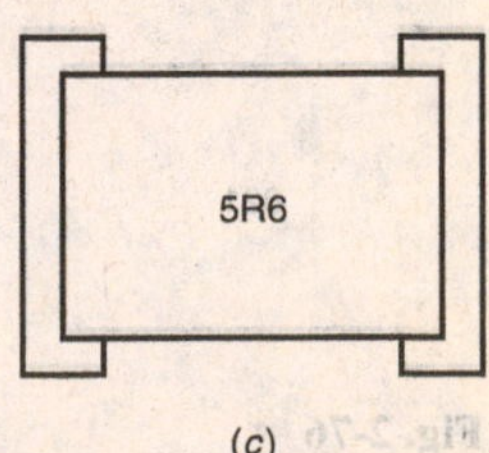

(*c*)

Fig. 2-59

Answer

The resistor in Fig. 2-59*a* shows a three-digit code. The first two digits, 6 and 8, represent the first and second digits of the resistance value. The third digit, 2, means add two zeros. This gives a resistance value of 6,800 Ω or 6.8 kΩ.

In Fig. 2-59*b*, a four-digit code is shown. The first three digits, 1, 0, and 0, represent the first three digits of the resistance value. The fourth digit, 2, means add two zeros. This gives a resistance value of 10,000 Ω or 10 kΩ.

In Fig. 2-59*c*, the letter *R* is used to signify a decimal point between the digits 5 and 6. This gives a resistance value of 5.6 Ω.

PRACTICE PROBLEMS

Sec. 2-3 For each resistor shown in Figs. 2-60 through 2-79, indicate the resistance value in ohms.

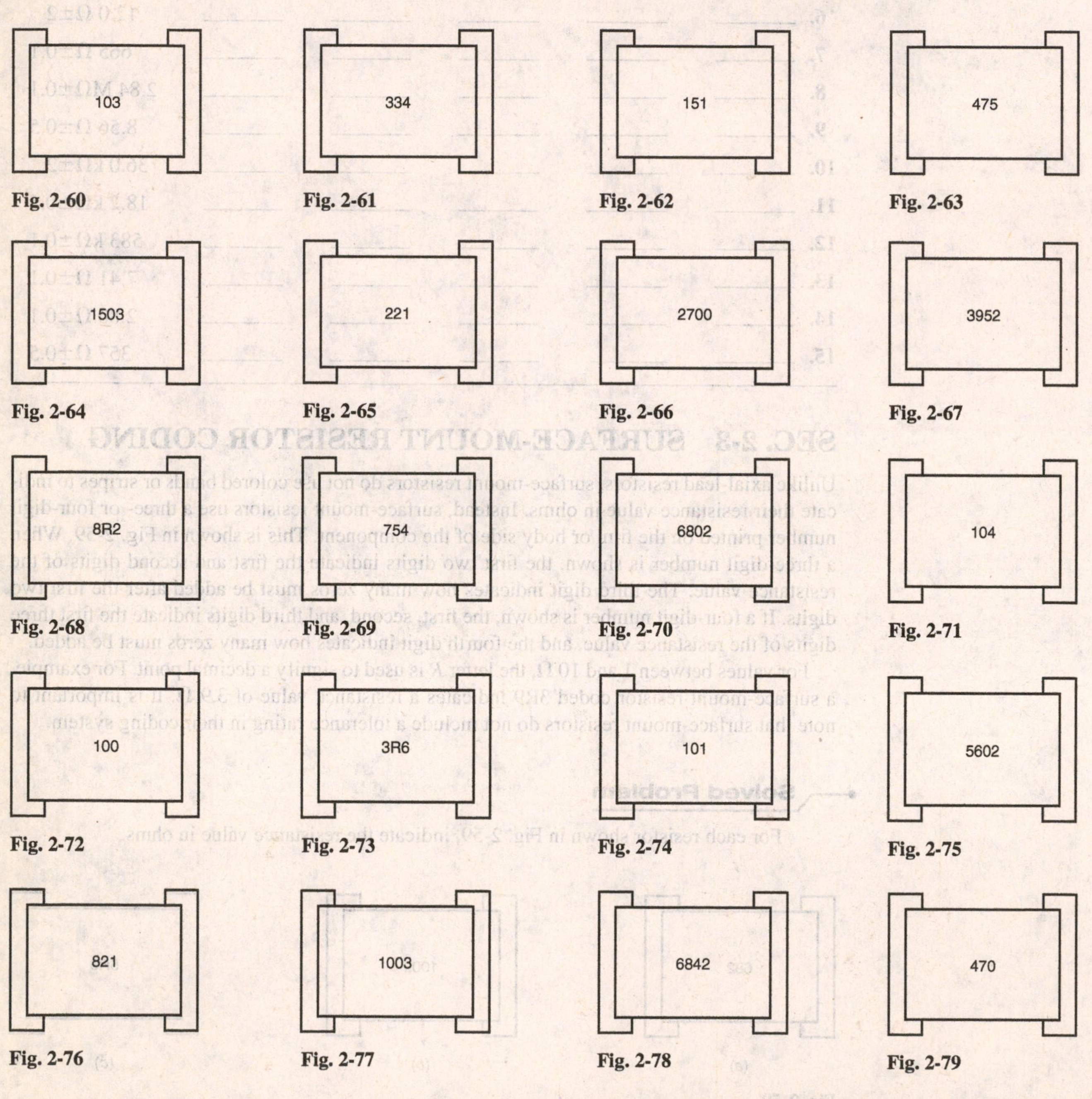

Fig. 2-60 Fig. 2-61 Fig. 2-62 Fig. 2-63

Fig. 2-64 Fig. 2-65 Fig. 2-66 Fig. 2-67

Fig. 2-68 Fig. 2-69 Fig. 2-70 Fig. 2-71

Fig. 2-72 Fig. 2-73 Fig. 2-74 Fig. 2-75

Fig. 2-76 Fig. 2-77 Fig. 2-78 Fig. 2-79

Chapter 2: Resistor Color Coding

Answer True or False.

1. The physical size of a resistor is an indication of its resistance value.
2. The physical size of a resistor is an indication of its power rating.
3. In the four-band resistor color code, a gold fourth band indicates a tolerance of ±5%.
4. When silver is used in the multiplier band, it indicates a fractional multiplier of 0.1.
5. In the resistor color-coding scheme, gold or silver never appear in the first or second bands.
6. A surface-mount resistor that is marked 5R6 has a resistance value of 5.6 Ω.
7. If a carbon resistor has only three colored bands or stripes, its tolerance is ±20%.
8. In the four-band resistor color code, a silver fourth band indicates a tolerance of ±10%.
9. Resistors with five colored bands or stripes provide more precise resistance values.
10. Orange in the multiplier band means multiply the first two or three digits by 10^3.
11. The main purpose of a resistor is to limit the current flow or to provide the desired drop in voltage.
12. If a 68 kΩ resistor, with a tolerance of ±10%, measures 77.3 kΩ, it is barely within tolerance.
13. A surface mount resistor that is marked 100 has a resistance value of 100 Ω.
14. In the five-band resistor color code, a blue fifth band indicates a tolerance of ±0.25%.
15. Gold in the multiplier band indicates a fractional multiplier of 0.1.

CHAPTER

Ohm's Law

In this chapter, you will begin examining the relationship between the voltage, current, and resistance in an electric circuit. You will also examine the formulas for calculating the power dissipation in a resistance. And finally, you will learn how to calculate the cost of energy consumption using typical electrical appliances found in our homes.

SEC. 3-1 USING OHM'S LAW

The relationship between voltage, current, and resistance in an electric circuit is very clearly expressed in a simple but extremely valuable law known as Ohm's law. This law states that the current in amperes (A) is equal to the voltage in volts (V) divided by the resistance in ohms (Ω). Ohm's law, expressed as an equation, is

$$I = \frac{V}{R}$$

where I = current in amperes
V = voltage in volts
R = resistance in ohms

If the voltage V and resistance R in a given circuit are known, the current I can be easily found. If the voltage and current are known, and the resistance is unknown, the resistance R is equal to V/I. When the voltage in a given circuit is unknown, it can be found by multiplying $I \times R$. The equations are shown in Table 3-1.

TABLE 3-1 OHM'S LAW EQUATIONS

$V = I \times R$	$I = \frac{V}{R}$
$R = \frac{V}{I}$	$G = \frac{1}{R} = \frac{I}{V}$

As you recall from Chap. 1, "Electricity," conductance is the opposite of resistance. The lower the resistance, the higher the conductance and vice versa. In a circuit, the conductance, G, can be calculated as:

$$G = \frac{1}{R}$$

or

$$G = \frac{1}{(V/I)}$$

$$G = \frac{I}{V}$$

These equations are also shown in Table 3-1.

Solved Problem

For each circuit shown in Fig. 3-1, solve for the unknown. Use the equations in Table 3-1.

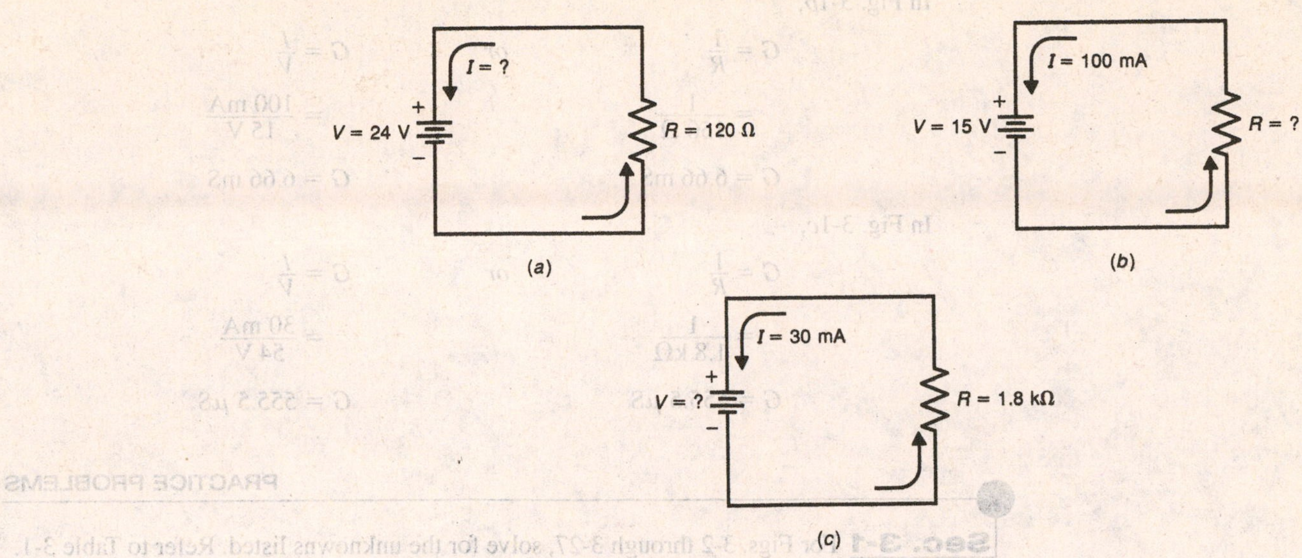

Fig. 3-1 Examples of using Ohm's law. (*a*) Finding the current when voltage and resistance are known, (*b*) finding the resistance when voltage and current are known, (*c*) finding the voltage when current and resistance are known.

Answer

In Fig. 3-1*a*, the voltage *V* and resistance *R* are both known. Since the current $I = V/R$, we have

$$I = \frac{V}{R}$$
$$= \frac{24 \text{ V}}{120 \ \Omega}$$
$$I = 200 \text{ mA}$$

In Fig. 3-1*b*, the voltage *V* and current *I* are known. Since the resistance $R = V/I$, we have

$$R = \frac{V}{I}$$
$$= \frac{15 \text{ V}}{100 \text{ mA}}$$
$$R = 150 \ \Omega$$

In Fig. 3-1*c*, the current *I* and resistance *R* are known. Since the voltage $V = I \times R$, we have

$$V = I \times R$$
$$= 30 \text{ mA} \times 1.8 \text{ k}\Omega$$
$$V = 54 \text{ V}$$

To calculate the conductance, *G*, for each circuit in Fig. 3-1, proceed as shown. In Fig. 3-1*a*,

$$G = \frac{1}{R} \quad \textit{or} \quad G = \frac{I}{V}$$
$$= \frac{1}{120 \ \Omega} \qquad = \frac{200 \text{ mA}}{24 \text{ V}}$$
$$G = 8.33 \text{ mS} \qquad G = 8.33 \text{ mS}$$

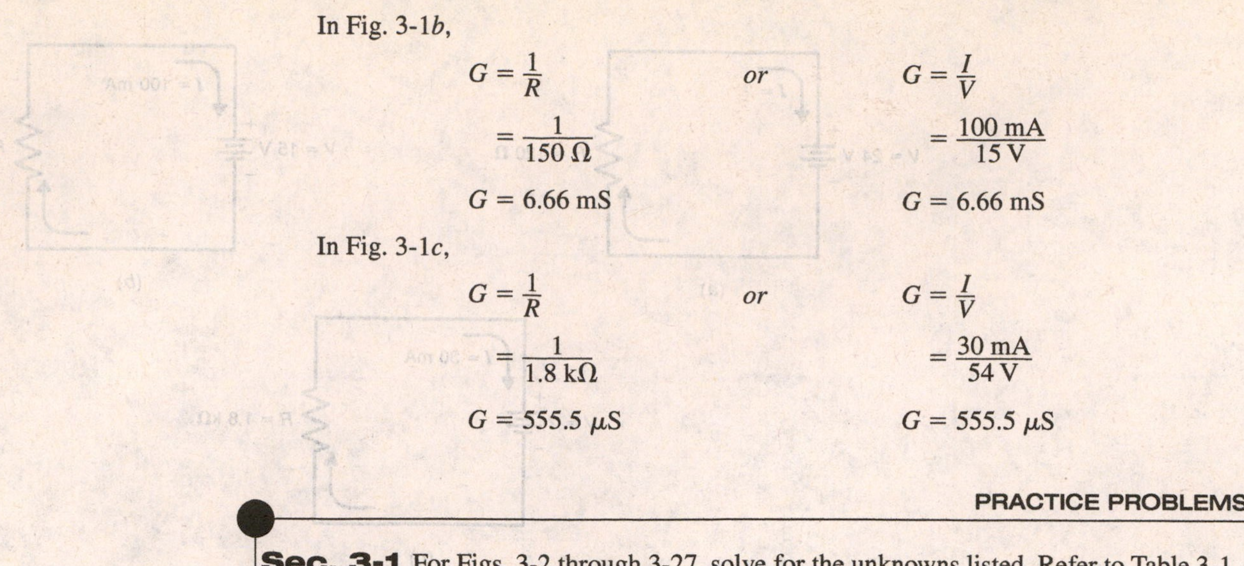

In Fig. 3-1*b*,

$$G = \frac{1}{R} \qquad or \qquad G = \frac{I}{V}$$

$$= \frac{1}{150\ \Omega} \qquad\qquad = \frac{100\text{ mA}}{15\text{ V}}$$

$$G = 6.66\text{ mS} \qquad\qquad G = 6.66\text{ mS}$$

In Fig. 3-1*c*,

$$G = \frac{1}{R} \qquad or \qquad G = \frac{I}{V}$$

$$= \frac{1}{1.8\text{ k}\Omega} \qquad\qquad = \frac{30\text{ mA}}{54\text{ V}}$$

$$G = 555.5\ \mu\text{S} \qquad\qquad G = 555.5\ \mu\text{S}$$

PRACTICE PROBLEMS

Sec. 3-1 For Figs. 3-2 through 3-27, solve for the unknowns listed. Refer to Table 3-1.

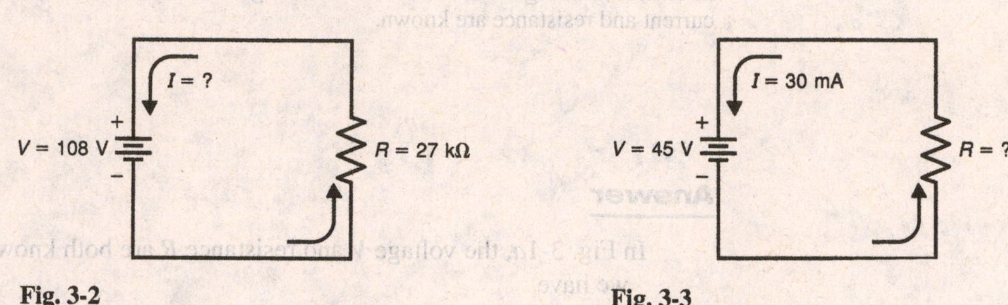

Fig. 3-2

Fig. 3-3

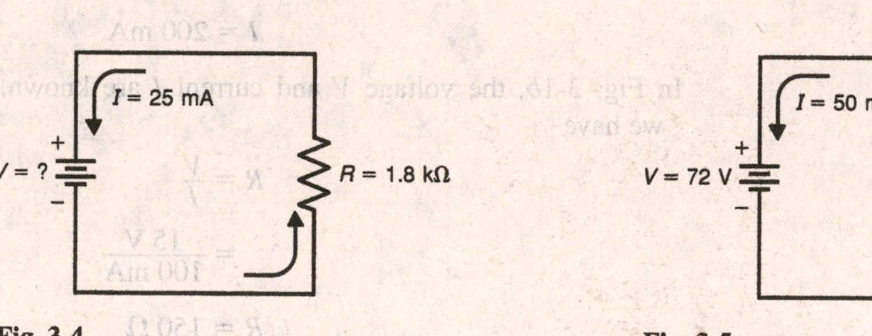

Fig. 3-4

Fig. 3-5

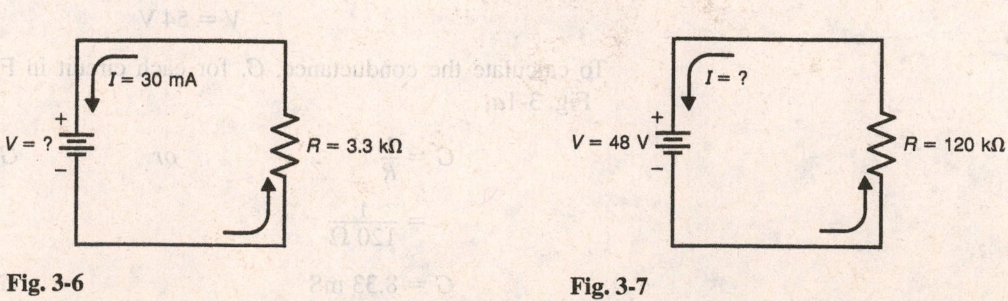

Fig. 3-6

Fig. 3-7

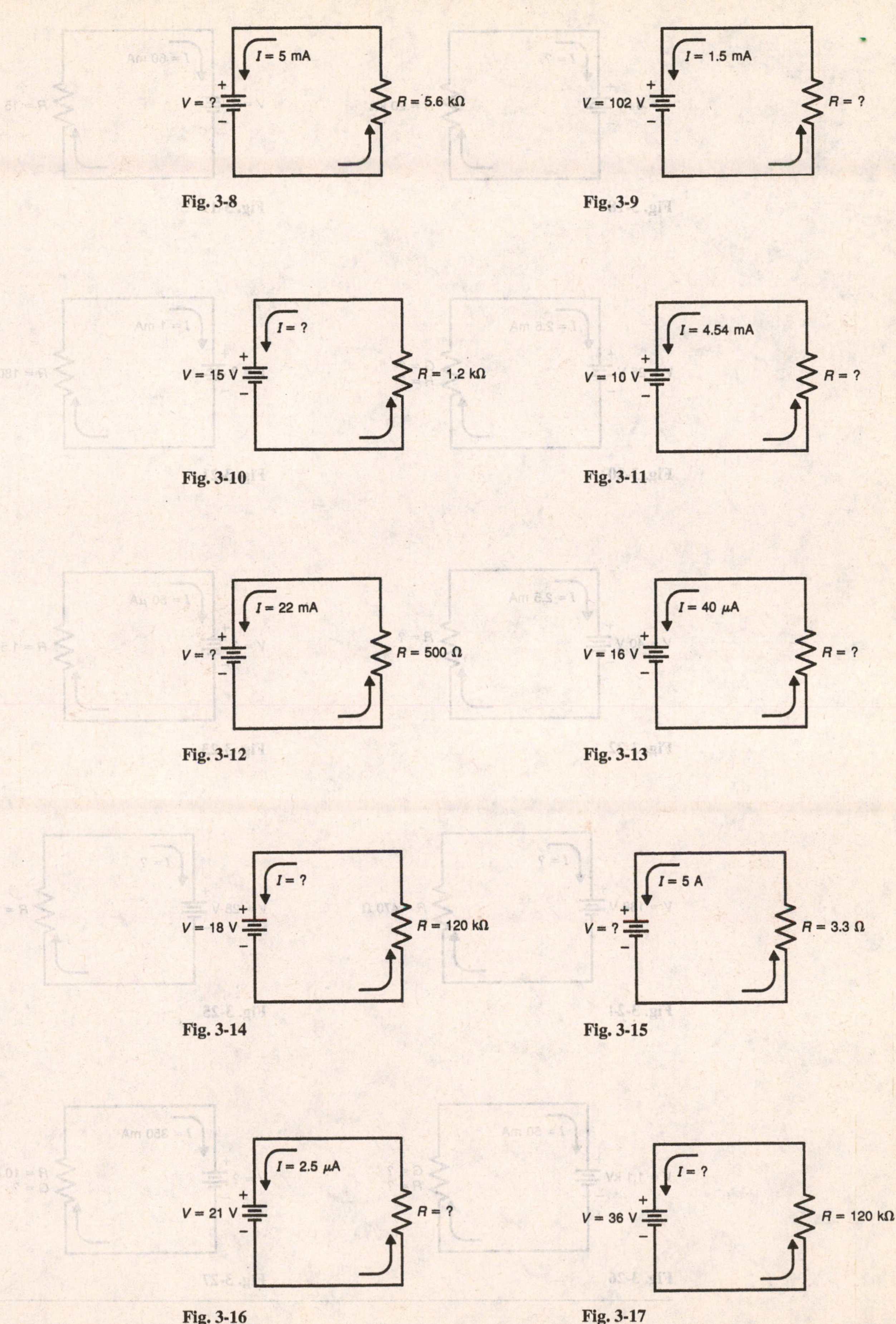

Fig. 3-8

Fig. 3-9

Fig. 3-10

Fig. 3-11

Fig. 3-12

Fig. 3-13

Fig. 3-14

Fig. 3-15

Fig. 3-16

Fig. 3-17

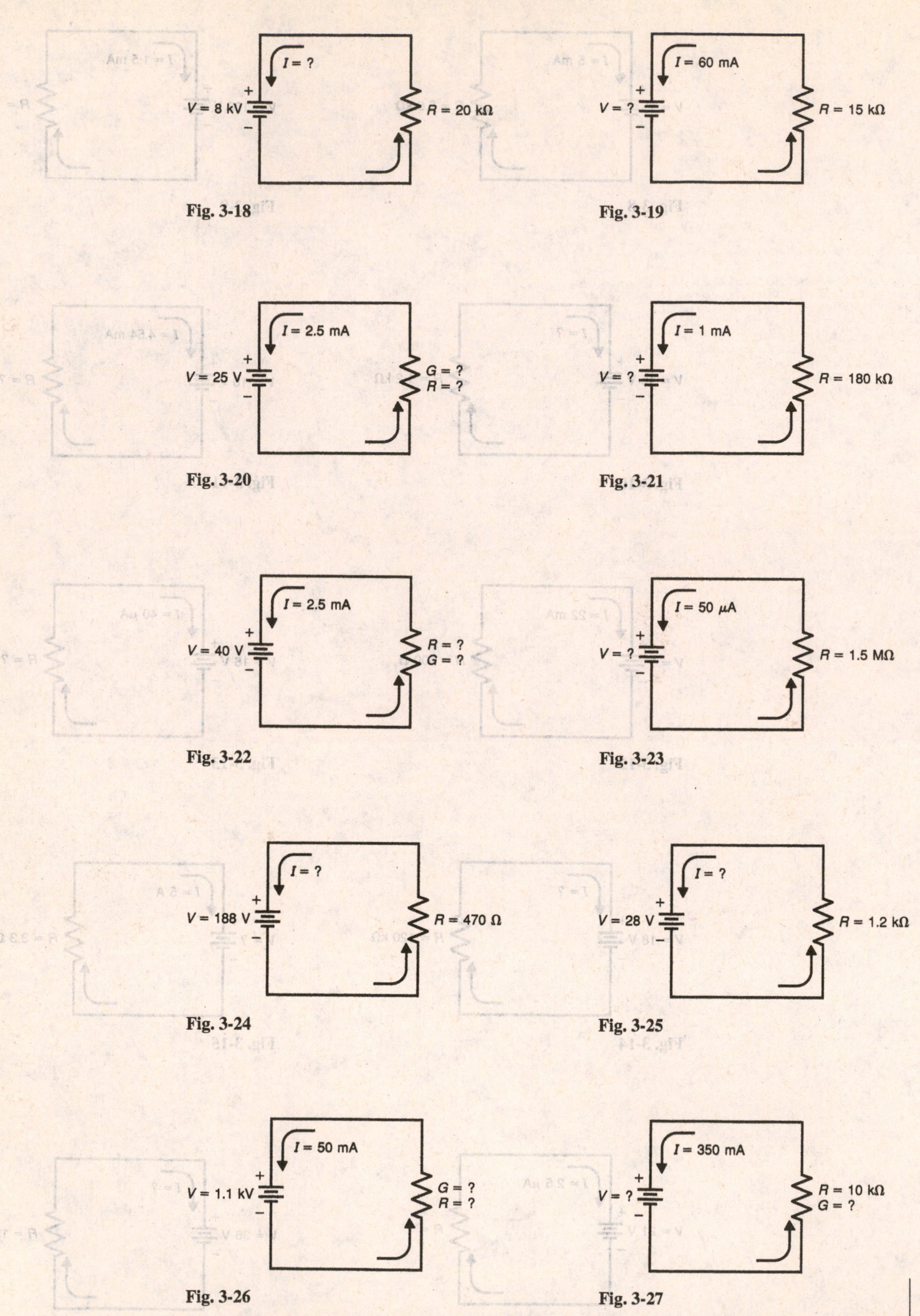

Fig. 3-18

Fig. 3-19

Fig. 3-20

Fig. 3-21

Fig. 3-22

Fig. 3-23

Fig. 3-24

Fig. 3-25

Fig. 3-26

Fig. 3-27

Sec. 3-1 Using the formulas listed in Table 3-1, solve the following word problems.

1. A relay operates at 48 V and draws 30 mA. What is its resistance?
2. What is the resistance of a buzzer if it draws 45 mA from a 9-V line?
3. A 1.5-MΩ resistor is connected across a 120-V line. What is the current flow through the resistor?
4. What current is carried by a 27-Ω line cord resistor if the voltage drop across it is 40.5 V?
5. What is the hot resistance of a lightbulb if it draws 750 mA from a 120-V line?
6. A 5-Ω heating element draws 15 A from the power line. Find the voltage drop across the heating element.
7. A 120-V line is protected with a 15-A fuse. Will the fuse carry a 6-Ω load?
8. A voltmeter has an internal resistance of 50 kΩ. What current will flow through the meter when it is placed across a 1.5-V battery?
9. An ammeter shows 5 A of current. If the current is drawn from a 12-V battery, what is the resistance of the circuit?
10. An ammeter has an internal resistance of 5 Ω. If 5 mA of current flows through the meter, find the voltage drop across it.
11. What voltage is developed across a 1-MΩ resistor when 6 μA of current is flowing through it?
12. An electromagnet draws 2.5 A from a 110-V line. What current will it draw from a 220-V line?
13. An emitter resistor in a transistor amplifier has 0.75 V across it with 7.5 mA of current through it. Calculate its resistance.
14. A 50-Ω load has 2 A of current through it. Calculate its voltage drop.
15. A motor drawing 20 A of current is connected to a 220-V line through leads that have a resistance of 0.2 Ω for each lead. What is the voltage available at the motor?
16. An insulator has 15 μA of current flowing through it with 4.5 kV across it. What is its resistance?
17. What is the resistance of an AM/FM radio if it draws 120 mA from a 13.6-V DC source?
18. The resistance of a circuit remains the same, but the applied voltage is doubled. What will happen to the current flowing through the circuit?
19. The resistance of a circuit is doubled, and the applied voltage is halved. What will happen to the current flowing through the circuit?
20. The conductance of a copper wire is 400 S. What is its resistance in ohms?

SEC. 3-2 ELECTRIC POWER

Power can be defined as the time rate of doing work. The symbol for power is *P*, and its unit is the watt (W). The power in watts equals the product of voltage and current.

$$\text{Power in watts} = \text{volts} \times \text{amperes}$$

$$P = V \times I$$

The other forms of this equation are

$$I = \frac{P}{V}$$

$$V = \frac{P}{I}$$

Since power is dissipated in the resistance of a circuit, it is convenient to express the power in terms of the resistance R. By substituting V/R for I, the formula $P = V \times I$ can be rearranged as

$$P = \frac{V^2}{R}$$

The other forms for this equation are

$$R = \frac{V^2}{P}$$

$$V = \sqrt{P \times R}$$

By substituting $I \times R$ for V, the formula $P = V \times I$ can also be arranged as

$$P = I^2 \times R$$

The other forms for this equation are

$$R = \frac{P}{I^2}$$

$$I = \sqrt{\frac{P}{R}}$$

Solved Problem

For each circuit shown in Fig. 3-28, calculate the power dissipated by the resistance R.

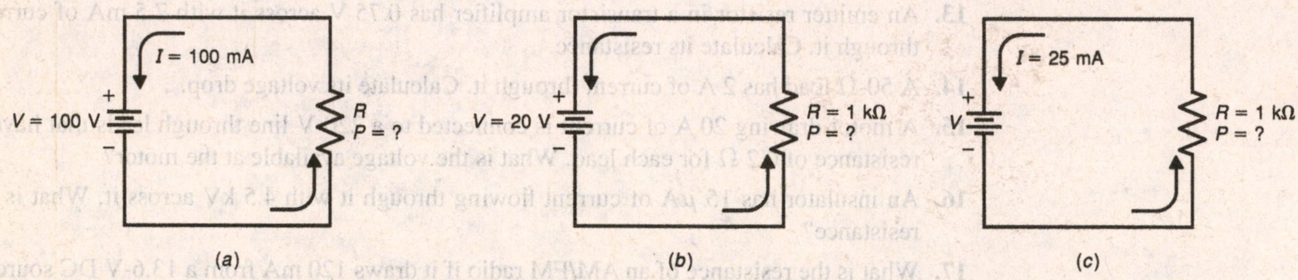

Fig. 3-28 Examples for calculating power. (*a*) Finding the power dissipated in the resistance when voltage and current are known, (*b*) finding the power dissipated in the resistance when voltage and resistance are known, (*c*) finding the power dissipated in the resistance when current and resistance are known.

Answer

In Fig. 3-28*a*, the voltage V and current I are both known. The simplest solution for power P is to multiply $V \times I$.

$$P = V \times I$$
$$= 100\text{ V} \times 100\text{ mA}$$
$$P = 10\text{ W}$$

In Fig. 3-28*b*, the voltage and resistance are both known. The simplest solution for power is to square the voltage and divide by the resistance.

$$P = \frac{V^2}{R}$$
$$= \frac{(20\text{ V})^2}{1\text{ k}\Omega}$$
$$P = 400\text{ mW}$$

In Fig. 3-28*c*, the current and resistance are both known. The simplest solution for power is to square the current and multiply by the resistance.

$$P = I^2 \times R$$
$$= (25\ \text{mA})^2 \times 1\ \text{k}\Omega$$
$$P = 625\ \text{mW}$$

PRACTICE PROBLEMS

Sec. 3-2 For Figs. 3-29 through 3-48, solve for the unknowns listed.

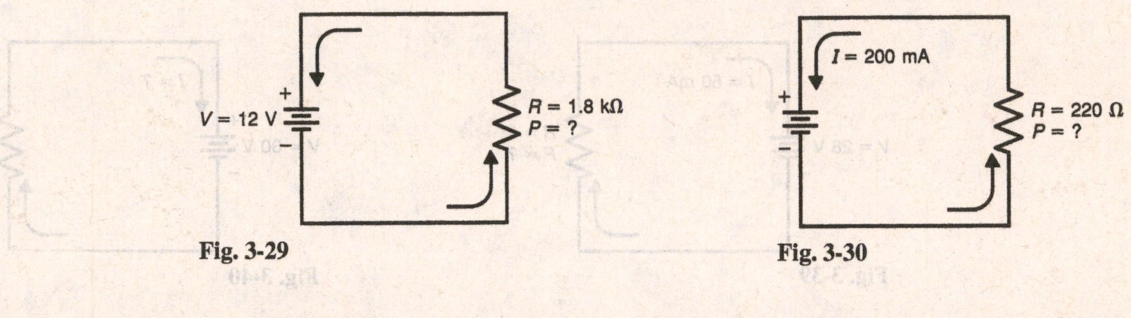

Fig. 3-29

Fig. 3-30

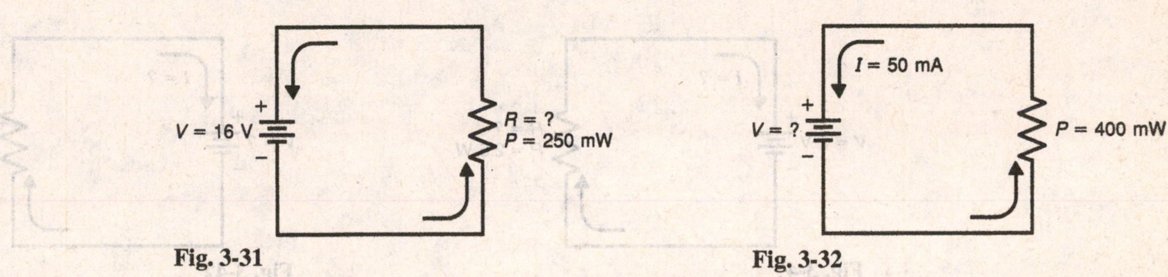

Fig. 3-31

Fig. 3-32

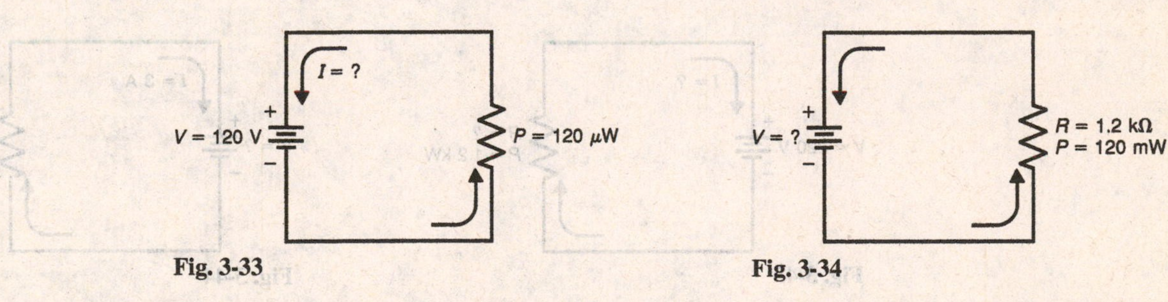

Fig. 3-33

Fig. 3-34

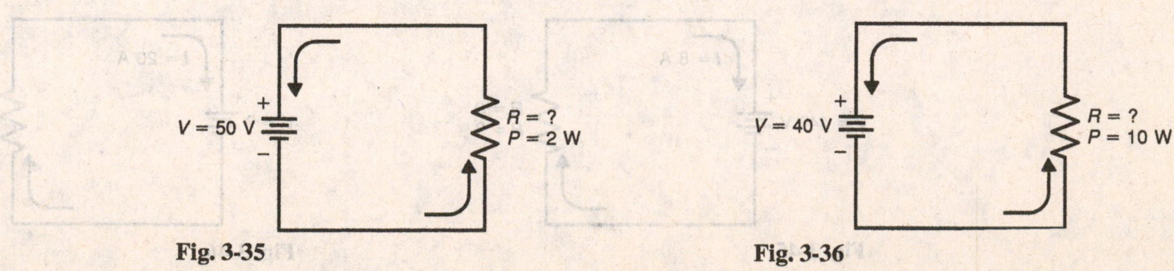

Fig. 3-35

Fig. 3-36

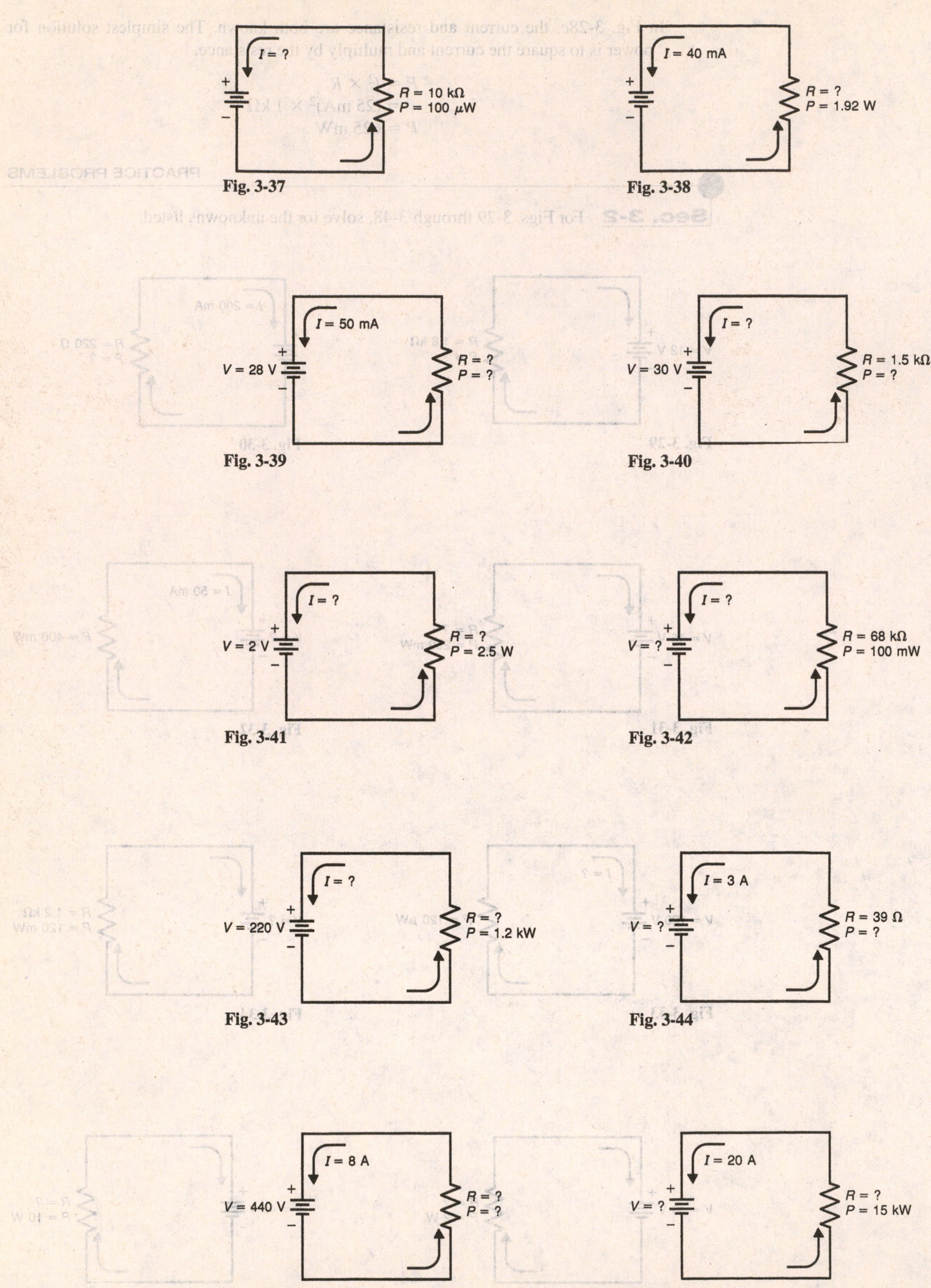

Fig. 3-37

Fig. 3-38

Fig. 3-39

Fig. 3-40

Fig. 3-41

Fig. 3-42

Fig. 3-43

Fig. 3-44

Fig. 3-45

Fig. 3-46

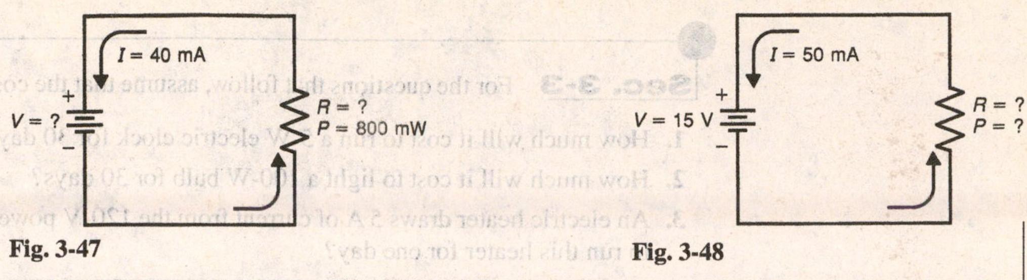

Fig. 3-47

Fig. 3-48

WORD PROBLEMS

Sec. 3-2 Solve the following word problems.

1. A clothes dryer draws 15 A from a 220-V line. Calculate the amount of power used.
2. How much current is supplied to a 6-W bulb if it is connected across a 12-V source?
3. How much current does a 30-W soldering iron draw from the 120-V power line?
4. How much current does a 75-W lightbulb draw from the 120-V power line?
5. A 50-Ω load dissipates 100 W of power. How much voltage is across the load?
6. An 8-Ω load dissipates 20 W of power. What is the current through the load?
7. A 4-Ω load dissipates 6 W of power. How much voltage is across the load?
8. A heater connected across the 120-V power line draws 5 A of current. Calculate the amount of power being used.
9. A 120-V source is connected across a 100-W lightbulb. Calculate the bulb's resistance.
10. A 27-Ω resistor has 20 mA of current flowing through it. Calculate its power dissipation.
11. A 1-kΩ resistor has 15 V across it. Calculate its power dissipation.
12. Calculate the maximum voltage drop a 1-kΩ $\frac{1}{4}$-W resistor can handle safely.
13. Calculate the maximum current a 150-Ω $\frac{1}{2}$-W resistor can have flowing through it safely.
14. Calculate the maximum voltage drop an 8-Ω 200-W load can handle safely.
15. A 100-ft copper wire carrying 50 A of current has a conductance value of 100 S. Calculate the power loss in the copper wire.

SEC. 3-3 ENERGY

Work and energy, with the symbol W, are essentially the same with identical units. The joule (J) is the basic unit of electrical energy: 1 J = 1 W × 1 s. It is seen from this formula that energy is power used during a period of time. In terms of charge and current, 1 J = 1 V × 1 C, and 1 W = 1 J/1 s.

A unit commonly used for larger amounts of electrical energy is the kilowatt-hour. As the unit's name implies, it is a product of power and time: 1 kWh = 1 kW × 1 h.

Solved Problem

How much electric energy [in kilowatt-hours (kWh)] is used by a 150-W lightbulb if it is on for 12 hours?

Answer

$$\begin{aligned} \text{kWh} &= P \times T \\ &= 150\text{ W} \times 12\text{ h} \\ \text{kWh} &= 1800\text{ watt-hours (Wh) } \textit{or } 1.8\text{ kWh} \end{aligned}$$

If the cost of energy is 10 cents/kWh, then the total cost would be 1.8 kWh × 0.10/kWh = 0.18 or 18 cents.

Sec. 3-3 For the questions that follow, assume that the cost of energy is 10 cents/kWh.

1. How much will it cost to run a 5-W electric clock for 30 days?
2. How much will it cost to light a 100-W bulb for 30 days?
3. An electric heater draws 5 A of current from the 120-V power line. How much will it cost to run this heater for one day?
4. A color television set draws 1.5 A from the 120-V power line when operating. If the set is on 50 percent of the time, what will it cost the user to operate it over a period of 30 days?
5. An air conditioner draws 15 A from the 220-V power line when running. How much will it cost to run this air conditioner for one week?
6. A quartz heater draws 12.5 A of current from the 120-V power line. How much will it cost to have this heater run for two days?
7. A light fixture containing three 60-W bulbs is left on for $2\frac{1}{2}$ days. How much will this cost the user?
8. A 1,050-W toaster is used for a total of 6 hours in a 1-year period. What is the cost per year? What is the average cost per month?
9. A water heater draws 20 A from the 220-V power line when heating. How much will it cost the user for a 30-day period if the water heater is on 10 percent of the time?
10. An electric fan draws 3 A of current from the 120-V power line. How much will it cost to run this fan for 2 weeks?

SEC. 3-4 EFFICIENCY

Efficiency is defined as the ratio of useful output power to total input power. Expressed as an equation,

$$\%\text{ efficiency} = \frac{\text{power output}}{\text{power input}} \times 100$$

Solved Problem

A 2-horsepower (hp) motor draws 11 A from the 220-V power line. Calculate its efficiency.

Answer

$$\text{Power output} = 2\text{ hp}$$
$$= 2 \times 746\text{ W}$$
$$\text{Power output} = 1{,}492\text{ W}$$

The power output is 1,492 W because 1 hp equals 746 W of electric power.

The amount of power input required to produce 2 hp is found by multiplying voltage by current.

$$P = V \times I$$
$$\text{Power in} = 220\text{ V} \times 11\text{ A}$$
$$\text{Power in} = 2{,}420\text{ W}$$

The % efficiency is calculated as

$$\%\text{ efficiency} = \frac{\text{power out}}{\text{power in}} \times 100$$
$$= \frac{1{,}492\text{ W}}{2{,}420\text{ W}} \times 100$$
$$\%\text{ efficiency} = 61.65\%$$

Sec. 3-4 Where applicable, assume that the cost of energy is 10 cents/kWh.

1. A 4-hp motor draws 20 A from the 220-V power line. What is the efficiency of the motor? How much will it cost to run the motor for 12 hours?
2. A 5-hp motor is 80 percent efficient. Calculate the current it draws from the 220-V power line.
3. Calculate the % efficiency of a 1-hp motor if it draws 9 A from the 120-V power line. How much will it cost the user if this motor runs 10 percent of the time over a 30-day period?
4. A $\frac{1}{4}$-hp motor draws 2 A of current from the 120-V power line. Calculate its efficiency.
5. A $\frac{3}{4}$-hp motor is 65 percent efficient. Calculate the current it draws from the 120-V line.
6. A $\frac{1}{2}$-hp motor is 70 percent efficient. Calculate the current it draws from the 120-V power line.
7. A 2-hp motor draws 8 A of current from the 220-V power line. Calculate the efficiency of the motor.
8. A 1-hp motor runs 15 percent of the time over a 30-day period. If the motor has an efficiency of 60 percent, how much will it cost the user?
9. A CB transmitter has 4 W of power output. If the transmitter is 80 percent efficient, how much current does it draw from a 12-V DC source?
10. An amateur radio transmitter draws 2 A of current from the 120-V power line when delivering 150 W of radio-frequency (RF) power to its antenna. Calculate the efficiency of the transmitter.

END OF CHAPTER TEST

Chapter 3: Ohm's Law Answer True or False.

1. If the voltage across a resistance is doubled, the current through the resistance will double.
2. If the voltage is held constant and the resistance increases, the current decreases.
3. Power can be defined as the time rate of doing work.
4. The joule (J) and kilowatt-hour (kWh) are both units of electrical energy.
5. 1 horsepower (hp) equals 746 W of electrical power.
6. If the current through a resistance doubles, its power dissipation also doubles.
7. When the power and resistance are known, the current, I, can be calculated using the equation $I = \frac{P}{R}$.
8. If the voltage across a resistance doubles, its power dissipation quadruples.
9. When the power and resistance are known, the voltage, V, can be calculated using the equation $V = \sqrt{P \times R}$.
10. A 100-W load operating from the 120-VAC power line draws more current than a 100-W load connected to a 12-VDC source.

CHAPTER

Series Circuits

A series circuit is any circuit that provides only one path for current flow. In this chapter, you will solve for the unknown values of voltage, current, resistance, and power in a series circuit.

SEC. 4-1 ANALYZING SERIES CIRCUITS

When components are connected in successive order, as shown in Fig. 4-1, they form a series circuit. The resistors R_1, R_2, and R_3 are in series with each other and with the battery. Perhaps the most important characteristic of series circuits is that the current, *I*, is the same at all points. This very important characteristic allows us to develop a list of equations that are applicable to series circuits. These equations are shown in Table 4-1.

TABLE 4-1 EQUATIONS USED FOR SERIES CIRCUITS

$R_T = R_1 + R_2 + R_3 + \cdots + R_n$	$V_T = V_1 + V_2 + V_3 + \cdots + V_n$
$R_T = \dfrac{V_T}{I_T}$	$V_T = I_T \times R_T$ $V_R = I_T \times R$ (voltage drop across any individual resistor)
$I_T = \dfrac{V_T}{R_T}$	$P_T = P_1 + P_2 + P_3 + \cdots + P_n$
$I_T = \dfrac{V_R}{R}$	$P_T = V_T \times I_T$ $P_R = V_R \times I_T$ (power dissipated by any individual resistor)

Solved Problem

For Fig. 4-1, find R_T, I_T, V_1, V_2, V_3, P_T, P_1, P_2, and P_3.

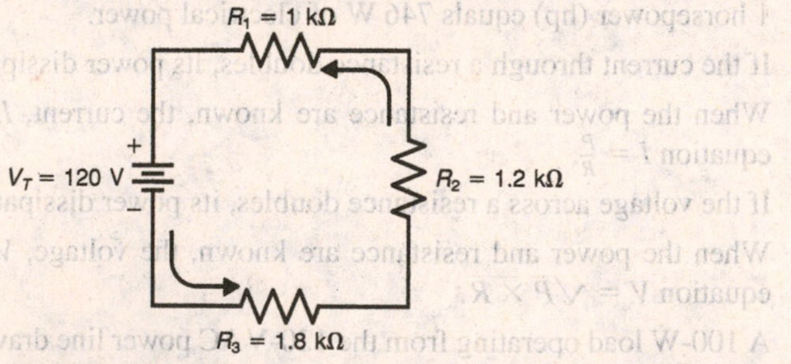

Fig. 4-1 Analyzing a series circuit to find R_T, I_T, V_1, V_2, V_3, P_T, P_1, P_2, and P_3.

Answer

Using the equations listed in Table 4-1, the procedure is as follows.

$$R_T = R_1 + R_2 + R_3$$

$$= 1\ \text{k}\Omega + 1.2\ \text{k}\Omega + 1.8\ \text{k}\Omega$$

$$R_T = 4\ \text{k}\Omega$$

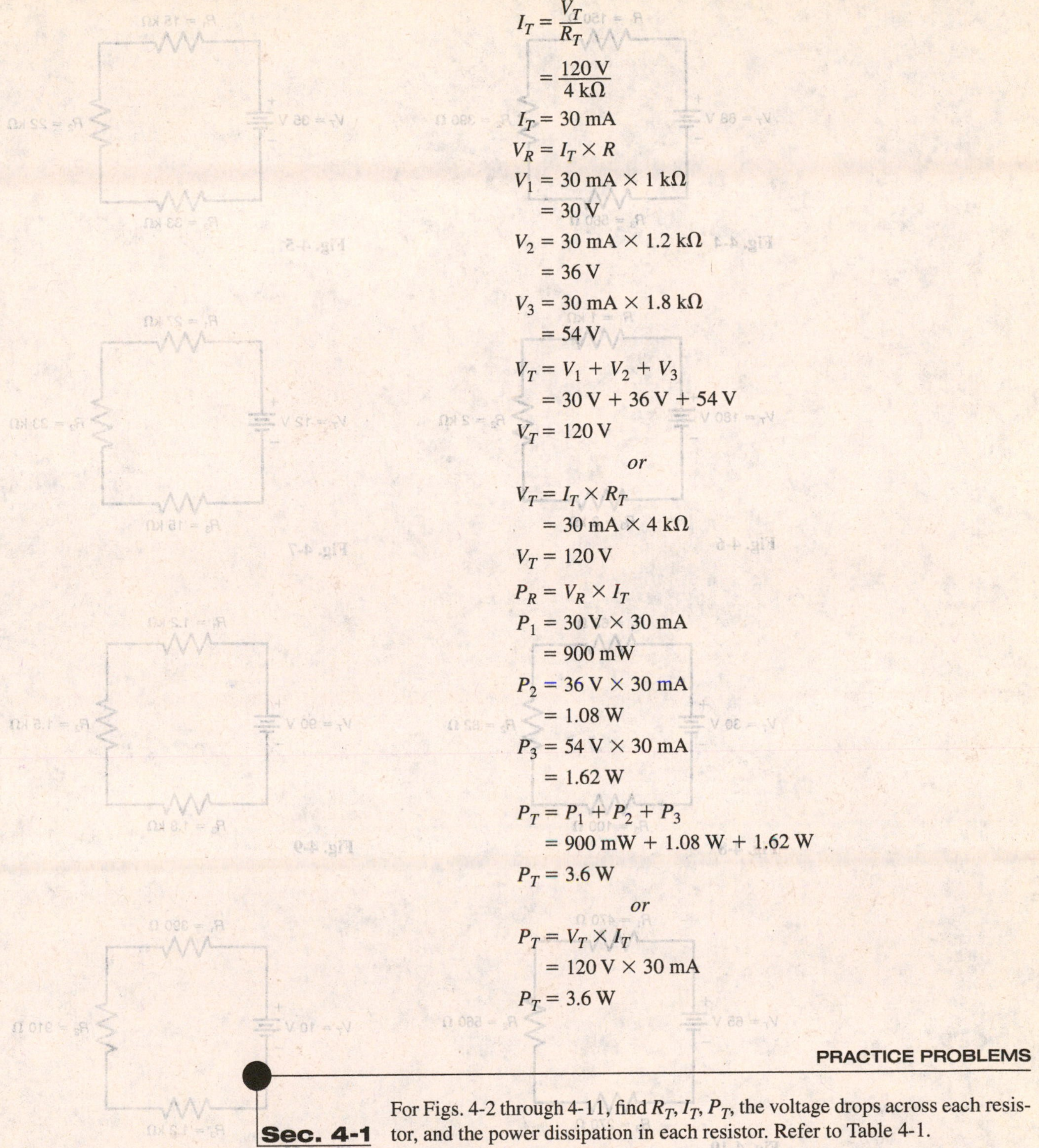

$$I_T = \frac{V_T}{R_T}$$

$$= \frac{120\text{ V}}{4\text{ k}\Omega}$$

$$I_T = 30\text{ mA}$$

$$V_R = I_T \times R$$

$$V_1 = 30\text{ mA} \times 1\text{ k}\Omega$$

$$= 30\text{ V}$$

$$V_2 = 30\text{ mA} \times 1.2\text{ k}\Omega$$

$$= 36\text{ V}$$

$$V_3 = 30\text{ mA} \times 1.8\text{ k}\Omega$$

$$= 54\text{ V}$$

$$V_T = V_1 + V_2 + V_3$$

$$= 30\text{ V} + 36\text{ V} + 54\text{ V}$$

$$V_T = 120\text{ V}$$

or

$$V_T = I_T \times R_T$$

$$= 30\text{ mA} \times 4\text{ k}\Omega$$

$$V_T = 120\text{ V}$$

$$P_R = V_R \times I_T$$

$$P_1 = 30\text{ V} \times 30\text{ mA}$$

$$= 900\text{ mW}$$

$$P_2 = 36\text{ V} \times 30\text{ mA}$$

$$= 1.08\text{ W}$$

$$P_3 = 54\text{ V} \times 30\text{ mA}$$

$$= 1.62\text{ W}$$

$$P_T = P_1 + P_2 + P_3$$

$$= 900\text{ mW} + 1.08\text{ W} + 1.62\text{ W}$$

$$P_T = 3.6\text{ W}$$

or

$$P_T = V_T \times I_T$$

$$= 120\text{ V} \times 30\text{ mA}$$

$$P_T = 3.6\text{ W}$$

PRACTICE PROBLEMS

Sec. 4-1 For Figs. 4-2 through 4-11, find R_T, I_T, P_T, the voltage drops across each resistor, and the power dissipation in each resistor. Refer to Table 4-1.

Fig. 4-2

Fig. 4-3

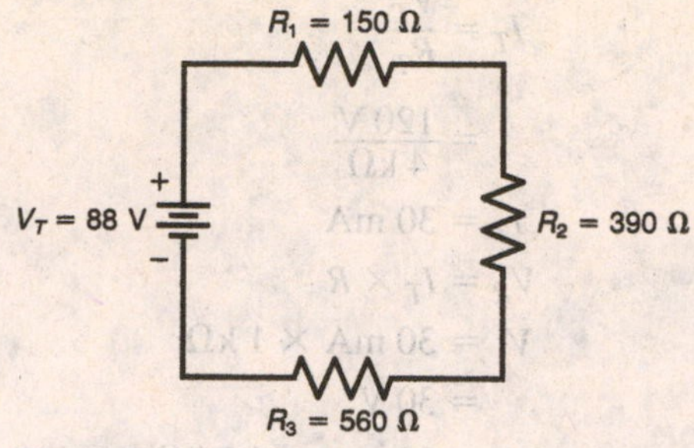

Fig. 4-4

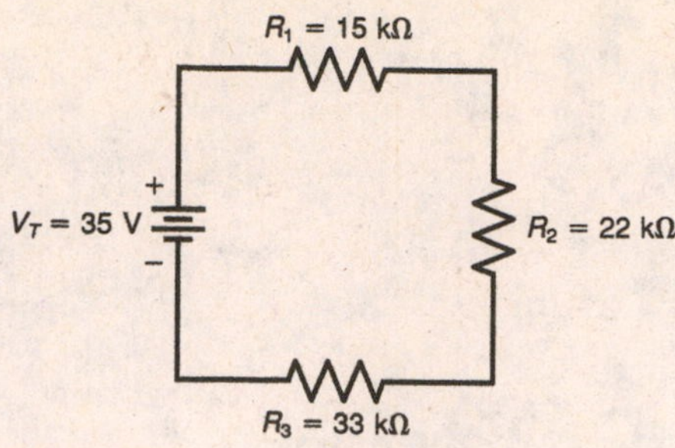

Fig. 4-5

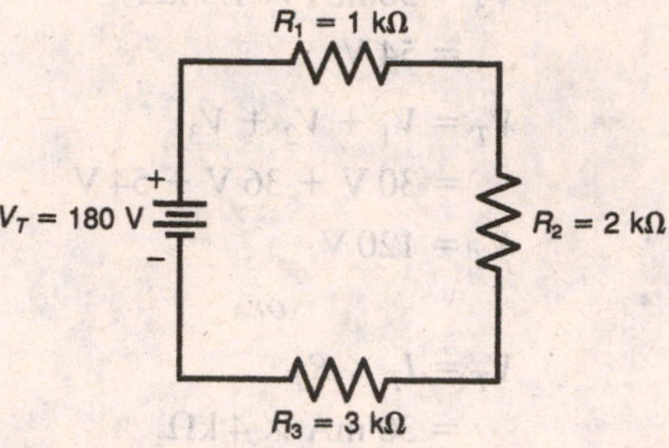

Fig. 4-6

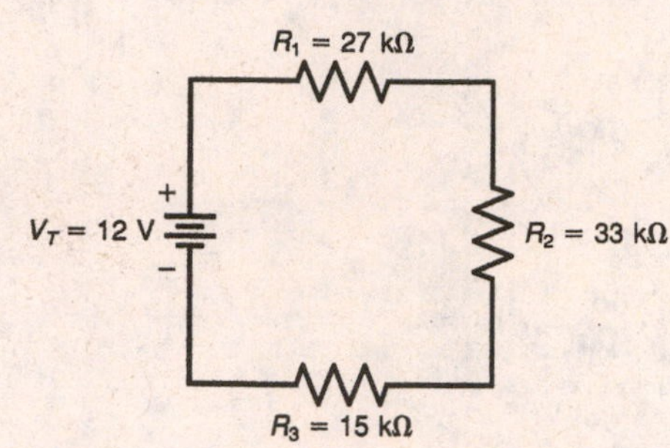

Fig. 4-7

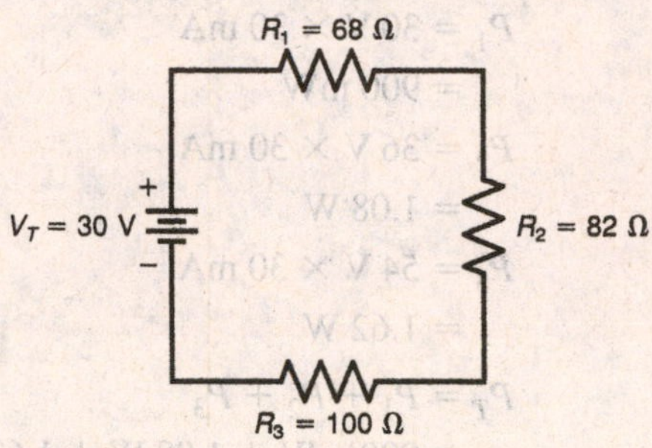

Fig. 4-8

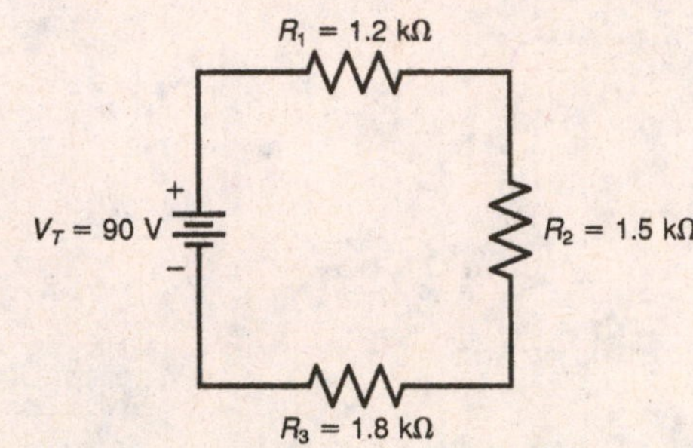

Fig. 4-9

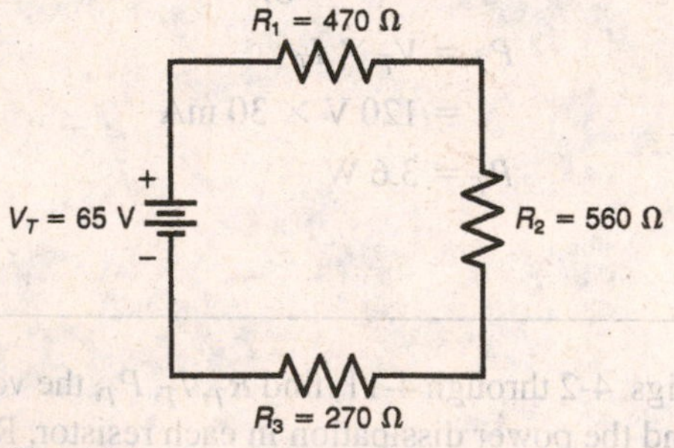

Fig. 4-10

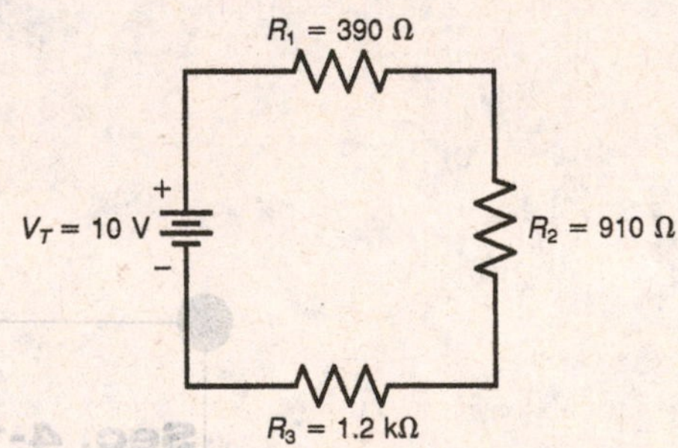

Fig. 4-11

WORD PROBLEMS

Sec. 4-1 Solve the following word problems.

1. A 9-V radio requires 200 mA to operate. In order to operate the radio from a 12-V battery at the same current, what size resistor must be connected in series with it?
2. A 12-Ω and 18-Ω resistor are in series. The current flow through the 18-Ω resistor is 150 mA. How much voltage exists across both resistors?

3. A 1.2-kΩ resistor is in series with an unknown resistor across a 20-V source. The current through the 1.2-kΩ resistor is 12.5 mA. Calculate the value of the unknown resistor.
4. Three resistors of 6 Ω, 12 Ω, and 24 Ω are connected in series across a DC voltage source. The voltage drop across the 12-Ω resistor is 9 V. What is the value of the applied voltage?
5. A 1-kΩ and 3-kΩ resistor are in series. The total power dissipated is 25 mW. Calculate the voltage dropped across the 1-kΩ resistor.
6. A 200-Ω and 600-Ω resistor are in series. The total power dissipated is 32 W. Calculate the voltage drop across the 600-Ω resistor.
7. A 50-Ω resistor is in series with an unknown resistor. The voltage drop across the unknown resistor is 25 V, and the power dissipated by the 50-Ω resistor is 0.5 W. Calculate the value of the unknown resistor.
8. A 6.8-kΩ and 8.2-kΩ resistor are in series. The total power dissipation of the series combination is 135 mW. Calculate the amount of applied voltage.
9. Three resistors of 1 kΩ, 2 kΩ, and 3 kΩ are in series across a 20-V source. If the 2-kΩ resistor is shorted, how much voltage exists across the 3-kΩ resistor?
10. Three resistors connected in series have a total resistance of 3.6 kΩ. The resistors are labeled R_1, R_2, and R_3. The resistor R_2 is twice the value of R_1, and R_3 is three times the value of R_1. Calculate the value of each resistor.
11. A 1-kΩ $^1/_4$-W resistor and a 1.5-kΩ $^1/_8$-W resistor are in series. What is the maximum amount of voltage that can be applied to this circuit without exceeding the wattage rating of either resistor?
12. A 120-Ω $^1/_2$-W resistor and a 180-Ω 1-W resistor are in series. What is the maximum total current that this series circuit can handle without exceeding the power dissipation rating of either resistor?
13. Four resistors connected in series have a total resistance of 6 kΩ. The resistors are labeled R_1, R_2, R_3, and R_4. The resistor R_2 is twice the value of R_1, R_3 is three times the value of R_1, and R_4 is twice the value of R_3. Calculate the value of each resistor.
14. Four series-connected resistors have a total resistance R_T of 24 kΩ. The resistor R_2 is twice the value of R_1, and R_3 is three times the value of R_2. The resistor $R_4 = 6$ kΩ. Calculate R_1, R_2, and R_3.
15. How much resistance must be added in series with a 150-Ω resistor to limit the current from a 9-V battery to 36 mA?

SEC. 4-2 SERIES CIRCUITS WITH RANDOM UNKNOWNS

Solved Problem

For Fig. 4-12, solve for I_T, R_T, V_2, V_3, R_3, P_T, P_1, P_2, and P_3.

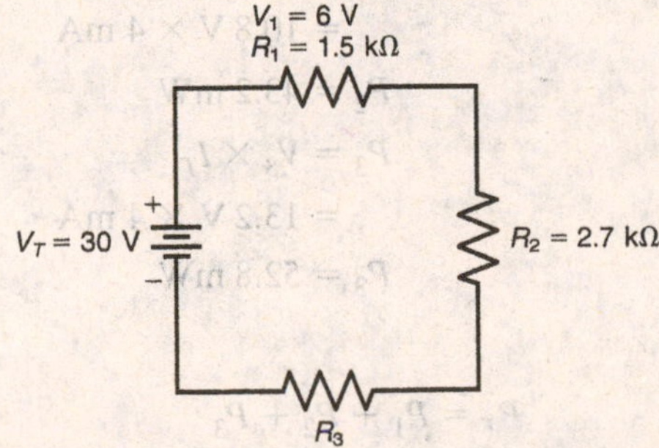

Fig. 4-12 Analyzing a series circuit to find I_T, R_T, V_2, V_3, R_3, P_T, P_1, P_2, and P_3.

Answer

When the current I is known for one component, this value can be used for the I in all the components because the current is the same in all parts of a series circuit. In Fig. 4-12, we have

$$I_T = \frac{V_1}{R_1}$$
$$= \frac{6\text{ V}}{1.5\text{ k}\Omega}$$
$$I_T = 4\text{ mA}$$

When I_T is known, R_T can be found.

$$R_T = \frac{V_T}{I_T}$$
$$= \frac{30\text{ V}}{4\text{ mA}}$$
$$R_T = 7.5\text{ k}\Omega$$

Knowing R_T, we can now solve for R_3. Since $R_T = R_1 + R_2 + R_3$,

$$R_3 = R_T - R_1 - R_2$$
$$= 7.5\text{ k}\Omega - 1.5\text{ k}\Omega - 2.7\text{ k}\Omega$$
$$R_3 = 3.3\text{ k}\Omega$$

The voltage drops across resistors R_2 and R_3 can be found as follows.

$$V_2 = I_T \times R_2$$
$$= 4\text{ mA} \times 2.7\text{ k}\Omega$$
$$V_2 = 10.8\text{ V}$$
$$V_3 = I_T \times R_3$$
$$= 4\text{ mA} \times 3.3\text{ k}\Omega$$
$$V_3 = 13.2\text{ V}$$

Power dissipation values are found by the methods shown.

$$P_T = V_T \times I_T$$
$$= 30\text{ V} \times 4\text{ mA}$$
$$P_T = 120\text{ mW}$$
$$P_1 = V_1 \times I_T$$
$$= 6\text{ V} \times 4\text{ mA}$$
$$P_1 = 24\text{ mW}$$
$$P_2 = V_2 \times I_T$$
$$= 10.8\text{ V} \times 4\text{ mA}$$
$$P_2 = 43.2\text{ mW}$$
$$P_3 = V_3 \times I_T$$
$$= 13.2\text{ V} \times 4\text{ mA}$$
$$P_3 = 52.8\text{ mW}$$

Also, note that

$$P_T = P_1 + P_2 + P_3$$
$$= 24\text{ mW} + 43.2\text{ mW} + 52.8\text{ mW}$$
$$P_T = 120\text{ mW}$$

Sec. 4-2 Solve for the unknowns listed in Figs. 4-13 through 4-32.

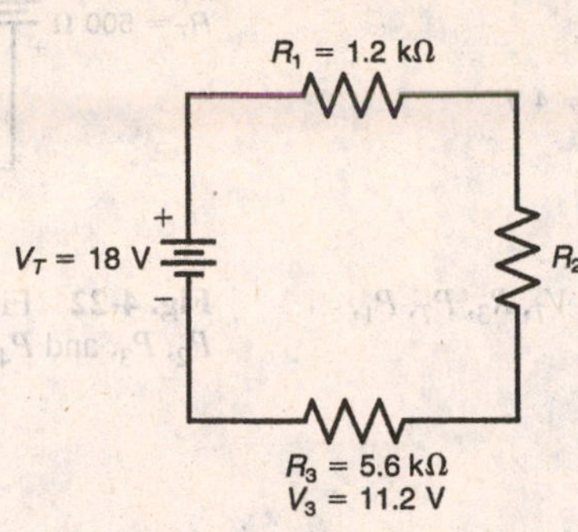

Fig. 4-13 Find R_T, I_T, R_2, V_1, V_2, P_T, P_1, P_2, and P_3.

R_1 = 1 kΩ
V_1 = 2.4 V
V_T
R_2 = 2.2 kΩ
R_3
V_3 = 16.32 V

Fig. 4-14 Find R_T, I_T, V_T, V_2, R_3, P_T, P_1, P_2, and P_3.

R_1
V_T = 90 V
V_2 = 30 V
R_2 = 20 kΩ
R_3 = 10 kΩ

Fig. 4-15 Find R_T, I_T, V_1, V_3, R_1, P_T, P_1, P_2, and P_3.

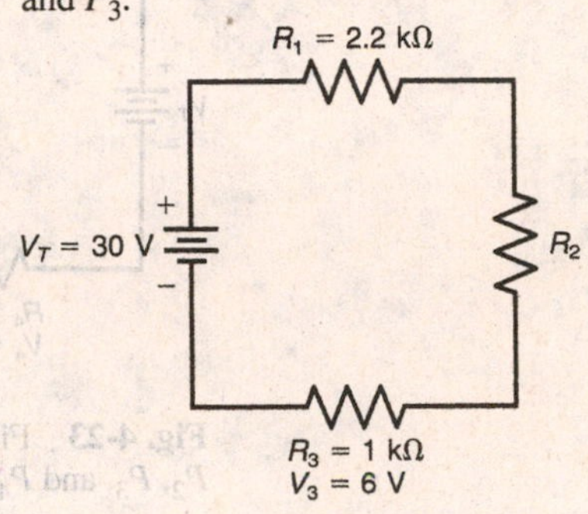

Fig. 4-16 Find R_T, I_T, V_1, V_2, R_2, P_T, P_1, P_2, and P_3.

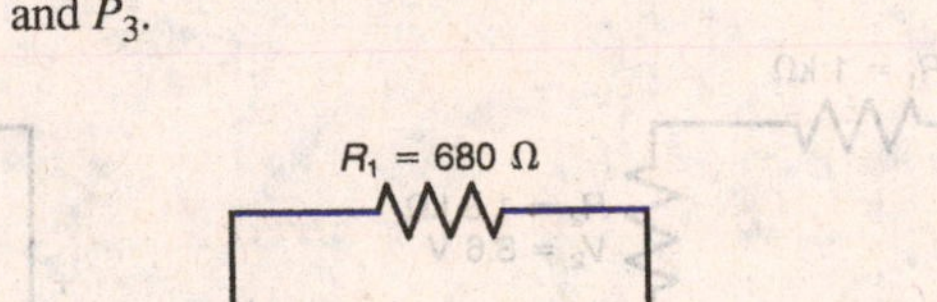
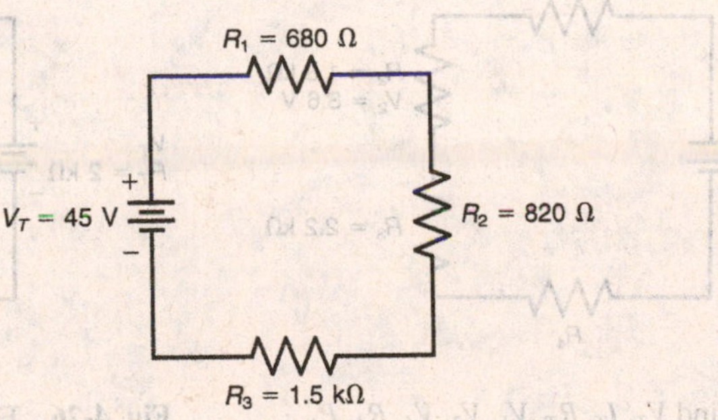

Fig. 4-17 Find R_T, I_T, V_1, V_2, V_3, P_T, P_1, P_2, and P_3.

R_1 = 1.2 kΩ
V_1 = 7.2 V
V_T = 120 V
R_2 = 6.8 kΩ
R_3

Fig. 4-18 Find R_T, I_T, V_2, V_3, R_3, P_T, P_1, P_2, and P_3.

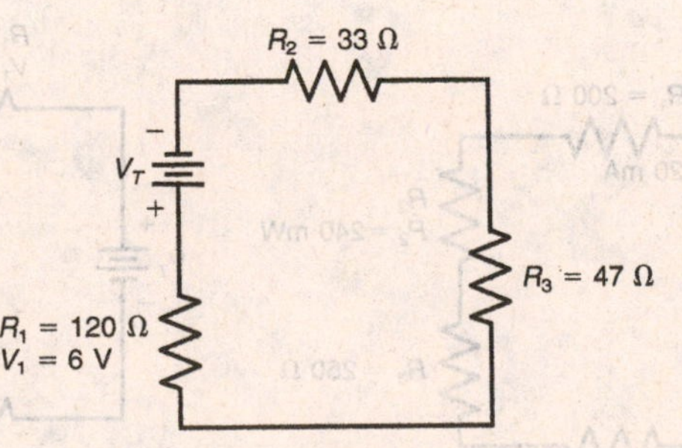

Fig. 4-19 Find V_T, R_T, I_T, V_2, V_3, P_T, P_1, P_2, and P_3.

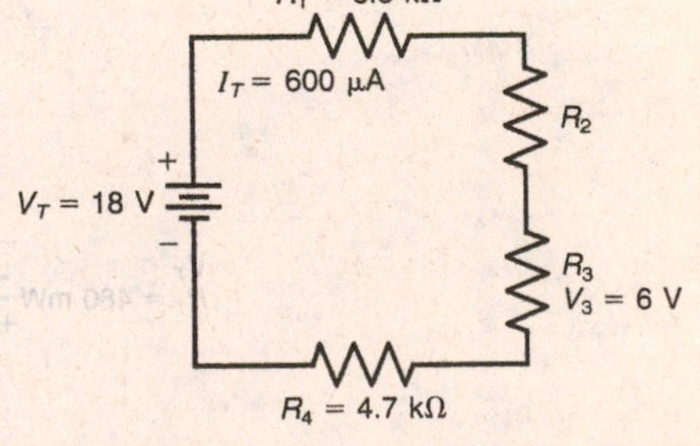

Fig. 4-20 Find R_T, V_1, V_2, V_4, R_2, R_3, P_T, P_1, P_2, P_3, and P_4.

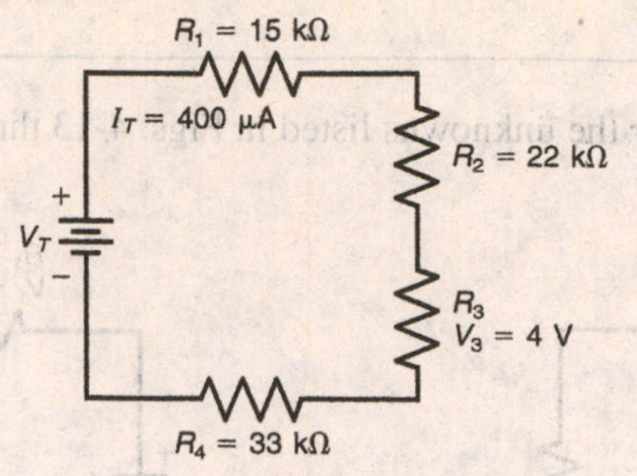

Fig. 4-21 Find R_T, V_1, V_2, V_4, V_T, R_3, P_T, P_1, P_2, P_3, and P_4.

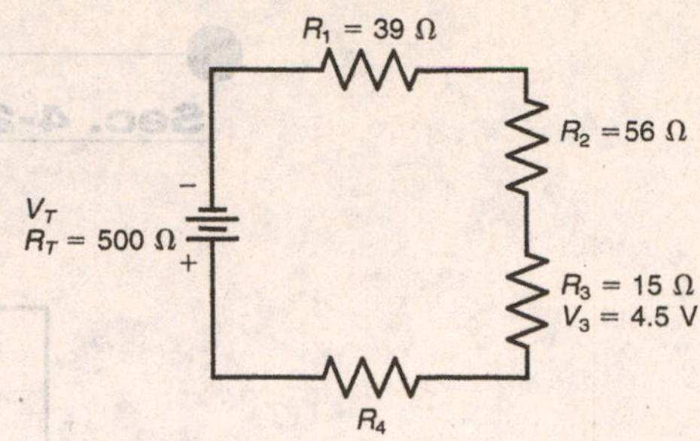

Fig. 4-22 Find V_T, I_T, V_1, V_2, V_4, R_4, P_T, P_1, P_2, P_3, and P_4.

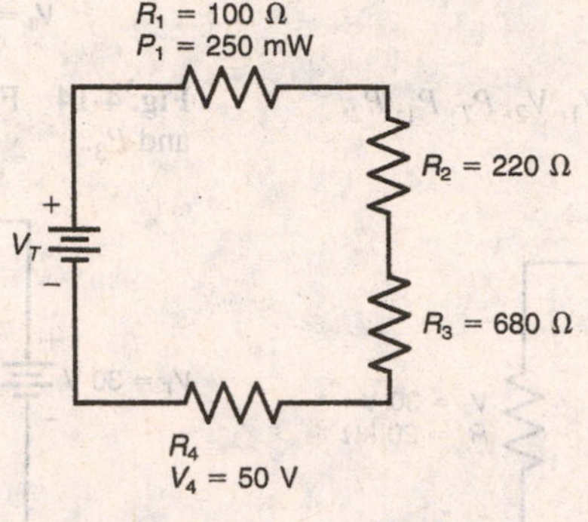

Fig. 4-23 Find V_T, I_T, R_T, V_1, V_2, V_3, R_4, P_T, P_2, P_3, and P_4.

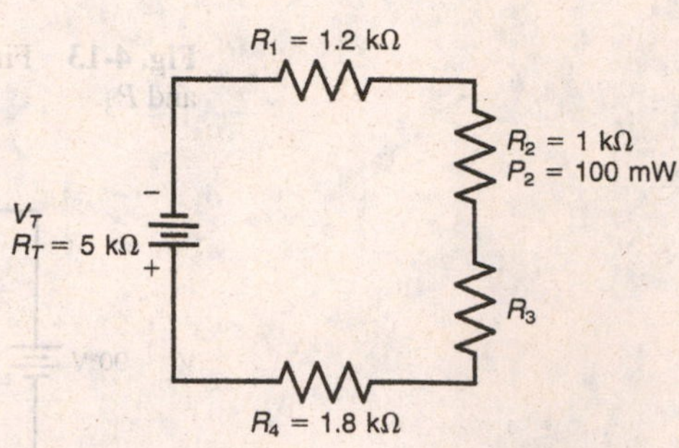

Fig. 4-24 Find V_T, I_T, V_1, V_2, V_3, V_4, R_3, P_T, P_1, P_3, and P_4.

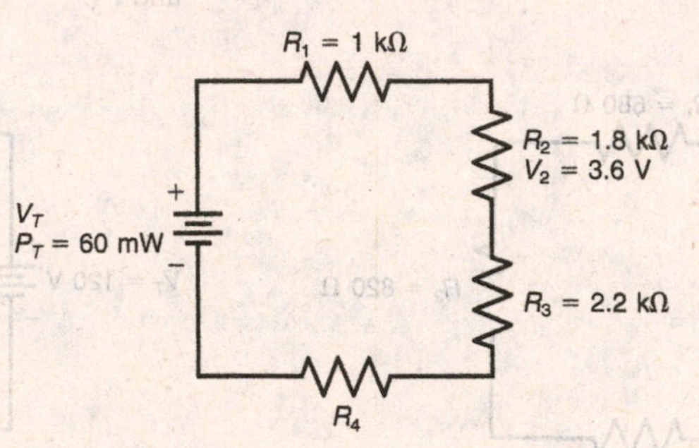

Fig. 4-25 Find V_T, I_T, R_T, V_1, V_3, V_4, R_4, P_1, P_2, P_3, and P_4.

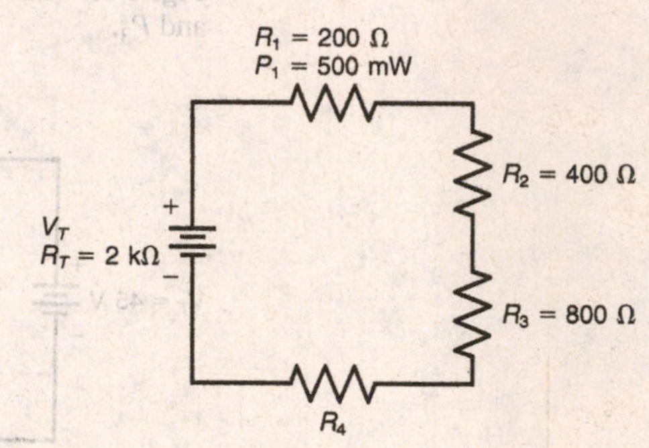

Fig. 4-26 Find V_T, I_T, V_1, V_2, V_3, V_4, R_4, P_T, P_2, P_3, and P_4.

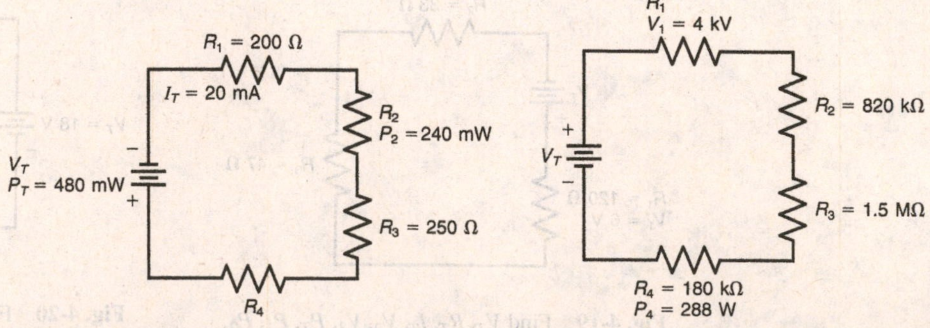

Fig. 4-27 Find V_T, R_T, V_1, V_2, V_3, V_4, R_4, R_2, P_1, P_3, and P_4.

Fig. 4-28 Find V_T, I_T, R_T, V_2, V_3, V_4, R_1, P_T, P_1, P_2, and P_3.

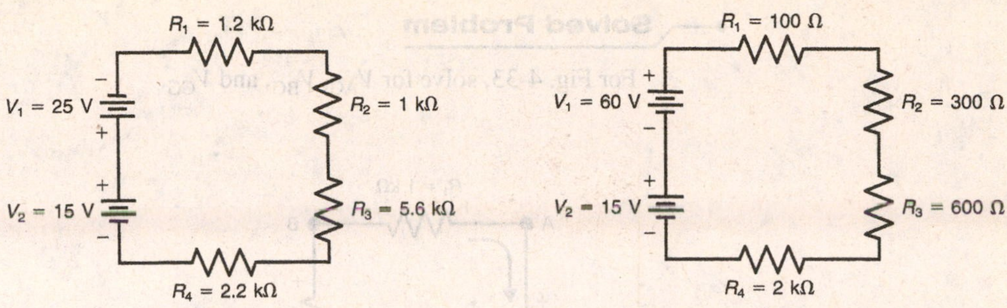

Fig. 4-29 Find V_T, I_T, R_T, V_{R_1}, V_{R_2}, V_{R_3}, V_{R_4}, P_T, P_{R_1}, P_{R_2}, P_{R_3}, and P_{R_4}.

Fig. 4-30 Find V_T, I_T, R_T, V_{R_1}, V_{R_2}, V_{R_3}, V_{R_4}, P_T, P_{R_1}, P_{R_2}, P_{R_3}, and P_{R_4}.

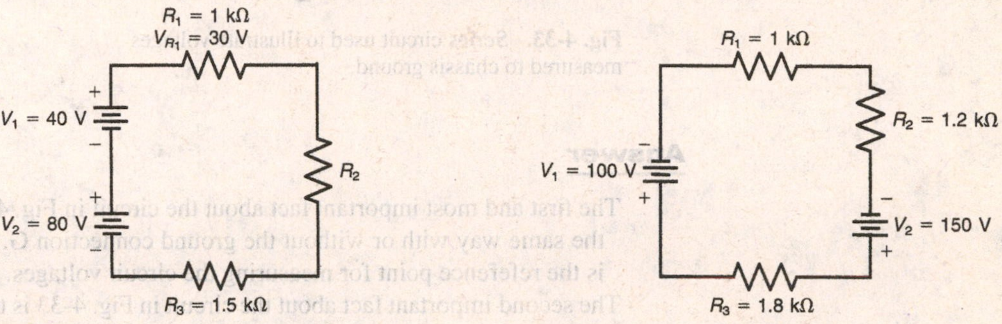

Fig. 4-31 Find V_T, I_T, R_T, V_{R_2}, V_{R_3}, R_2, P_T, P_{R_1}, P_{R_2}, and P_{R_3}.

Fig. 4-32 Find V_T, I_T, R_T, V_{R_1}, V_{R_2}, V_{R_3}, P_T, P_{R_1}, P_{R_2}, and P_{R_3}.

SEC. 4-3 VOLTAGES MEASURED TO CHASSIS GROUND

In practical circuits, one side of the voltage source is connected to ground (usually chassis ground). The purpose is to simplify the wiring. When a circuit has the chassis as a common return, we generally measure voltages with respect to this chassis ground. Some people have the misconception that ground is always negative, but this is not necessarily so. In fact, in electronic circuits, voltages can be either positive or negative with respect to chassis ground. To determine whether a voltage is positive or negative with respect to chassis ground, we must first understand how the polarity of a voltage drop is determined.

Polarities of Voltage Drops

When an *IR* voltage drop exists across a resistance, one end must be either positive or negative with respect to the other end. The polarity of this *IR* voltage can be associated with the direction of the current through *R*. In brief, we label the side of the resistor in which electrons enter as negative and the side of the resistor in which electrons leave as positive.

Double Subscript Notation

Double subscript notation is very common when voltages in electronic circuits are specified. The second letter in the subscript denotes the point of reference, and the first letter indicates the point in the circuit at which the measurement is being taken. For example, V_{AG} would be the voltage at point A in the circuit with respect to point G.

The voltage drop of an individual resistor is not specified as being positive or negative because one end is not being specified as the reference.

Solved Problem

For Fig. 4-33, solve for V_{AG}, V_{BG}, and V_{CG}.

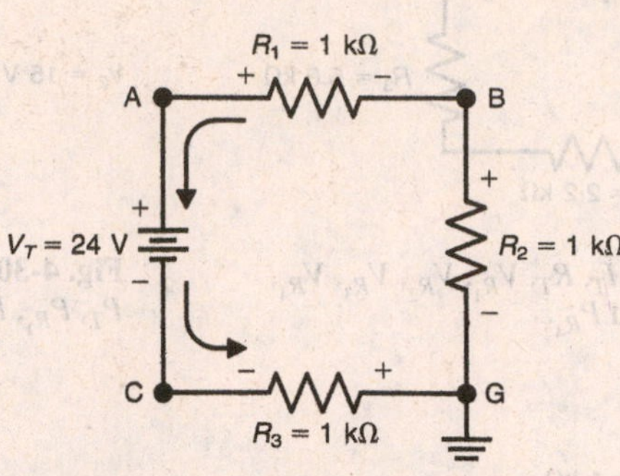

Fig. 4-33 Series circuit used to illustrate voltages measured to chassis ground.

Answer

The first and most important fact about the circuit in Fig. 4-33 is that it operates exactly the same way with or without the ground connection G. The only factor that changes is the reference point for measuring the circuit voltages.

The second important fact about the circuit in Fig. 4-33 is the polarity of each individual *IR* voltage drop. As mentioned earlier, we label the side of the resistor in which electrons enter as negative, and the side of the resistor in which electrons leave as positive.

Knowing these two factors, we solve for V_{AG}, V_{BG}, and V_{CG} as follows.

First we solve for R_T.

$$R_T = R_1 + R_2 + R_3$$
$$= 1\ k\Omega + 1\ k\Omega + 1\ k\Omega$$
$$= 3\ k\Omega$$

Next we solve for I_T.

$$I_T = \frac{V_T}{R_T}$$
$$= \frac{24\ V}{3\ k\Omega}$$
$$= 8\ mA$$

Next we solve for the individual *IR* voltage drops.

$$V_1 = I_T \times R_1$$
$$= 8\ mA \times 1\ k\Omega$$
$$= 8\ V$$

Since $R_1 = R_2 = R_3$, V_2 and V_3 also equal 8 V.

Next we find V_{AG}, V_{BG}, and V_{CG}.

$$V_{CG} = -V_3$$
$$= -8\ V$$

Notice that V_{CG} is a negative voltage. This is because electrons enter R_3 at point C and leave R_3 at point G, thereby making the voltage drop negative at point C with respect to point G.

Next we find V_{BG}.

$$V_{BG} = V_2$$
$$= +8\ V$$

Since the top of R_2 is positive with respect to point G, V_{BG} is a positive voltage. Finally, we find V_{AG} as follows:

$$V_{AG} = V_1 + V_2$$
$$= 8\text{ V} + 8\text{ V}$$
$$= +16\text{ V}$$

Since the top of R_1 is positive with respect to point G, V_{AG} is also a positive voltage.

When the voltages in Fig. 4-33 are measured, the voltmeter's common (black) lead is connected to point G and left there. Then the voltmeter's probe (red) lead is connected to points A, B, or C for the desired reading.

PRACTICE PROBLEMS

Sec. 4-3 For the circuits shown in Figs. 4-34 through 4-43, solve for the voltages V_{AG}, V_{BG}, V_{CG}, and V_{DG}.

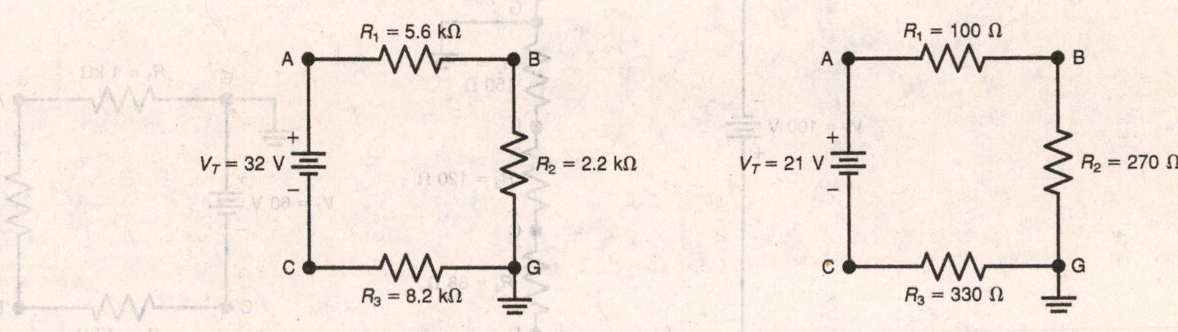

Fig. 4-34

Fig. 4-35

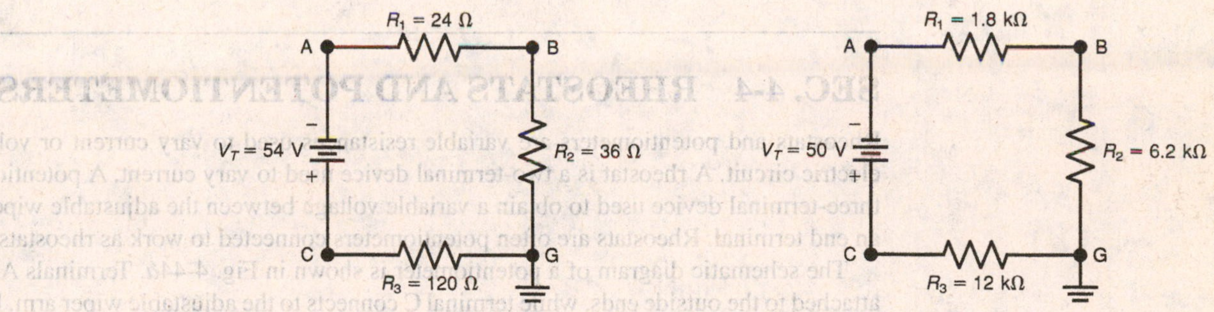

Fig. 4-36

Fig. 4-37

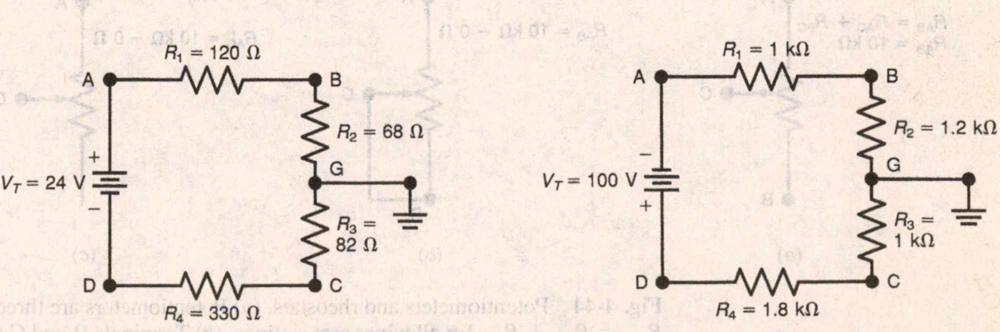

Fig. 4-38

Fig. 4-39

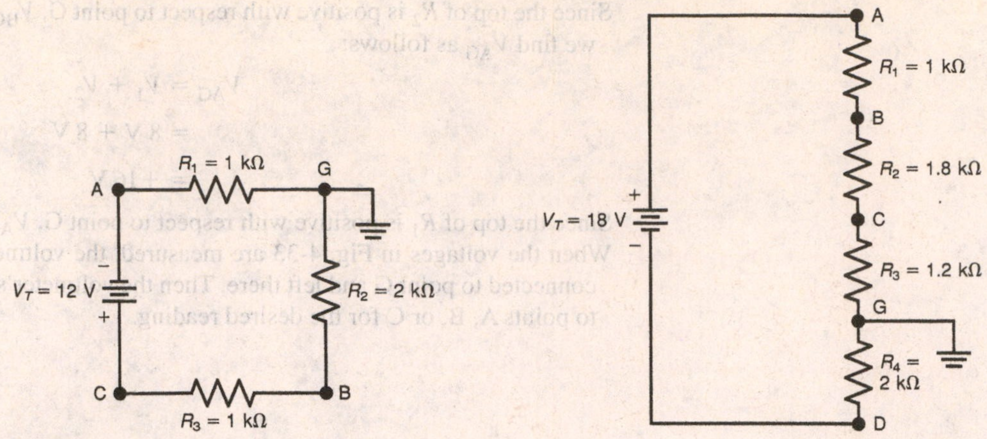

Fig. 4-40

Fig. 4-41

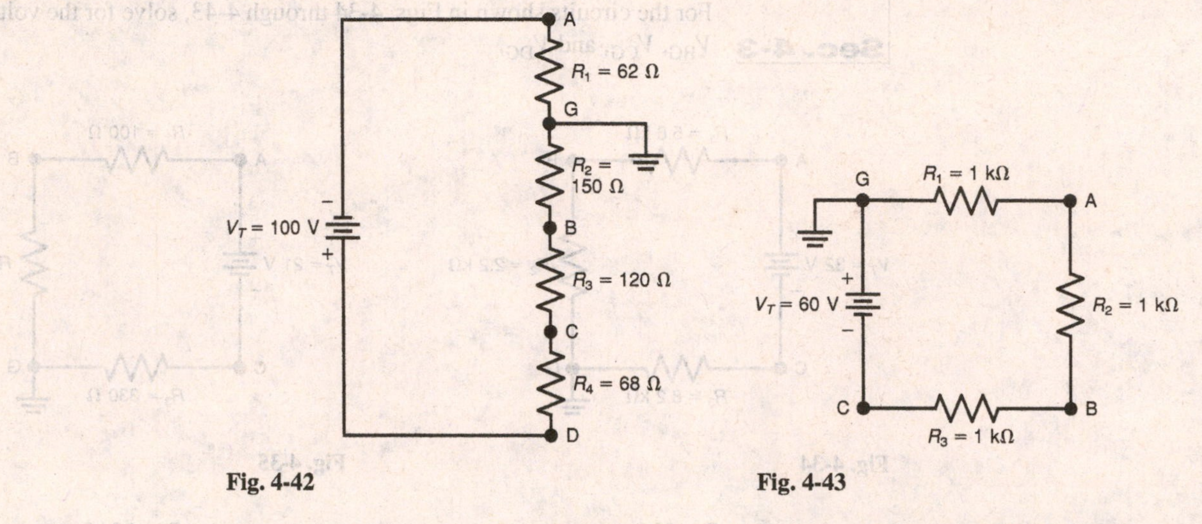

Fig. 4-42

Fig. 4-43

SEC. 4-4 RHEOSTATS AND POTENTIOMETERS

Rheostats and potentiometers are variable resistances used to vary current or voltage in an electric circuit. A rheostat is a two-terminal device used to vary current. A potentiometer is a three-terminal device used to obtain a variable voltage between the adjustable wiper arm and an end terminal. Rheostats are often potentiometers connected to work as rheostats.

The schematic diagram of a potentiometer is shown in Fig. 4-44*a*. Terminals A and B are attached to the outside ends, while terminal C connects to the adjustable wiper arm. Resistance

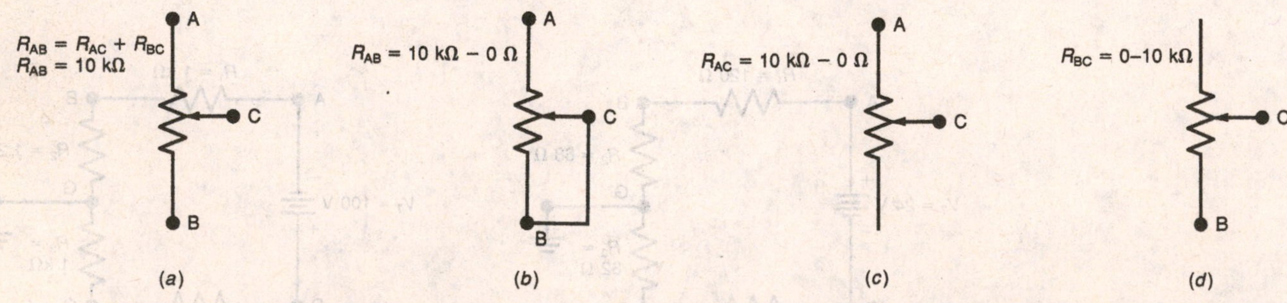

Fig. 4-44 Potentiometers and rheostats. (*a*) Potentiometers are three-terminal devices. $R_{AB} = R_{AC} + R_{BC}$ for all wiper arm settings. (*b*) Terminals B and C are shorted to make a rheostat, which is a two-terminal device. (*c*) End terminal B is left disconnected to form a rheostat using terminals A and C. (*d*) End terminal A is left disconnected to form a rheostat using terminals B and C.

R_{AB} remains fixed for all the settings of the wiper arm. The resistance between terminals A and C is variable as is the resistance between terminals B and C. Regardless of the wiper arm setting, $R_{AB} = R_{AC} + R_{BC}$.

For our analysis here, assume that clockwise (CW) shaft rotation moves the wiper arm, connected to terminal C, toward terminal A. Counterclockwise (CCW) rotation, therefore, will move the wiper arm toward terminal B. For clockwise shaft rotation, R_{AC} decreases from a maximum value of 10 kΩ to a minimum value of 0 Ω. For the same clockwise shaft rotation, R_{BC} increases from a minimum value of 0 Ω to a maximum value of 10 kΩ.

The schematic diagrams used to represent potentiometers connected as rheostats are shown in Fig. 4-44*b*, *c*, and *d*. The resistance range is specified for clockwise rotation. In Fig. 4-44*b*, terminals A and B serve as the two leads of the rheostat. With clockwise rotation, some portion of resistance R_{AB} will be shorted out. With the shaft rotated fully clockwise, $R_{AB} = 0\ \Omega$ because the wire jumper across terminals B and C is connected across terminals A and B. With the shaft rotated completely counterclockwise, resistance $R_{AB} = 10\ \text{k}\Omega$, since no portion of resistance R_{AB} is being shorted. In order to produce a variation of resistance just opposite to that described, terminal C should be shorted to terminal A. Resistance R_{AB} will then vary in a manner opposite to that described for Fig. 4-44*b*. The diagrams shown in Fig. 4-44*c* and *d* are also used to represent potentiometers connected as rheostats. Notice the variation of resistance for clockwise rotation in each case.

Analyzing Circuits Containing Potentiometers

Solved Problem

For Fig. 4-45, find R_T, I_T, individual voltage drops, P_T, and individual power dissipation values. Also, find the voltage range available from the wiper arm of R_2 to ground, as indicated by the voltmeter V.

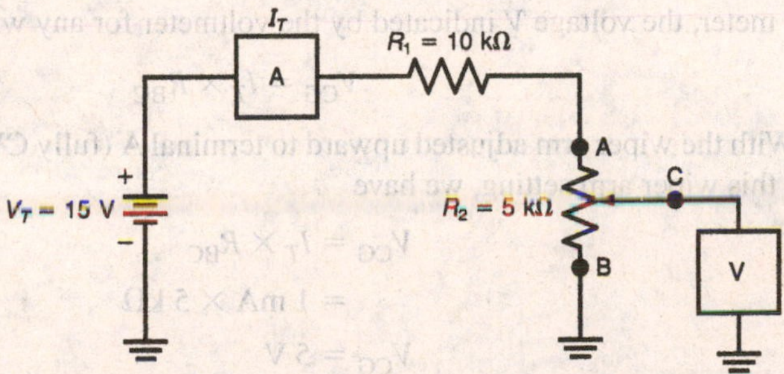

Fig. 4-45 Potentiometer used to vary voltage. Voltage V_{CG} varies as the wiper arm is moved up and down.

Answer

In Fig. 4-45, we see that the voltmeter V is connected to the wiper arm at terminal C. As the wiper arm is varied, however, R_{AB} remains fixed at 5 kΩ. With R_{AB} fixed, I_T also remains fixed. In Fig. 4-45, we have

$$R_T = R_1 + R_{AB}$$
$$= 10\ \text{k}\Omega + 5\ \text{k}\Omega$$
$$R_T = 15\ \text{k}\Omega$$
$$I_T = \frac{V_T}{R_T}$$
$$= \frac{15\ \text{V}}{15\ \text{k}\Omega}$$
$$I_T = 1\ \text{mA}$$

$$V_{R_1} = I_T \times R_1$$
$$= 1 \text{ mA} \times 10 \text{ k}\Omega$$
$$V_{R_1} = 10 \text{ V}$$
$$V_{R_2} = V_{AB}$$

Therefore,

$$V_{R_2} = I_T \times R_{AB}$$
$$= 1 \text{ mA} \times 5 \text{ k}\Omega$$
$$V_{R_2} = 5 \text{ V}$$
$$P_T = V_T \times I_T$$
$$= 15 \text{ V} \times 1 \text{ mA}$$
$$P_T = 15 \text{ mW}$$
$$P_1 = V_1 \times I_T$$
$$= 10 \text{ V} \times 1 \text{ mA}$$
$$P_1 = 10 \text{ mW}$$
$$P_2 = V_2 \times I_T$$
$$= 5 \text{ V} \times 1 \text{ mA}$$
$$P_2 = 5 \text{ mW}$$

Also, notice that

$$P_T = P_1 + P_2$$
$$= 10 \text{ mW} + 5 \text{ mW}$$
$$P_T = 15 \text{ mW}$$

Since the voltmeter V is actually connected across terminals B and C of the potentiometer, the voltage V indicated by the voltmeter for any wiper arm setting is

$$V_{CG} = I_T \times R_{BC}$$

With the wiper arm adjusted upward to terminal A (fully CW), $R_{BC} = R_{AB} = 5 \text{ k}\Omega$. For this wiper arm setting, we have

$$V_{CG} = I_T \times R_{BC}$$
$$= 1 \text{ mA} \times 5 \text{ k}\Omega$$
$$V_{CG} = 5 \text{ V}$$

The voltmeter indicates 5 V.

With the wiper arm adjusted downward to terminal B (fully CCW), $R_{BC} = 0 \ \Omega$. For this wiper arm setting, we have

$$V_{CG} = I_T \times R_{BC}$$
$$= 1 \text{ mA} \times 0 \ \Omega$$
$$V_{CG} = 0 \text{ V}$$

Now the voltmeter indicates 0 V.

With the wiper arm adjusted midway between terminals A and B, $R_{BC} = 2.5 \text{ k}\Omega$. For this wiper arm setting, we have

$$V_{CG} = I_T \times R_{BC}$$
$$= 1 \text{ mA} \times 2.5 \text{ k}\Omega$$
$$V_{CG} = 2.5 \text{ V}$$

The voltmeter indicates 2.5 V.

The voltage range as indicated by the voltmeter V, therefore, is 0 to 5 V for CW rotation.

Sec. 4-4

For Figs. 4-46 through 4-53, find R_T, I_T, P_T, individual voltage drops, individual power dissipation values, and the voltage range available from the wiper arm of the potentiometer to ground. Indicate the voltage range of V_{CG} as the shaft of the potentiometer is varied from its CCW-most position to its CW-most position.

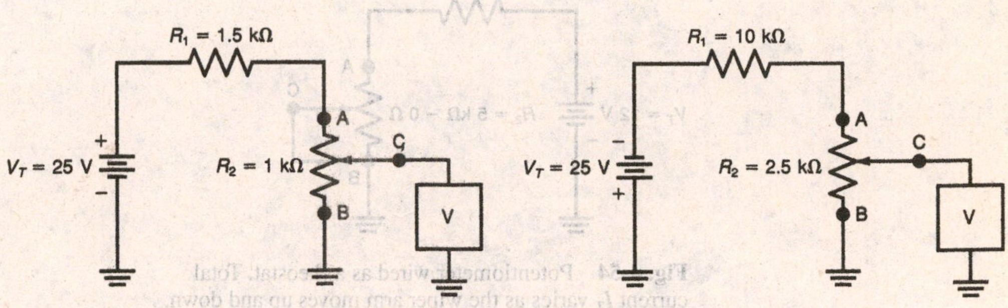

Fig. 4-46

Fig. 4-47

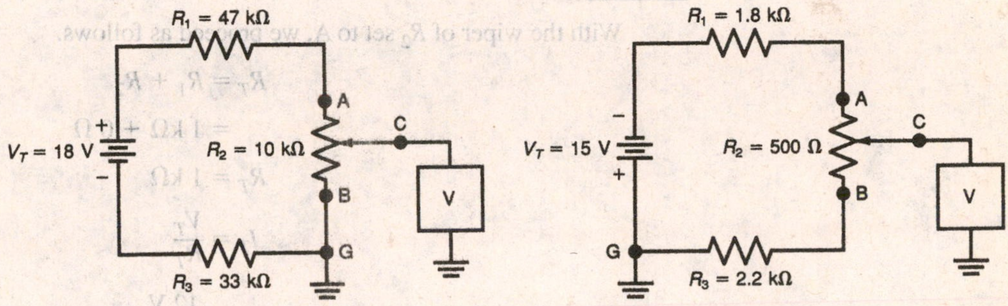

Fig. 4-48

Fig. 4-49

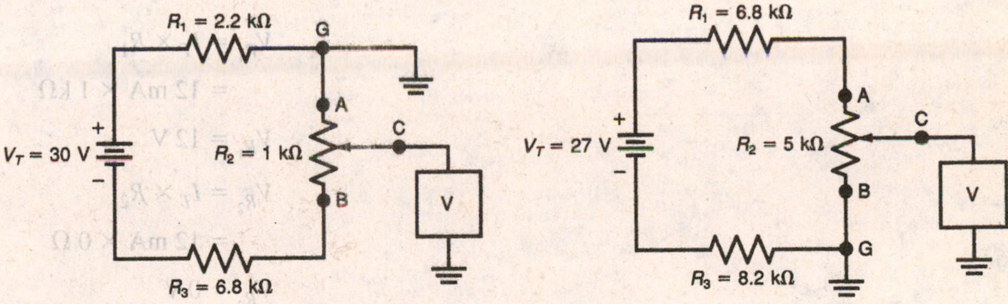

Fig. 4-50

Fig. 4-51

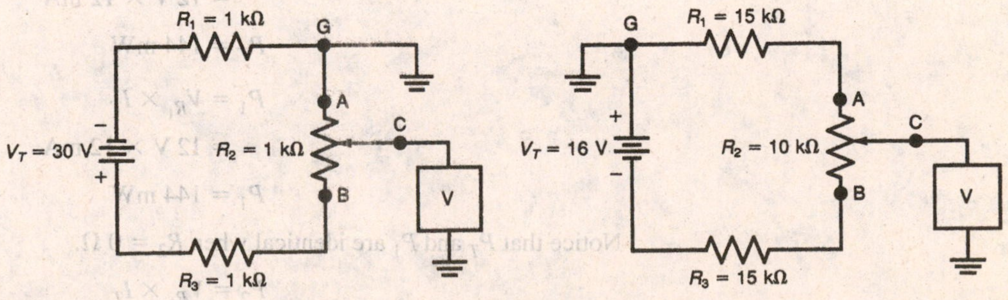

Fig. 4-52

Fig. 4-53

Analyzing Circuits Containing Rheostats

Solved Problem

Refer to Fig. 4-54. Find R_T, I_T, individual voltage drops, P_T, and individual power dissipation values for each extreme setting of R_2.

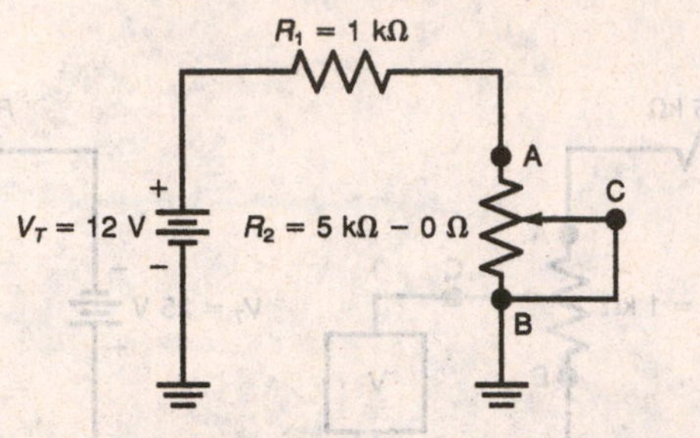

Fig. 4-54 Potentiometer wired as a rheostat. Total current I_T varies as the wiper arm moves up and down.

Answer

With the wiper of R_2 set to A, we proceed as follows.

$$R_T = R_1 + R_2$$
$$= 1\ \text{k}\Omega + 0\ \Omega$$
$$R_T = 1\ \text{k}\Omega$$
$$I_T = \frac{V_T}{R_T}$$
$$= \frac{12\ \text{V}}{1\ \text{k}\Omega}$$
$$I_T = 12\ \text{mA}$$
$$V_{R_1} = I_T \times R_1$$
$$= 12\ \text{mA} \times 1\ \text{k}\Omega$$
$$V_{R_1} = 12\ \text{V}$$
$$V_{R_2} = I_T \times R_2$$
$$= 12\ \text{mA} \times 0\ \Omega$$
$$V_{R_2} = 0\ \text{V}$$
$$P_T = V_T \times I_T$$
$$= 12\ \text{V} \times 12\ \text{mA}$$
$$P_T = 144\ \text{mW}$$
$$P_1 = V_{R_1} \times I_T$$
$$= 12\ \text{V} \times 12\ \text{mA}$$
$$P_1 = 144\ \text{mW}$$

Notice that P_T and P_1 are identical when $R_2 = 0\ \Omega$.

$$P_2 = V_{R_2} \times I_T$$
$$= 0\ \text{V} \times 12\ \text{mA}$$
$$P_2 = 0\ \text{W}$$

When the wiper of R_2 is set to B, we have

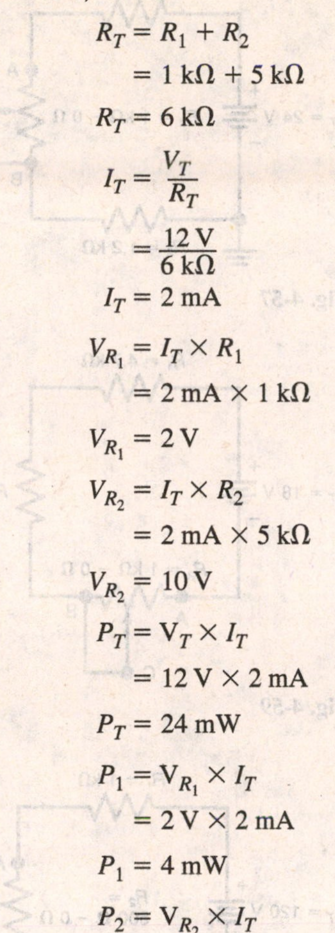

$$R_T = R_1 + R_2$$
$$= 1\text{ k}\Omega + 5\text{ k}\Omega$$
$$R_T = 6\text{ k}\Omega$$
$$I_T = \frac{V_T}{R_T}$$
$$= \frac{12\text{ V}}{6\text{ k}\Omega}$$
$$I_T = 2\text{ mA}$$
$$V_{R_1} = I_T \times R_1$$
$$= 2\text{ mA} \times 1\text{ k}\Omega$$
$$V_{R_1} = 2\text{ V}$$
$$V_{R_2} = I_T \times R_2$$
$$= 2\text{ mA} \times 5\text{ k}\Omega$$
$$V_{R_2} = 10\text{ V}$$
$$P_T = \text{V}_T \times I_T$$
$$= 12\text{ V} \times 2\text{ mA}$$
$$P_T = 24\text{ mW}$$
$$P_1 = \text{V}_{R_1} \times I_T$$
$$= 2\text{ V} \times 2\text{ mA}$$
$$P_1 = 4\text{ mW}$$
$$P_2 = \text{V}_{R_2} \times I_T$$
$$= 10\text{ V} \times 2\text{ mA}$$
$$P_2 = 20\text{ W}$$

Also, notice

$$P_T = P_1 + P_2$$
$$= 4\text{ mW} + 20\text{ mW}$$
$$P_T = 24\text{ mW}$$

MORE PRACTICE PROBLEMS

Sec. 4-4

For Figs. 4-55 through 4-62, find R_T, I_T, individual voltage drops, P_T, and individual power dissipation values for each extreme setting of R_2. List these values with the wiper for R_2 set to point A first, and then the values that exist with the wiper set to point B.

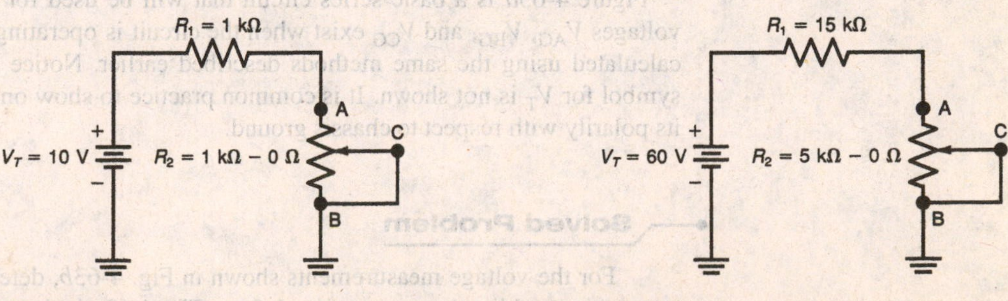

Fig. 4-55

Fig. 4-56

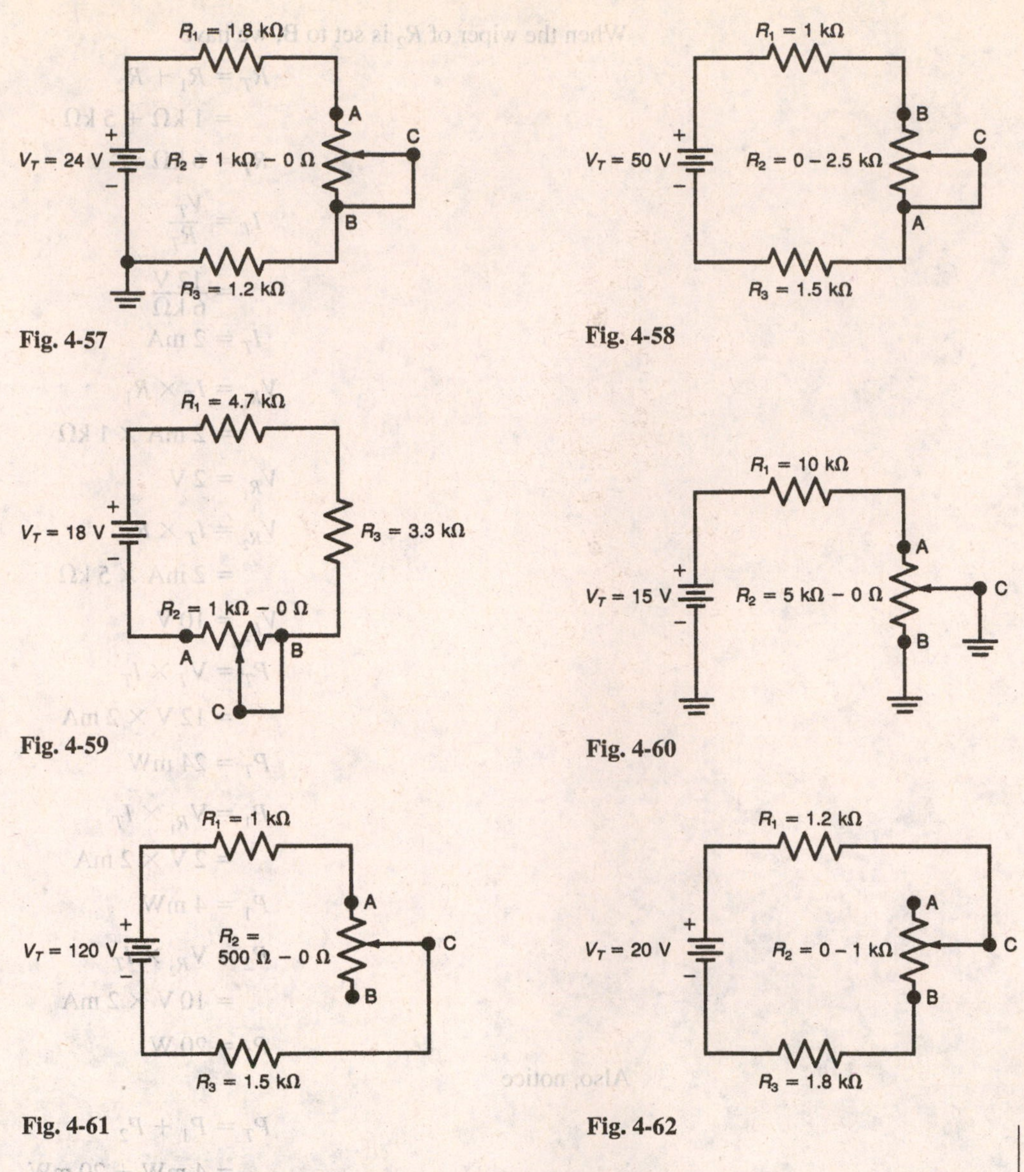

Fig. 4-57

Fig. 4-58

Fig. 4-59

Fig. 4-60

Fig. 4-61

Fig. 4-62

SEC. 4-5 TROUBLESHOOTING SERIES CIRCUITS

The two most common troubles in series circuits are opens and shorts. It is most common for resistors either to open or to change value due to excessive heat over a long period of time. Resistors, on the other hand, cannot short-circuit internally. They can, however, be short-circuited by some other part of the circuit such as a solder splash on the bottom side of a printed circuit board.

Figure 4-63*a* is a basic series circuit that will be used for troubleshooting analysis. The voltages V_{AG}, V_{BG}, and V_{CG} exist when the circuit is operating normally. These voltages are calculated using the same methods described earlier. Notice in Fig. 4-63*a* that the battery symbol for V_T is not shown. It is common practice to show only the potential difference and its polarity with respect to chassis ground.

Solved Problem

For the voltage measurements shown in Fig. 4-63*b*, determine which resistor is defective and list the nature of its defect. (Fig. 4-63*a* is the normal circuit.)

Answer

In Fig. 4-63*b*, notice that V_{BG} and V_{CG} are both +6 V. This indicates that R_2 is dropping 0 V. Therefore, R_2 must be shorted! Resistors R_1 and R_3 then form a two-resistor voltage divider that divides V_T evenly.

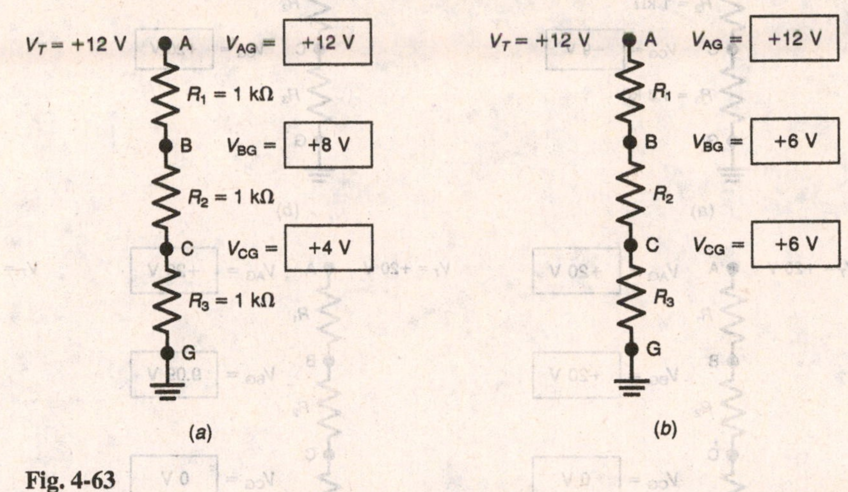

Fig. 4-63

PRACTICE PROBLEMS

Sec. 4-5

For the voltage measurements shown in Fig. 4-64*b* through *f*, Fig. 4-65*b* through *g*, and Fig. 4-66*b* through *j*, determine which resistor is defective and list the nature of its defect. For each figure number, *a* is the normal circuit. Only one defect exists for each circuit shown.

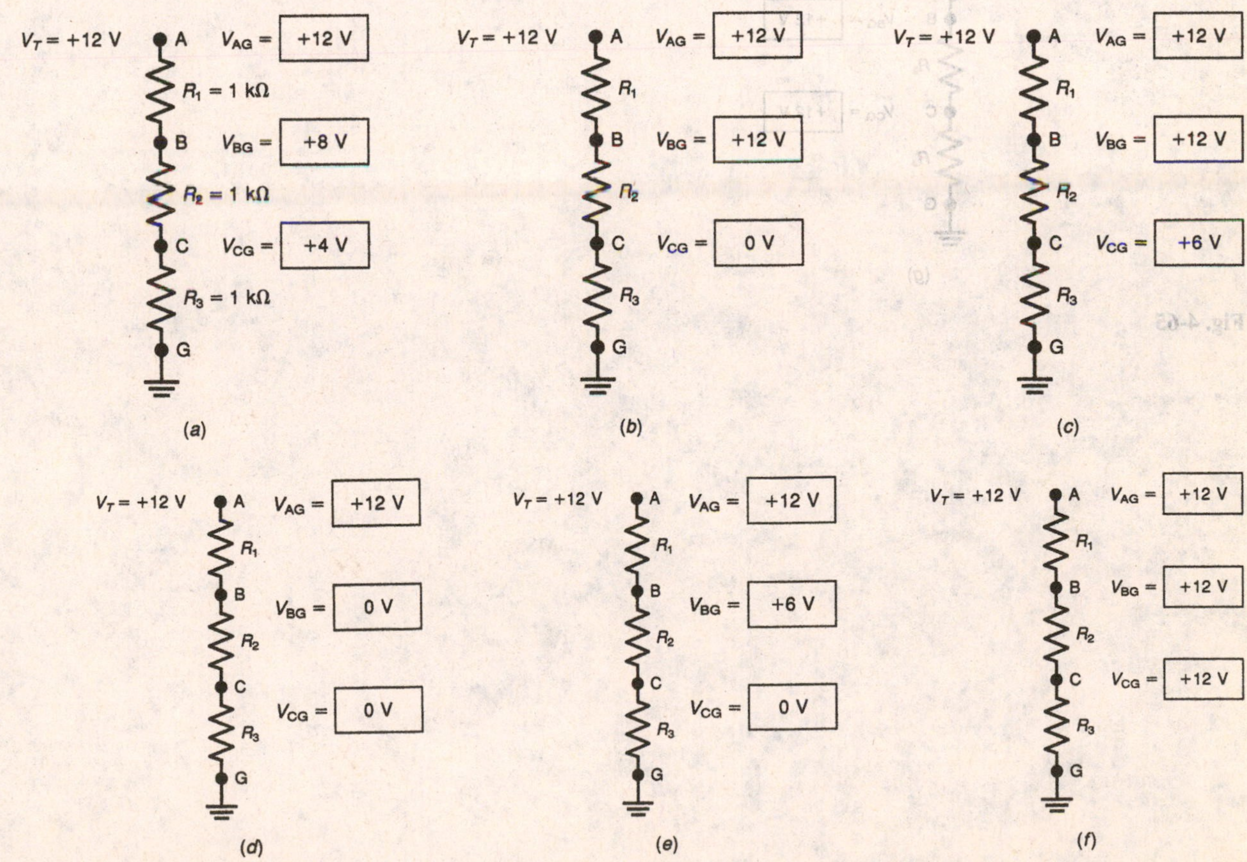

Fig. 4-64

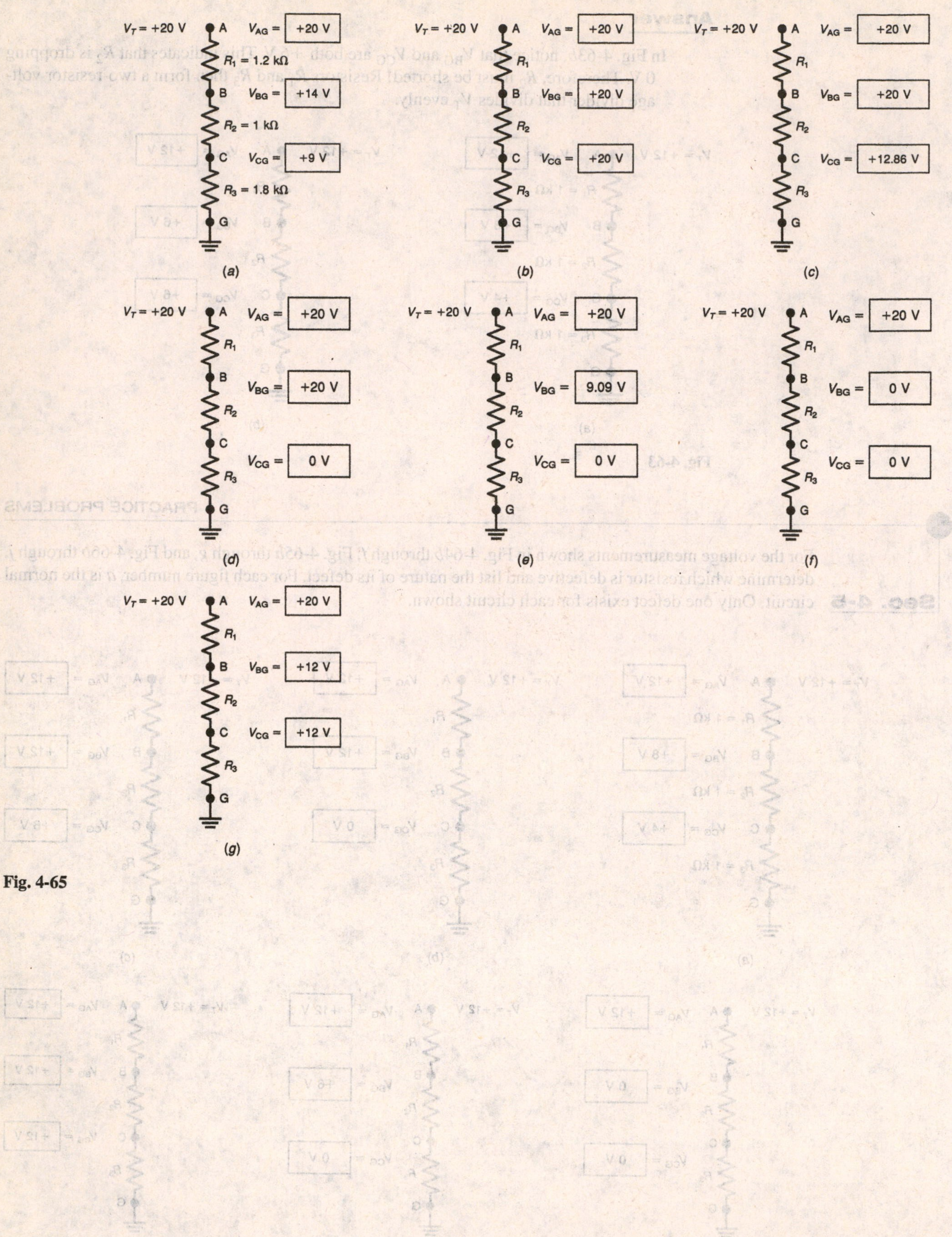

V_T = +20 V
A
V_{AG} = +20 V
R_1 = 1.2 kΩ
B
V_{BG} = +14 V
R_2 = 1 kΩ
C
V_{CG} = +9 V
R_3 = 1.8 kΩ
G
(a)
V_T = +20 V
A
V_{AG} = +20 V
R_1
B
V_{BG} = +20 V
R_2
C
V_{CG} = +20 V
R_3
G
(b)
V_T = +20 V
A
V_{AG} = +20 V
R_1
B
V_{BG} = +20 V
R_2
C
V_{CG} = +12.86 V
R_3
G
(c)
V_T = +20 V
A
V_{AG} = +20 V
R_1
B
V_{BG} = +20 V
R_2
C
V_{CG} = 0 V
R_3
G
(d)
V_T = +20 V
A
V_{AG} = +20 V
R_1
B
V_{BG} = 9.09 V
R_2
C
V_{CG} = 0 V
R_3
G
(e)
V_T = +20 V
A
V_{AG} = +20 V
R_1
B
V_{BG} = 0 V
R_2
C
V_{CG} = 0 V
R_3
G
(f)
V_T = +20 V
A
V_{AG} = +20 V
R_1
B
V_{BG} = +12 V
R_2
C
V_{CG} = +12 V
R_3
G
(g)

Fig. 4-65

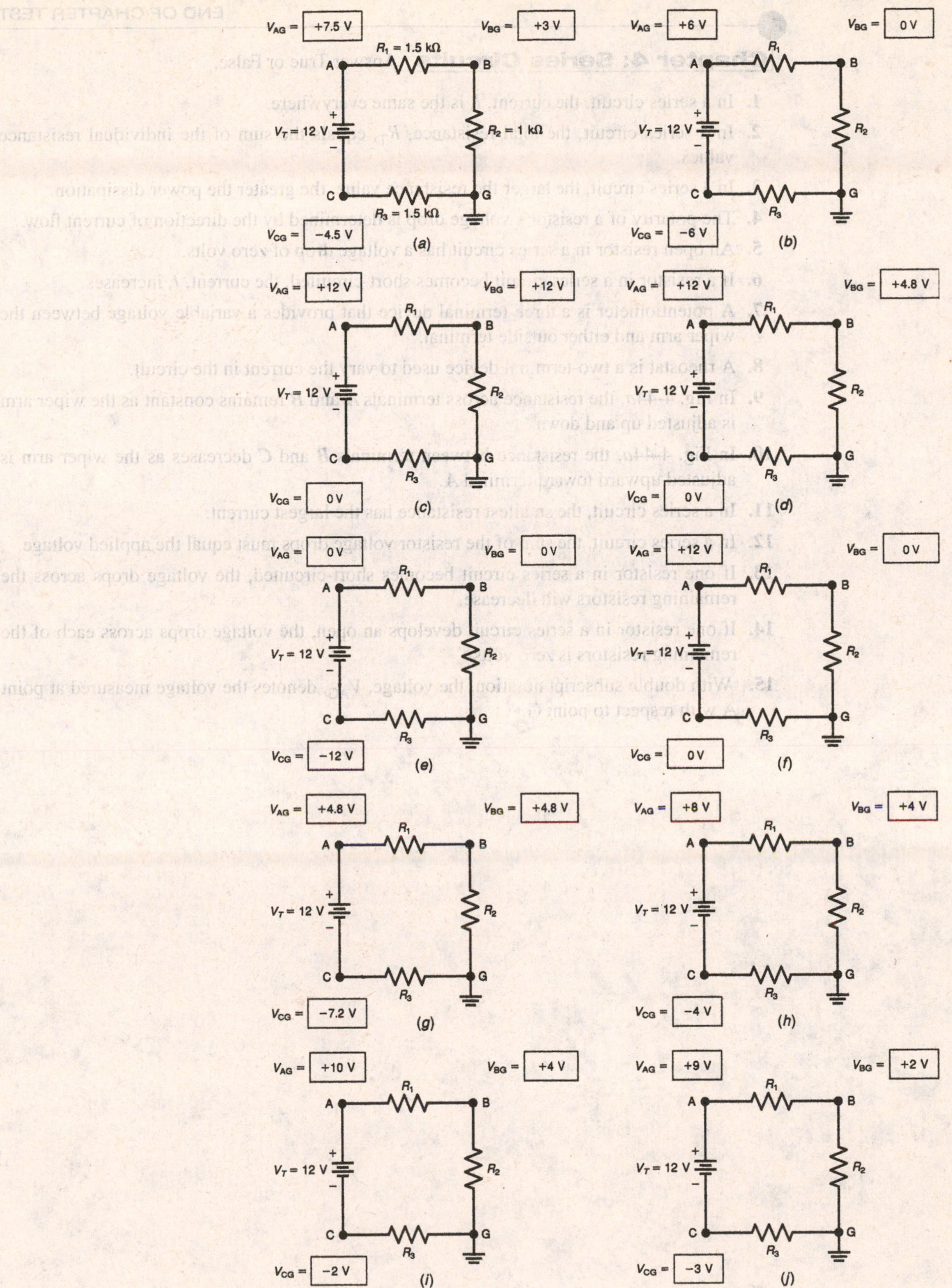
V_{AG} = +7.5 V
V_{BG} = +3 V
R_1 = 1.5 kΩ
A
B
V_T = 12 V
R_2 = 1 kΩ
C
G
R_3 = 1.5 kΩ
V_{CG} = −4.5 V
(a)
V_{AG} = +6 V
V_{BG} = 0 V
R_1
R_2
R_3
V_T = 12 V
V_{CG} = −6 V
(b)
V_{AG} = +12 V
V_{BG} = +12 V
V_T = 12 V
V_{CG} = 0 V
(c)
V_{AG} = +12 V
V_{BG} = +4.8 V
V_T = 12 V
V_{CG} = 0 V
(d)
V_{AG} = 0 V
V_{BG} = 0 V
V_T = 12 V
V_{CG} = −12 V
(e)
V_{AG} = +12 V
V_{BG} = 0 V
V_T = 12 V
V_{CG} = 0 V
(f)
V_{AG} = +4.8 V
V_{BG} = +4.8 V
V_T = 12 V
V_{CG} = −7.2 V
(g)
V_{AG} = +8 V
V_{BG} = +4 V
V_T = 12 V
V_{CG} = −4 V
(h)
V_{AG} = +10 V
V_{BG} = +4 V
V_T = 12 V
V_{CG} = −2 V
(i)
V_{AG} = +9 V
V_{BG} = +2 V
V_T = 12 V
V_{CG} = −3 V
(j)

Fig. 4-66

Chapter 4: Series Circuits

Answer True or False.

1. In a series circuit, the current, I, is the same everywhere.
2. In a series circuit, the total resistance, R_T, equals the sum of the individual resistance values.
3. In a series circuit, the larger the resistance value, the greater the power dissipation.
4. The polarity of a resistor's voltage drop is determined by the direction of current flow.
5. An open resistor in a series circuit has a voltage drop of zero volts.
6. If a resistor in a series circuit becomes short-circuited, the current, I, increases.
7. A potentiometer is a three-terminal device that provides a variable voltage between the wiper arm and either outside terminal.
8. A rheostat is a two-terminal device used to vary the current in the circuit.
9. In Fig. 4-44*a*, the resistance across terminals ***A*** and ***B*** remains constant as the wiper arm is adjusted up and down.
10. In Fig. 4-44*a*, the resistance between terminals ***B*** and ***C*** decreases as the wiper arm is adjusted upward toward terminal ***A***.
11. In a series circuit, the smallest resistance has the largest current.
12. In a series circuit, the sum of the resistor voltage drops must equal the applied voltage.
13. If one resistor in a series circuit becomes short-circuited, the voltage drops across the remaining resistors will decrease.
14. If one resistor in a series circuit develops an open, the voltage drops across each of the remaining resistors is zero volts.
15. With double subscript notation, the voltage, V_{AG}, denotes the voltage measured at point A with respect to point G.

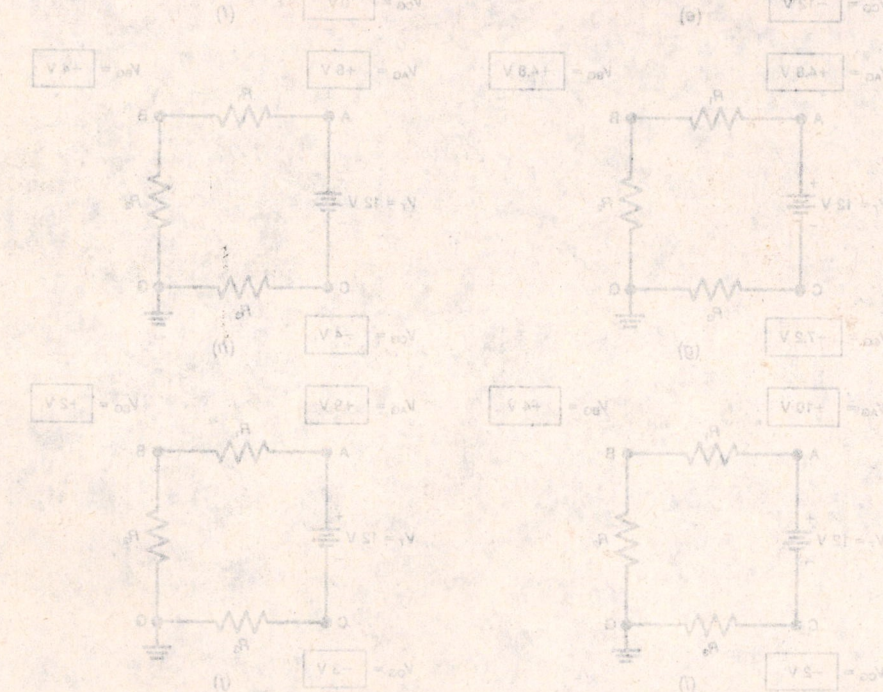

CHAPTER

5

Parallel Circuits

A circuit that provides one common voltage across all the components is called a parallel circuit. In this chapter, you will solve for the unknown values of voltage, current, resistance, and power in a parallel circuit.

SEC. 5-1 ANALYZING PARALLEL CIRCUITS

A parallel circuit is formed when two or more components are connected across a voltage source, as shown in Fig. 5-1. In this figure, R_1, R_2, and R_3 are in parallel with each other and

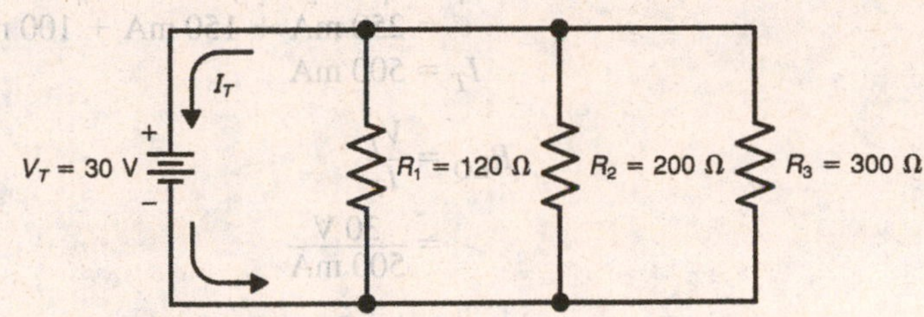

Fig. 5-1 Resistances in parallel. The current in each parallel branch equals the applied voltage V_T divided by each branch resistance R.

with the battery. Perhaps the most important characteristic of parallel circuits is that the voltage V is the same across each separate branch. This characteristic allows us to develop a list of equations that are applicable to parallel circuits. These equations are shown in Table 5-1.

TABLE 5-1 EQUATIONS USED FOR PARALLEL CIRCUITS

$I_T = I_1 + I_2 + I_3 + \cdots + I_n$	$V_T = I_T \times R_{EQ}$
$I_T = \frac{V_T}{R_{EQ}}$	$V_T = I_R \times R$
$I_R = \frac{V_T}{R}$ (current through any individual resistor)	
$R_{EQ} = \frac{1}{1/R_1 + 1/R_2 + 1/R_3 + \cdots + 1/R_n}$	$P_T = P_1 + P_2 + P_3 + \cdots + P_n$
$R_{EQ} = \frac{V_T}{I_T}$	$P_T = V_T \times I_T$
$R_{EQ} = \frac{R_1 \times R_2}{R_1 + R_2}$ (for just two resistors)	$P_R = V_T \times I_R$

Solved Problem

For Fig. 5-1, find R_{EQ}, I_T, I_1, I_2, I_3, P_T, P_1, P_2, and P_3.

Answer

Using the equations listed in Table 5-1, we proceed as follows.

$$I_R = \frac{V_T}{R}$$

$$I_1 = \frac{30\text{ V}}{120\ \Omega}$$

$$I_1 = 250\text{ mA}$$

$$I_2 = \frac{30\text{ V}}{200\ \Omega}$$

$$I_2 = 150\text{ mA}$$

$$I_3 = \frac{30\text{ V}}{300\ \Omega}$$

$$I_3 = 100\text{ mA}$$

$$\begin{aligned} I_T &= I_1 + I_2 + I_3 + \cdots + I_n \\ &= 250\text{ mA} + 150\text{ mA} + 100\text{ mA} \\ I_T &= 500\text{ mA} \end{aligned}$$

$$\begin{aligned} R_{EQ} &= \frac{V_T}{I_T} \\ &= \frac{30\text{ V}}{500\text{ mA}} \\ R_{EQ} &= 60\ \Omega \end{aligned}$$

or

$$\begin{aligned} R_{EQ} &= \frac{1}{1/R_1 + 1/R_2 + 1/R_3 + \cdots + 1/R_n} \\ &= \frac{1}{1/120\ \Omega + 1/200\ \Omega + 1/300\ \Omega} \\ R_{EQ} &= 60\ \Omega \end{aligned}$$

$$\begin{aligned} P_R &= V_T \times I_R \\ P_1 &= 30\text{ V} \times 250\text{ mA} \\ P_1 &= 7.5\text{ W} \end{aligned}$$

$$\begin{aligned} P_2 &= 30\text{ V} \times 150\text{ mA} \\ P_2 &= 4.5\text{ W} \end{aligned}$$

$$\begin{aligned} P_3 &= 30\text{ V} \times 100\text{ mA} \\ P_3 &= 3\text{ W} \end{aligned}$$

$$\begin{aligned} P_T &= P_1 + P_2 + P_3 + \cdots + P_n \\ &= 7.5\text{ W} + 4.5\text{ W} + 3\text{ W} \\ P_T &= 15\text{ W} \end{aligned}$$

or

$$\begin{aligned} P_T &= V_T \times I_T \\ &= 30\text{ V} \times 500\text{ mA} \\ P_T &= 15\text{ W} \end{aligned}$$

Sec. 5-1 For Figs. 5-2 through 5-11, find I_T, R_{EQ}, individual branch currents, P_T, and individual power dissipation values. Refer to Table 5-1.

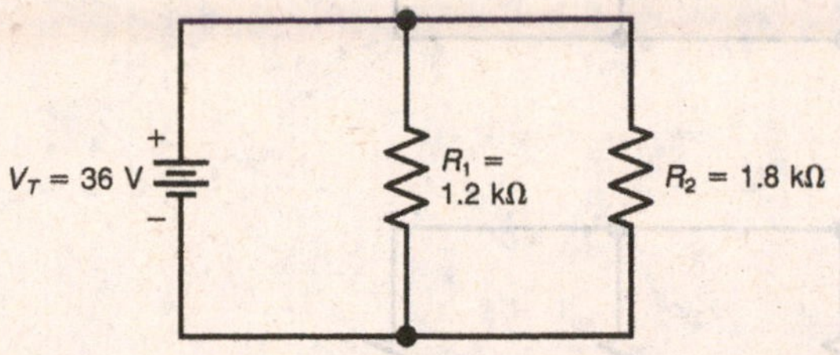

V_T = 36 V
R_1 = 1.2 kΩ
R_2 = 1.8 kΩ

Fig. 5-2

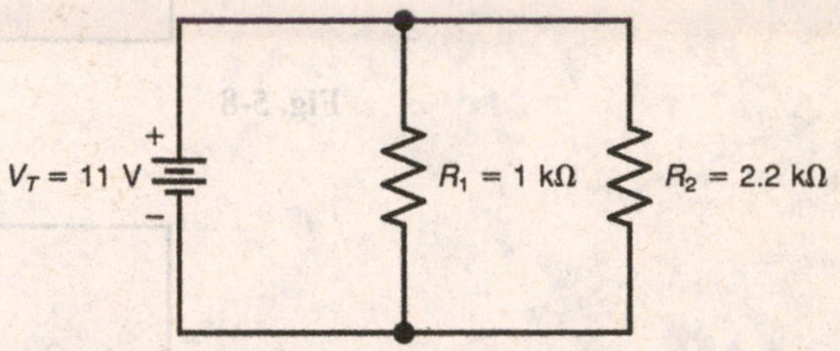

V_T = 11 V
R_1 = 1 kΩ
R_2 = 2.2 kΩ

Fig. 5-3

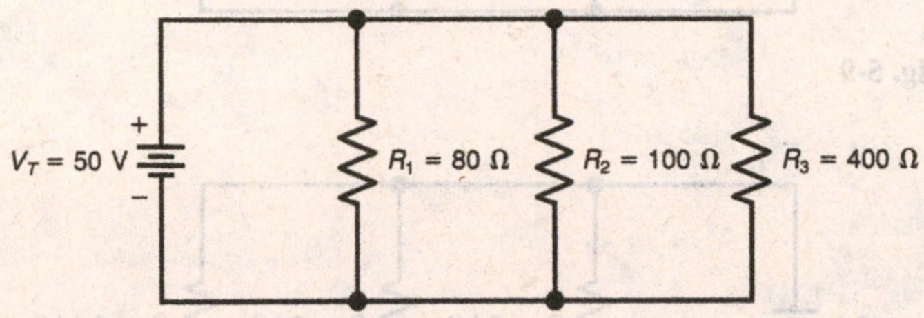

V_T = 50 V
R_1 = 80 Ω
R_2 = 100 Ω
R_3 = 400 Ω

Fig. 5-4

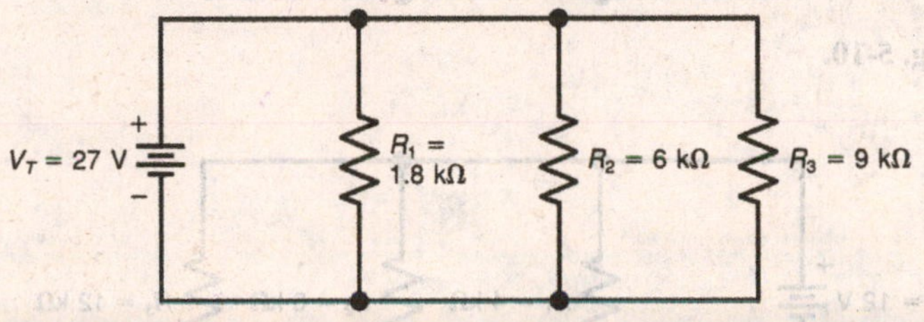

V_T = 27 V
R_1 = 1.8 kΩ
R_2 = 6 kΩ
R_3 = 9 kΩ

Fig. 5-5

V_T = 20 V
R_1 = 5 kΩ
R_2 = 10 kΩ
R_3 = 10 kΩ

Fig. 5-6

V_T = 45 V
R_1 = 1.5 kΩ
R_2 = 2 kΩ
R_3 = 6 kΩ

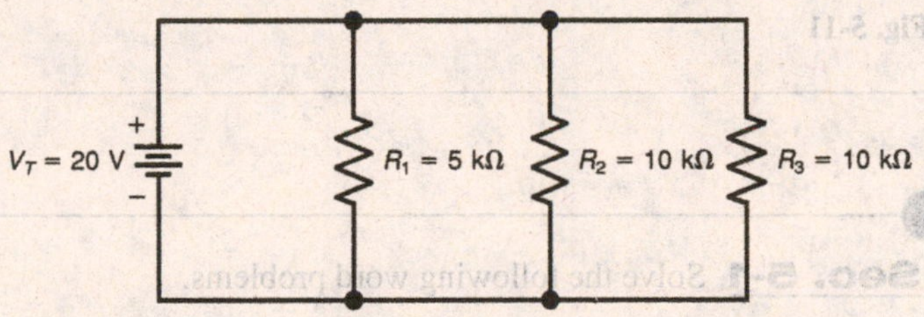

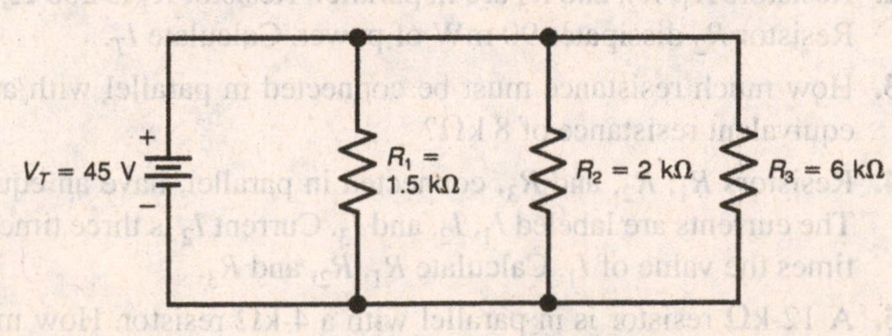

Fig. 5-7

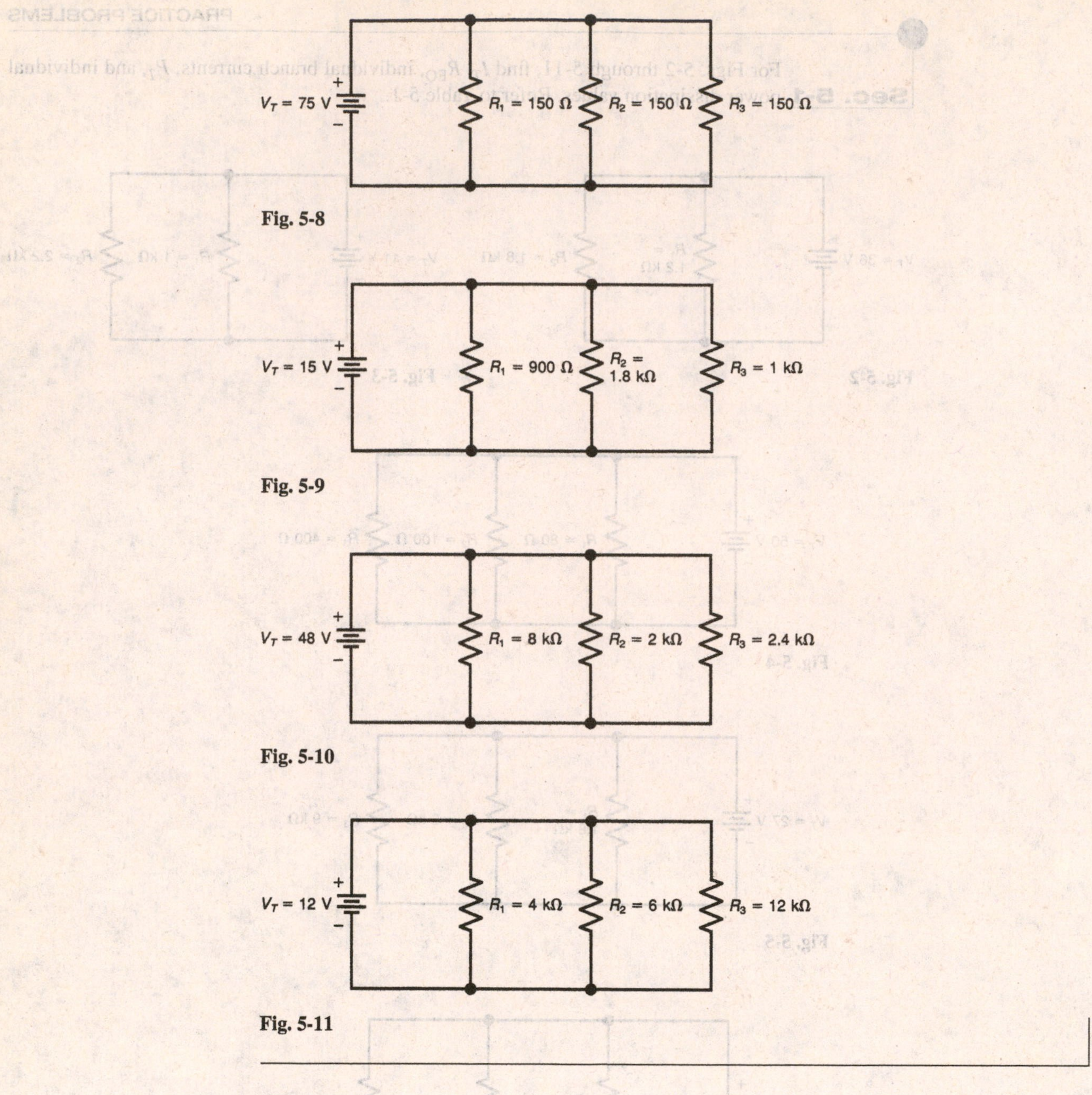

Fig. 5-8

Fig. 5-9

Fig. 5-10

Fig. 5-11

WORD PROBLEMS

Sec. 5-1 Solve the following word problems.

1. Three resistors are connected in parallel to a 120-V line. Total current is 12 mA. Resistor R_1 is 20 kΩ. Resistor R_2 is 30 kΩ. Calculate the current in resistor R_3.
2. Resistors R_1, R_2, and R_3 are in parallel. Resistor R_1 is 200 Ω, R_2 is 400 Ω, and R_3 is 1 kΩ. Resistor R_2 dissipates 90 mW of power. Calculate I_T.
3. How much resistance must be connected in parallel with a 12-kΩ resistor to obtain an equivalent resistance of 8 kΩ?
4. Resistors R_1, R_2, and R_3, connected in parallel, have an equivalent resistance of 720 Ω. The currents are labeled I_1, I_2, and I_3. Current I_2 is three times the value of I_1, and I_3 is six times the value of I_1. Calculate R_1, R_2, and R_3.
5. A 12-kΩ resistor is in parallel with a 4-kΩ resistor. How much resistance must be connected in parallel with this combination to obtain an equivalent resistance of 2 kΩ?

6. Resistors R_1, R_2, and R_3 are in parallel. Resistor R_1 is 20 Ω, R_2 is 25 Ω, and R_3 is 100 Ω. The total current passed by this combination is 2.5 A. What is the current through each branch?
7. Three 1.2-kΩ resistors are connected in parallel, with each other and a 24-V DC source. If the value of DC voltage from the source is tripled, what effect does it have on R_{EQ}? Why?
8. A 1-kΩ 0.5-W resistor is in parallel with a 100-Ω 1-W resistor. What is the maximum voltage that can be applied across this circuit without exceeding the wattage rating of either resistor?
9. Three lightbulbs connected to the 120-V power line are rated at 60 W, 100 W, and 150 W, respectively. Calculate the hot resistance of this parallel combination.
10. A 220-Ω 0.5-W resistor is in parallel with a 1-kΩ 0.25-W and a 150-Ω 0.5-W resistor. What is the maximum total current that this parallel combination can handle without exceeding the wattage rating of any resistor?
11. Three equal resistances in parallel have a combined equivalent resistance, R_{EQ}, of 90 Ω. What will be the value of R_{EQ} if one resistance is disconnected?
12. How many 330-Ω resistors must be connected in parallel to obtain an R_{EQ} of 5 Ω?
13. A 10-kΩ resistor in parallel with a potentiometer draws 4.5 mA from a 15-V DC source. What is the resistance of the potentiometer?
14. Three resistors in parallel draw a total current, I_T, of 120 mA from a 90-V DC source. If $I_2 = 2I_1$ and $I_3 = 2I_2$, calculate the values for R_1, R_2, and R_3.
15. Three resistors in parallel with the 120-V power line dissipate a total power, P_T, of 288 W. If $R_2 = 3R_1$ and $R_2 = 2R_3$, calculate the values for R_1, R_2, and R_3.

For Probs. 16–25, refer to Fig. 5-12.

16. Calculate the current drawn by the lights and each kitchen appliance when they are connected to the 120-V AC power line.
17. Will the fuse, F_1, blow when all switches are closed?
18. Will the fuse, F_1, blow when only switches S_1, S_4, and S_5 are closed?
19. Will the fuse, F_1, blow when only switches S_1, S_2, and S_4 are closed?
20. Will the fuse, F_1, blow if only switches S_1, S_2, S_3, and S_5 are closed?
21. What is the resistance of the (a) toaster, (b) microwave, and (c) dishwasher?
22. How much is the current between points B and C if only S_1, S_2, and S_4 are closed?
23. How much is the current between points I and J if only switches S_1, S_4, and S_5 are closed?
24. How much is the current between points D and E if only switches S_3 and S_5 are closed?
25. Assume only switches S_1, S_3, S_4, and S_5 are closed. Which items will still operate if the wire between points H and I develops an open?

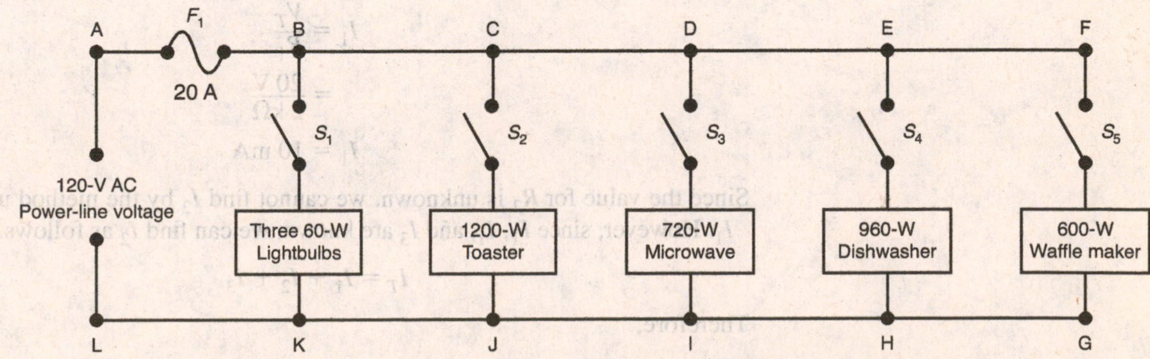

Fig. 5-12 Lights and kitchen appliances operated from the 120-V AC power line.

SEC. 5-2 PARALLEL CIRCUITS WITH RANDOM UNKNOWNS

Solved Problem

For Fig. 5-13, solve for V_T, R_{EQ}, I_1, I_2, R_2, P_T, P_1, P_2, and P_3.

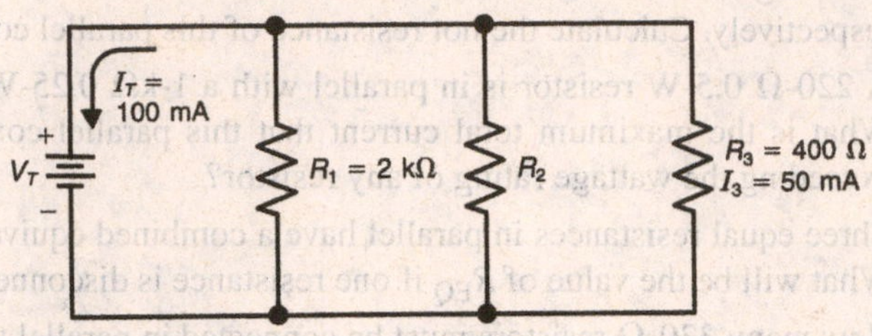

Fig. 5-13 Analyzing a parallel circuit to find V_T, R_{EQ}, I_1, I_2, R_2, P_T, P_1, P_2, and P_3.

Answer

When you know the voltage across one branch, this voltage is across all branches. For Fig. 5-13, we can find the voltage across R_3, since both current and resistance are known.

$$V_3 = I_3 \times R_3$$
$$= 50 \text{ mA} \times 400\ \Omega$$
$$V_3 = 20 \text{ V}$$

This voltage exists across all parallel branches and is the voltage that is being supplied by the battery. Therefore,

$$V_T = V_1 = V_2 = V_3$$
$$V_T = 20 \text{ V}$$

When both V_T and I_T are known, we can find R_{EQ}.

$$R_{EQ} = \frac{V_T}{I_T}$$
$$= \frac{20 \text{ V}}{100 \text{ mA}}$$
$$R_{EQ} = 200\ \Omega$$

The branch currents I_1 and I_2 can be found as follows.

$$I_1 = \frac{V_T}{R_1}$$
$$= \frac{20 \text{ V}}{2 \text{ k}\Omega}$$
$$I_1 = 10 \text{ mA}$$

Since the value for R_2 is unknown, we cannot find I_2 by the method used to determine I_1. However, since I_T, I_1, and I_3 are known, we can find I_2 as follows.

$$I_T = I_1 + I_2 + I_3$$

Therefore,

$$I_2 = I_T - I_1 - I_3$$
$$= 100 \text{ mA} - 10 \text{ mA} - 50 \text{ mA}$$
$$I_2 = 40 \text{ mA}$$

This tells us that the current flowing in branch resistance R_2 is equal to the total current minus the currents flowing in all other branches.

With I_2 known, we can solve for R_2.

$$R_2 = \frac{V_T}{I_2}$$
$$= \frac{20\text{ V}}{40\text{ mA}}$$
$$R_2 = 200\ \Omega$$

Power dissipation values are found by the methods shown.

$$P_T = V_T \times I_T$$
$$= 20\text{ V} \times 100\text{ mA}$$
$$P_T = 2\text{ W}$$

$$P_1 = V_T \times I_1$$
$$= 20\text{ V} \times 10\text{ mA}$$
$$P_1 = 200\text{ mW}$$

$$P_2 = V_T \times I_2$$
$$= 20\text{ V} \times 40\text{ mA}$$
$$P_2 = 800\text{ mW}$$

$$P_3 = V_T \times I_3$$
$$= 20\text{ V} \times 50\text{ mA}$$
$$P_3 = 1\text{ W}$$

Also, notice that

$$P_T = P_1 + P_2 + P_3$$
$$= 200\text{ mW} + 800\text{ mW} + 1\text{ W}$$
$$P_T = 2\text{ W}$$

PRACTICE PROBLEMS

Sec. 5-2 Solve for the unknowns listed in Figs. 5-14 through 5-32.

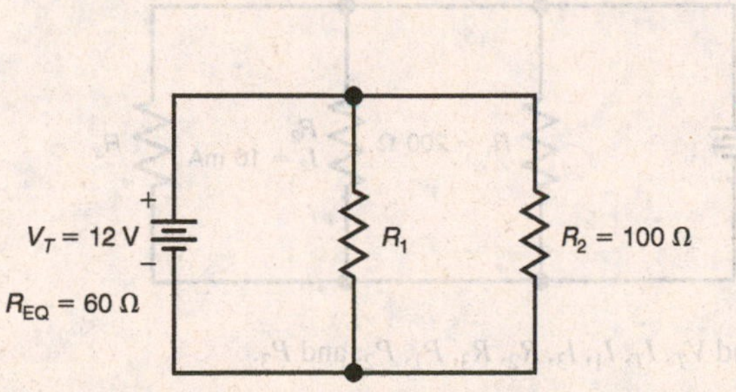

Fig. 5-14 Find I_T, I_1, I_2, R_1, P_T, P_1, and P_2.

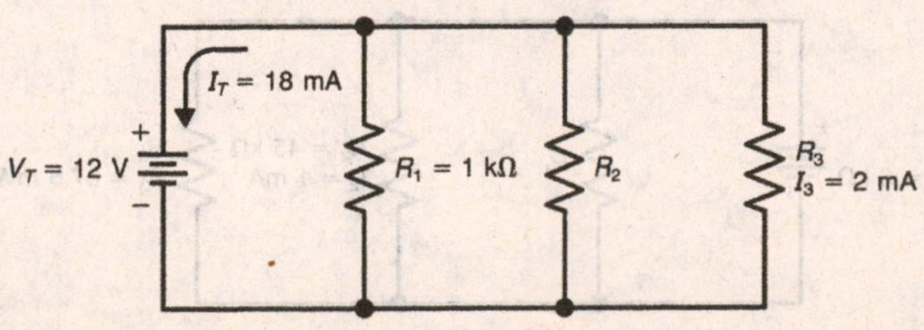

Fig. 5-15 Find R_{EQ}, I_1, I_2, R_2, R_3, P_T, P_1, P_2, and P_3.

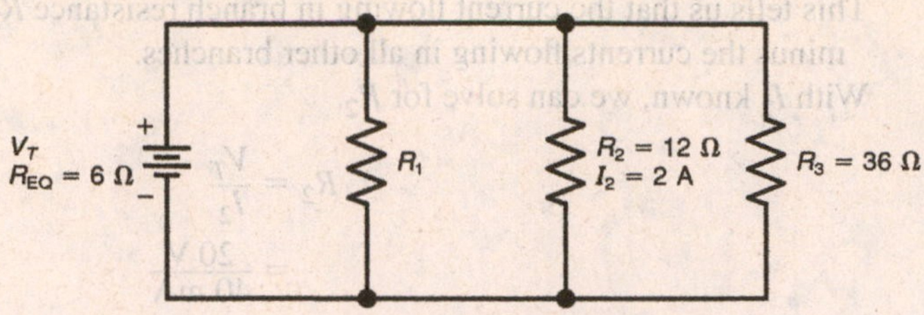

Fig. 5-16 Find V_T, I_T, I_1, I_3, R_1, P_T, P_1, P_2, and P_3.

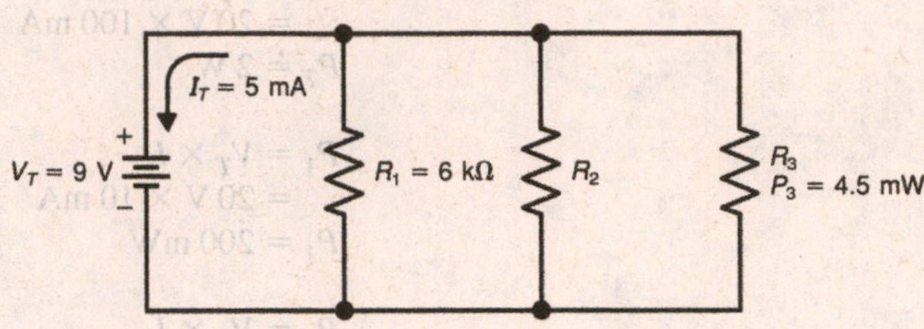

Fig. 5-17 Find R_{EQ}, I_1, I_2, I_3, R_2, R_3, P_T, P_1, and P_2.

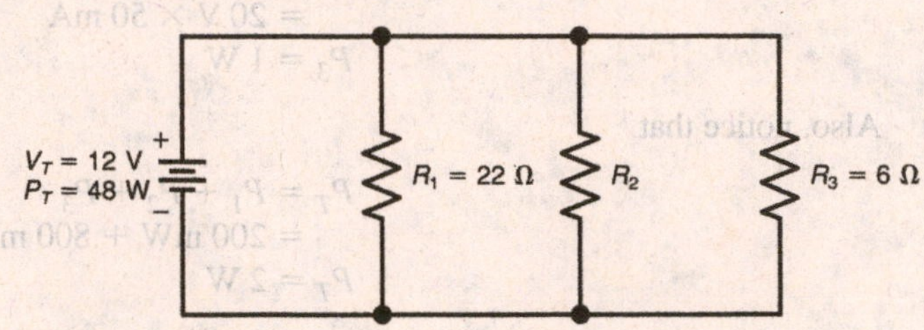

Fig. 5-18 Find R_{EQ}, I_T, I_1, I_2, I_3, R_2, P_1, P_2, and P_3.

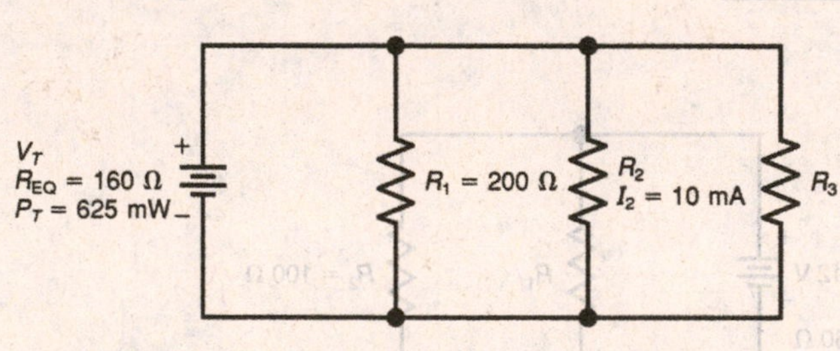

Fig. 5-19 Find V_T, I_T, I_1, I_3, R_2, R_3, P_1, P_2, and P_3.

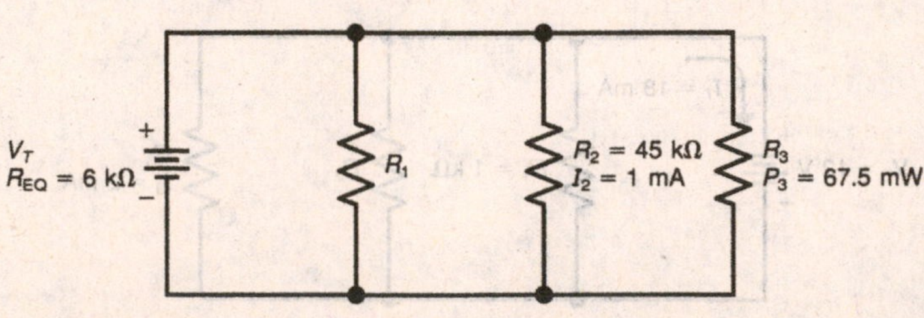

Fig. 5-20 Find V_T, I_T, I_1, I_3, R_1, R_3, P_T, P_1, and P_2.

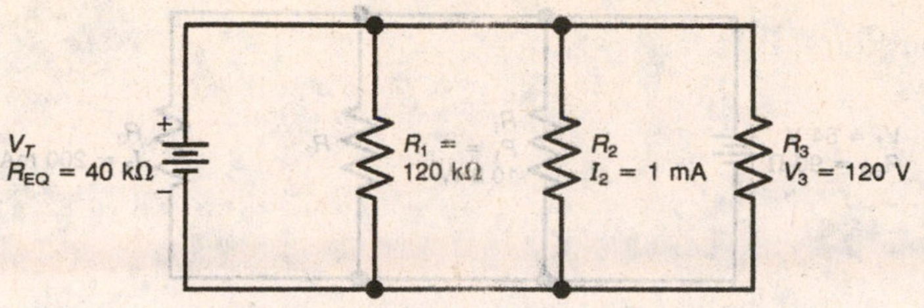

Fig. 5-21 Find V_T, I_T, I_1, I_3, R_2, R_3, P_T, P_1, P_2, and P_3.

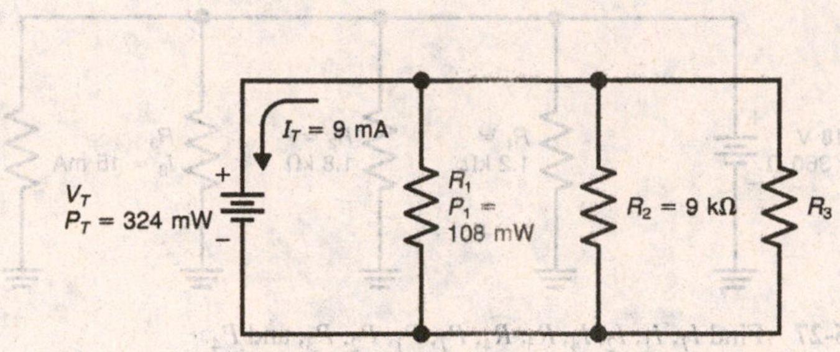

Fig. 5-22 Find V_T, R_{EQ}, I_1, I_2, I_3, R_1, R_3, P_2, and P_3.

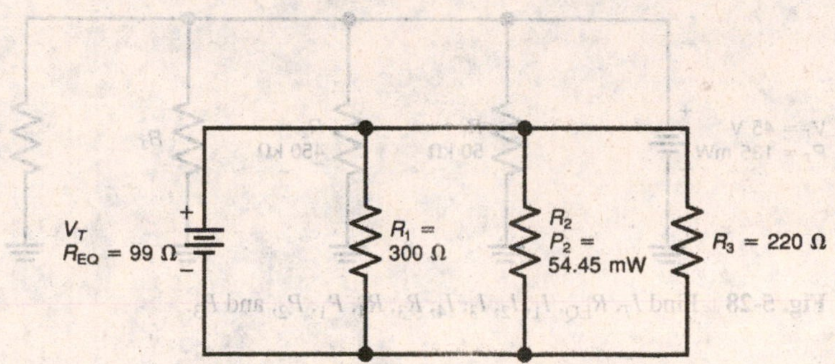

Fig. 5-23 Find V_T, I_T, I_1, I_2, I_3, R_2, P_T, P_1, and P_3.

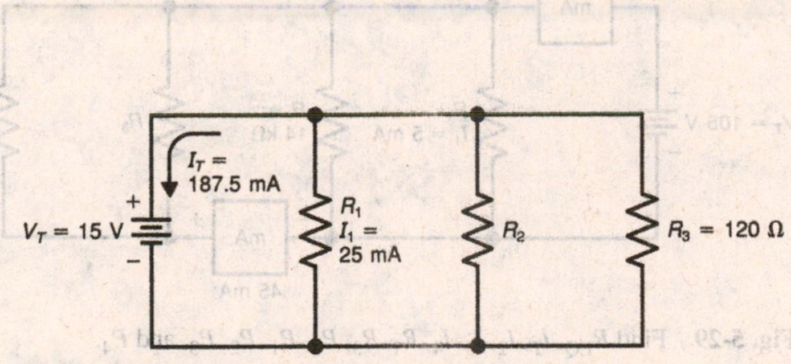

Fig. 5-24 Find R_{EQ}, I_2, I_3, R_1, R_2, P_T, P_1, P_2, and P_3.

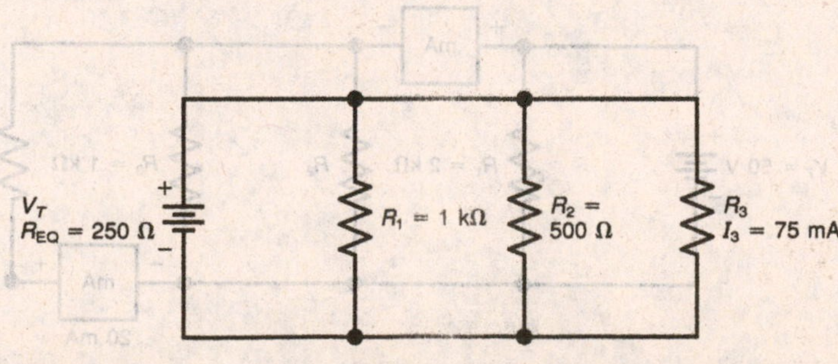

Fig. 5-25 Find V_T, I_T, I_1, I_2, R_3, P_T, P_1, P_2, and P_3.

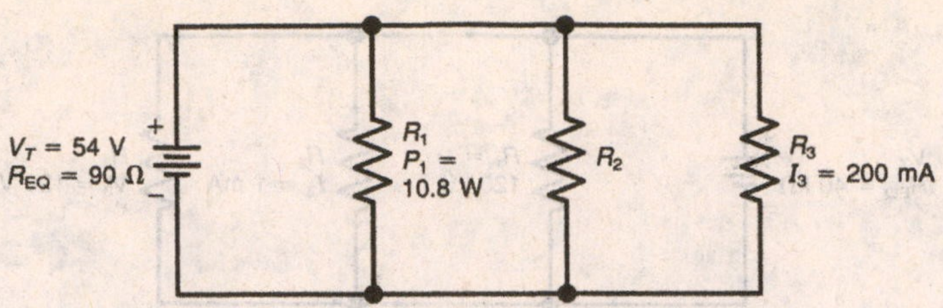

Fig. 5-26 Find I_T, I_1, I_2, R_1, R_2, R_3, P_T, P_2, and P_3.

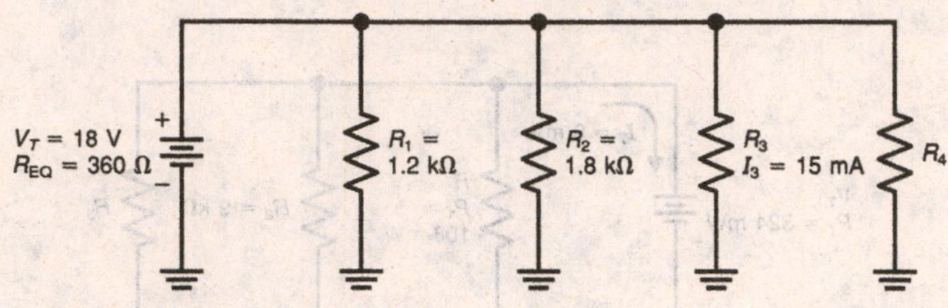

Fig. 5-27 Find I_T, I_1, I_2, I_4, R_3, R_4, P_T, P_1, P_2, P_3, and P_4.

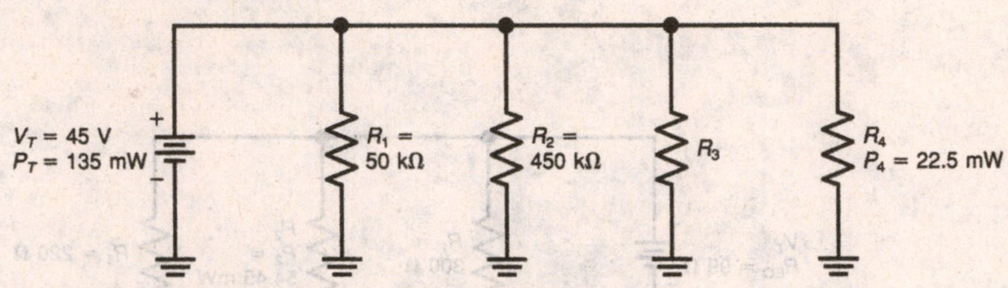

Fig. 5-28 Find I_T, R_{EQ}, I_1, I_2, I_3, I_4, R_3, R_4, P_1, P_2, and P_3.

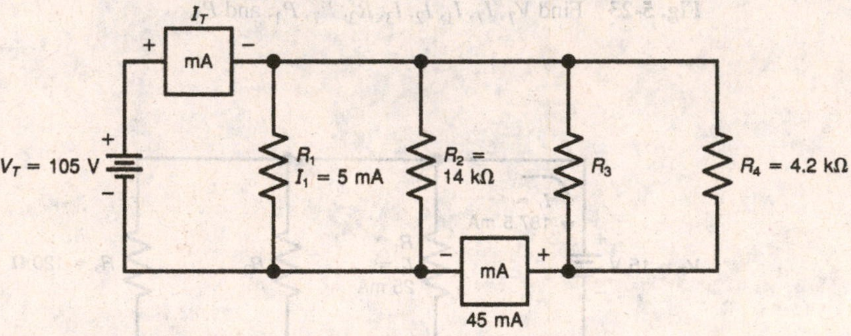

Fig. 5-29 Find R_{EQ}, I_T, I_2, I_3, I_4, R_1, R_3, P_T, P_1, P_2, P_3, and P_4.

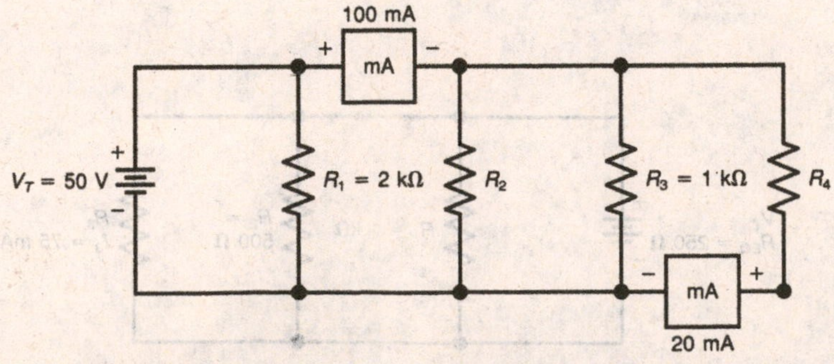

Fig. 5-30 Find R_{EQ}, I_T, I_1, I_2, I_3, R_2, R_4, P_T, P_1, P_2, P_3, and P_4.

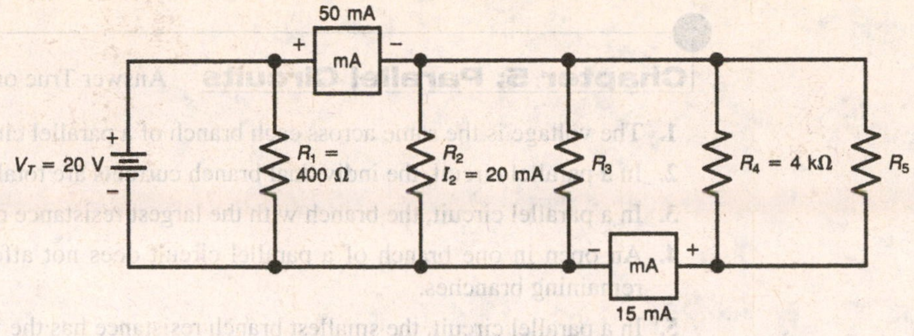

Fig. 5-31 Find R_{EQ}, I_T, I_1, I_3, I_4, I_5, R_2, R_3, R_5, P_T, P_1, P_2, P_3, P_4, and P_5.

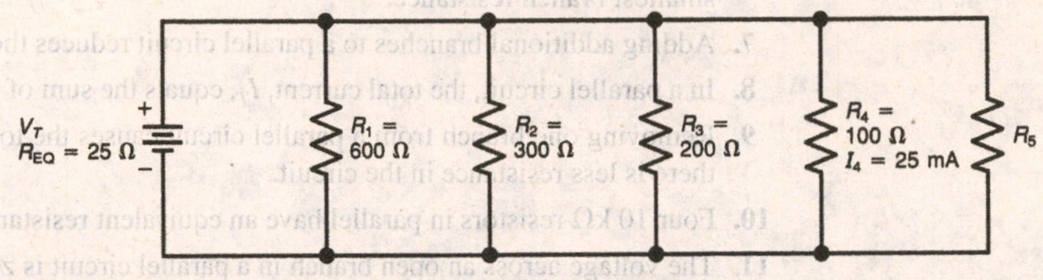

Fig. 5-32 Find I_T, V_T, I_1, I_2, I_3, I_5, R_5, P_T, P_1, P_2, P_3, P_4, and P_5.

SEC. 5-3 TROUBLESHOOTING PARALLEL CIRCUITS

The currents shown in Fig. 5-33 exist when the circuit is operating normally.

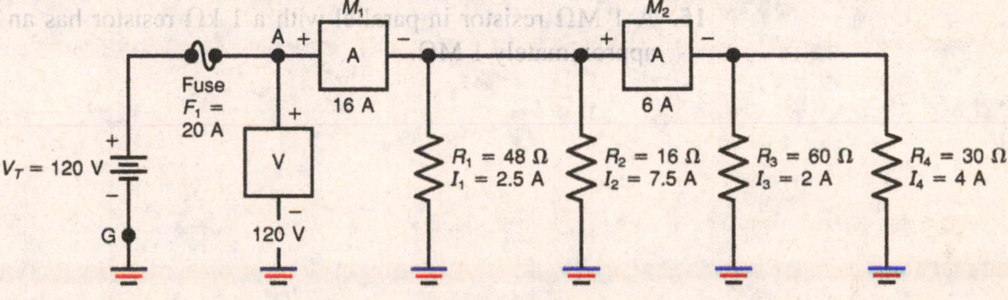

Fig. 5-33 Parallel circuit used for troubleshooting analysis.

PRACTICE PROBLEMS

Sec. 5-3

Refer to Fig. 5-33. For each set of measurements shown in Probs. 1–5, identify the defective component. Indicate how the component is defective, and describe a procedure that would lead you to the defective component.

1. $V_{AG} = 120$ V; M_1 reads 8.5 A; M_2 reads 6 A.
2. $V_{AG} = 120$ V; M_1 reads 14 A; M_2 reads 4 A.
3. $V_{AG} = 120$ V; M_1 reads 12 A; M_2 reads 2 A.
4. $V_{AG} = 120$ V; M_1 reads 13.5 A; M_2 reads 6 A.
5. $V_{AG} = 0$ V; M_1 reads 0 A; M_2 reads 0 A.
6. How much voltage would be measured across fuse F_1 in Fig. 5-33 if it were open?
7. How much voltage would be measured across fuse F_1 if it were good?
8. How would you measure the resistance of an individual branch in Fig. 5-33? What precautions must be taken when measuring resistance?
9. If, in Fig. 5-33, R_3 opens, what voltage will be measured across it? How will this open branch affect the other branch currents?

Chapter 5: Parallel Circuits

Answer True or False.

1. The voltage is the same across each branch of a parallel circuit.
2. In a parallel circuit, the individual branch currents are totally independent of each other.
3. In a parallel circuit, the branch with the largest resistance dissipates the most power.
4. An open in one branch of a parallel circuit does not affect the current in each of the remaining branches.
5. In a parallel circuit, the smallest branch resistance has the largest branch current.
6. In a parallel circuit, the combined equivalent resistance, R_{EQ}, is always smaller than the smallest branch resistance.
7. Adding additional branches to a parallel circuit reduces the equivalent resistance, R_{EQ}.
8. In a parallel circuit, the total current, I_T, equals the sum of the individual branch currents.
9. Removing one branch from a parallel circuit causes the total current to increase because there is less resistance in the circuit.
10. Four 10 kΩ resistors in parallel have an equivalent resistance, R_{EQ}, of 2.5 kΩ.
11. The voltage across an open branch in a parallel circuit is zero volts.
12. If the fuse in the main line of a parallel circuit opens, the voltage across all parallel connected branches is zero volts.
13. If one branch of a parallel circuit shorts, all the branches are effectively short-circuited.
14. Removing one branch from a parallel circuit causes the remaining branch currents to increase.
15. A 1 MΩ resistor in parallel with a 1 kΩ resistor has an equivalent resistance, R_{EQ}, of approximately 1 MΩ.

CHAPTER 6

Series-Parallel Circuits

A series-parallel circuit, sometimes called a combination circuit, is any circuit that combines both series and parallel connections. In this chapter, you will solve for the unknown values of voltage, current, resistance, and power in a variety of series-parallel circuits.

SEC. 6-1 ANALYZING SERIES-PARALLEL CIRCUITS

Almost all electronic circuits contain a combination of series and parallel connections. In analyzing these circuits, the techniques of series and parallel circuits are applied individually to produce a much simpler overall circuit.

Solved Problem

For Fig. 6-1, solve for R_T, I_T, V_1, V_2, V_3, V_4, I_3, and I_4.

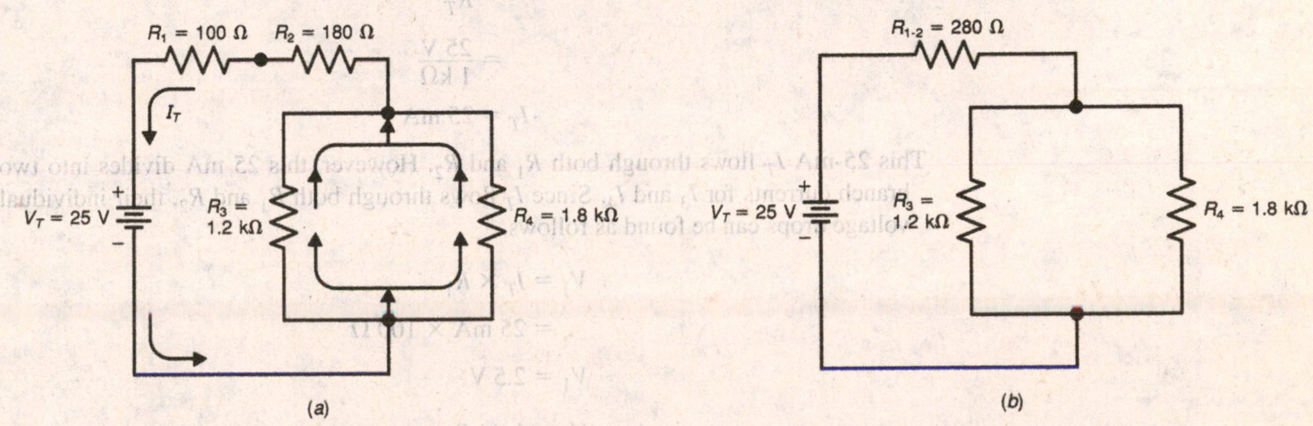

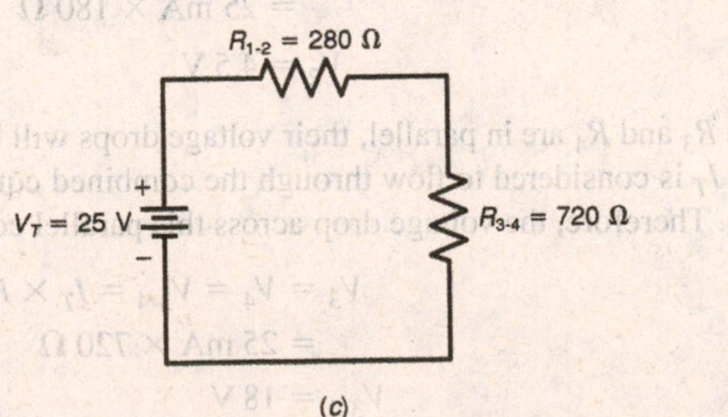

Fig. 6-1 Series-parallel circuit. (*a*) Schematic diagram before simplification. (*b*) Resistors R_1 and R_2 are combined as one and are represented as $R_{1\text{-}2}$. (*c*) Resistors R_3 and R_4 are combined as one and are represented as $R_{3\text{-}4}$.

Answer

To find R_T, add the series resistances R_1 and R_2 and combine the parallel resistances R_3 and R_4.

In Fig. 6-1*b*, the 100-Ω R_1 and 180-Ω R_2 in series total 280 Ω for $R_{1\text{-}2}$. The calculation is

$$R_{1\text{-}2} = R_1 + R_2$$
$$= 100\ \Omega + 180\ \Omega$$
$$R_{1\text{-}2} = 280\ \Omega$$

Also, the 1.2-kΩ R_3 in parallel with the 1.8-kΩ R_4 can be combined for an equivalent resistance of 720 Ω for $R_{3\text{-}4}$, as shown in Fig. 6-1*c*. The calculation is

$$R_{3\text{-}4} = \frac{R_3 \times R_4}{R_3 + R_4}$$
$$= \frac{1.2\ \text{k}\Omega \times 1.8\ \text{k}\Omega}{1.2\ \text{k}\Omega + 1.8\ \text{k}\Omega}$$
$$R_{3\text{-}4} = 720\ \Omega$$

This parallel $R_{3\text{-}4}$ combination of 720 Ω is then added to the series $R_{1\text{-}2}$ combination for the final R_T of 1 kΩ. The calculation for R_T is

$$R_T = R_{1\text{-}2} + R_{3\text{-}4}$$
$$= 280\ \Omega + 720\ \Omega$$
$$R_T = 1\ \text{k}\Omega$$

With R_T known to be 1 kΩ, we can solve for I_T.

$$I_T = \frac{V_T}{R_T}$$
$$= \frac{25\ \text{V}}{1\ \text{k}\Omega}$$
$$I_T = 25\ \text{mA}$$

This 25-mA I_T flows through both R_1 and R_2. However, this 25 mA divides into two branch currents for I_3 and I_4. Since I_T flows through both R_1 and R_2, their individual voltage drops can be found as follows.

$$V_1 = I_T \times R_1$$
$$= 25\ \text{mA} \times 100\ \Omega$$
$$V_1 = 2.5\ \text{V}$$

$$V_2 = I_T \times R_2$$
$$= 25\ \text{mA} \times 180\ \Omega$$
$$V_2 = 4.5\ \text{V}$$

Since R_3 and R_4 are in parallel, their voltage drops will be identical. Notice in Fig. 6-1*c* that I_T is considered to flow through the combined equivalent resistance of 720 Ω for $R_{3\text{-}4}$. Therefore, the voltage drop across this parallel combination is found as follows.

$$V_3 = V_4 = V_{3\text{-}4} = I_T \times R_{3\text{-}4}$$
$$= 25\ \text{mA} \times 720\ \Omega$$
$$V_{3\text{-}4} = 18\ \text{V}$$

Knowing this voltage allows us to find currents I_3 and I_4.

$$I_3 = \frac{V_3}{R_3}$$
$$= \frac{18\ \text{V}}{1.2\ \text{k}\Omega}$$
$$I_3 = 15\ \text{mA}$$

$$I_4 = \frac{V_4}{R_4}$$

$$= \frac{18 \text{ V}}{1.8 \text{ k}\Omega}$$

$$I_4 = 10 \text{ mA}$$

PRACTICE PROBLEMS

Sec. 6-1 For Figs. 6-2 through 6-23, find R_T, I_T, and all individual voltage and current values.

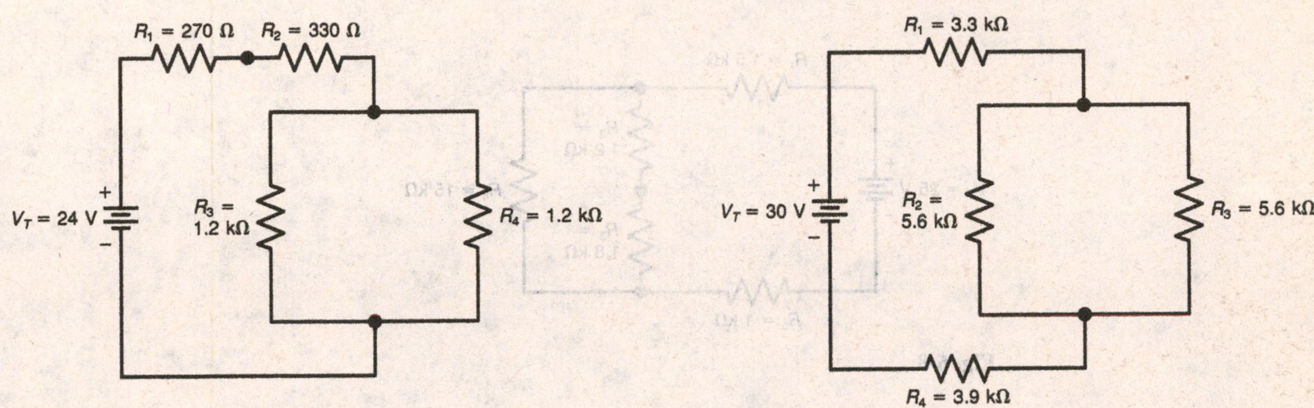

Fig. 6-2

Fig. 6-3

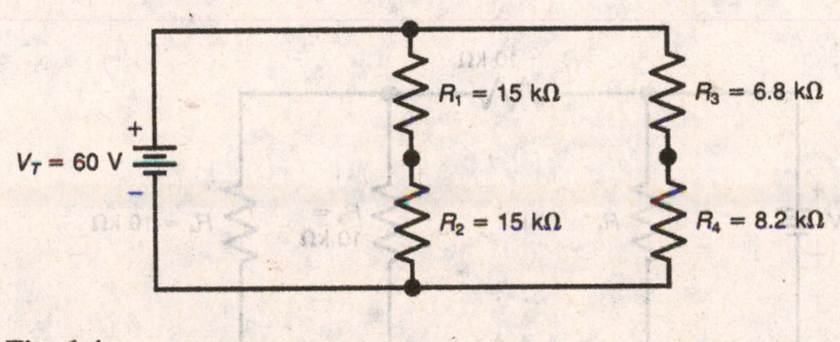

Fig. 6-4

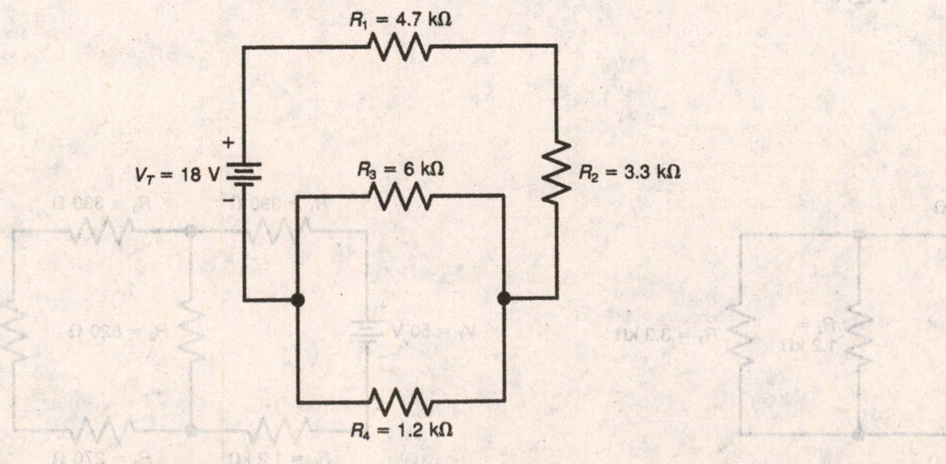

Fig. 6-5

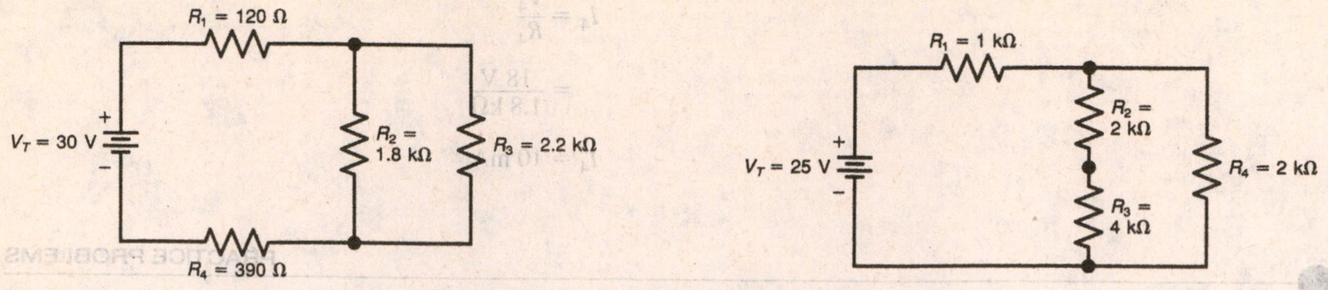

Fig. 6-6

Fig. 6-7

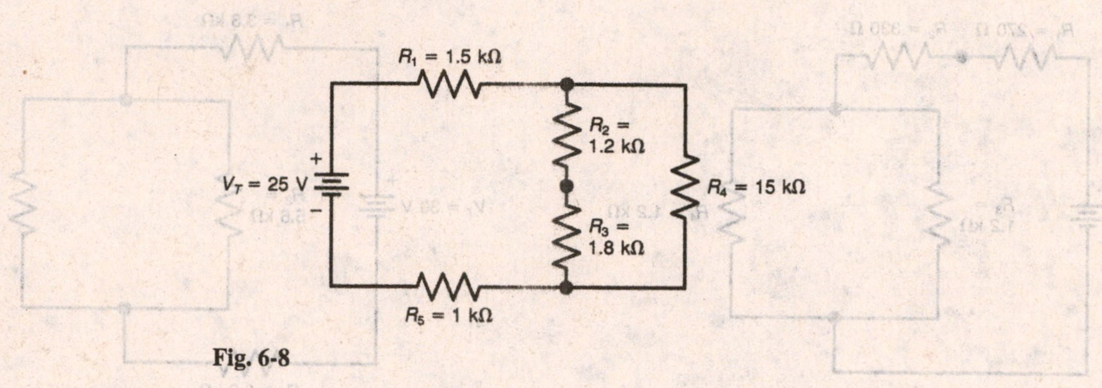

Fig. 6-8

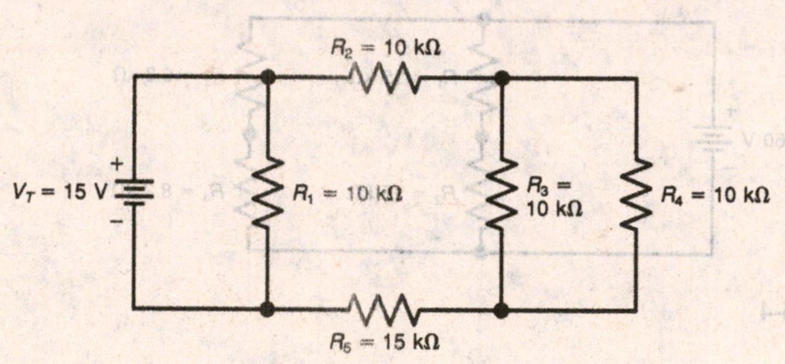

Fig. 6-9

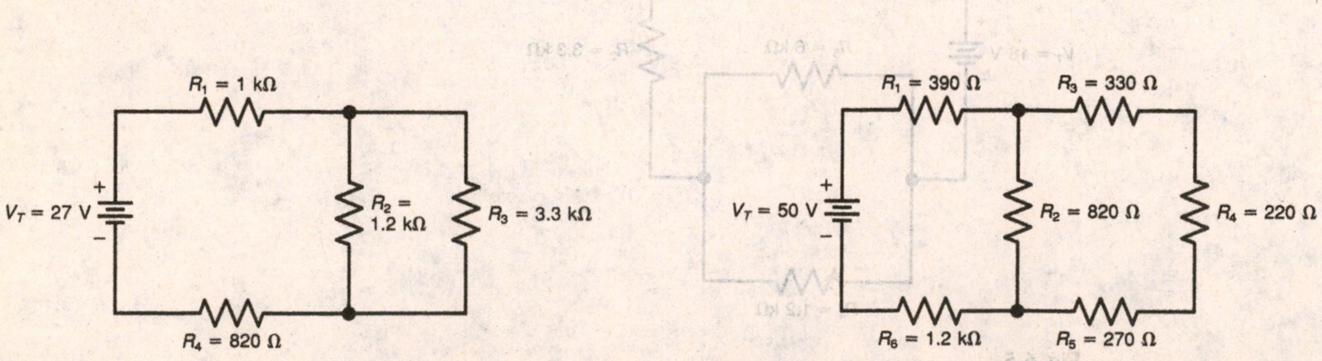

Fig. 6-10

Fig. 6-11

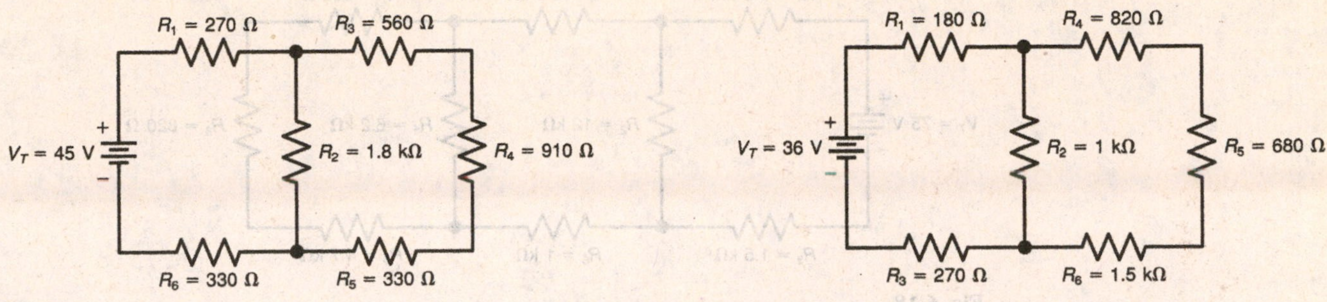

Fig. 6-12

Fig. 6-13

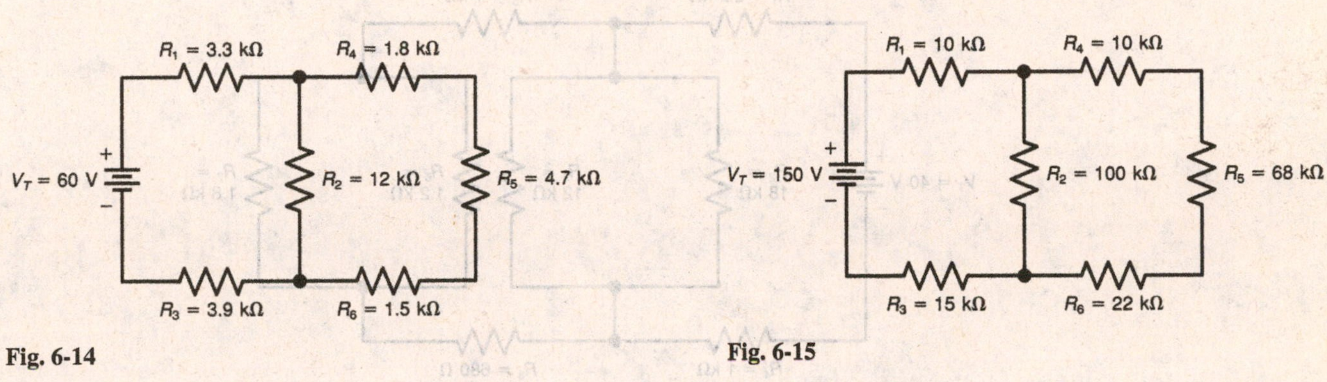

Fig. 6-14

Fig. 6-15

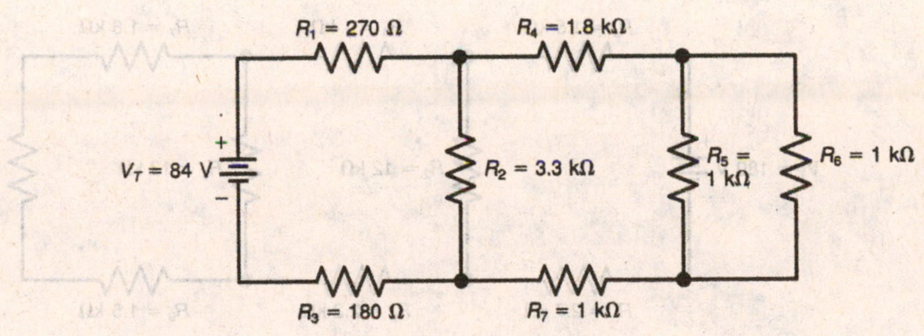

Fig. 6-16

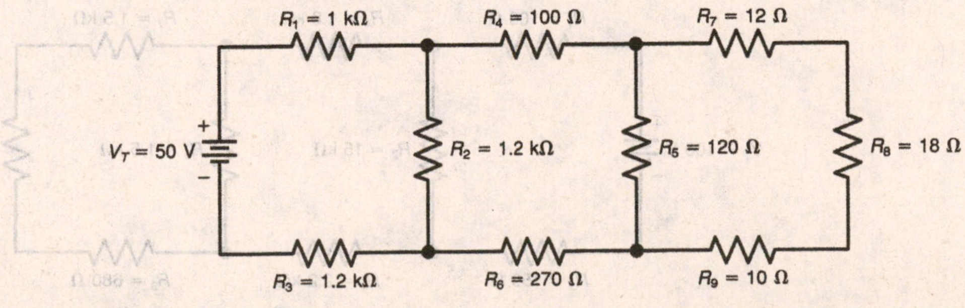

Fig. 6-17

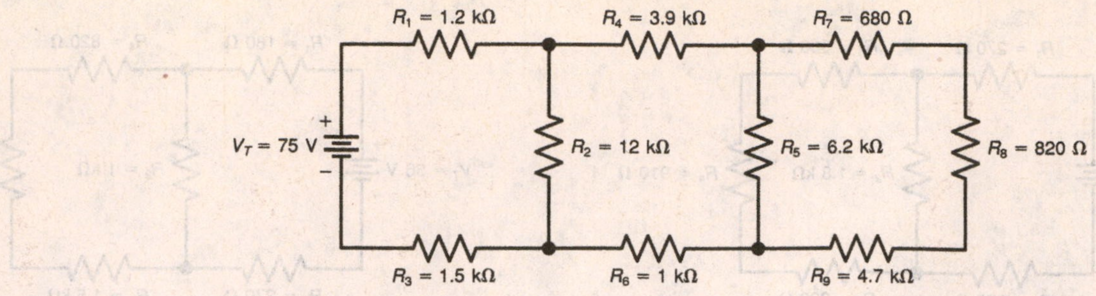

Fig. 6-18

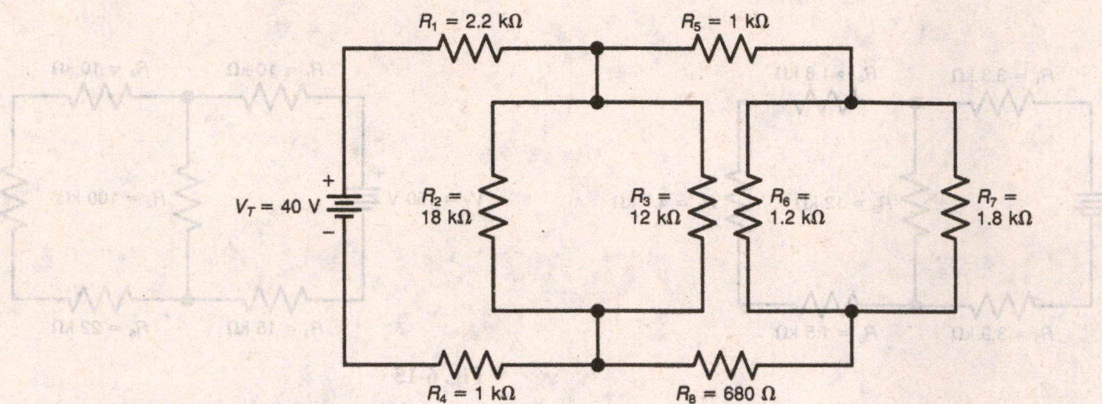

Fig. 6-19

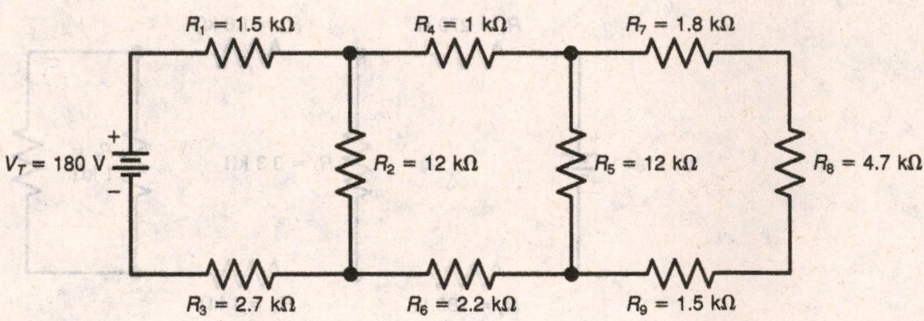

Fig. 6-20

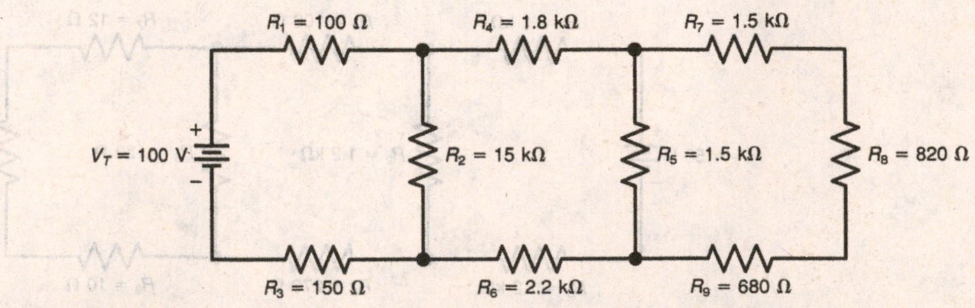

Fig. 6-21

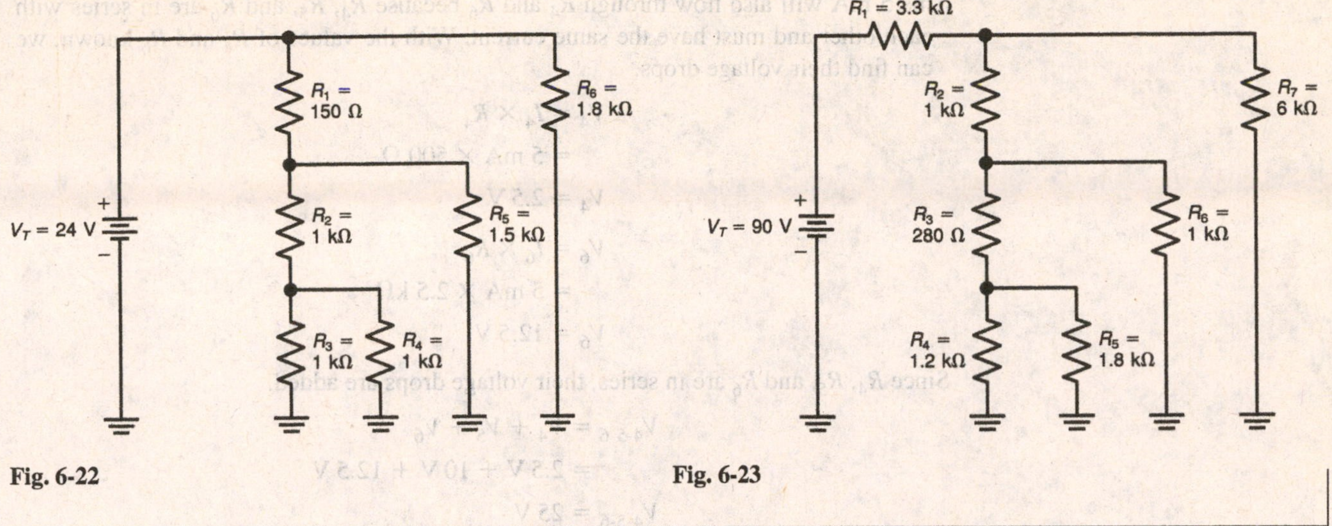

Fig. 6-22

Fig. 6-23

SEC. 6-2 SOLVING SERIES-PARALLEL CIRCUITS WITH RANDOM UNKNOWNS

Solved Problem

For Fig. 6-24, solve for V_T, R_T, V_1, V_2, V_3, V_4, V_6, I_T, and I_2.

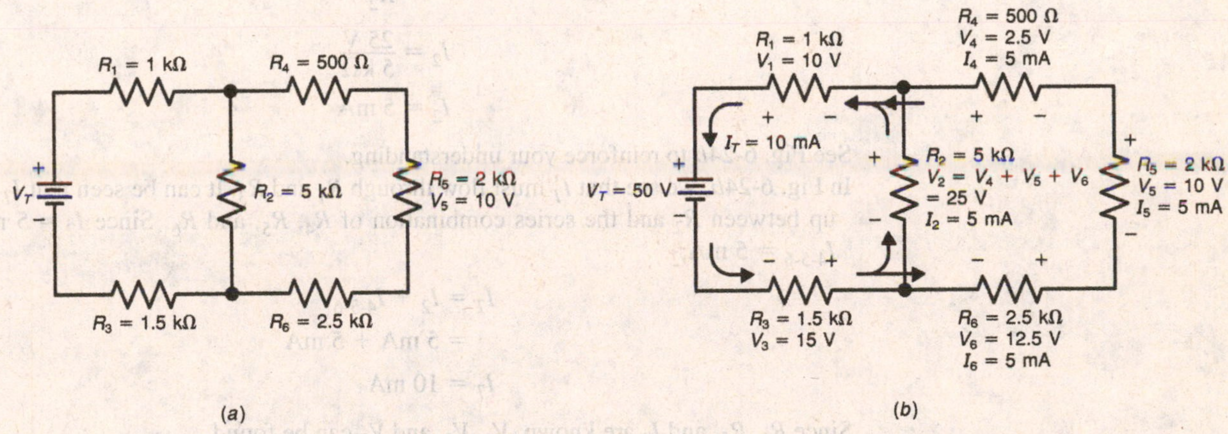

Fig. 6-24 Analyzing series-parallel circuits. (*a*) Series-parallel circuit with V_T unknown. (*b*) Solved circuit showing all voltage and current values.

Answer

To find V_T in Fig. 6-24*a*, we must work toward the applied voltage from R_5. Since V_5 and R_5 are known, we can find I_5.

$$I_5 = \frac{V_5}{R_5}$$

$$= \frac{10\text{ V}}{2\text{ k}\Omega}$$

$$I_5 = 5\text{ mA}$$

This 5 mA will also flow through R_4 and R_6 because R_4, R_5, and R_6 are in series with each other and must have the same current. With the values of R_4 and R_6 known, we can find their voltage drops.

$$V_4 = I_4 \times R_4$$
$$= 5 \text{ mA} \times 500\ \Omega$$
$$V_4 = 2.5 \text{ V}$$
$$V_6 = I_6 \times R_6$$
$$= 5 \text{ mA} \times 2.5 \text{ k}\Omega$$
$$V_6 = 12.5 \text{ V}$$

Since R_4, R_5, and R_6 are in series, their voltage drops are added.

$$V_{4\text{-}5\text{-}6} = V_4 + V_5 + V_6$$
$$= 2.5 \text{ V} + 10 \text{ V} + 12.5 \text{ V}$$
$$V_{4\text{-}5\text{-}6} = 25 \text{ V}$$

This 25 V is across the series string of R_4, R_5, and R_6. Since this series string is connected across R_2, V_2 must have a voltage drop that equals the sum of V_4, V_5, and V_6. This is shown as

$$V_2 = V_4 + V_5 + V_6$$
$$= 2.5 \text{ V} + 10 \text{ V} + 12.5 \text{ V}$$
$$V_2 = 25 \text{ V}$$

(See Fig. 6-24*b*.)

Since V_2 and R_2 are known, I_2 can be found as shown.

$$I_2 = \frac{V_2}{R_2}$$
$$I_2 = \frac{25 \text{ V}}{5 \text{ k}\Omega}$$
$$I_2 = 5 \text{ mA}$$

See Fig. 6-24*b* to reinforce your understanding.

In Fig. 6-24*b*, we see that I_T must flow through R_1 and R_3. It can be seen that I_T divides up between R_2 and the series combination of R_4, R_5, and R_6. Since $I_2 = 5$ mA and $I_{4\text{-}5\text{-}6} = 5$ mA,

$$I_T = I_2 + I_{4\text{-}5\text{-}6}$$
$$= 5 \text{ mA} + 5 \text{ mA}$$
$$I_T = 10 \text{ mA}$$

Since R_1, R_3, and I_T are known, V_1, V_3, and V_T can be found.

$$V_1 = I_T \times R_1$$
$$= 10 \text{ mA} \times 1 \text{ k}\Omega$$
$$V_1 = 10 \text{ V}$$
$$V_3 = I_T \times R_3$$
$$= 10 \text{ mA} \times 1.5 \text{ k}\Omega$$
$$V_3 = 15 \text{ V}$$

To find the total voltage V_T, we must realize that R_2 is in parallel with the series combination of R_4, R_5, and R_6. It is the voltage drop across this series-parallel combination that is added to V_1 and V_3 to find V_T. Refer to Fig. 6-24*b*. Notice that V_2 is the voltage across the series-parallel connection, and realize too that $V_4 + V_5 + V_6 = V_2$. Therefore, the following is true.

$$V_T = V_1 + V_2 + V_3$$
$$= 10\text{ V} + 25\text{ V} + 15\text{ V}$$
$$V_T = 50\text{ V}$$

Since $V_2 = V_4 + V_5 + V_6$, V_T can also be shown as

$$V_T = V_1 + V_4 + V_5 + V_6 + V_3$$
$$= 10\text{ V} + 2.5\text{ V} + 10\text{ V} + 12.5\text{ V} + 15\text{ V}$$
$$V_T = 50\text{ V}$$

It is important to realize that V_T is *not* equal to $V_1 + V_2 + V_3 + V_4 + V_5 + V_6$ because of the series-parallel connections.

With I_T and V_T known, R_T can be found.

$$R_T = \frac{V_T}{I_T}$$
$$= \frac{50\text{ V}}{10\text{ mA}}$$
$$R_T = 5\text{ k}\Omega$$

It should also be pointed out that R_T can be found by combining resistances alone. For Fig. 6-24,

$$R_T = R_1 + [R_2 \| (R_4 + R_5 + R_6)] + R_3$$
$$= 1\text{ k}\Omega + [5\text{ k}\Omega \| (500\ \Omega + 2\text{ k}\Omega + 2.5\text{ k}\Omega)] + 1.5\text{ k}\Omega$$
$$R_T = 5\text{ k}\Omega$$

where "$\|$" represents R_2 in parallel with the series combination of R_4, R_5, and R_6.

PRACTICE PROBLEMS

Sec. 6-2 For Figs. 6-25 through 6-32, find R_T, I_T, and all individual voltage and current values.

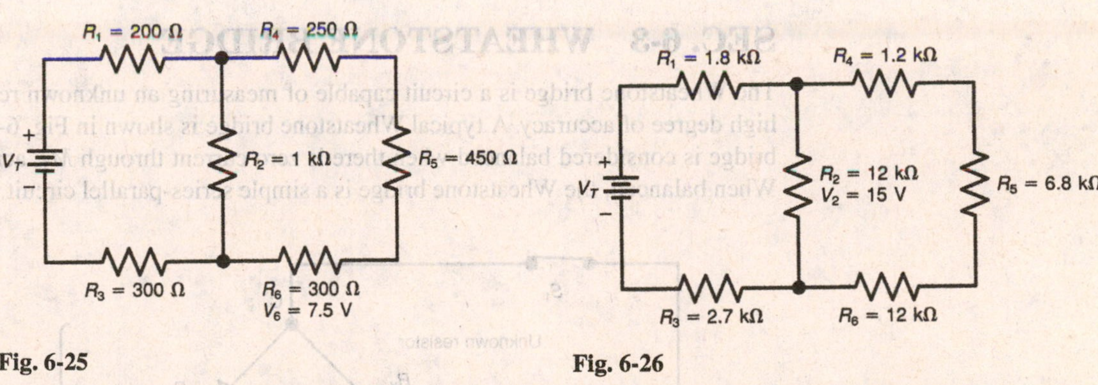

Fig. 6-25

Fig. 6-26

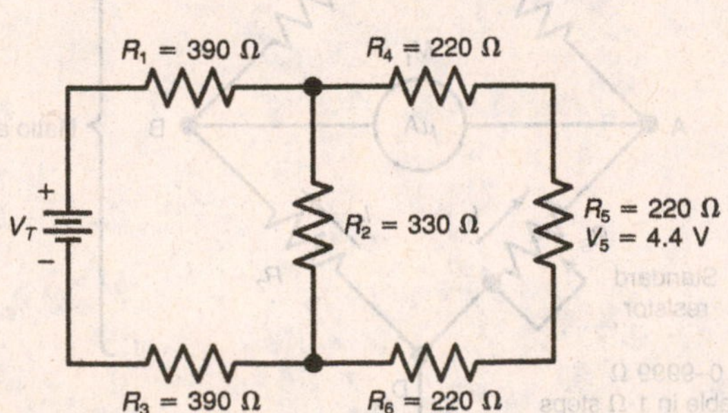

Fig. 6-27

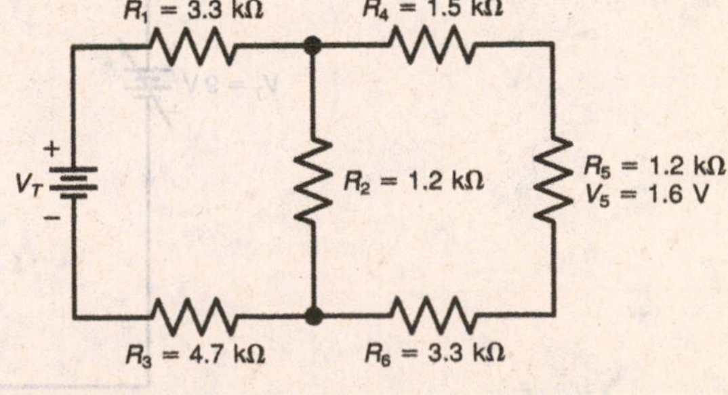

Fig. 6-28

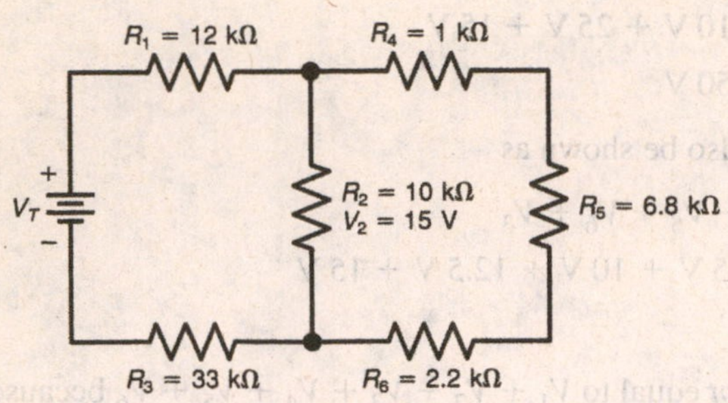

Fig. 6-29

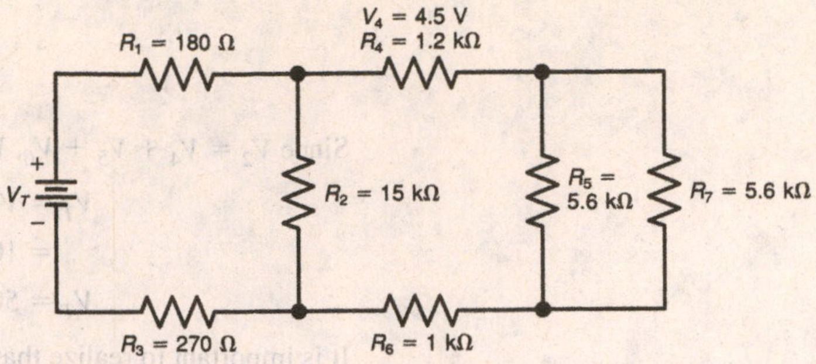

Fig. 6-30

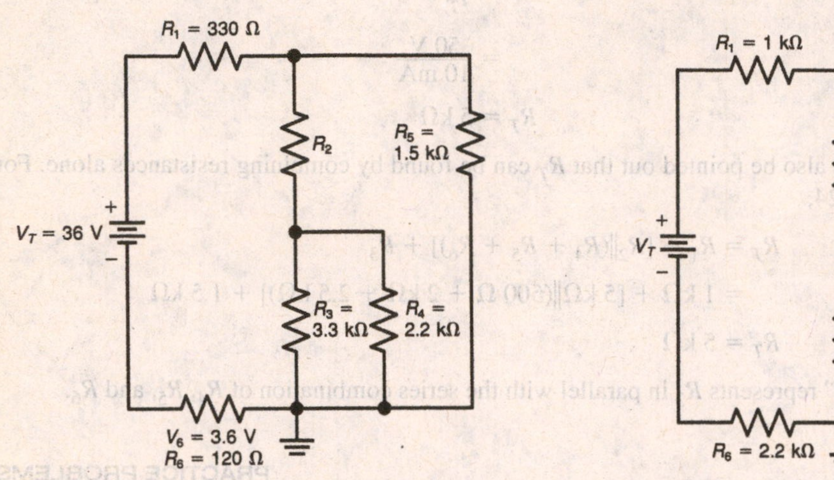

Fig. 6-31

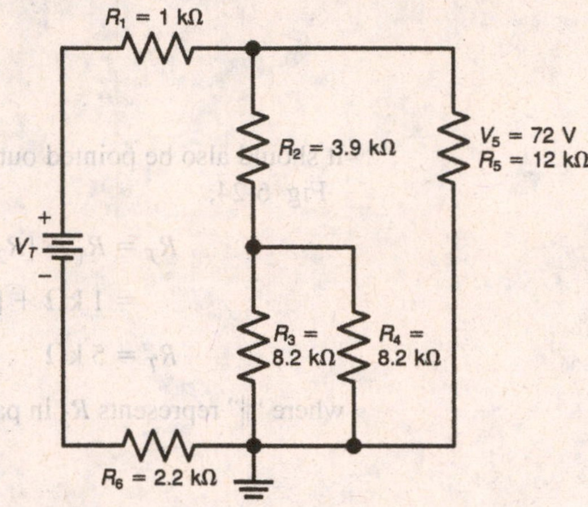

Fig. 6-32

SEC. 6-3 WHEATSTONE BRIDGE

The Wheatstone bridge is a circuit capable of measuring an unknown resistance with a very high degree of accuracy. A typical Wheatstone bridge is shown in Fig. 6-33. The Wheatstone bridge is considered balanced when there is zero current through M_1, across points A and B. When balanced, the Wheatstone bridge is a simple series-parallel circuit.

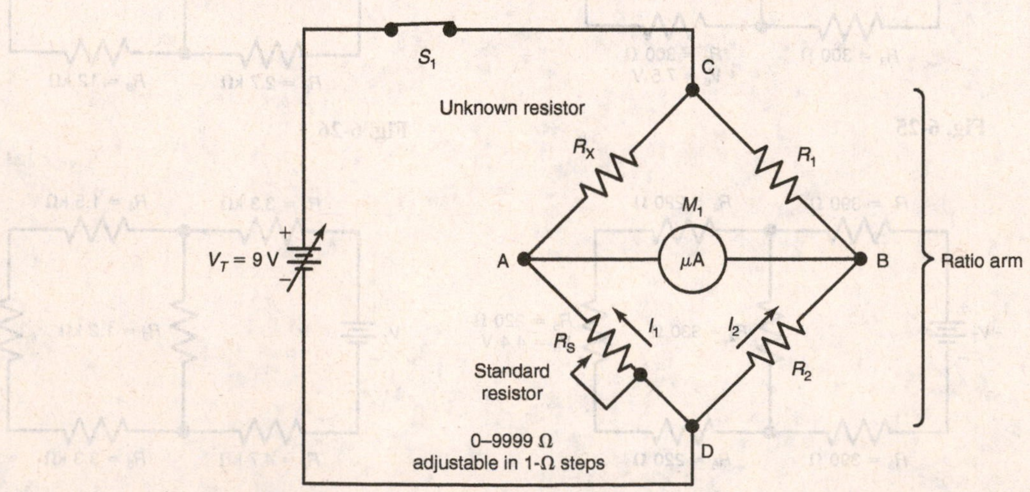

Fig. 6-33 Wheatstone bridge circuit.

In Fig. 6-33, notice the Wheatstone bridge has four terminals: two for input voltage and two for output voltage. Terminals C and D are the input terminals, whereas terminals A and B are the output terminals. The microammeter M_1 placed across the output terminals A and B will have zero current flowing through it when the bridge is balanced. Resistor R_S is varied to obtain balance. Resistor R_X is the unknown resistance being measured while R_1 and R_2 form a series voltage divider for V_T. Resistors R_1 and R_2 are referred to as the ratio arm.

When the voltage division in the ratio arm R_1 and R_2 is identical to that in the series string of R_X and R_S, the potential difference across M_1 is 0, thereby resulting in zero current. At balance, the equal voltage ratios can be stated as

$$\frac{I_1 \times R_X}{I_1 \times R_S} = \frac{I_2 \times R_1}{I_2 \times R_2}$$

or

$$\frac{R_X}{R_S} = \frac{R_1}{R_2}$$

Solving for R_X, we have

$$R_X = R_S \times \frac{R_1}{R_2}$$

The maximum unknown resistance that can be measured for a specific ratio in the ratio arm is

$$R_{X(\text{max})} = R_{S(\text{max})} \times \frac{R_1}{R_2}$$

(In Fig. 6-33, $R_{S(\text{max})} = 9999\ \Omega$.)

When measuring unknown resistors, always use the lowest possible R_1/R_2 ratio setting to obtain greatest accuracy. The ratio arm fraction determines the placement accuracy of resistor R_X.

Solved Problem

In Fig. 6-33, assume the bridge has been balanced by adjusting R_S to 827 Ω. If the ratio arm, $R_1/R_2 = \frac{10}{1}$, calculate the value of the unknown resistor, R_X.

Answer

$$R_X = R_S \times \frac{R_1}{R_2}$$

$$= 827\ \Omega \times \frac{10}{1}$$

$$= 8270\ \Omega$$

PRACTICE PROBLEMS

Sec. 6-3

Refer to Fig. 6-33. In Probs. 1–10, solve for the unknown. Resistor R_S is variable from 0 Ω to 9999 Ω in 1-Ω steps. The bridge is at balance for all problems.

1. $R_X = ?;\ R_S = 1236\ \Omega;\ \frac{R_1}{R_2} = \frac{1}{100}$

2. $R_X = ?;\ R_S = 9217\ \Omega;\ \frac{R_1}{R_2} = \frac{10}{1}$

3. $R_X = ?; R_S = 4166\ \Omega; \frac{R_1}{R_2} = \frac{1}{100}$

4. $R_X = ?; R_S = 2647\ \Omega; \frac{R_1}{R_2} = \frac{1}{10}$

5. $R_X = ?; R_S = 2854\ \Omega; \frac{R_1}{R_2} = \frac{100}{1}$

6. $R_X = 2.71\ \text{k}\Omega; R_S = ?; \frac{R_1}{R_2} = \frac{1}{1}$

7. $R_X = 684.3\ \Omega; R_S = ?; \frac{R_1}{R_2} = \frac{1}{10}$

8. $R_X = 18.64\ \Omega; R_S = ?; \frac{R_1}{R_2} = \frac{1}{100}$

9. $R_X = 14.97\ \Omega; R_S = 1497\ \Omega; \frac{R_1}{R_2} = ?$

10. $R_X = 1.459\ \Omega; R_S = 1459\ \Omega; \frac{R_1}{R_2} = ?$

Problems 11–15 also refer to the Wheatstone bridge in Fig. 6-33.

11. Calculate the maximum unknown resistance, $R_{X(\text{max})}$, that can be measured for the following ratio arm values: (a) $\frac{R_1}{R_2} = \frac{1}{10}$, (b) $\frac{R_1}{R_2} = \frac{1}{100}$, (c) $\frac{R_1}{R_2} = \frac{10}{1}$, (d) $\frac{R_1}{R_2} = \frac{100}{1}$.

12. How much voltage will be measured across points A and B when the bridge is balanced?

13. Assume the bridge is balanced when the ratio arm fraction, $\frac{R_1}{R_2} = \frac{100\ \Omega}{1\ \text{k}\Omega}$ and $R_S = 2{,}775\ \Omega$.
    a. How much voltage will be measured across points A and D and points B and D?
    b. How much is the total current, I_T, flowing to and from the terminals of the voltage source, V_T?

14. In reference to Prob. 13, which direction (A to B or B to A) will electrons flow through M_1 if R_S is reduced in value?

15. Assume the same unknown resistance, R_X, is measured using different ratio arm fractions. In each case, the standard resistor, R_S, is adjusted to provide the balanced condition. The values for each measurement are:

    a. $R_S = 47\ \Omega$ and $\frac{R_1}{R_2} = \frac{1}{1}$

    b. $R_S = 474\ \Omega$ and $\frac{R_1}{R_2} = \frac{1}{10}$

    c. $R_S = 4736\ \Omega$ and $\frac{R_1}{R_2} = \frac{1}{100}$

    Calculate the value of the unknown resistor, R_X, for each measurement. Which ratio arm fraction provides the greatest accuracy?

END OF CHAPTER TEST

Chapter 6: Series-Parallel Circuits

Answer True or False.

1. In Fig. 6-6, resistors R_1 and R_2 are in series.
2. In Fig. 6-6, resistors R_2 and R_3 are in parallel.
3. In Fig. 6-6, resistors R_2 and R_3 are in parallel with the total voltage, V_T of 30 V.
4. In Fig. 6-6, R_1 and R_4 are in series.
5. In Fig. 6-6, I_2 and I_3 add to equal I_T.

6. In Fig. 6-6, $V_T = V_1 + V_2 + V_3 + V_4$.
7. In Fig. 6-6, $V_2 = V_3$.
8. In Fig. 6-33, the Wheatstone bridge is balanced when the currents I_1 and I_2 are equal.
9. In Fig. 6-33, the Wheatstone bridge is balanced when $\frac{V_x}{V_s} = \frac{V_1}{V_2}$.
10. In Fig. 6-33, the voltage between points A and B is zero when the Wheatstone bridge is balanced.

CHAPTER

Voltage and Current Dividers

In this chapter, you will learn how to calculate the voltage drops in a series circuit without knowing the value of the series current. Similarly, you will learn how to calculate the branch currents in a parallel circuit without knowing the value of the applied voltage. And finally, you will learn how to calculate the voltage, current, and resistance values in a loaded voltage divider.

SEC. 7-1 UNLOADED VOLTAGE DIVIDERS

Any series circuit is a voltage divider in which the individual *IR* voltage drops are directly proportional to the series resistance values. This means that the higher resistance values will drop a larger percentage of the applied voltage.

In a series circuit, remember that each individual *IR* voltage drop is calculated as $V = I_T \times R$. Since we also know that $I_T = V_T/R_T$, we can substitute this for I_T in the equation for V. Then we have the following:

$$V = \frac{V_T}{R_T} \times R \text{ or } V = \frac{R}{R_T} \times V_T$$

This formula shows us that the voltage drops in a series circuit are proportional to the series resistance values. The formula $V = \frac{R}{R_T} \times V_T$ is commonly referred to as the voltage divider rule (VDR).

Solved Problem

For the circuit shown in Fig. 7-1, calculate the voltage at point A with respect to ground for each position of the switch, S_1.

Answer

First, we calculate R_T by adding the individual resistance values.

$$\begin{aligned} R_T &= R_1 + R_2 + R_3 + R_4 + R_5 \\ &= 1\text{ k}\Omega + 1\text{ k}\Omega + 3\text{ k}\Omega + 2.5\text{ k}\Omega + 2.5\text{ k}\Omega \\ &= 10\text{ k}\Omega \end{aligned}$$

Next, using the formula $V = R/R_T \times V_T$, we calculate the voltage V_{AG} for each switch position as follows:

$$\begin{aligned} V_{AG}\text{ (Position 1)} &= \frac{R_1}{R_T} \times V_T \\ &= \frac{1\text{ k}\Omega}{10\text{ k}\Omega} \times 100\text{ V} \\ &= 10\text{ V} \end{aligned}$$

$$\begin{aligned} V_{AG}\text{ (Position 2)} &= \frac{R_1 + R_2}{R_T} \times 100\text{ V} \\ &= \frac{2\text{ k}\Omega}{10\text{ k}\Omega} \times 100\text{ V} \\ &= 20\text{ V} \end{aligned}$$

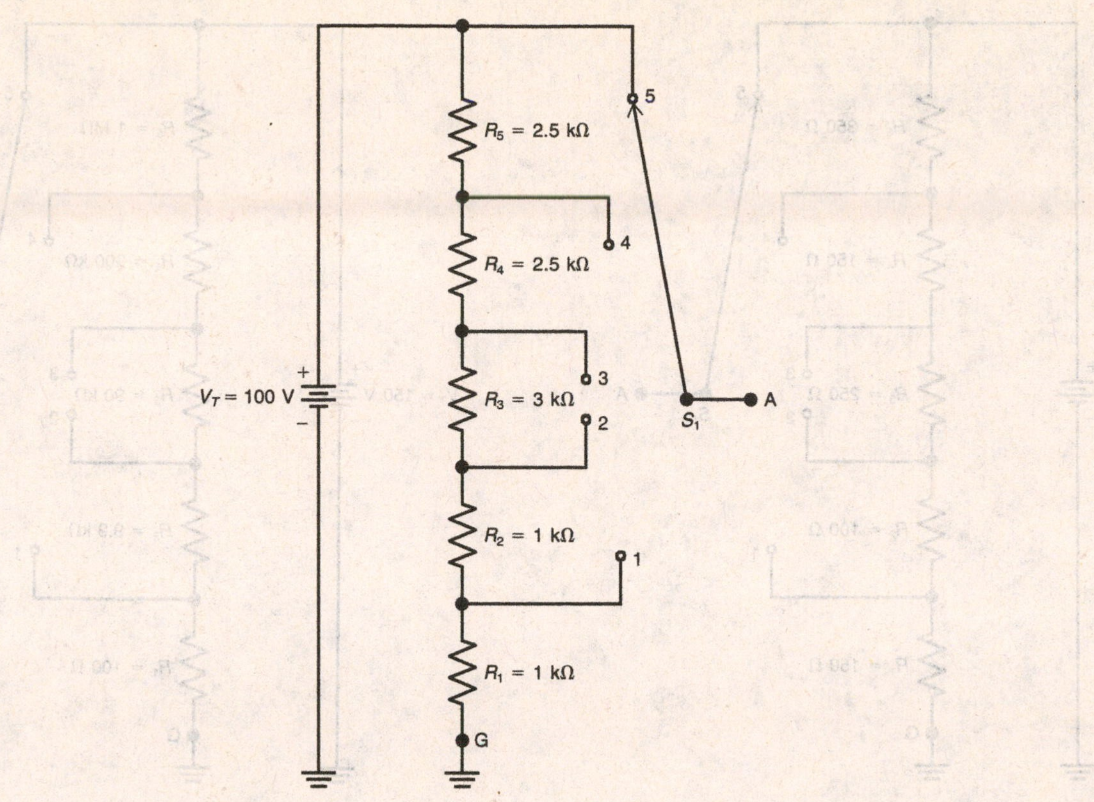

Fig. 7-1

$$V_{AG} \text{ (Position 3)} = \frac{R_1 + R_2 + R_3}{R_T} \times 100 \text{ V}$$

$$= \frac{5 \text{ k}\Omega}{10 \text{ k}\Omega} \times 100 \text{ V}$$

$$= 50 \text{ V}$$

$$V_{AG} \text{ (Position 4)} = \frac{R_1 + R_2 + R_3 + R_4}{R_T} \times 100 \text{ V}$$

$$= \frac{7.5 \text{ k}\Omega}{10 \text{ k}\Omega} \times 100 \text{ V}$$

$$= 75 \text{ V}$$

Finally,

$$V_{AG} \text{ (Position 5)} = \frac{R_1 + R_2 + R_3 + R_4 + R_5}{R_T} \times 100 \text{ V}$$

$$= \frac{10 \text{ k}\Omega}{10 \text{ k}\Omega} \times 100 \text{ V}$$

$$= 100 \text{ V}$$

Notice that for the higher switch positions, the voltage is being measured across more than just one resistance. For example, when S_1 is in position 2, the voltmeter is measuring V across both resistors R_1 and R_2. Since $R_1 + R_2$ makes up 20% of the total resistance R_T, the voltage V_{AG} will also be 20% of the applied voltage.

PRACTICE PROBLEMS

Sec. 7-1

For the voltage dividers shown in Figs. 7-2 through 7-11, indicate the voltage V_{AG} for each position of the switch, S_1.

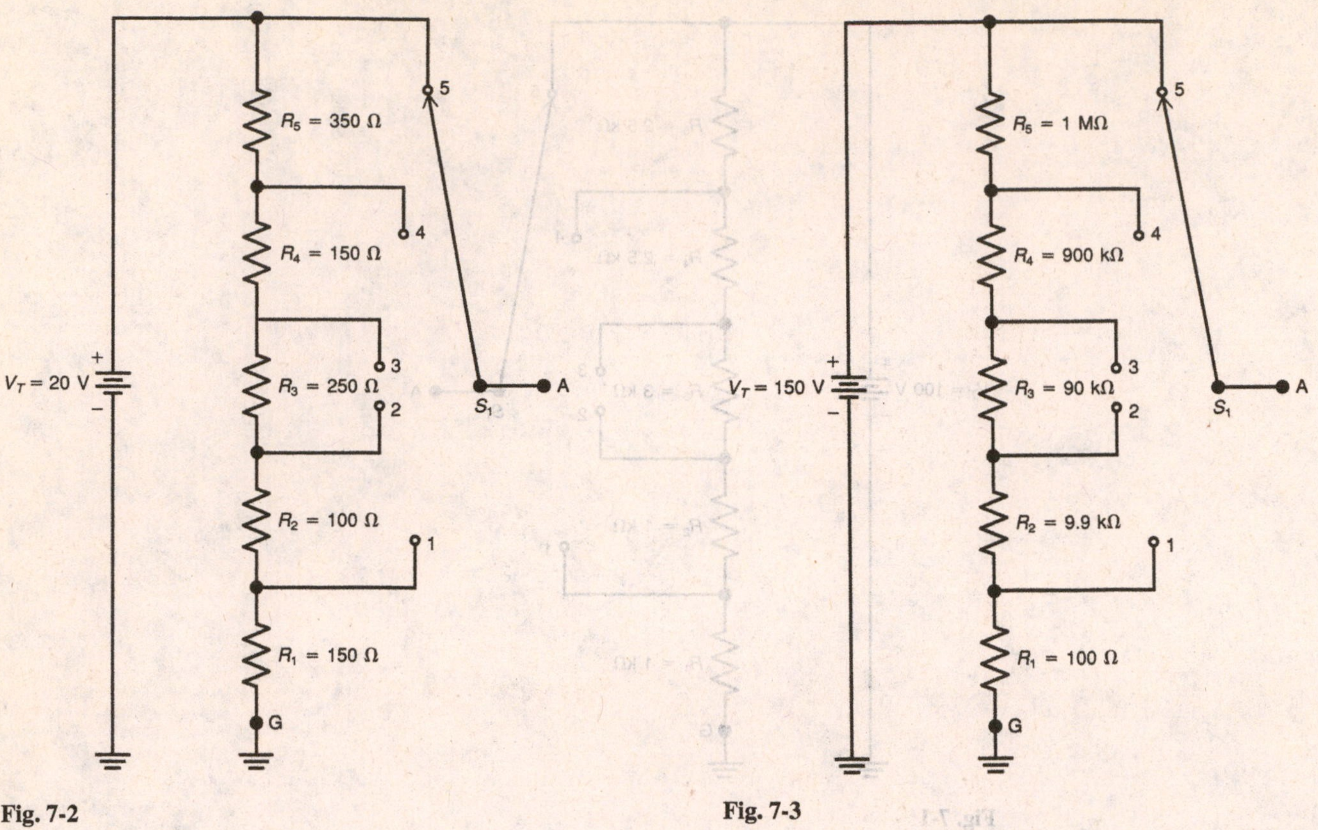

Fig. 7-2

Fig. 7-3

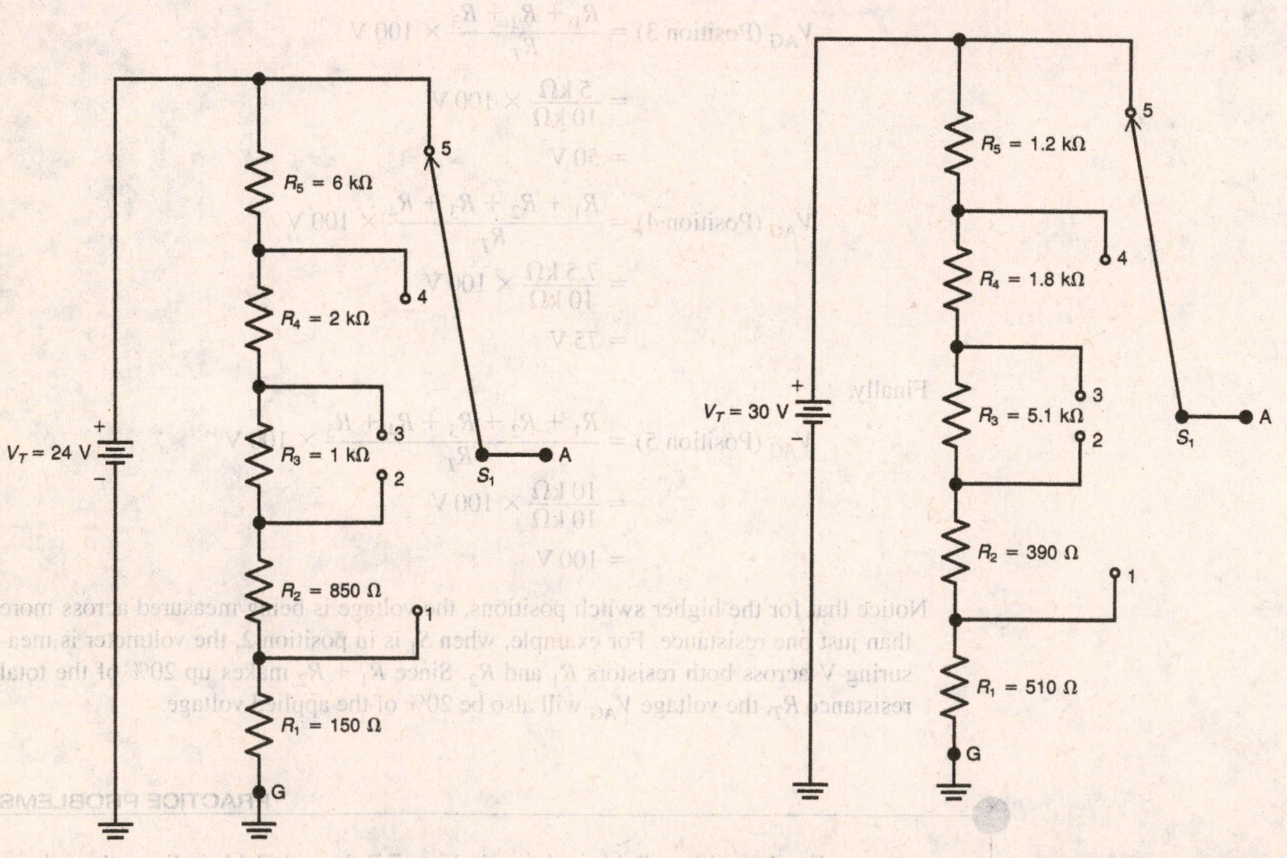

Fig. 7-4

Fig. 7-5

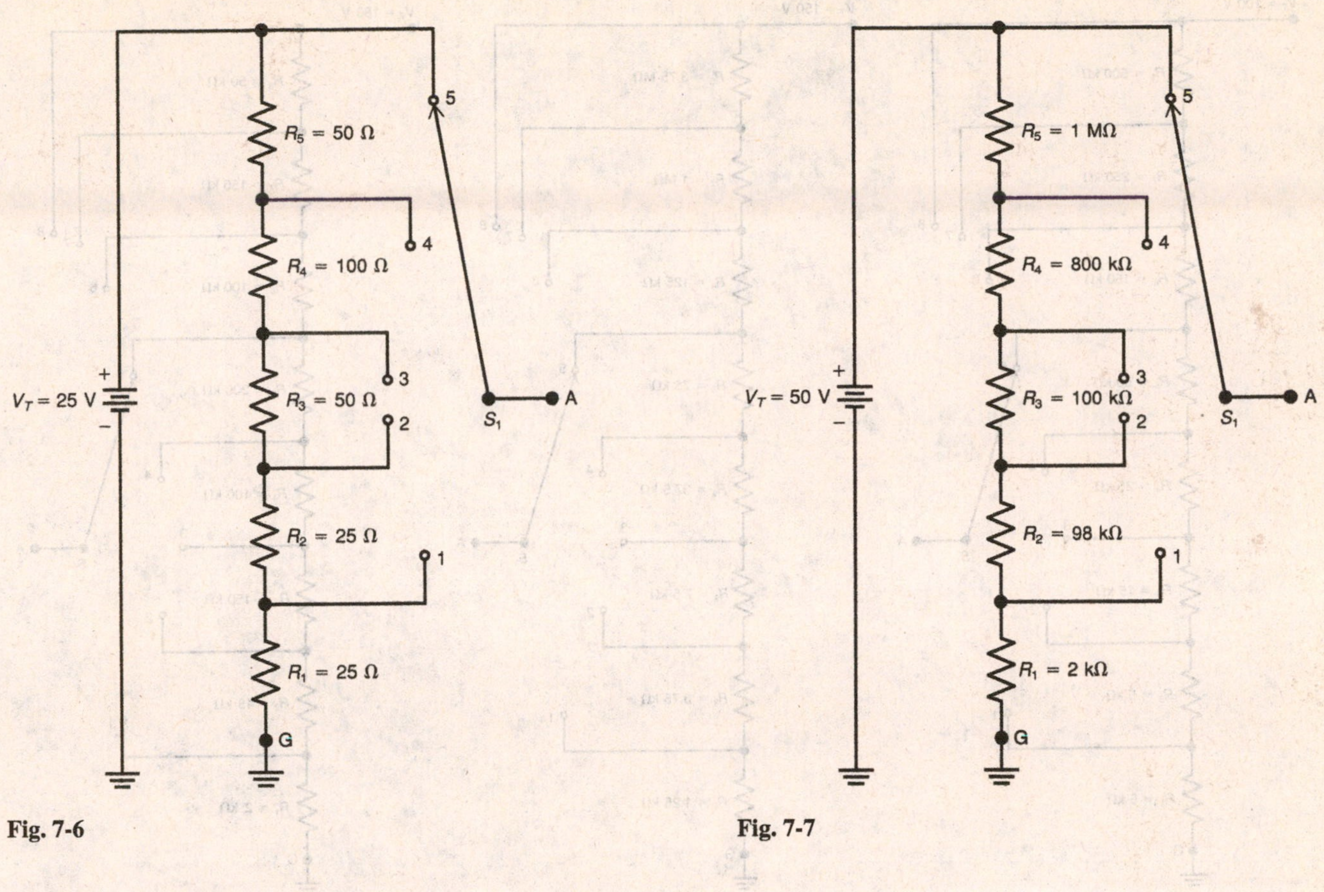

Fig. 7-6

Fig. 7-7

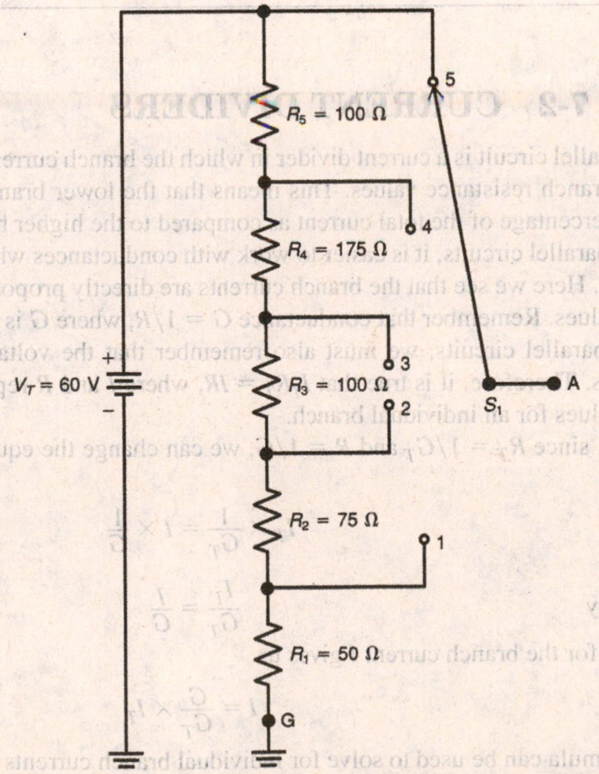

5
R_5 = 100 Ω
4
R_4 = 175 Ω
+
V_T = 60 V
−
3
R_3 = 100 Ω
2
S_1
A
R_2 = 75 Ω
1
R_1 = 50 Ω
G

Fig. 7-8

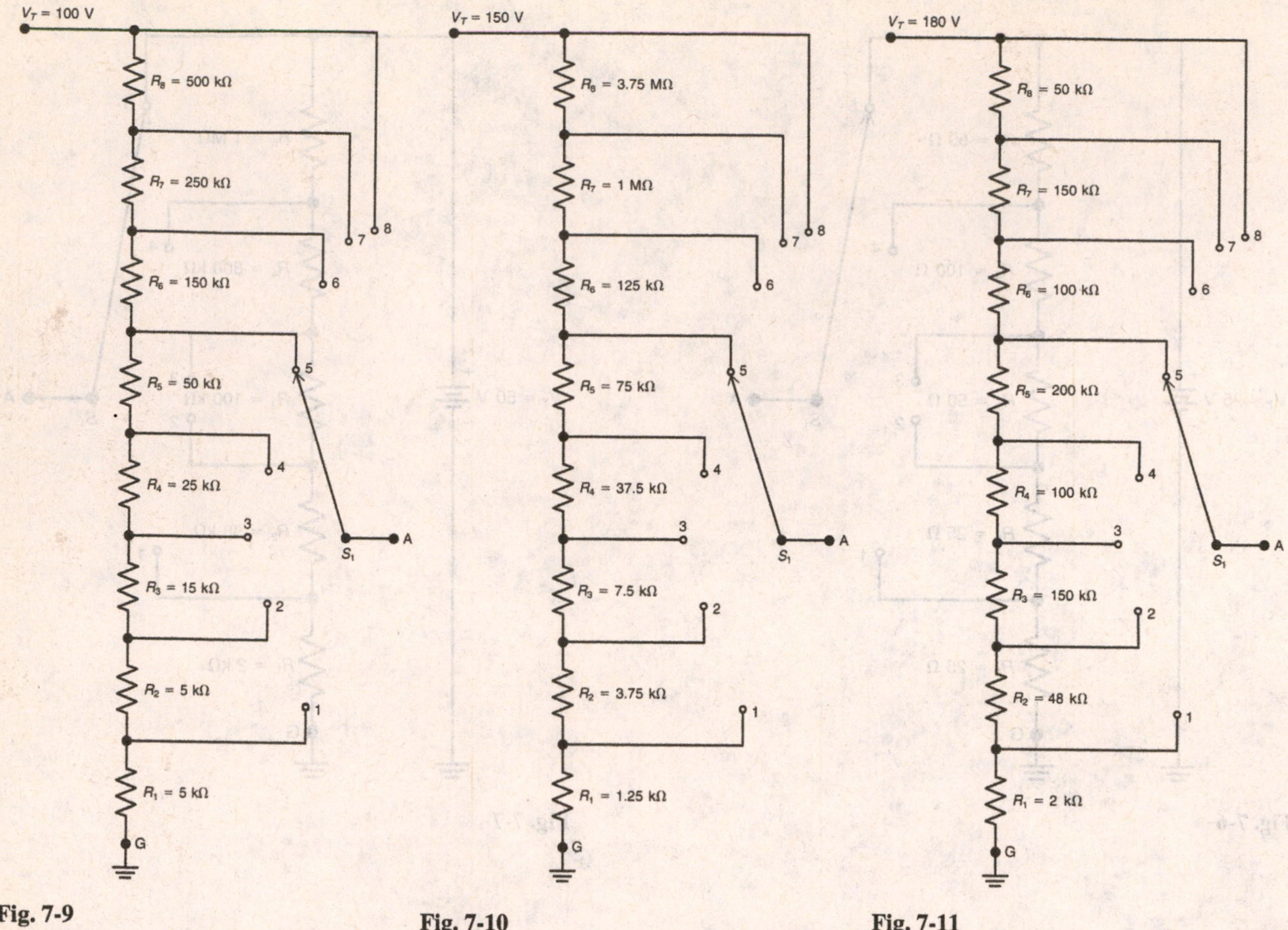

Fig. 7-9 **Fig. 7-10** **Fig. 7-11**

SEC. 7-2 CURRENT DIVIDERS

Any parallel circuit is a current divider in which the branch currents are inversely proportional to the branch resistance values. This means that the lower branch resistances will receive a larger percentage of the total current as compared to the higher branch resistance values.

For parallel circuits, it is easier to work with conductances when finding individual branch currents. Here we see that the branch currents are directly proportional to the branch conductance values. Remember that conductance $G = 1/R$, where G is measured in siemens (S).

For parallel circuits, we must also remember that the voltage V is the same across all branches. Therefore, it is true that $I_T R_T = IR$, where I and R represent the current and resistance values for an individual branch.

Also, since $R_T = 1/G_T$ and $R = 1/G$, we can change the equation $I_T R_T = IR$ to look like this:

$$I_T \times \frac{1}{G_T} = I \times \frac{1}{G}$$

or simply

$$\frac{I_T}{G_T} = \frac{I}{G}$$

Solving for the branch current I gives us

$$I = \frac{G}{G_T} \times I_T$$

This formula can be used to solve for individual branch currents in any parallel circuit.

The formula $I = \frac{G}{G_T} \times I_T$ is commonly referred to as the current divider rule (CDR).

It is important to note that G is calculated as $1/R$ and that the smaller resistance values correspond to higher conductance values.

Solved Problem

For the circuit shown in Fig. 7-12, find the branch currents I_1, I_2, I_3, and I_4.

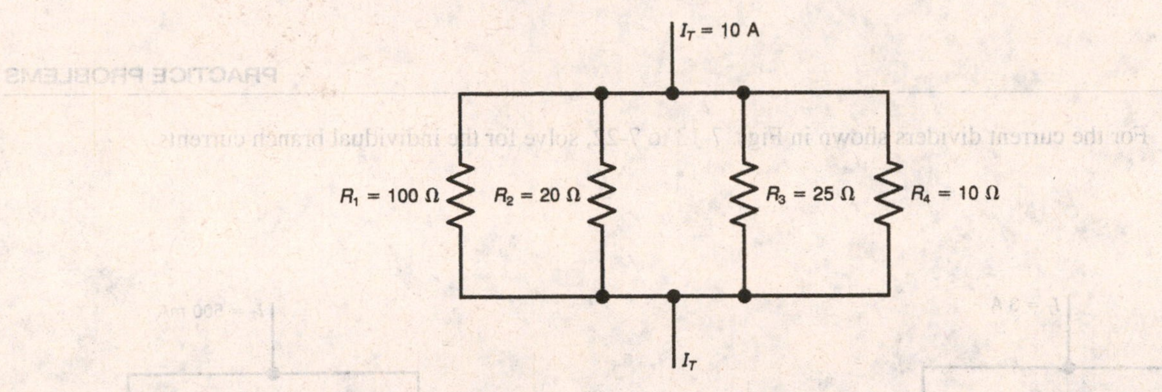

Fig. 7-12

Answer

First, we calculate G_1, G_2, G_3, G_4, and G_T. We proceed as follows:

$$G_1 = \frac{1}{R_1} = \frac{1}{100\ \Omega} = 10\ \text{mS}$$

$$G_2 = \frac{1}{R_2} = \frac{1}{20\ \Omega} = 50\ \text{mS}$$

$$G_3 = \frac{1}{R_3} = \frac{1}{25\ \Omega} = 40\ \text{mS}$$

$$G_4 = \frac{1}{R_4} = \frac{1}{10\ \Omega} = 100\ \text{mS}$$

$$\begin{aligned} G_T &= G_1 + G_2 + G_3 + G_4 \\ &= 10\ \text{mS} + 50\ \text{mS} + 40\ \text{mS} + 100\ \text{mS} \\ &= 200\ \text{mS} \end{aligned}$$

Notice that branch conductance values in parallel circuits are added just like resistance values in series circuits.

Using the formula $I = G/G_T \times I_T$, we calculate each branch current as follows:

$$\begin{aligned} I_1 &= \frac{G_1}{G_T} \times I_T \\ &= \frac{10\ \text{mS}}{200\ \text{mS}} \times 10\ \text{A} \\ &= 0.5\ \text{A} \end{aligned}$$

$$\begin{aligned} I_2 &= \frac{G_2}{G_T} \times I_T \\ &= \frac{50\ \text{mS}}{200\ \text{mS}} \times 10\ \text{A} \\ &= 2.5\ \text{A} \end{aligned}$$

$$I_3 = \frac{40 \text{ mS}}{200 \text{ mS}} \times 10 \text{ A}$$
$$= 2 \text{ A}$$

$$I_4 = \frac{100 \text{ mS}}{200 \text{ mS}} \times 10 \text{ A}$$
$$= 5 \text{ A}$$

Notice again that the branch currents are greater for those branches with the lowest resistance values; this is analogous to saying that the branch currents are directly proportional to the branch conductance values.

PRACTICE PROBLEMS

Sec. 7-2 For the current dividers shown in Figs. 7-13 to 7-22, solve for the individual branch currents.

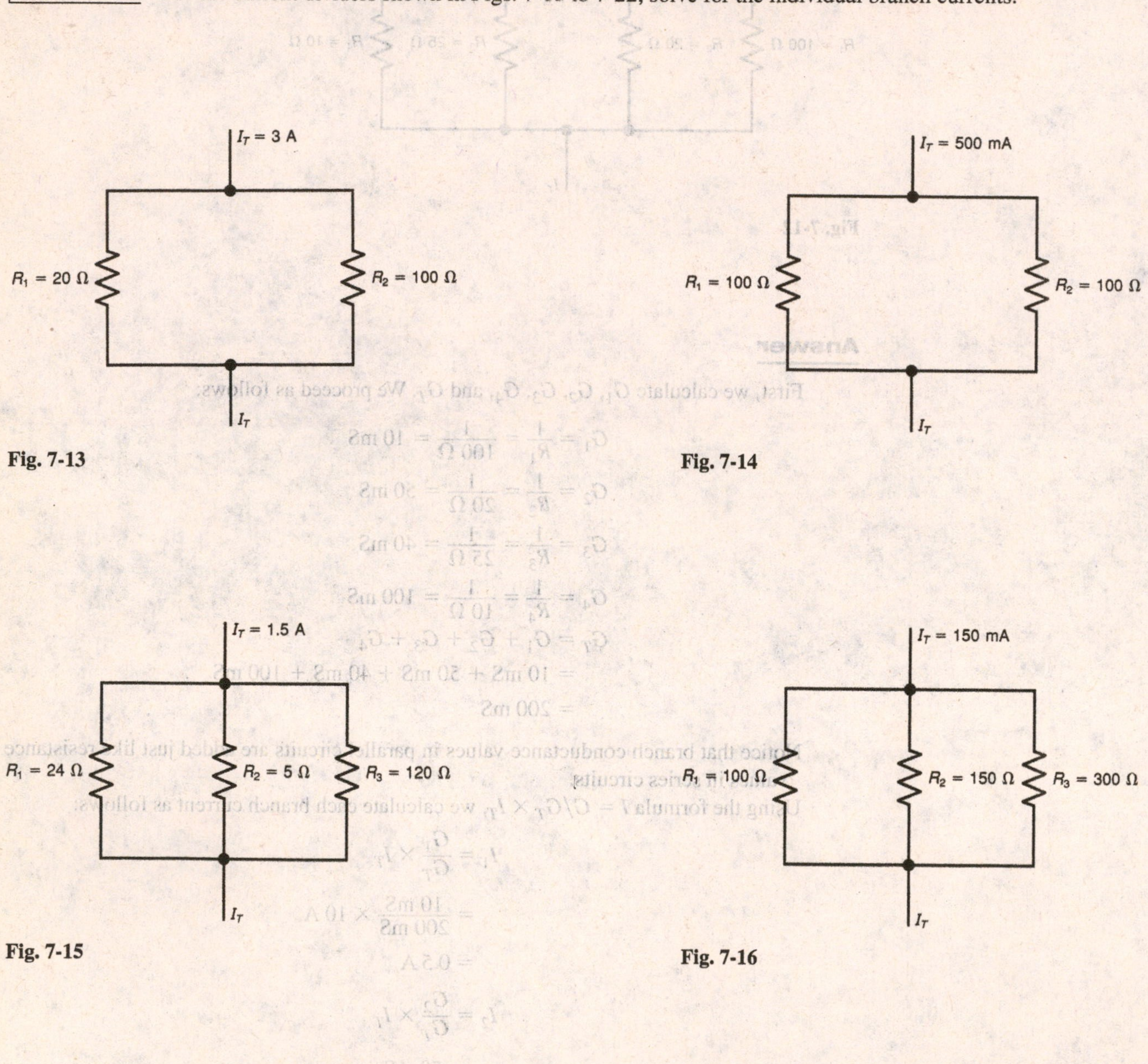

Fig. 7-13

Fig. 7-14

Fig. 7-15

Fig. 7-16

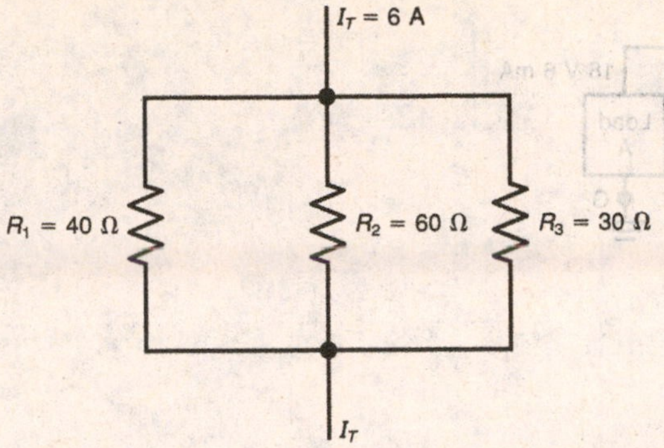

I_T = 6 A
R_1 = 40 Ω
R_2 = 60 Ω
R_3 = 30 Ω
I_T

Fig. 7-17

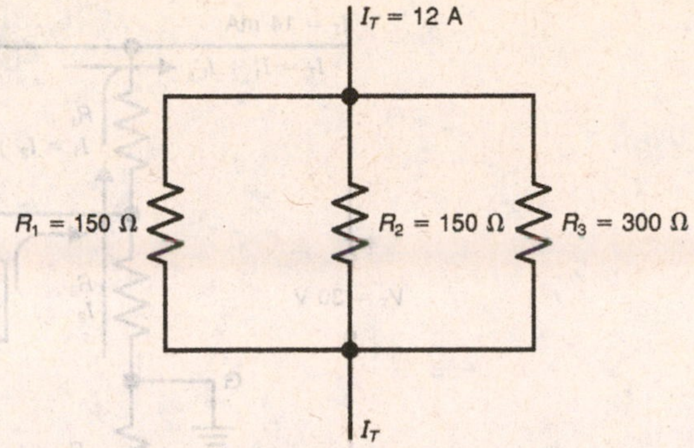

I_T = 12 A
R_1 = 150 Ω
R_2 = 150 Ω
R_3 = 300 Ω
I_T

Fig. 7-18

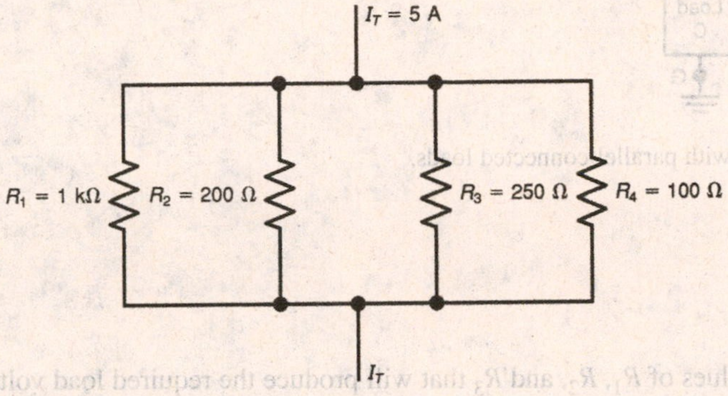

I_T = 5 A
R_1 = 1 kΩ
R_2 = 200 Ω
R_3 = 250 Ω
R_4 = 100 Ω
I_T

Fig. 7-19

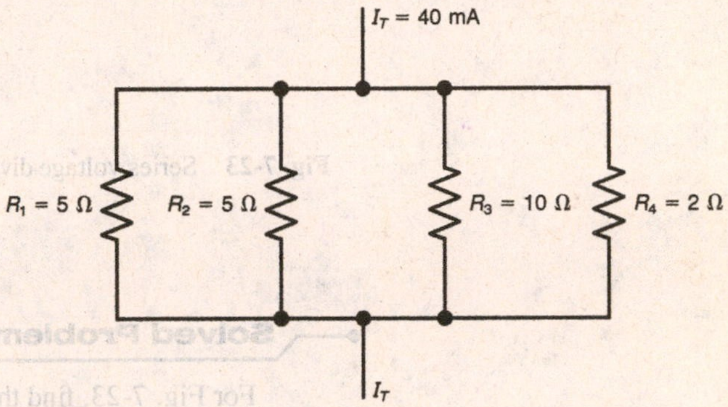

I_T = 40 mA
R_1 = 5 Ω
R_2 = 5 Ω
R_3 = 10 Ω
R_4 = 2 Ω
I_T

Fig. 7-20

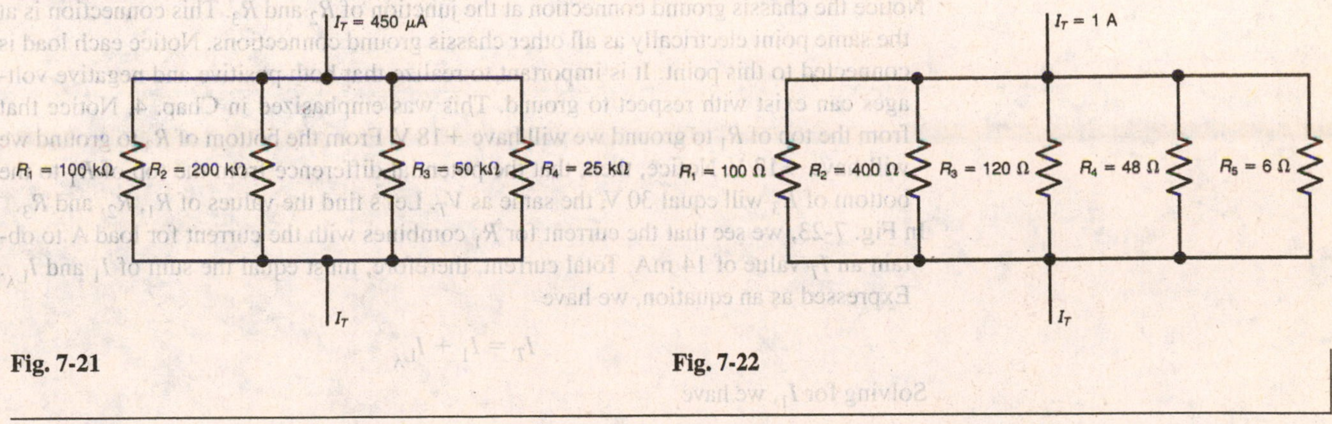

I_T = 450 μA
R_1 = 100 kΩ
R_2 = 200 kΩ
R_3 = 50 kΩ
R_4 = 25 kΩ
I_T

Fig. 7-21

I_T = 1 A
R_1 = 100 Ω
R_2 = 400 Ω
R_3 = 120 Ω
R_4 = 48 Ω
R_5 = 6 Ω
I_T

Fig. 7-22

SEC. 7-3 SERIES VOLTAGE DIVIDERS WITH PARALLEL-CONNECTED LOADS

The voltage dividers in Sec. 7-1 illustrate just a series string without any branch currents. In most cases, a voltage divider is used to tap off part of the applied voltage V_T for a load that needs less voltage than V_T. A series string with parallel-connected loads is often referred to as a loaded voltage divider. An example of a loaded voltage divider is shown in Fig. 7-23. Each load is represented by a box. The actual load could be something other than a resistor, and often is. Notice that V_T is given, but the battery symbol is not shown. This is because V is often obtained from an electronic circuit called a *power supply.*

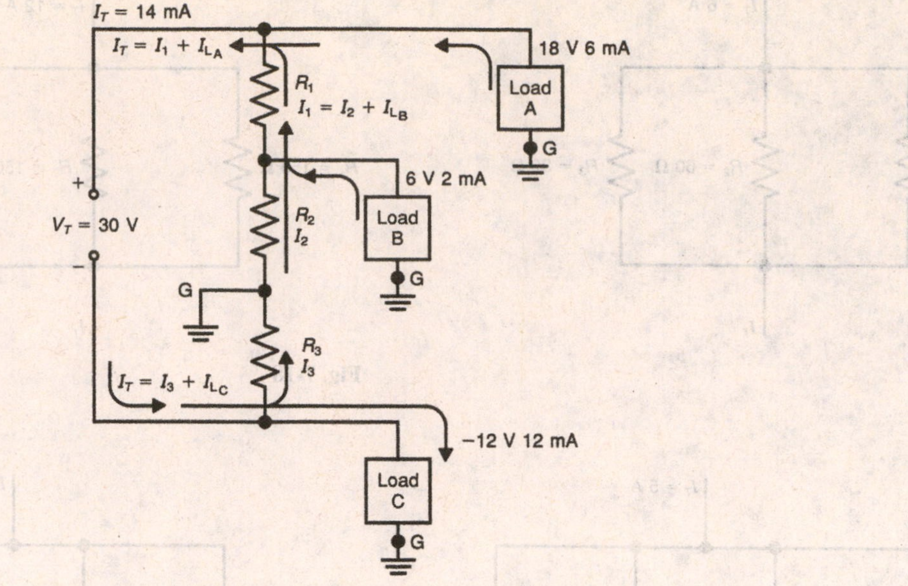

Fig. 7-23 Series voltage divider with parallel-connected loads.

Solved Problem

For Fig. 7-23, find the values of R_1, R_2, and R_3 that will produce the required load voltages and currents shown.

Answer

Notice the chassis ground connection at the junction of R_2 and R_3. This connection is at the same point electrically as all other chassis ground connections. Notice each load is connected to this point. It is important to realize that both positive and negative voltages can exist with respect to ground. This was emphasized in Chap. 4. Notice that from the top of R_1 to ground we will have +18 V. From the bottom of R_3 to ground we will have −12 V. Notice, then, that the potential difference from the top of R_1 to the bottom of R_3 will equal 30 V, the same as V_T. Let's find the values of R_1, R_2, and R_3.

In Fig. 7-23, we see that the current for R_1 combines with the current for load A to obtain an I_T value of 14 mA. Total current, therefore, must equal the sum of I_1 and I_{L_A}. Expressed as an equation, we have

$$I_T = I_1 + I_{L_A}$$

Solving for I_1, we have

$$\begin{aligned} I_1 &= I_T - I_{L_A} \\ &= 14\text{ mA} - 6\text{ mA} \\ I_1 &= 8\text{ mA} \end{aligned}$$

To solve for R_1, we must find the voltage drop for R_1. Examining Fig. 7-23, we see that the potential difference across R_1 is 12 V. With +18 V at the top of R_1 to ground, and +6 V at the bottom of R_1 to ground, V_1 = 18 V − 6 V or 12 V. Solving for R_1, we have

$$\begin{aligned} R_1 &= \frac{V_1}{I_1} \\ &= \frac{12\text{ V}}{8\text{ mA}} \\ R_1 &= 1.5\text{ k}\Omega \end{aligned}$$

In solving for R_2, find its voltage and current values. From Fig. 7-23 it can be seen that $V_2 = 6$ V. Notice that load B and R_2 are in parallel and, therefore, must have the same voltage.

The current through R_1 is the sum of the current for load B and the current for R_2. Expressed as an equation, we have

$$I_1 = I_2 + I_{L_B}$$

Solving for I_2, we have

$$I_2 = I_1 - I_{L_B}$$
$$= 8 \text{ mA} - 2 \text{ mA}$$
$$I_2 = 6 \text{ mA}$$

Solving for R_2, we have

$$R_2 = \frac{V_2}{I_2}$$
$$= \frac{6 \text{ V}}{6 \text{ mA}}$$
$$R_2 = 1 \text{ k}\Omega$$

In solving for R_3, we must realize that total current I_T divides between R_3 and load C. Since $I_{\text{load C}} = 12$ mA, we have the following.

$$I_T = I_3 + I_{L_C}$$

Solving for I_3, we have

$$I_3 = I_T - I_{L_C}$$
$$= 14 \text{ mA} - 12 \text{ mA}$$
$$I_3 = 2 \text{ mA}$$

The voltage across R_3 is the same as the voltage across load C. Therefore, $V_3 = 12$ V. Solving for R_3, we have

$$R_3 = \frac{V_3}{I_3}$$
$$= \frac{12 \text{ V}}{2 \text{ mA}}$$
$$R_3 = 6 \text{ k}\Omega$$

The resistance of each load can be determined using Ohm's law.

$$R_{\text{load A}} = \frac{V_{L_A}}{I_{L_A}}$$
$$= \frac{18 \text{ V}}{6 \text{ mA}}$$
$$R_{\text{load A}} = 3 \text{ k}\Omega$$
$$R_{\text{load B}} = \frac{V_{L_B}}{I_{L_B}}$$
$$= \frac{6 \text{ V}}{2 \text{ mA}}$$
$$R_{\text{load B}} = 3 \text{ k}\Omega$$
$$R_{\text{load C}} = \frac{V_{L_C}}{I_{L_C}}$$
$$= \frac{12 \text{ V}}{12 \text{ mA}}$$
$$R_{\text{load C}} = 1 \text{ k}\Omega$$

A loaded voltage divider is actually a practical application of series-parallel circuits. By choosing the correct resistances, the specified load voltages and currents can be obtained.

PRACTICE PROBLEMS

Sec. 7-3 For Figs. 7-24 through 7-33, solve for the unknowns listed.

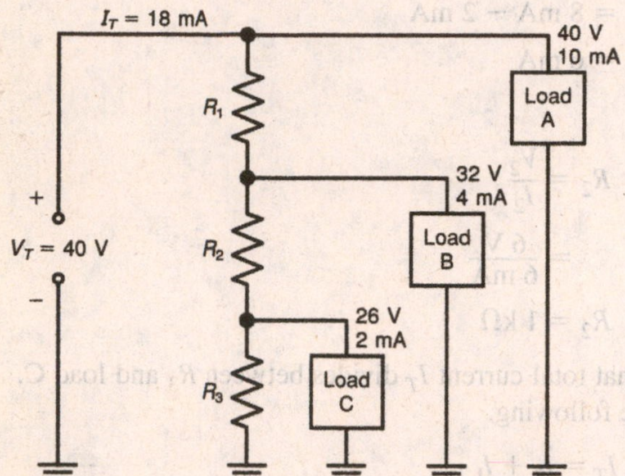

Fig. 7-24 Find R_1, R_2, R_3, R_{L_A}, R_{L_B}, R_{L_C}, P_1, P_2, and P_3.

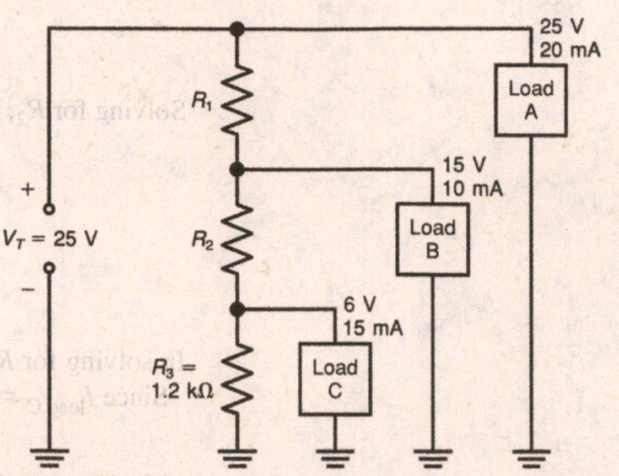

Fig. 7-25 Find R_1, R_2, I_T, R_{L_A}, R_{L_B}, R_{L_C}, P_1, P_2, and P_3.

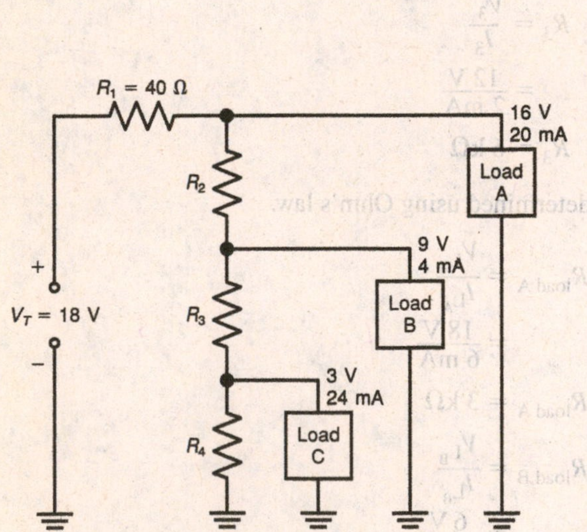

Fig. 7-26 Find R_2, R_3, R_4, I_T, R_{L_A}, R_{L_B}, R_{L_C}, P_1, P_2, P_3, and P_4.

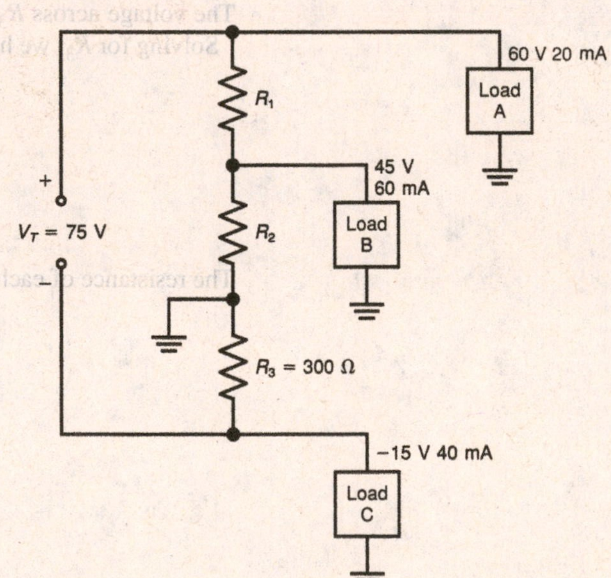

Fig. 7-27 Find R_1, R_2, I_T, R_{L_A}, R_{L_B}, R_{L_C}, P_1, P_2, and P_3.

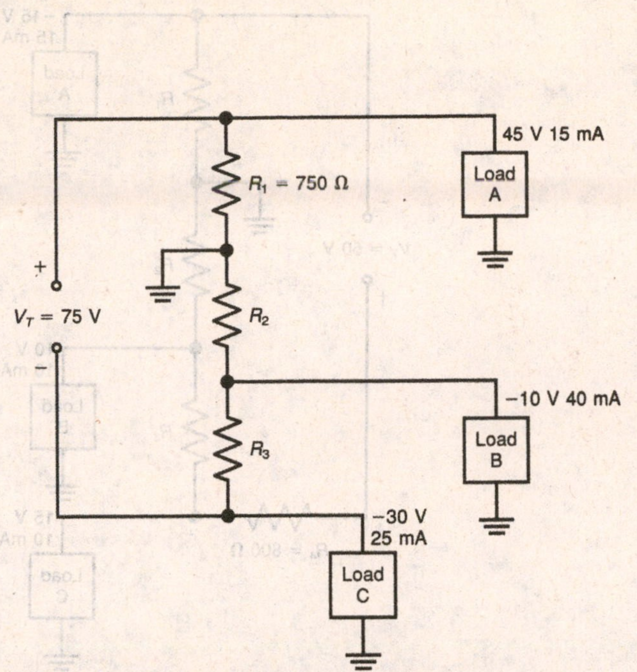

Fig. 7-28 Find R_2, R_3, I_T, R_{L_A}, R_{L_B}, R_{L_C}, P_1, P_2, and P_3.

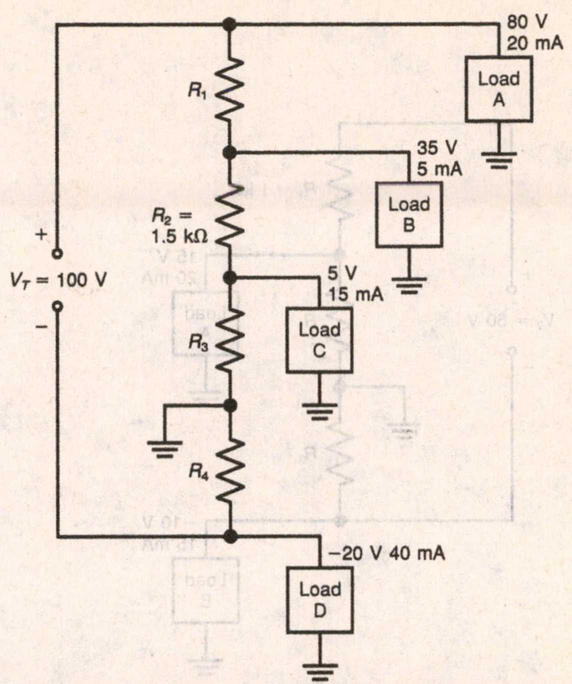

Fig. 7-29 Find R_1, R_3, R_4, I_T, R_{L_A}, R_{L_B}, R_{L_C}, R_{L_D}, P_1, P_2, P_3, and P_4.

Fig. 7-30 Find R_1, R_2, R_3, R_{L_A}, R_{L_B}, R_{L_C}, P_1, P_2, and P_3.

Fig. 7-31 Find R_1, R_2, R_3, R_{L_A}, R_{L_B}, R_{L_C}, P_1, P_2, and P_3.

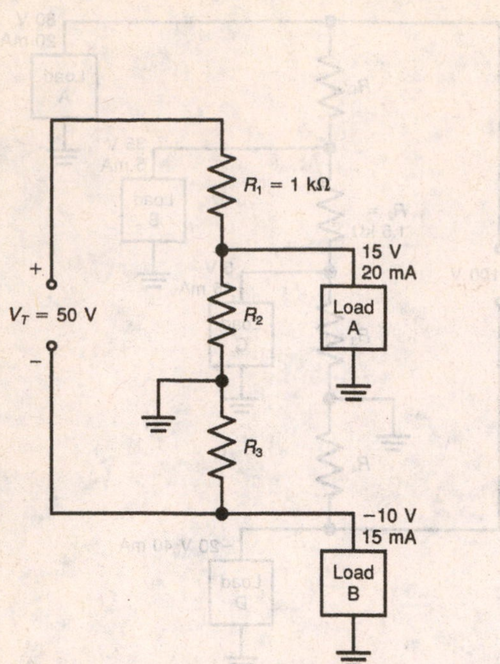

Fig. 7-32 Find I_T, R_2, R_3, R_{L_A}, R_{L_B}, P_1, P_2, and P_3.

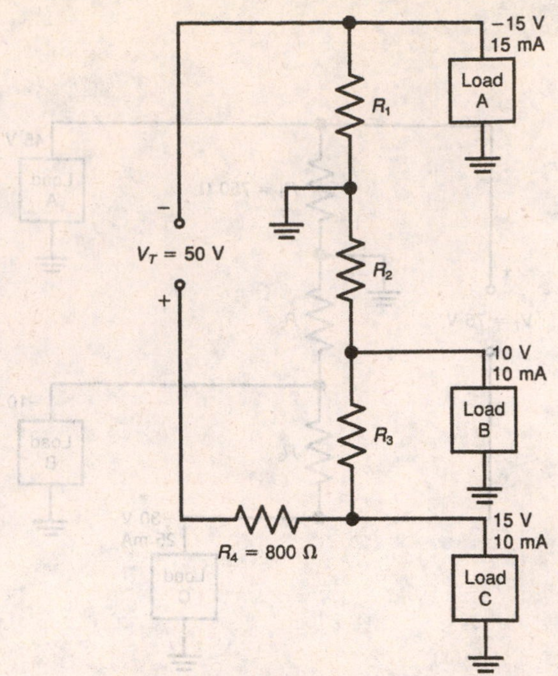

Fig. 7-33 Find R_1, R_2, R_3, I_T, R_{L_A}, R_{L_B}, R_{L_C}, P_1, P_2, P_3, and P_4.

SEC. 7-4 TROUBLESHOOTING VOLTAGE DIVIDERS

The voltages and currents shown in Fig. 7-34 exist when the circuit is operating normally.

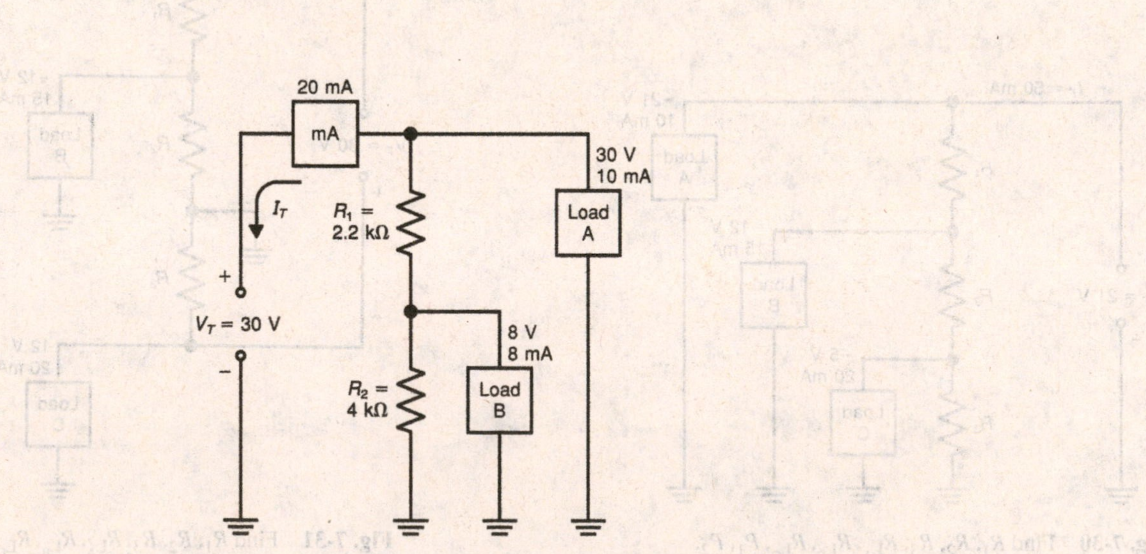

Fig. 7-34 Loaded voltage divider used for troubleshooting analysis.

Sec. 7-4 For each set of measurements shown below, identify the defective component. Indicate how the component is defective.

1. $V_{L_A} = 30$ V; $V_{L_B} = 8$ V; $I_T = 10$ mA
2. $V_{L_A} = 30$ V; $V_{L_B} = 0$ V; $I_T = 10$ mA
3. $V_{L_A} = 30$ V; $V_{L_B} = 19.35$ V; $I_T = 14.84$ mA
4. $V_{L_A} = 30$ V; $V_{L_B} = 30$ V; $I_T = 47.5$ mA
5. $V_{L_A} = 30$ V; $V_{L_B} = 9.375$ V; $I_T = 19.375$ mA

END OF CHAPTER TEST

Chapter 7: Voltage and Current Dividers

Answer True or False.

1. Any series circuit is a voltage divider.
2. Any parallel circuit is a current divider.
3. In a series circuit, the individual resistor voltage drops are proportional to the series resistance values.
4. In a parallel circuit, the individual branch currents are inversely proportional to the branch resistance values.
5. In a series circuit, it is not possible to calculate the individual resistor voltage drops without knowing the value of the series current.
6. In a parallel circuit, the individual branch currents can be calculated without knowing the value of the applied voltage.
7. In a parallel circuit, the individual branch currents are directly proportional to the branch conductance values.
8. A series circuit with parallel connected loads is called a loaded voltage divider.
9. In Fig. 7-24, $I_2 = I_3 + I_{L_C}$.
10. In Fig. 7-24, $V_2 = 32$ V.

CHAPTER 8

Direct Current Meters

Voltage, current, and resistance measurements are often made with a combination volt-ohm-milliammeter (VOM). In this chapter, we will examine the circuitry associated with the basic moving-coil meter movement that makes it capable of measuring voltage, current, and resistance. The two most important characteristics of a moving-coil meter movement are its internal resistance, r_M, and its full-scale deflection current, I_M. The resistance, r_M, is the resistance of the wire of the moving coil. The full-scale deflection current, I_M, is the amount of current needed in the moving-coil to deflect the pointer all the way to the right to the last mark on the printed scale. Meters that use a moving pointer and a printed scale are called analog meters. Most analog meters are built using a moving-coil meter movement.

SEC. 8-1 CURRENT METERS

It is often desirable to measure currents greater than the meter's I_M value. In order to do this, we must bypass or "shunt" a specific fraction of the circuit's current around the meter movement. The combination then provides us with a current meter having an extended range. This is illustrated in the solved problem that follows.

Solved Problem

Refer to Fig. 8-1. Calculate the shunt resistance, R_{sh}, needed to extend the range of the meter movement shown to 10 mA.

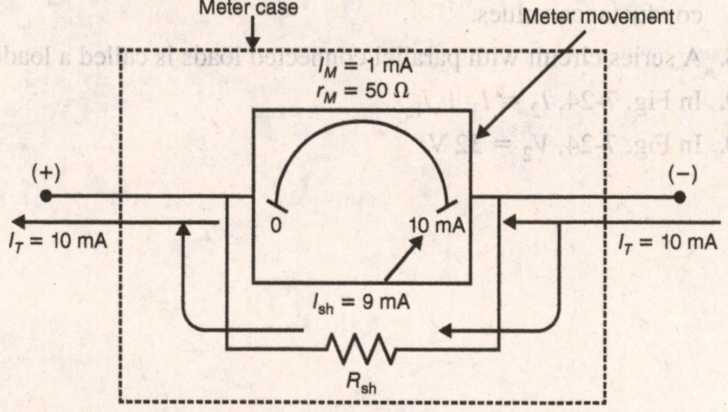

Fig. 8-1 Calculating the shunt resistance, R_{sh}, required to extend the range of the 1-mA meter movement to 10 mA.

Answer

In Fig. 8-1, 10 mA of total meter current must produce full-scale meter deflection; 9 mA must be shunted around the meter movement through resistor R_{sh}, and 1 mA must flow through the meter movement. Expressed as an equation, we have

$$I_T = I_{sh} + I_M$$

where I_T = total meter current
I_{sh} = current through shunt resistor R_{sh}
I_M = current required in the meter movement to produce full-scale deflection

As can be seen, the current through the shunt resistor, R_{sh}, is the difference between the total current, I_T, and the current through the meter movement, I_M. Or, solving for I_{sh}, we have

$$I_{sh} = I_T - I_M$$
$$= 10 \text{ mA} - 1 \text{ mA}$$
$$I_{sh} = 9 \text{ mA}$$

With 1 mA flowing through the 50-Ω meter movement, there will be a 50-mV drop across it. The voltage drop across the meter, designated V_M, can be expressed as

$$V_M = I_M \times r_M$$
$$= 1 \text{ mA} \times 50\ \Omega$$
$$V_M = 50 \text{ mV}$$

Since the meter movement is in parallel with R_{sh}, their voltage drops will be identical. Thus, R_{sh} can be found as follows:

$$R_{sh} = \frac{V_M}{I_{sh}}$$
$$= \frac{50 \text{ mV}}{9 \text{ mA}}$$
$$R_{sh} = 5.55\ \Omega$$

The parallel combination of R_{sh} and r_M produce a total meter resistance, R_M, of 5 Ω.

$$R_M = \frac{R_{sh} \times r_M}{R_{sh} + r_M}$$
$$= \frac{5.55\ \Omega \times 50\ \Omega}{5.55\ \Omega + 50\ \Omega}$$
$$R_M = 5\ \Omega$$

It is important to realize that converting the basic meter movement to a higher current range lowers the total meter resistance, R_M. This is because smaller values of R_{sh} are required to bypass the required current. For Fig. 8-1, R_M can also be calculated as

$$R_M = \frac{V_M}{I_T}$$
$$= \frac{50 \text{ mV}}{10 \text{ mA}}$$
$$R_M = 5\ \Omega$$

PRACTICE PROBLEMS

Sec. 8-1 For Figs. 8-2 through 8-7, calculate the shunt resistances needed to extend the range of the meter movements shown to the current values listed. Also, calculate the total meter resistance for each range.

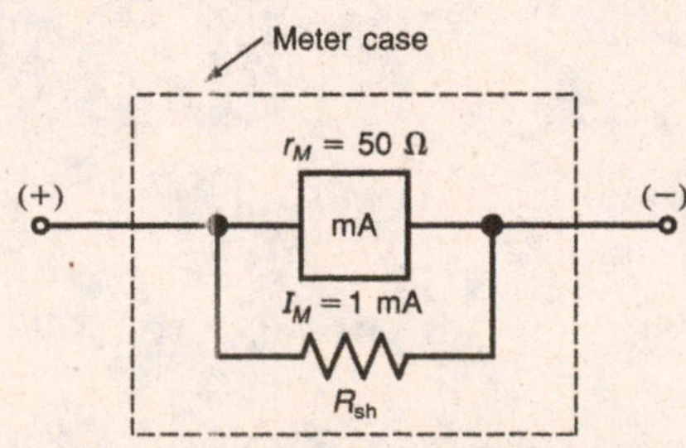

Fig. 8-2 Current ranges: 2 mA, 5 mA, 10 mA, 20 mA, 50 mA, and 100 mA.

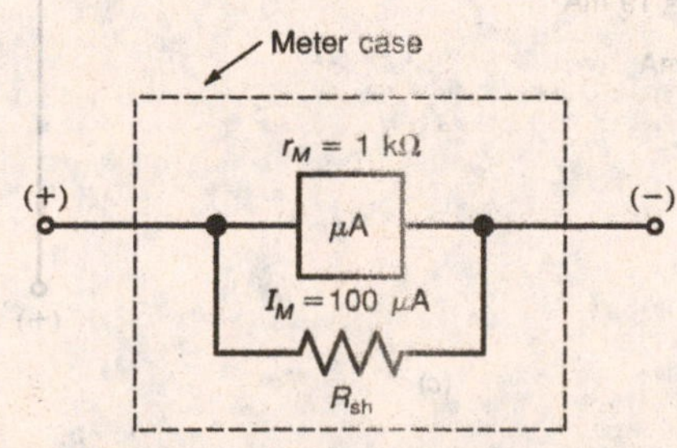

Fig. 8-3 Current ranges: 1 mA, 2 mA, 5 mA, 10 mA, 20 mA, and 50 mA.

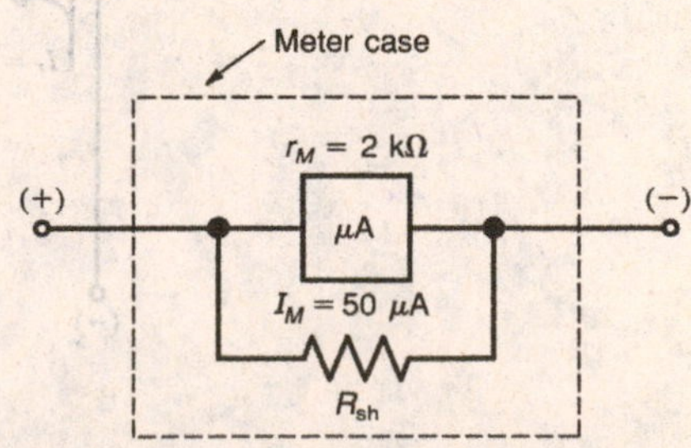

Fig. 8-4 Current ranges: 100 μA, 1 mA, 2 mA, 5 mA, 10 mA, and 50 mA.

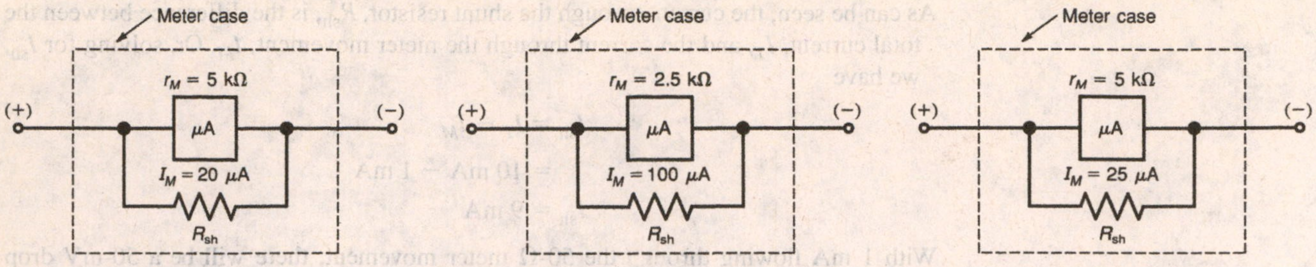

Fig. 8-5 Current ranges: 100 μA, 1 mA, 2 mA, 5 mA, and 10 mA.

Fig. 8-6 Current ranges: 1 mA, 2.5 mA, 10 mA, and 25 mA.

Fig. 8-7 Current ranges: 250 μA, 1 mA, 5 mA, and 25 mA.

The Universal Shunt Current Meter

Solved Problem

For Fig. 8-8*a*, find R_1, R_2, and R_3 for the current ranges listed.

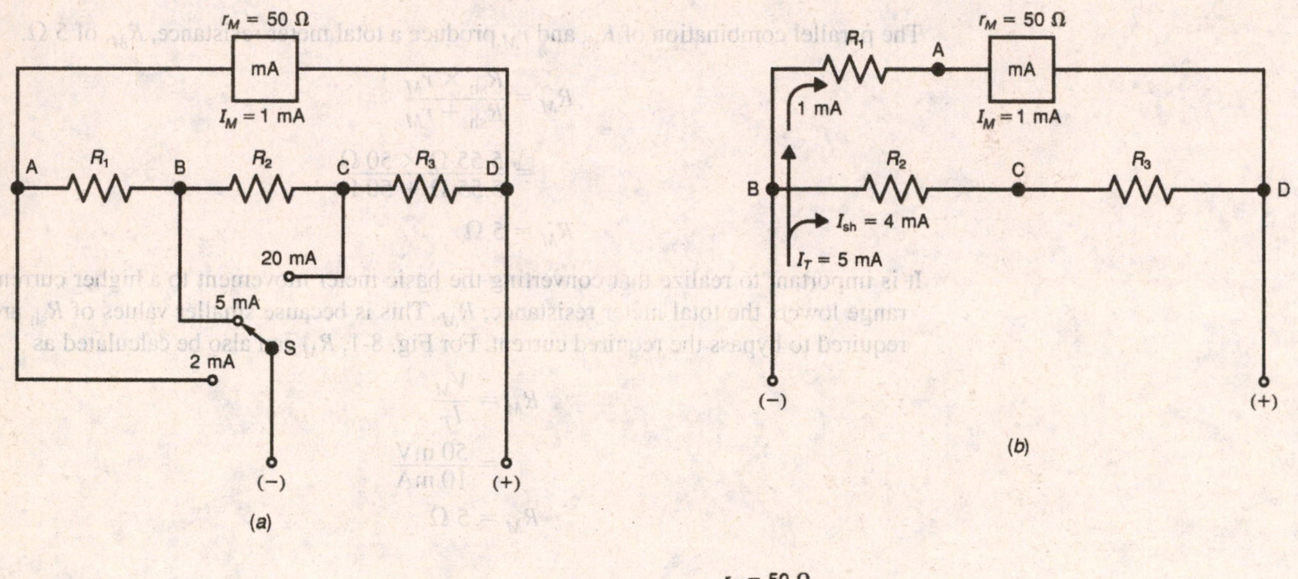

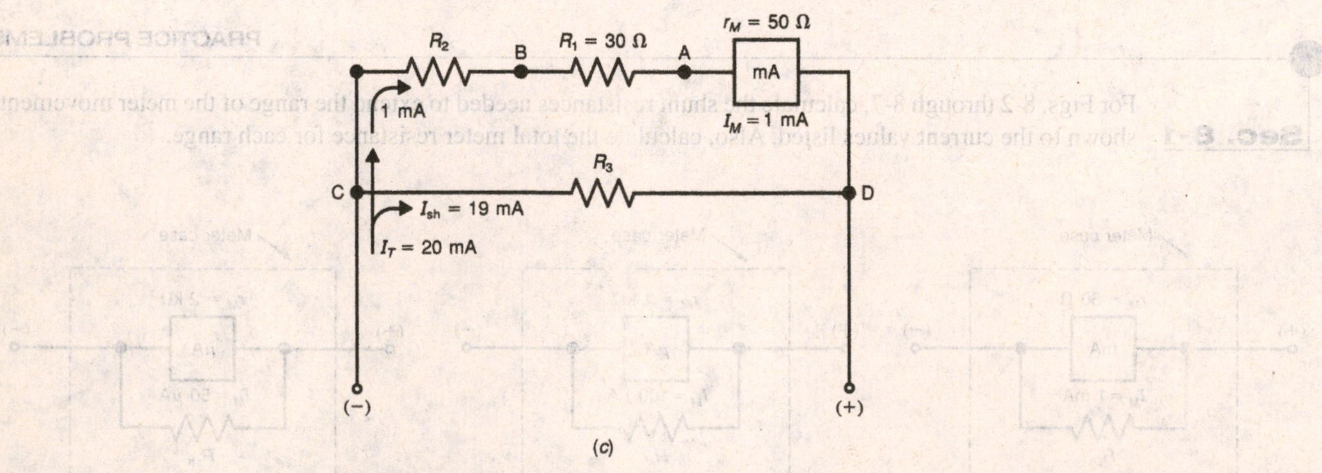

Fig. 8-8 Universal or Ayrton shunt for three current ranges. The *I* values are shown for full-scale deflection. (*a*) Actual circuit with switch *S* to choose different ranges. (*b*) Circuit for 5-mA range. (*c*) Circuit for 20-mA range.

Answer

With the meter set to the 2-mA range in Fig. 8-8*a*, the shunt current, I_{sh}, must flow through the series combination of R_1, R_2, and R_3. Since $I_T = 2$ mA and $I_M = 1$ mA, we have

$$I_{sh} = I_T - I_M$$
$$= 2\text{ mA} - 1\text{ mA}$$
$$I_{sh} = 1\text{ mA}$$

Notice that the meter is in parallel with the series combination R_1, R_2, and R_3. The voltage drop across each branch, therefore, is the same.

With $r_M = 50\ \Omega$, the voltage across the meter movement with full-scale deflection is

$$V_M = I_M \times r_M$$
$$= 1\text{ mA} \times 50\ \Omega$$
$$V_M = 50\text{ mV}$$

Since the shunt resistance on the 2-mA range consists of R_1, R_2, and R_3, we designate it as the total shunt resistance, R_{st}. Its value is found as follows:

$$R_{st} = \frac{V_M}{I_{sh}}$$
$$= \frac{50\text{ mV}}{1\text{ mA}}$$
$$R_{st} = 50\ \Omega$$

Therefore,

$$R_{st} = R_1 + R_2 + R_3$$
$$R_{st} = 50\ \Omega$$

With the meter set to the 5-mA range in Fig. 8-8*b*, the shunt current, I_{sh}, must flow through R_2 and R_3 in series; I_M flows through both R_1 and the meter movement. In Fig. 8-8*b*, $I_{sh} = 4$ mA and $I_M = 1$ mA.

To calculate R_1, we must write a basic equation.

$$1\text{ mA}(R_1 + r_M) = 4\text{ mA}(R_2 + R_3)$$

This tell us that the voltage drop across each branch is the same.

We know that r_M is 50 Ω and that R_{st} is 50 Ω. We do not know R_1, R_2, or R_3, but we can calculate them knowing that $(R_2 + R_3) = (R_{st} - R_1)$ or $(50\ \Omega - R_1)$. Knowing this, we can make the following substitutions:

$$1\text{ mA}(R_1 + 50\ \Omega) = 4\text{ mA}(50\ \Omega - R_1)$$

Solving for R_1, we have

$$1\text{ mA } R_1 + 50\text{ mV} = 200\text{ mV} - 4\text{ mA } R_1$$
$$5\text{ mA } R_1 = 150\text{ mV}$$
$$R_1 = 30\ \Omega$$

Not only do we know that $R_1 = 30\ \Omega$, but we also know that $R_2 + R_3$ must equal 20 Ω. This must be true because

$$R_2 + R_3 = R_{st} - R_1$$
$$= 50\ \Omega - 30\ \Omega$$
$$R_2 + R_3 = 20\ \Omega$$

This 20-Ω value is used in the next step of the calculation.

For the 20-mA range in Fig. 8-8*c*, the shunt current, I_{sh}, must flow through R_3; I_M flows through the series combination of R_1, R_2, and r_M. In Fig. 8-8*c*, I_{sh} = 19 mA and I_M = 1 mA. To calculate R_2, a basic equation must be written.

$$1 \text{ mA}(R_1 + R_2 + r_M) = 19 \text{ mA } R_3$$

Since $R_1 = 30\ \Omega$ and $(R_2 + R_3) = 20\ \Omega$,

$$1 \text{ mA}(30\ \Omega + R_2 + 50\ \Omega) = 19 \text{ mA } R_3$$

Also, if $(R_2 + R_3) = 20\ \Omega$, then

$$R_3 = (20\ \Omega - R_2)$$

Substituting $(20\ \Omega - R_2)$ for R_3 in the original equation, we have

$$1 \text{ mA}(80\ \Omega + R_2) = 19 \text{ mA}(20\ \Omega - R_2)$$

$$80 \text{ mV} + 1 \text{ mA } R_2 = 380 \text{ mV} - 19 \text{ mA } R_2$$

$$20 \text{ mA } R_2 = 300 \text{ mV}$$

$$R_2 = 15\ \Omega$$

Finally, R_3 must equal 5 Ω since $R_{st} = R_1 + R_2 + R_3$. Then

$$R_3 = R_{st} - R_1 - R_2$$

$$= 50\ \Omega - 30\ \Omega - 15\ \Omega$$

$$R_3 = 5\ \Omega$$

MORE PRACTICE PROBLEMS

Sec. 8-1 For Figs. 8-9 through 8-18, find R_1, R_2, and R_3 for the current ranges listed.

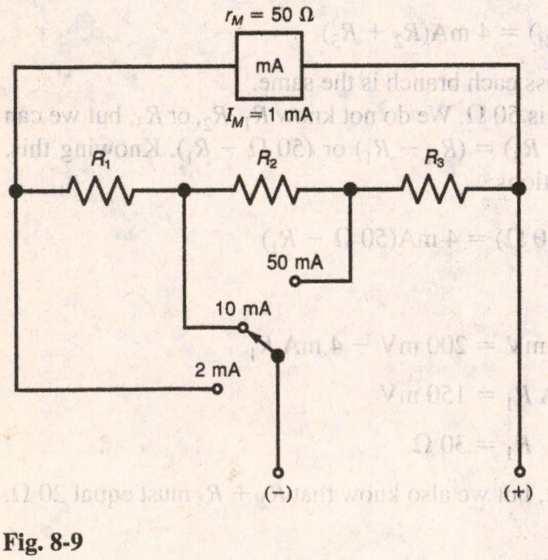

Fig. 8-9

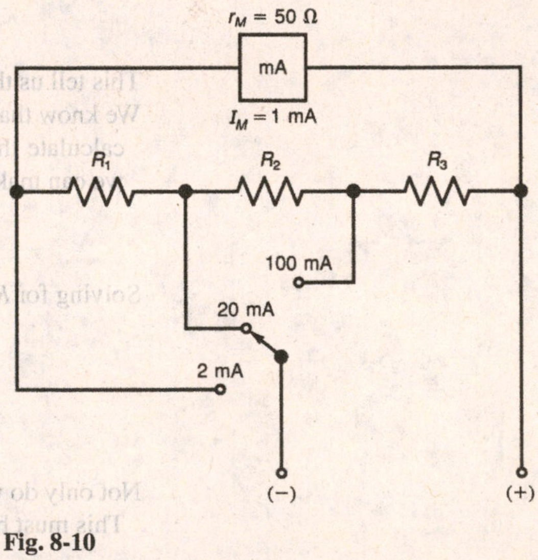

Fig. 8-10

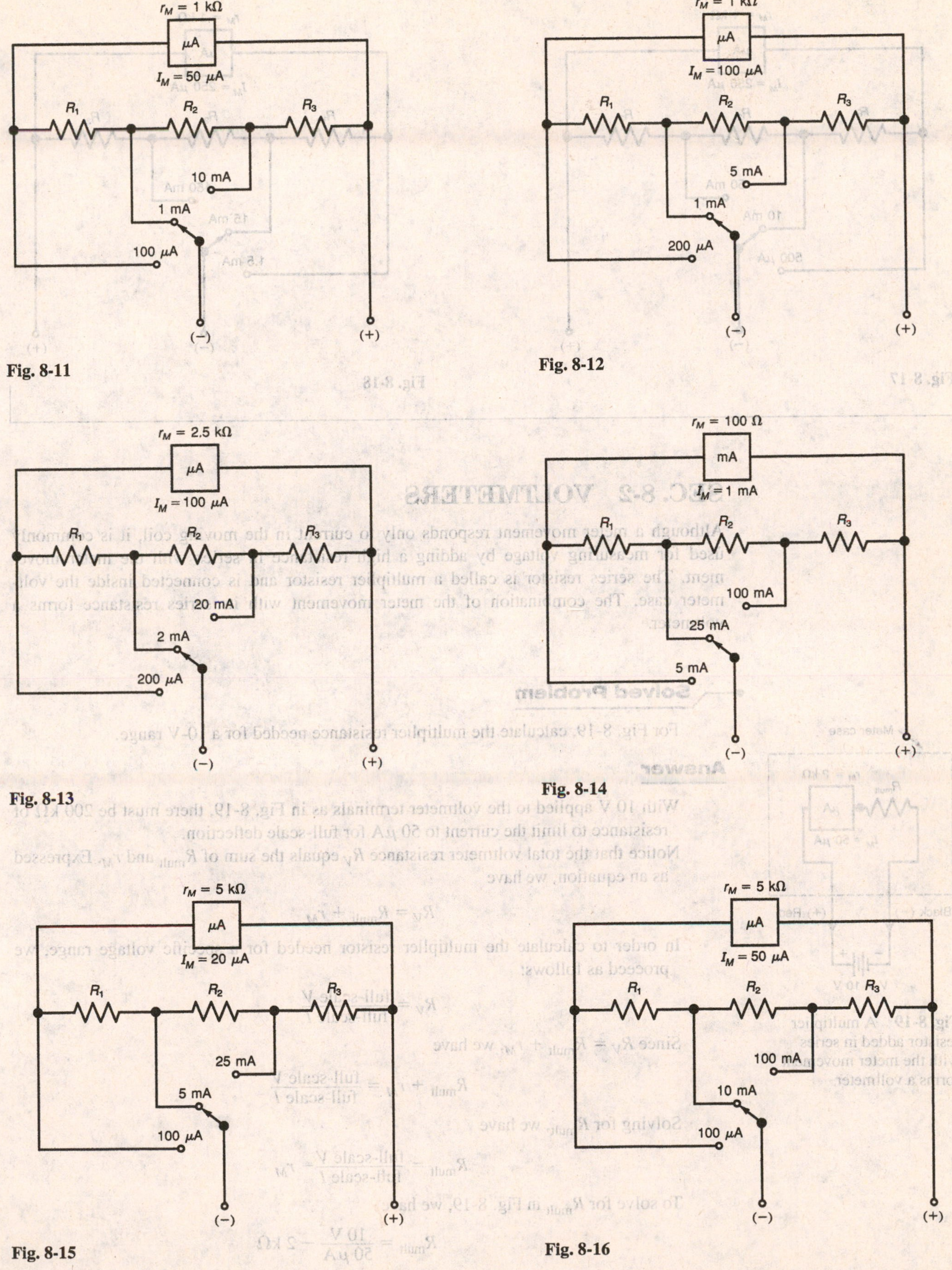

Fig. 8-11

Fig. 8-12

Fig. 8-13

Fig. 8-14

Fig. 8-15

Fig. 8-16

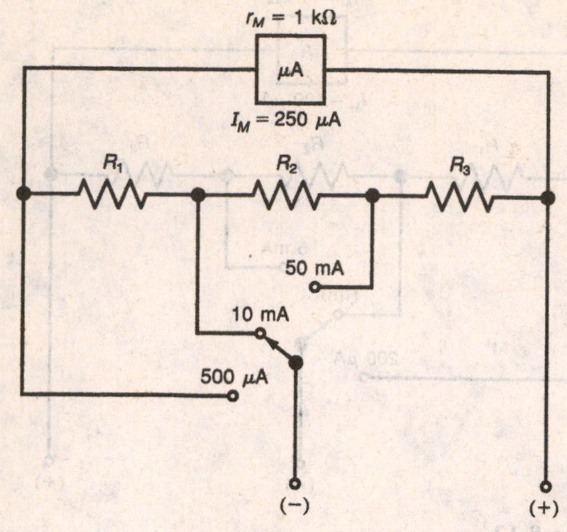

Fig. 8-17

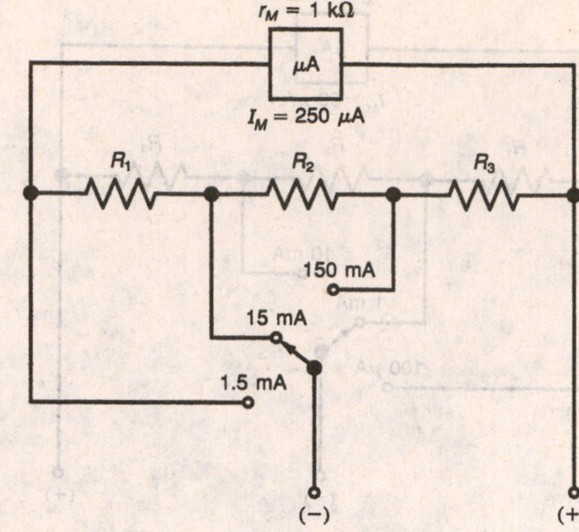

Fig. 8-18

SEC. 8-2 VOLTMETERS

Although a meter movement responds only to current in the moving coil, it is commonly used for measuring voltage by adding a high resistance in series with the meter movement. The series resistor is called a multiplier resistor and is connected inside the voltmeter case. The combination of the meter movement with its series resistance forms a voltmeter.

Solved Problem

For Fig. 8-19, calculate the multiplier resistance needed for a 10-V range.

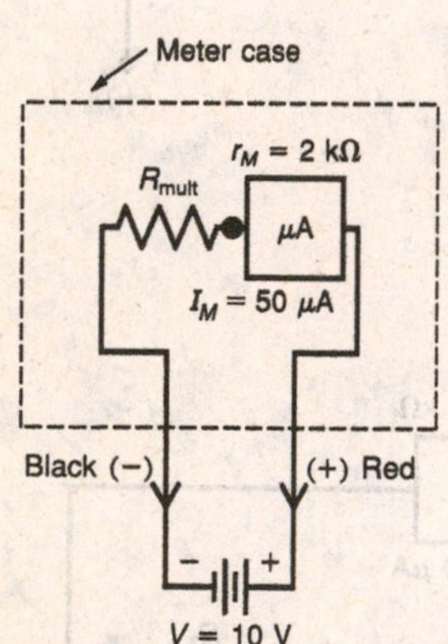

Fig. 8-19 A multiplier resistor added in series with the meter movement forms a voltmeter.

Answer

With 10 V applied to the voltmeter terminals as in Fig. 8-19, there must be 200 kΩ of resistance to limit the current to 50 μA for full-scale deflection.

Notice that the total voltmeter resistance R_V equals the sum of R_{mult} and r_M. Expressed as an equation, we have

$$R_V = R_{mult} + r_M$$

In order to calculate the multiplier resistor needed for a specific voltage range, we proceed as follows:

$$R_V = \frac{\text{full-scale } V}{\text{full-scale } I}$$

Since $R_V = R_{mult} + r_M$, we have

$$R_{mult} + r_M = \frac{\text{full-scale } V}{\text{full-scale } I}$$

Solving for R_{mult}, we have

$$R_{mult} = \frac{\text{full-scale } V}{\text{full-scale } I} - r_M$$

To solve for R_{mult} in Fig. 8-19, we have

$$R_{mult} = \frac{10 \text{ V}}{50 \ \mu\text{A}} - 2 \text{ k}\Omega$$

$$R_{mult} = 198 \text{ k}\Omega$$

With 5 V connected across the meter terminals in Fig. 8-19, we have 25 μA of current and, therefore, half-scale deflection. Notice, however, that R_V remains unchanged, regardless of the voltage connected to the meter terminals.

To indicate the voltmeter's resistance independently of the voltage range, voltmeters are rated in ohms of resistance needed for 1 V of deflection. This is the ohms-per-volt rating of the meter expressed as Ω/V. In Fig. 8-19, we have

$$\frac{R_V}{V_{\text{range}}} = \frac{\Omega}{\text{V}}$$

$$\frac{200\ \text{k}\Omega}{10\ \text{V}} = \frac{20\ \text{k}\Omega}{1\ \text{V}}$$

$$\frac{R_V}{V_{\text{range}}} = \frac{20\ \text{k}\Omega}{\text{V}}$$

This tells us there must be 20 kΩ of resistance for every 1 V of input voltage to limit the current through the meter movement to 50 μA. The Ω/V rating is the same for all ranges.

The Ω/V rating can also be found by taking the reciprocal of I_M. In Fig. 8-19, we have

$$\frac{\Omega}{\text{V}}\ \text{rating} = \frac{1}{I_M}$$

$$= \frac{1}{50\ \mu\text{A}}$$

$$\frac{\Omega}{\text{V}}\ \text{rating} = \frac{20\ \text{k}\Omega}{\text{V}}$$

The resistance R_V of a voltmeter can be found when the Ω/V rating is known. In Fig. 8-19, R_V can also be found, as shown.

$$R_V = \Omega/\text{V rating} \times V_{\text{range}}$$

$$= 20\ \text{k}\Omega/\text{V} \times 10\ \text{V}$$

$$R_V = 200\ \text{k}\Omega$$

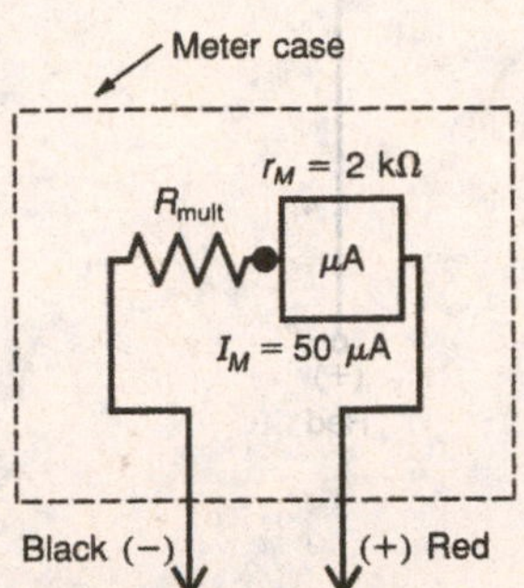

Fig. 8-20 Voltage ranges: 2.5 V, 10 V, 50 V, 100 V, and 250 V.

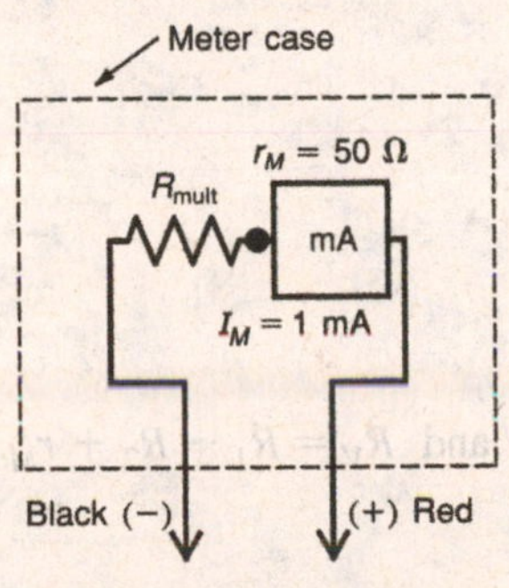

Fig. 8-21 Voltage ranges: 2.5 V, 10 V, 50 V, 100 V, and 250 V.

PRACTICE PROBLEMS

Sec. 8-2

For Figs. 8-20 through 8-25, calculate the multiplier resistors needed for the ranges listed. Also, calculate the total voltmeter resistance R_V for each range and the meter's Ω/V rating for all ranges.

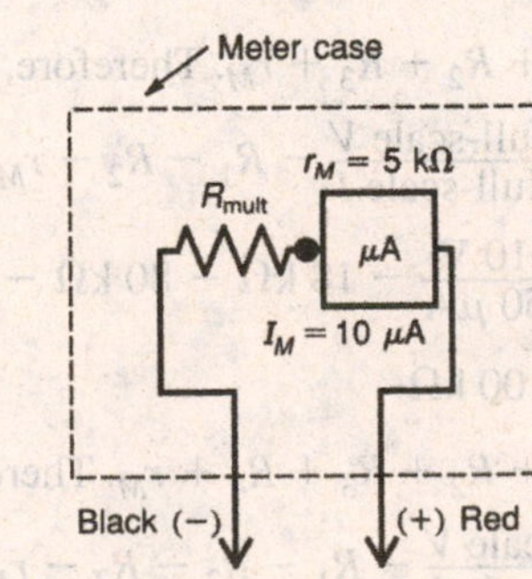

Fig. 8-22 Voltage ranges: 1 V, 5 V, 25 V, 100 V, and 500 V.

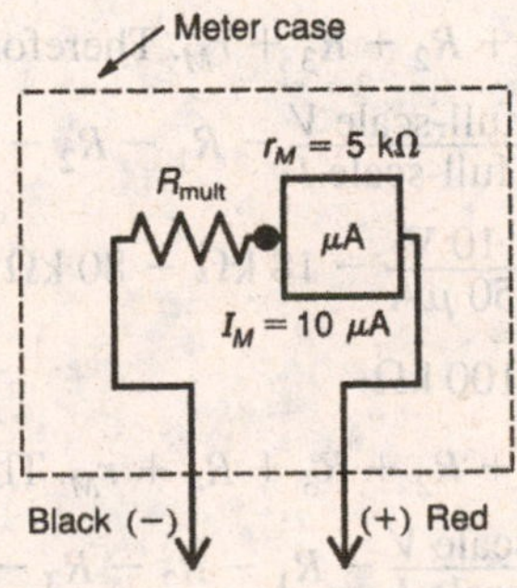

Fig. 8-23 Voltage ranges: 1 V, 5 V, 25 V, 100 V, and 500 V.

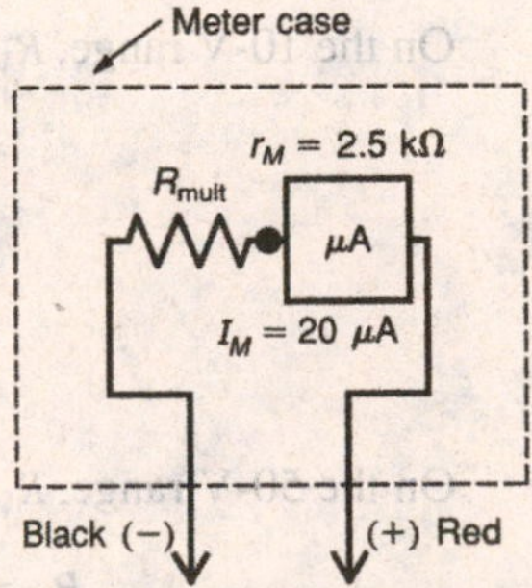

Fig. 8-24 Voltage ranges: 1 V, 2.5 V, 10 V, 25 V, and 100 V.

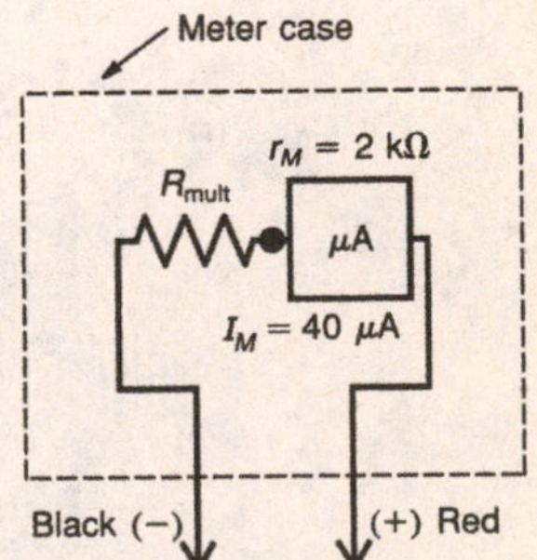

Fig. 8-25 Voltage ranges: 3 V, 10 V, 30 V, 100 V, and 300 V.

Multiple-Range Voltmeters

Solved Problem

Find R_1, R_2, R_3, R_4, and R_5 for the different voltmeter ranges shown in Fig. 8-26.

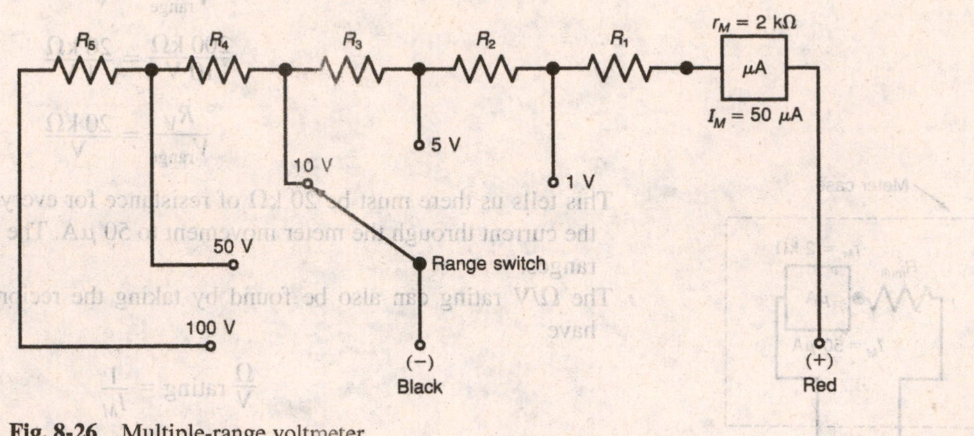

Fig. 8-26 Multiple-range voltmeter.

Answer

On the 1-V range, R_1 is found as follows:

$$R_1 = \frac{\text{full-scale } V}{\text{full-scale } I} - r_M$$

$$= \frac{1\ \text{V}}{50\ \mu\text{A}} - 2\ \text{k}\Omega$$

$$R_1 = 18\ \text{k}\Omega$$

On the 5-V range, $R_1 + R_2$ form the series multiplier, and $R_V = R_1 + R_2 + r_M$. Therefore, R_2 is found as follows:

$$R_2 = \frac{\text{full-scale } V}{\text{full-scale } I} - R_1 - r_M$$

$$= \frac{5\ \text{V}}{50\ \mu\text{A}} - 18\ \text{k}\Omega - 2\ \text{k}\Omega$$

$$R_2 = 80\ \text{k}\Omega$$

On the 10-V range, $R_V = R_1 + R_2 + R_3 + r_M$. Therefore,

$$R_3 = \frac{\text{full-scale } V}{\text{full-scale } I} - R_1 - R_2 - r_M$$

$$= \frac{10\ \text{V}}{50\ \mu\text{A}} - 18\ \text{k}\Omega - 80\ \text{k}\Omega - 2\ \text{k}\Omega$$

$$R_3 = 100\ \text{k}\Omega$$

On the 50-V range, $R_V = R_1 + R_2 + R_3 + R_4 + r_M$. Therefore,

$$R_4 = \frac{\text{full-scale } V}{\text{full-scale } I} - R_1 - R_2 - R_3 - r_M$$

$$= \frac{50\ \text{V}}{50\ \mu\text{A}} - 18\ \text{k}\Omega - 80\ \text{k}\Omega - 100\ \text{k}\Omega - 2\ \text{k}\Omega$$

$$R_4 = 800\ \text{k}\Omega$$

On the 100-V range, $R_V = R_1 + R_2 + R_3 + R_4 + R_5 + r_M$. Therefore,

$$R_5 = \frac{\text{full-scale } V}{\text{full-scale } I} - R_1 - R_2 - R_3 - R_4 - r_M$$

$$= \frac{100 \text{ V}}{50\ \mu\text{A}} - 18 \text{ k}\Omega - 80 \text{ k}\Omega - 100 \text{ k}\Omega - 800 \text{ k}\Omega - 2 \text{ k}\Omega$$

$$R_5 = 1 \text{ M}\Omega$$

The Ω/V rating can be found as follows:

$$\frac{\Omega}{\text{V}} \text{ rating} = \frac{1}{I_M}$$

$$= \frac{1}{50\ \mu\text{A}}$$

$$\frac{\Omega}{\text{V}} \text{ rating} = \frac{20 \text{ k}\Omega}{\text{V}}$$

This rating is the same for all the voltmeter ranges.

MORE PRACTICE PROBLEMS

Sec. 8-2 For Figs. 8-27 through 8-36, find R_1, R_2, R_3, R_4, and R_5 for the voltmeter ranges listed. Also, calculate total voltmeter resistance R_V for each range.

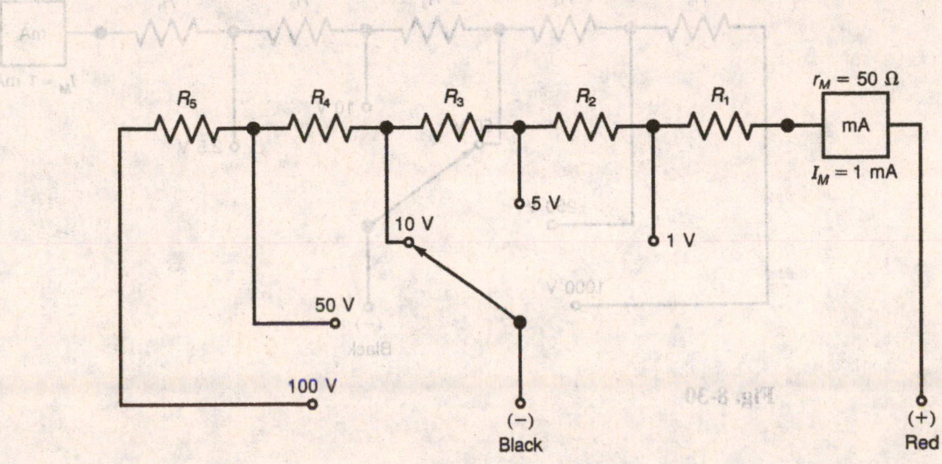

Fig. 8-27

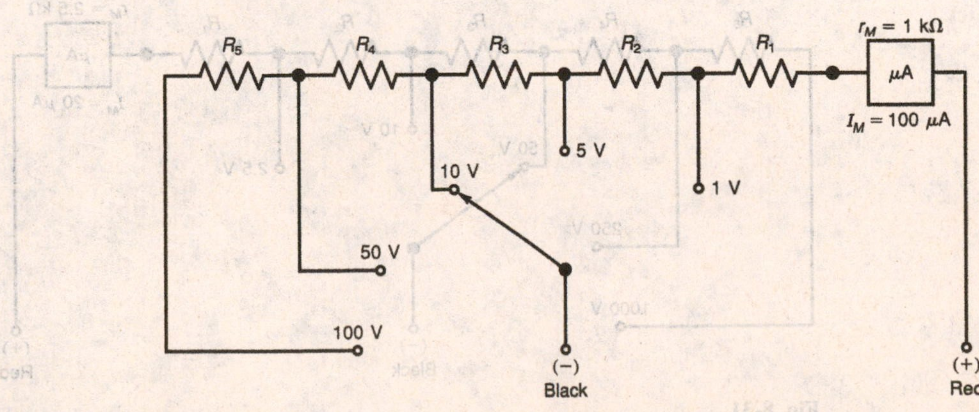

Fig. 8-28

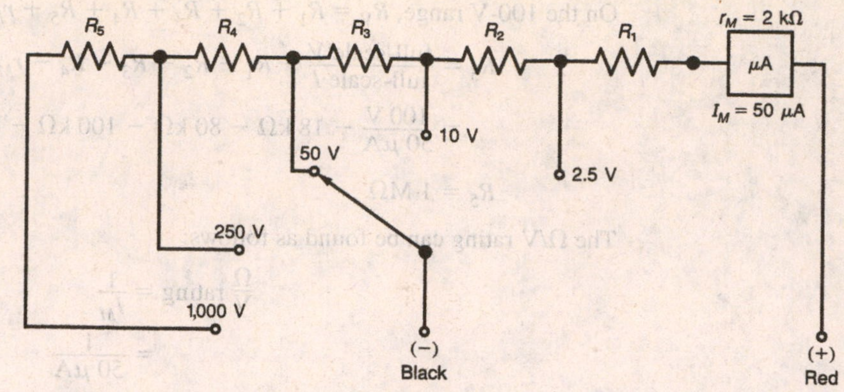

Fig. 8-29

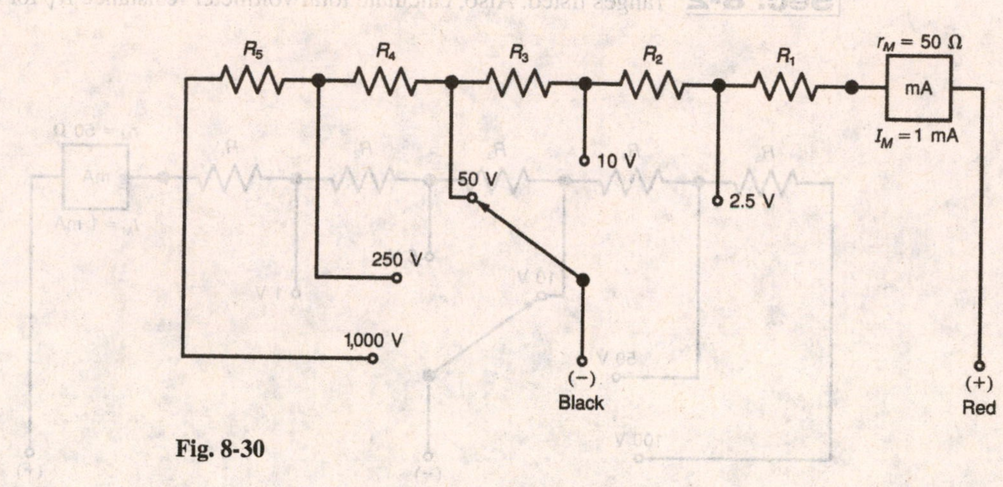

Fig. 8-30

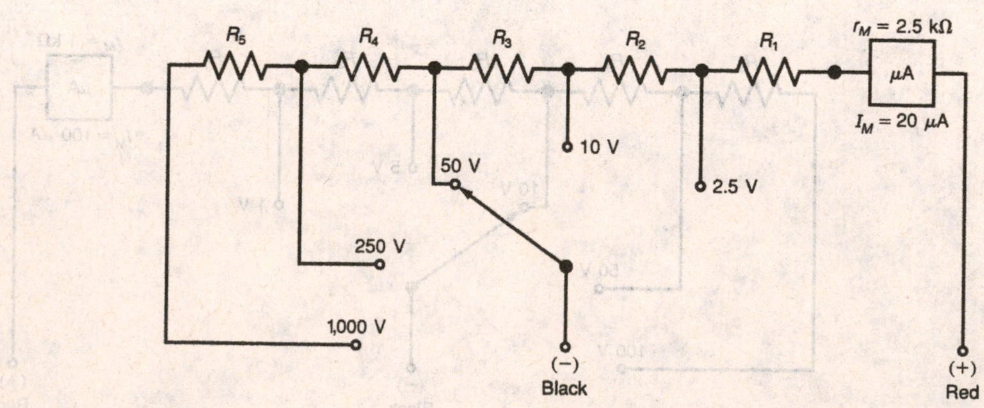

Fig. 8-31

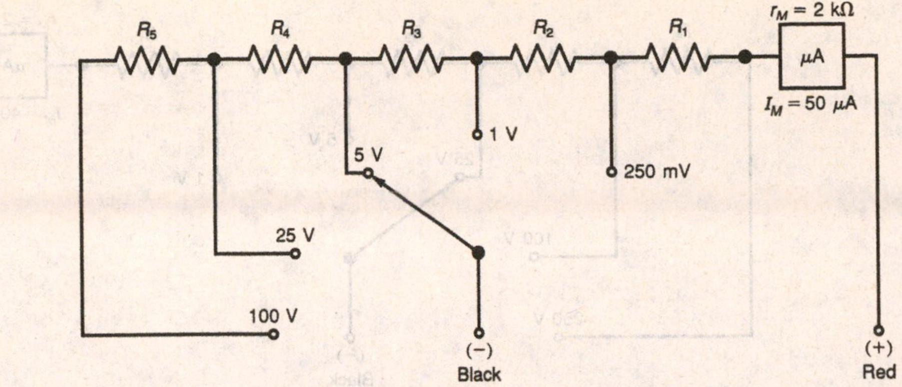

Fig. 8-32

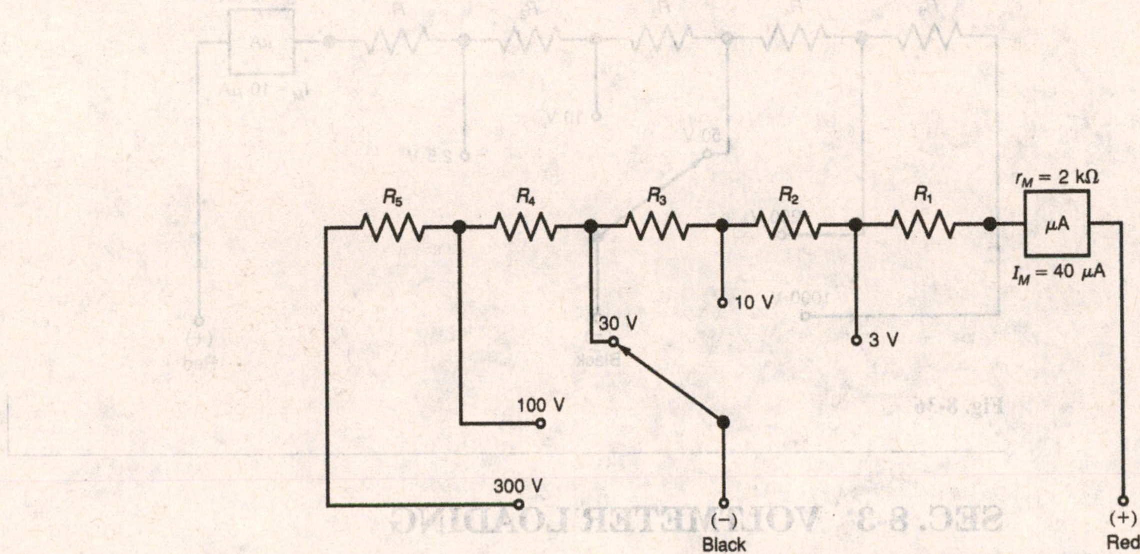

Fig. 8-33

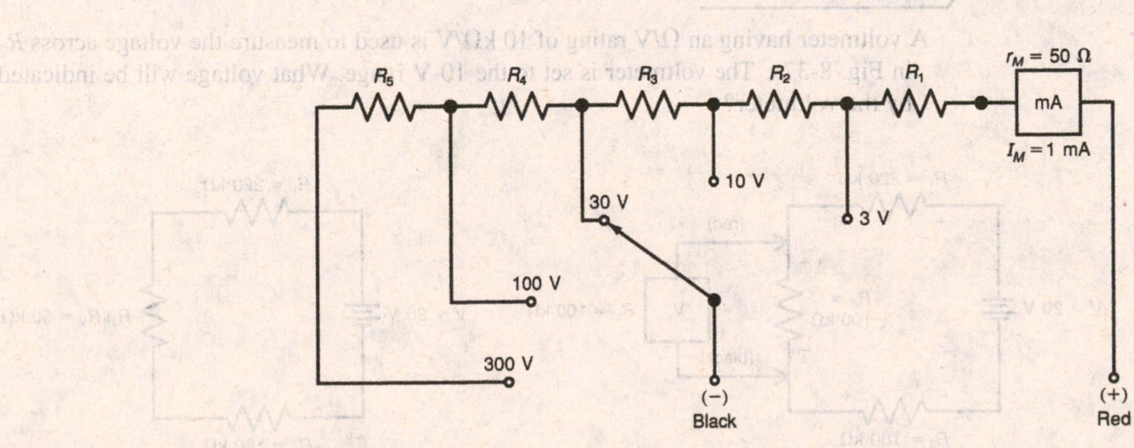

Fig. 8-34

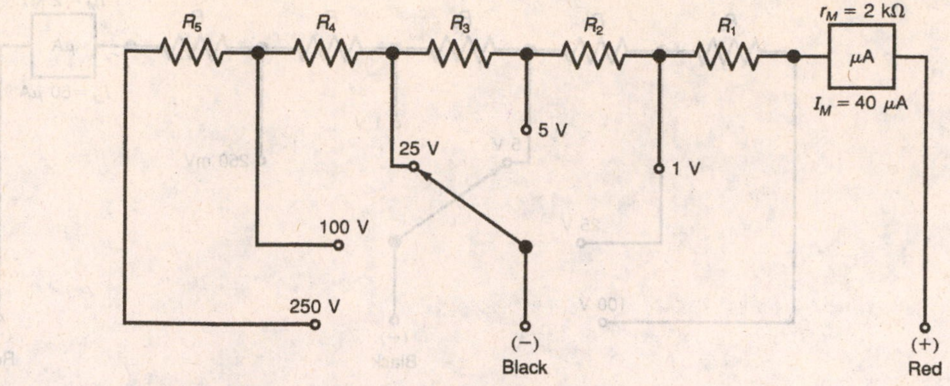

Fig. 8-35

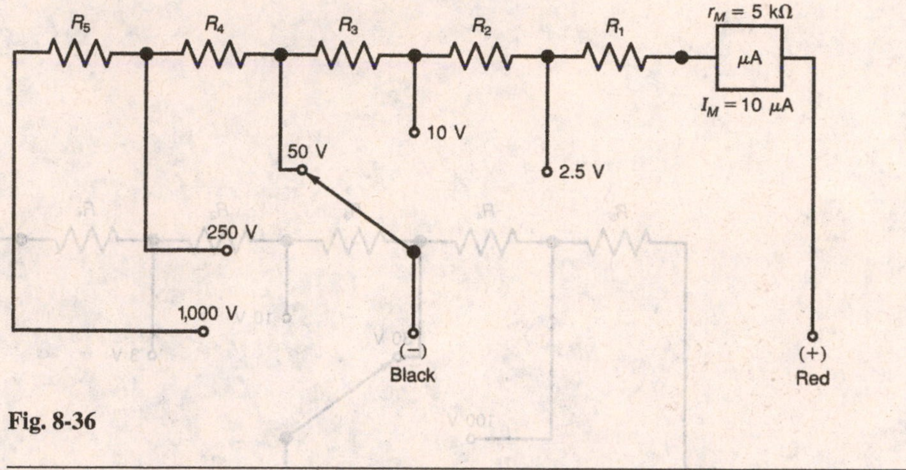

Fig. 8-36

SEC. 8-3 VOLTMETER LOADING

When the voltmeter resistance R_V is not high enough, connecting it across the circuit can reduce the measured voltage, as compared to the voltage present without the voltmeter. This is called voltmeter loading because the measured voltage decreases due to the additional current required for the meter.

Solved Problem

A voltmeter having an Ω/V rating of 10 kΩ/V is used to measure the voltage across R_2 in Fig. 8-37*a*. The voltmeter is set to the 10-V range. What voltage will be indicated by the voltmeter?

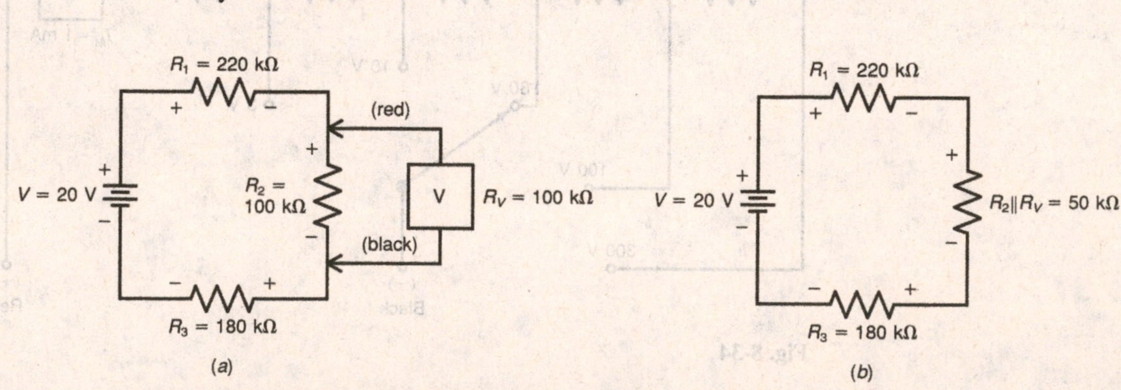

Fig. 8-37 Example of voltmeter loading. (*a*) Voltmeter *V* connected across R_2. (*b*) Resistor R_2 in parallel with R_V equals 50 kΩ.

Answer

The voltage drop across R_2 without the voltmeter connected is found by the methods used in the solution of series circuits.

$$I_T = \frac{V_T}{R_T}$$

$$= \frac{20\text{ V}}{500\text{ k}\Omega}$$

$$I_T = 40\ \mu\text{A}$$

$$V_2 = I_T \times R_2$$

$$= 40\ \mu\text{A} \times 100\text{ k}\Omega$$

$$V_2 = 4\text{ V}$$

With the voltmeter connected, I_T will change due to the paralleling effect of the voltmeter. The parallel combination of R_V and R_2 produces a combined equivalent resistance of 50 kΩ. Now R_T is 450 kΩ. This is shown in Fig. 8-37*b*. Remember that the voltmeter resistance $R_V = \Omega/\text{V rating} \times V_{\text{range}}$. In Fig. 8-37*b*, $R_V = 100\text{ k}\Omega$. To solve for V_2 with the meter connected, we proceed as follows:

$$I_T = \frac{V_T}{R_T}$$

$$= \frac{20\text{ V}}{450\text{ k}\Omega}$$

$$I_T = 44.44\ \mu\text{A}$$

The voltage drop across the meter and its parallel connection R_2 are found as follows:

$$V_2 = V_{\text{meter}}$$

$$V_{\text{meter}} = I_T \times (R_2 \| R_V)$$

$$= 44.44\ \mu\text{A} \times 50\text{ k}\Omega$$

$$V_2 = 2.22\text{ V}$$

Notice that this is almost half what it actually is without the voltmeter connected. This is severe voltmeter loading! The effect can be minimized considerably by using a voltmeter with a higher Ω/V rating. It is commonplace to use a digital multimeter (DMM) to measure voltages in electronic circuitry. These meters usually have 10 MΩ of resistance for all voltage ranges—a tremendous advantage!

PRACTICE PROBLEMS

Sec. 8-3 Problems 1 – 5 refer to Fig. 8-38.

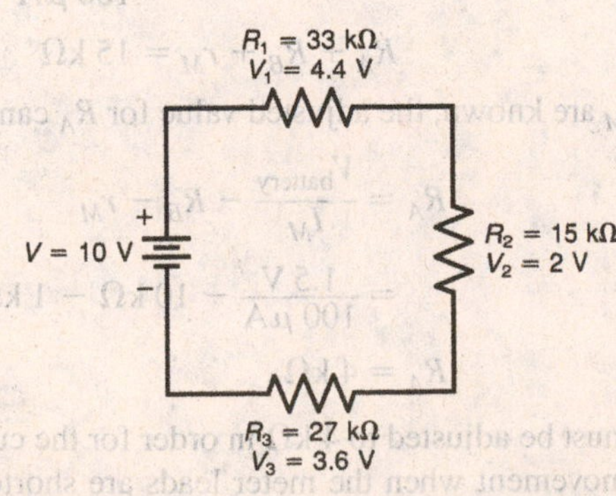

Fig. 8-38 Series circuit used for Probs. 1–5 in Sec. 8-3. The voltage drops shown exist without a meter connected.

1. Using the voltmeter shown in Fig. 8-27, determine the voltage that would be indicated when measuring V_1, V_2, and V_3 in Fig. 8-38. The voltmeter is set on the 10-V range.
2. Using the voltmeter shown in Fig. 8-29, determine the voltage that would be indicated when measuring V_1, V_2, and V_3 in Fig. 8-38. The voltmeter is set on the 10-V range.
3. Determine the voltage that would be indicated when measuring V_1, V_2, and V_3 in Fig. 8-38 if a digital multimeter having a 10-MΩ input resistance were used.
4. Using the voltmeter shown in Fig. 8-32, determine the voltage that would be indicated when measuring V_1, V_2, and V_3 in Fig. 8-38. The voltmeter is set on the 5-V range.
5. Using the voltmeter shown in Fig. 8-35, determine the voltage that would be indicated when measuring V_1, V_2, and V_3 in Fig. 8-38. The voltmeter is set on the 5-V range.
6. Which meter produced the most severe case of voltmeter loading in Probs. 1–3? Why?
7. Which meter in Probs. 1–3 provided the least amount of voltmeter loading? Why?

SEC. 8-4 OHMMETERS

The ohmmeter is used to measure the resistance of resistors and other electronic components. Here we will analyze a typical ohmmeter circuit found in multimeters. The meter-movements scale is calibrated in ohms. Full-scale deflection corresponds to 0 Ω, whereas no deflection of the needle corresponds to a resistance of infinity (∞).

Solved Problem

For Fig. 8-39*a*, determine the value to which R_A must be adjusted to produce full-scale deflection when the meter leads are shorted. Once the meter has been zeroed, what value of resistance connected across the meter leads produces half-scale deflection?

Answer

With the meter leads shorted, as in Fig. 8-39*b*, we see that the internal voltage of 1.5 V is connected across two separate parallel branches. One branch consists of the series connection of R_A, R_B, and the meter movement. The other branch consists of resistor R_S. With 1.5 V connected across each branch, R_A must be adjusted to a value that will limit current flow to 100 μA. To find the total resistance required in this branch, we proceed as follows:

$$R_A + R_B + r_M = \frac{V_{\text{battery}}}{I_M}$$

$$= \frac{1.5\text{ V}}{100\ \mu\text{A}}$$

$$R_A + R_B + r_M = 15\text{ k}\Omega$$

Since R_B and r_M are known, the adjusted value for R_A can be determined.

$$R_A = \frac{V_{\text{battery}}}{I_M} - R_B - r_M$$

$$= \frac{1.5\text{ V}}{100\ \mu\text{A}} - 10\text{ k}\Omega - 1\text{ k}\Omega$$

$$R_A = 4\text{ k}\Omega$$

Therefore, R_A must be adjusted to 4 kΩ in order for the current to be limited to 100 μA in the meter movement when the meter leads are shorted. If the battery ages and its voltage output reduces for a specific load, the value for R_A can be reduced to increase the current to the value required to produce full-scale deflection.

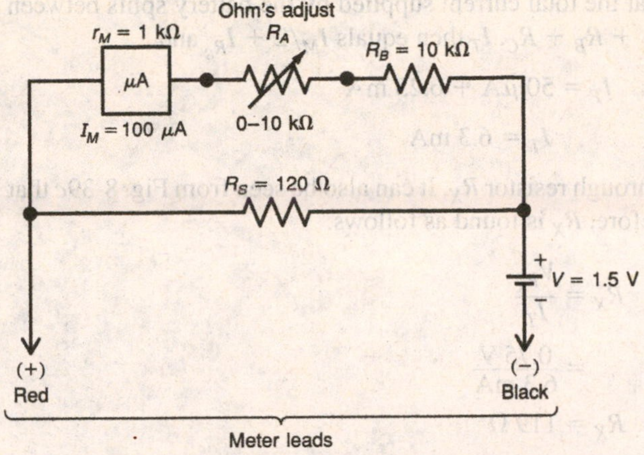

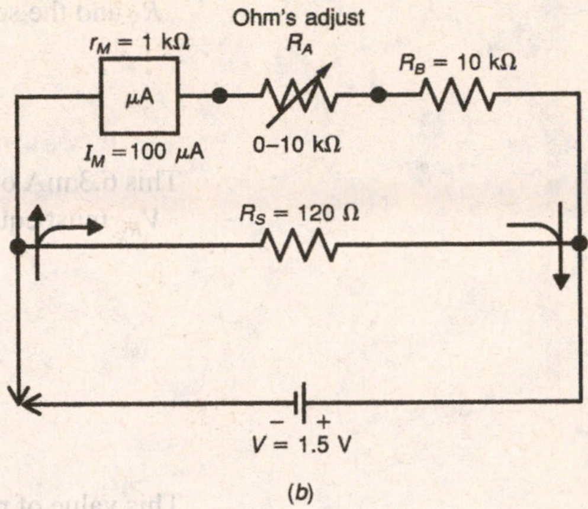

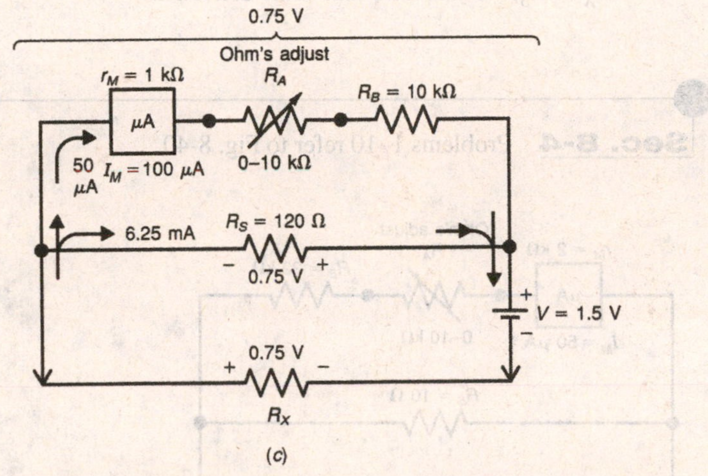

Fig. 8-39 Typical ohmmeter circuit. (*a*) Circuit before 0-Ω adjustment. (*b*) Meter leads short-circuited to adjust for 0 Ω. (*c*) Measuring an external resistance R_X.

Next, we need to determine what value of resistance R_X across the ohmmeter leads produces half-scale deflection. Refer to Fig. 8-39*c*.

First, we must realize that half-scale deflection corresponds to a meter-movement current of 50 μA. This 50-μA current produces a 0.75-V drop across the series combination of $R_A + R_B + r_M$.

$$V = \frac{I_M}{2}(R_A + R_B + r_M)$$

$$= \frac{100\ \mu\text{A}}{2}(4\ \text{k}\Omega + 10\ \text{k}\Omega + 1\ \text{k}\Omega)$$

$$V = 0.75\ \text{V}$$

Since R_S is in parallel with R_A, R_B, and r_M, V_{R_S} also equals 0.75 V. Since V_{R_S} and R_S are known, I_{R_S} can be found.

$$I_{R_S} = \frac{V_{R_S}}{R_S}$$

$$= \frac{0.75\ \text{V}}{120\ \Omega}$$

$$I_{R_S} = 6.25\ \text{mA}$$

It can be seen in Fig. 8-39*c* that the total current supplied by the battery splits between R_S and the series string of $R_A + R_B + R_C$. I_T then equals $I_M/2 + I_{R_S}$ and

$$I_T = 50\ \mu\text{A} + 6.25\ \text{mA}$$

$$I_T = 6.3\ \text{mA}$$

This 6.3 mA of current flows through resistor R_X. It can also be seen from Fig. 8-39*c* that V_{R_X} must equal 0.75 V. Therefore, R_X is found as follows:

$$R_X = \frac{V_{R_X}}{I_T}$$

$$= \frac{0.75\ \text{V}}{6.3\ \text{mA}}$$

$$R_X = 119\ \Omega$$

This value of resistance connected to the ohmmeter leads will produce half-scale deflection. For all practical purposes, R_S is so much smaller than $R_A + R_B + r_M$ in Fig. 8-39 that the battery voltage division is determined solely by R_S and R_X. Therefore, when $R_X = R_S$, we will have half-scale deflection.

PRACTICE PROBLEMS

Sec. 8-4 Problems 1–10 refer to Fig. 8-40.

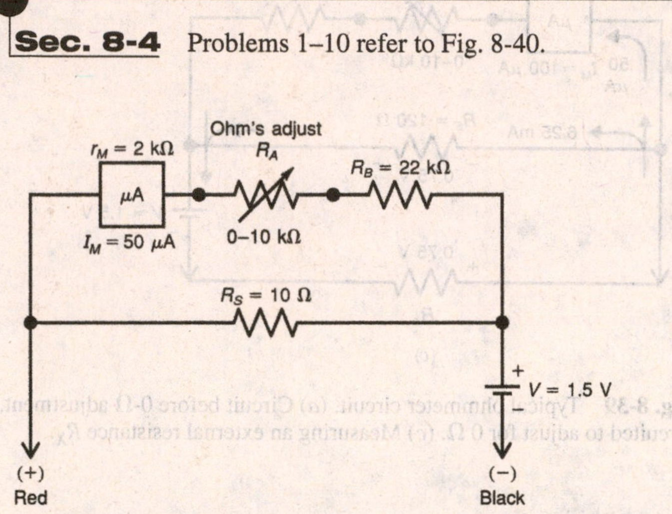

Fig. 8-40 Ohmmeter circuit used for Probs. 1–10 in Sec. 8-4.

1. When the ohmmeter is zeroed, what is the resistance of R_A?
2. What resistance across the ohmmeter leads will produce ⅓ full-scale deflection?
3. What resistance across the ohmmeter leads will produce ⅔ full-scale deflection?
4. What resistance across the ohmmeter leads will produce ¼ full-scale deflection?
5. What resistance across the ohmmeter leads will produce ¾ full-scale deflection?
6. What resistance across the ohmmeter leads will produce ⅕ full-scale deflection?
7. What component must be changed, and to what value, in order for half-scale deflection to correspond to 100 Ω?
8. If $V = 1.3$ V when the ohmmeter leads are shorted, to what value must R_A be adjusted to produce full-scale deflection?
9. What is the minimum battery voltage that will produce full-scale deflection when the meter leads are shorted?
10. With the ohmmeter leads left open, zero current flows through the meter movement. How would the meter face be marked for this condition?

Chapter 8: Direct Current Meters

Answer True or False.

1. To extend the current measuring capabilities of a basic moving-coil meter movement, a shunt resistor is placed in parallel with the meter movement.
2. To make a voltmeter using a basic moving-coil meter movement, a series resistor, called a multiplier resistor, is added in series with the meter movement.
3. In a current meter, the shunt current, I_{sh}, is the difference between I_T and I_M.
4. In a current meter, the shunt resistance, R_{sh}, is less for higher current ranges.
5. If a voltmeter uses a moving-coil meter movement, the total meter resistance does not change as the voltage range is increased.
6. If the meter movement in a voltmeter has a full-scale deflection current, I_M, of 1 mA, the $\frac{\Omega}{V}$ rating of the voltmeter is $\frac{1\ k\Omega}{V}$.
7. A voltmeter with a $\frac{\Omega}{V}$ rating of $\frac{20\ k\Omega}{V}$ has a voltmeter resistance of 1 MΩ on the 50-V range.
8. Voltmeter loading occurs when the voltmeter's resistance is 100 or more times larger than the resistance across which the voltage is measured.
9. A typical DMM has a voltmeter resistance of 10 MΩ on all voltage ranges.
10. With an analog ohmmeter, full-scale meter deflection corresponds to zero ohms, whereas no meter deflection corresponds to infinite ohms.

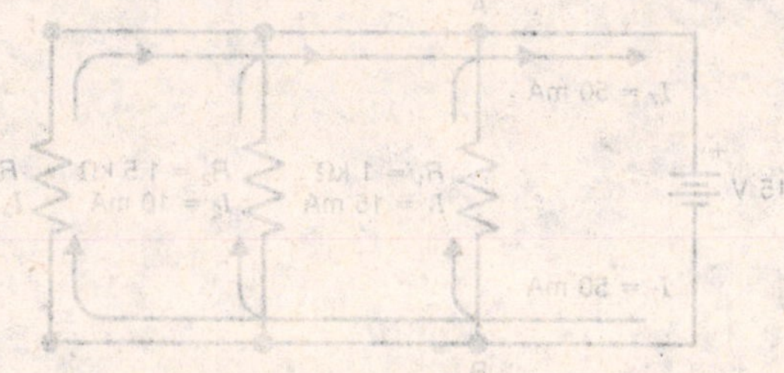

CHAPTER

Kirchhoff's Laws

In this chapter, you will learn how to solve for the unknown values of voltage and current in circuits containing components that are not connected in series, in parallel, or in series-parallel. To solve for the unknown values in these types of circuits, we can apply Kirchhoff's voltage and current laws. As you will see, all types of circuits can be solved with the use of Kirchhoff's laws because these laws do not depend on whether components are connected in series or parallel with each other.

SEC. 9-1 KIRCHHOFF'S CURRENT LAW

Kirchhoff's current law (KCL) states that the algebraic sum of the currents entering and leaving a point must total zero. Consider all currents flowing into a point as positive and all currents leaving that point as negative. Kirchhoff's current law can be applied to the circuit shown in Fig. 9-1.

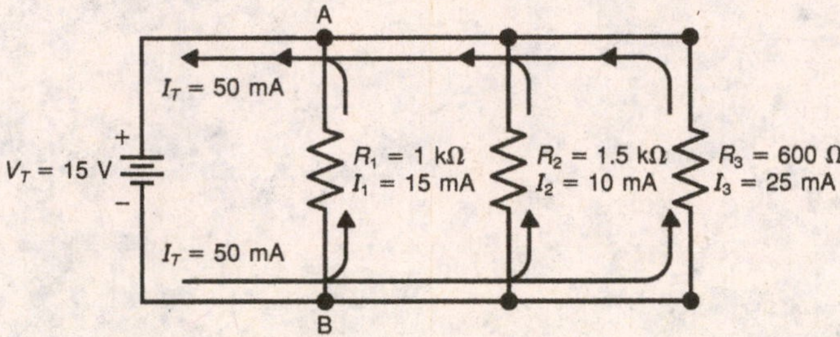

Fig. 9-1 Parallel circuit used to illustrate Kirchhoff's current law.

The 50-mA I_T into point B divides into the 15-mA I_1, 10-mA I_2, and 25-mA I_3. Since 50 mA flows into point B, it is considered positive. The individual branch currents of 15 mA, 10 mA, and 25 mA are considered negative because they leave point B. This can be expressed as an equation.

$$I_T - I_1 - I_2 - I_3 = 0$$

Substituting the values from Fig. 9-1, we have

$$50 \text{ mA} - 15 \text{ mA} - 10 \text{ mA} - 25 \text{ mA} = 0$$

For the opposite direction, refer to point A at the top of Fig. 9-1. Here the branch currents of 15 mA, 10 mA, and 25 mA flow into point A and are considered positive. Now the total current of 50 mA is directed out from point A and therefore is considered negative. The algebraic equation is

$$I_1 + I_2 + I_3 - I_T = 0$$

Substituting again with the values from Fig. 9-1, we have

$$15 \text{ mA} + 10 \text{ mA} + 25 \text{ mA} - 50 \text{ mA} = 0$$

Adding the I_T value of 50 mA to both sides of the equation results in the following:

$$15 \text{ mA} + 10 \text{ mA} + 25 \text{ mA} = 50 \text{ mA}$$

or

$$I_1 + I_2 + I_3 = I_T$$

Notice that this is a basic rule for parallel circuits. The sum of the individual branch currents must equal the total current I_T.

Solved Problem

For Fig. 9-2, find the value and direction of all unknown currents. (This is an exercise in understanding Kirchhoff's current law. It is of no interest here to determine voltage and resistance values.)

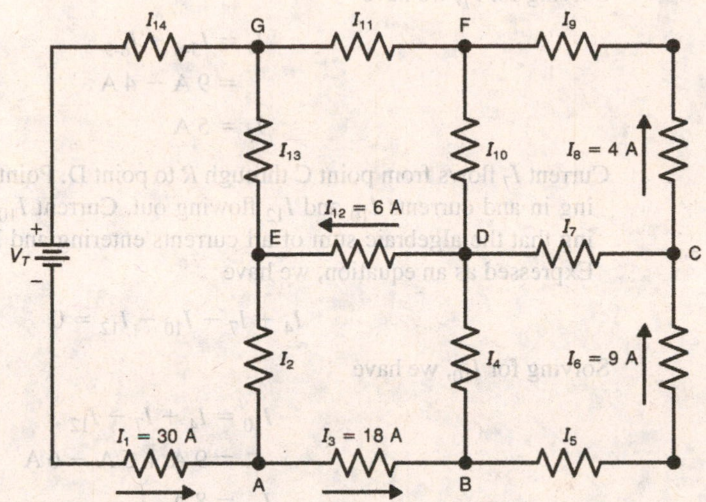

Fig. 9-2 Complex resistor network used to illustrate Kirchhoff's current law.

Answer

In Fig. 9-2, we must remember that all currents flowing into any point are considered positive, whereas all currents flowing away from any point are considered negative. The algebraic sum of all currents flowing into and out of any point must total zero.

The I_1 value of 30 A flowing into point A is positive. The I_3 value of 18 A flowing away from point A is negative. Since the algebraic sum of currents entering and leaving any point must total zero, I_2 must also flow away from point A and, therefore, is considered negative. To determine the value of I_2, we write an equation that will satisfy Kirchhoff's current law.

$$I_1 - I_2 - I_3 = 0$$

Solving for I_2, we have

$$I_2 = I_1 - I_3$$
$$= 30\text{ A} - 18\text{ A}$$
$$I_2 = 12\text{ A}$$

Current I_2 must flow from point A through R to point E.

The I_3 value of 18 A flowing into point B is positive. The current I_5 must be the same as I_6 because the resistors carrying the current are in series. The $I_{5\text{-}6}$ value of 9 A flows away from point B and, therefore, is negative. To determine the value of I_4, we can write an equation that satisfies Kirchhoff's current law.

$$I_3 - I_{5\text{-}6} - I_4 = 0$$

Solving for I_4, we have

$$I_4 = I_3 - I_{5\text{-}6}$$
$$= 18\text{ A} - 9\text{ A}$$
$$I_4 = 9\text{ A}$$

For the current equation at point B, we know that I_4 must be negative because only 9 A of the 18 A coming into point B flow away as $I_{5\text{-}6}$. In order for Kirchhoff's current law to be satisfied, 9 A of current must also flow away from B as I_4. The I_4 value of 9 A flows from point B through R to point D.

The $I_{5\text{-}6}$ value of 9 A flowing into point C is positive. Since the $I_{8\text{-}9}$ value of 4 A flows away from point C, 5 A must flow away from point C as current I_7. Expressed as an equation, we have

$$I_{5\text{-}6} - I_{8\text{-}9} - I_7 = 0$$

Solving for I_7, we have

$$\begin{aligned} I_7 &= I_{5\text{-}6} - I_{8\text{-}9} \\ &= 9\text{ A} - 4\text{ A} \\ I_7 &= 5\text{ A} \end{aligned}$$

Current I_7 flows from point C through R to point D. Point D has currents I_4 and I_7 flowing in and currents I_{10} and I_{12} flowing out. Current I_{10} can be solved by remembering that the algebraic sum of all currents entering and leaving point D must be zero. Expressed as an equation, we have

$$I_4 + I_7 - I_{10} - I_{12} = 0$$

Solving for I_{10}, we have

$$\begin{aligned} I_{10} &= I_4 + I_7 - I_{12} \\ &= 9\text{ A} + 5\text{ A} - 6\text{ A} \\ I_{10} &= 8\text{ A} \end{aligned}$$

It is known here that I_{10} flows away from point D because only 6 A of the 14 A coming in flows away from point D as I_{12}. Current I_{10} flows from point D through R to point F. Since I_{10} and $I_{8\text{-}9}$ flow into point F, they are considered positive. Current I_{11} must leave point F and, therefore, is negative.

$$I_{8\text{-}9} + I_{10} - I_{11} = 0$$

Solving for I_{11}, we have

$$\begin{aligned} I_{11} &= I_{8\text{-}9} + I_{10} \\ &= 4\text{ A} + 8\text{ A} \\ I_{11} &= 12\text{ A} \end{aligned}$$

Current I_{11} flows from point F through R to point G. The I_2 value of 12 A and the I_{12} value of 6 A flow into point E and, therefore, are positive. Current I_{13} must flow away from point E to satisfy Kirchhoff's current law. Expressed as an equation, we have

$$I_2 + I_{12} - I_{13} = 0$$

Solving for I_{13}, we have

$$\begin{aligned} I_{13} &= I_2 + I_{12} \\ &= 12\text{ A} + 6\text{ A} \\ I_{13} &= 18\text{ A} \end{aligned}$$

The last current to solve for is I_{14}. Since the I_{13} value of 18 A and the I_{11} value of 12 A flow into point G, they are positive. Current I_{14} must flow away from point E and, therefore, is considered negative. Expressed as an equation, we have

$$I_{11} + I_{13} - I_{14} = 0$$

$$\begin{aligned} I_{14} &= I_{11} + I_{13} \\ &= 12\text{ A} + 18\text{ A} \\ I_{14} &= 30\text{ A} \end{aligned}$$

This value is the same as I_1 because these currents flow to and from the source voltage terminals in the main line.

After doing the practice problems that follow, you will have a firm grip on Kirchhoff's current law.

PRACTICE PROBLEMS

Sec. 9-1 For Figs. 9-3 through 9-12, find the value and direction of all unknown currents. Indicate the current direction for all solved currents with an arrow. *Example:* $I_{12} = 12\text{ A}\uparrow$, $I_7 = 5\text{ A}\leftarrow$, and so on.

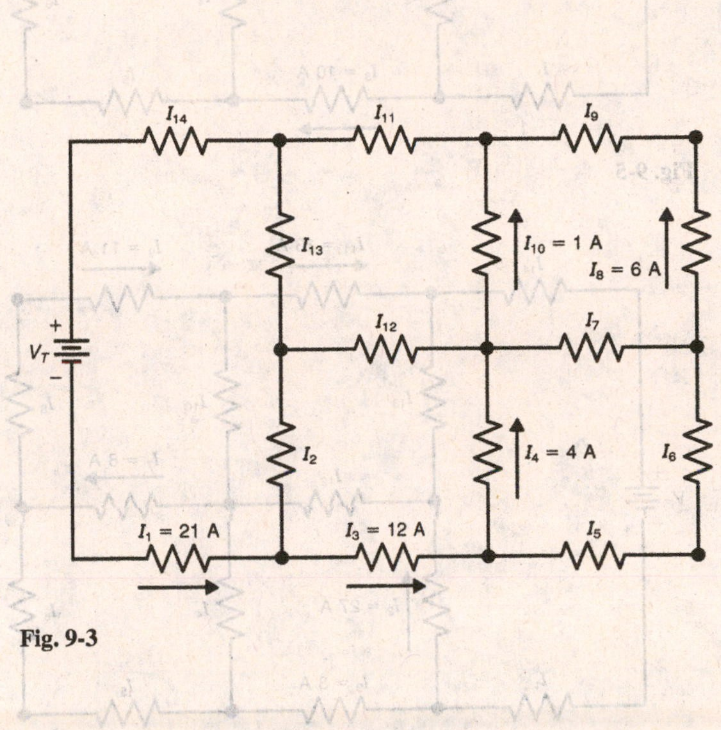

Fig. 9-3

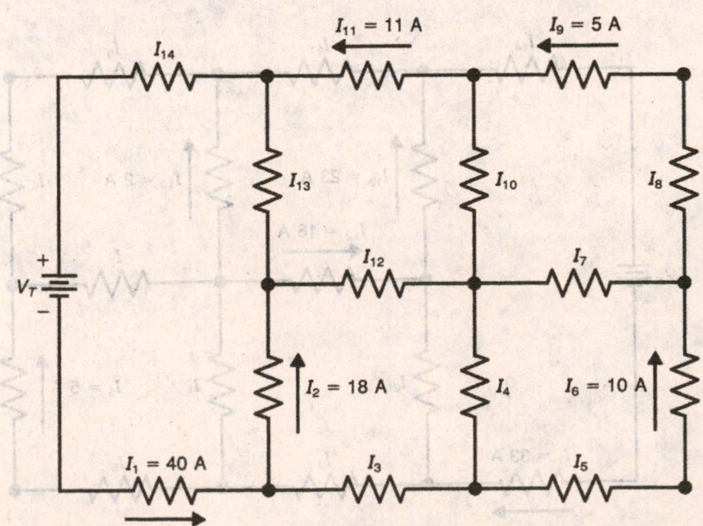

Fig. 9-4

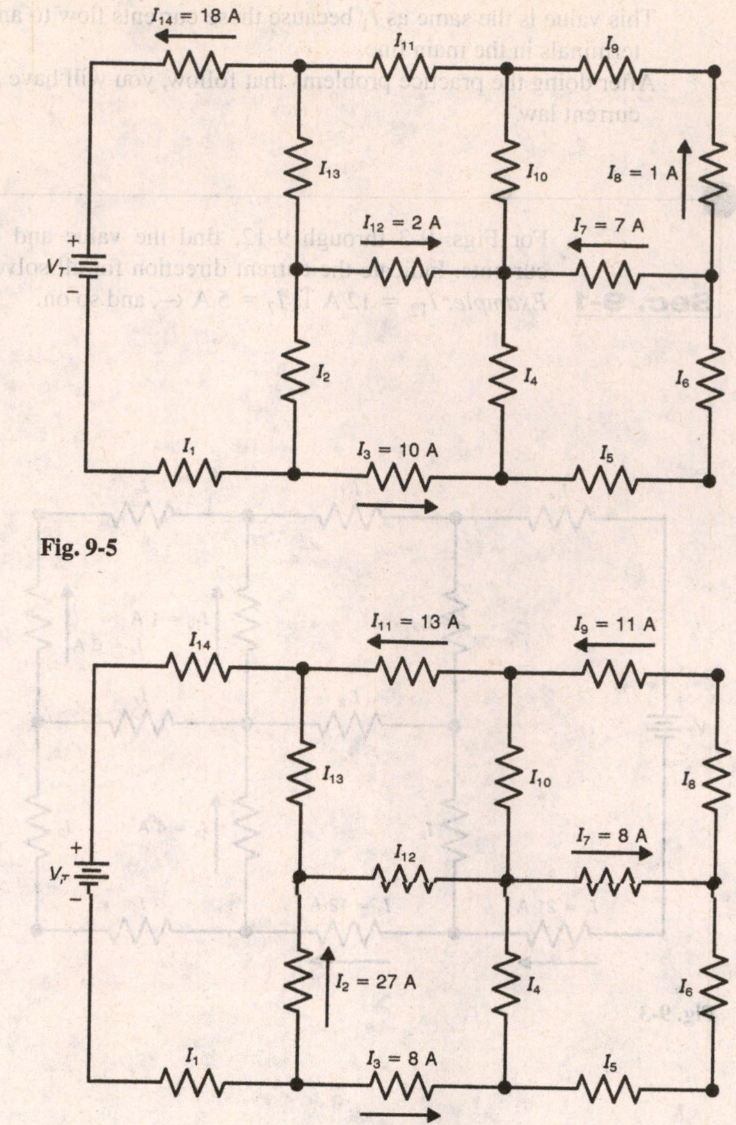

Fig. 9-5

Fig. 9-6

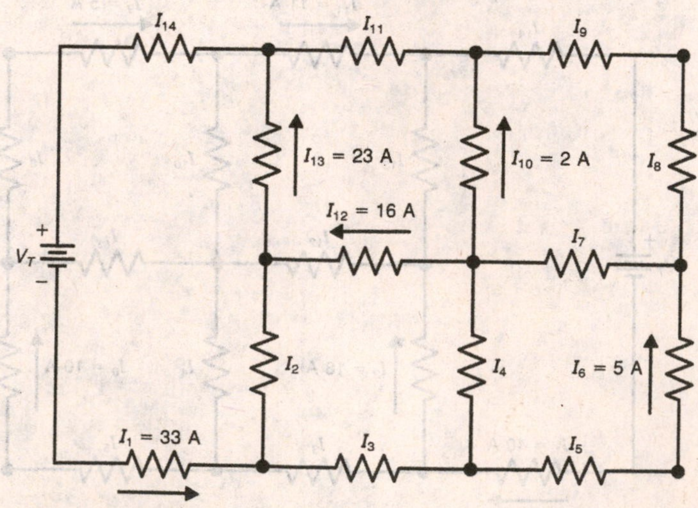

Fig. 9-7

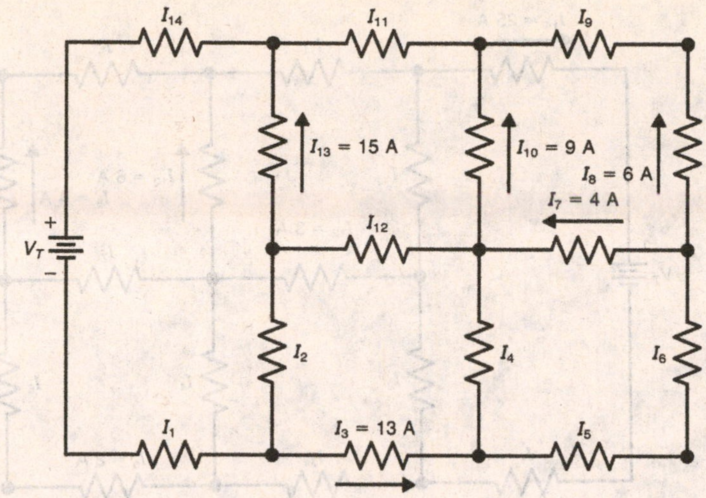

Fig. 9-8

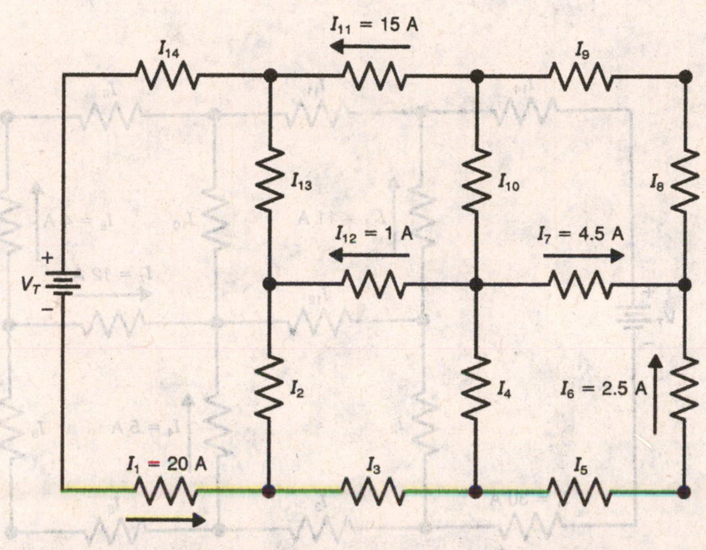

Fig. 9-9

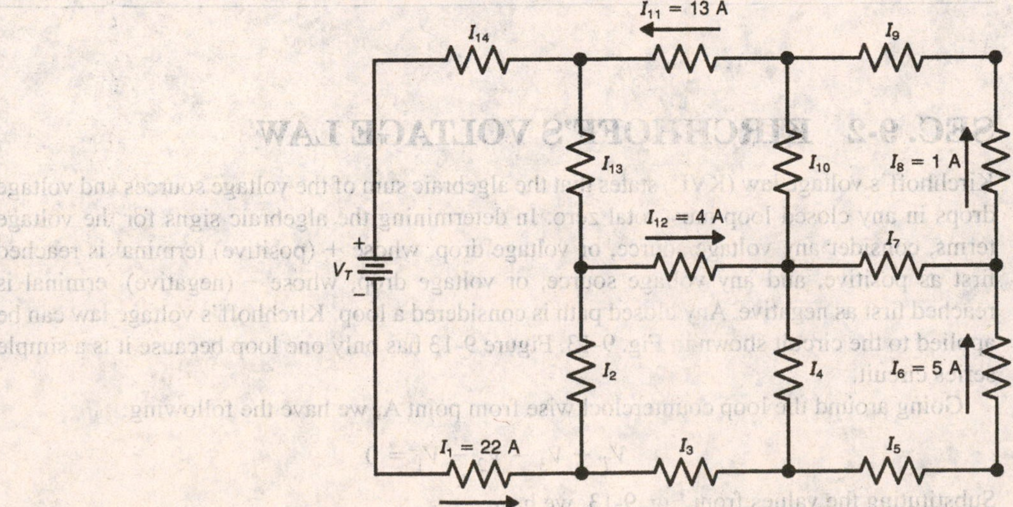

Fig. 9-10

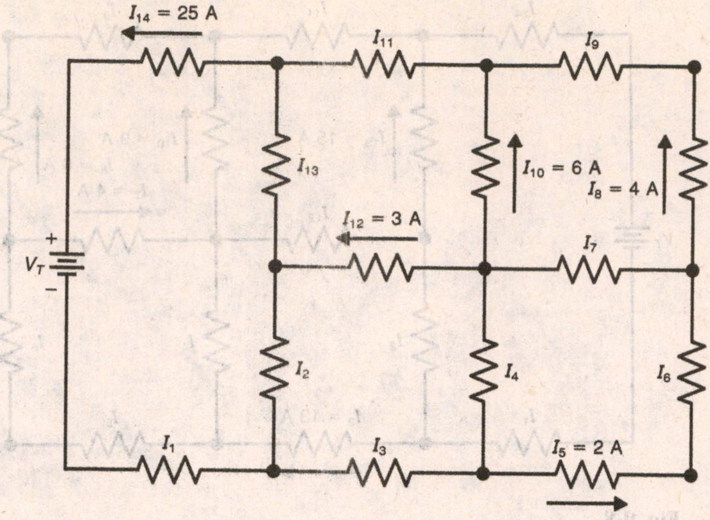

Fig. 9-11

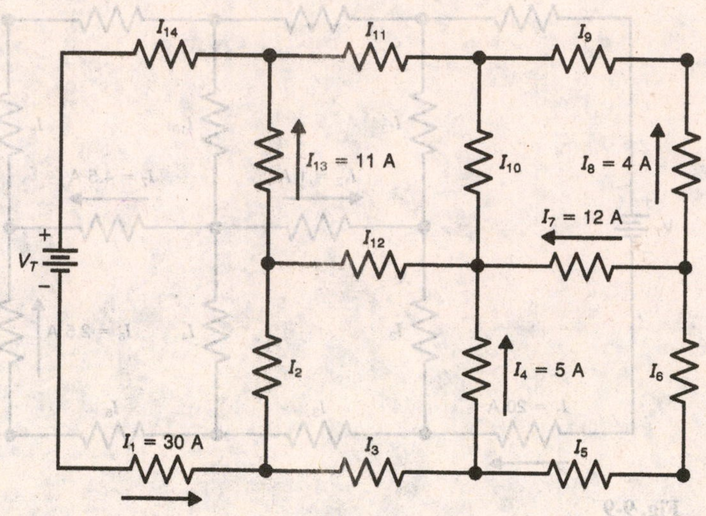

Fig. 9-12

SEC. 9-2 KIRCHHOFF'S VOLTAGE LAW

Kirchhoff's voltage law (KVL) states that the algebraic sum of the voltage sources and voltage drops in any closed loop must total zero. In determining the algebraic signs for the voltage terms, consider any voltage source, or voltage drop, whose + (positive) terminal is reached first as positive, and any voltage source, or voltage drop, whose − (negative) terminal is reached first as negative. Any closed path is considered a loop. Kirchhoff's voltage law can be applied to the circuit shown in Fig. 9-13. Figure 9-13 has only one loop because it is a simple series circuit.

Going around the loop counterclockwise from point A, we have the following:

$$V_T - V_3 - V_2 - V_1 = 0$$

Substituting the values from Fig. 9-13, we have

$$50\text{ V} - 30\text{ V} - 8.25\text{ V} - 11.75\text{ V} = 0$$

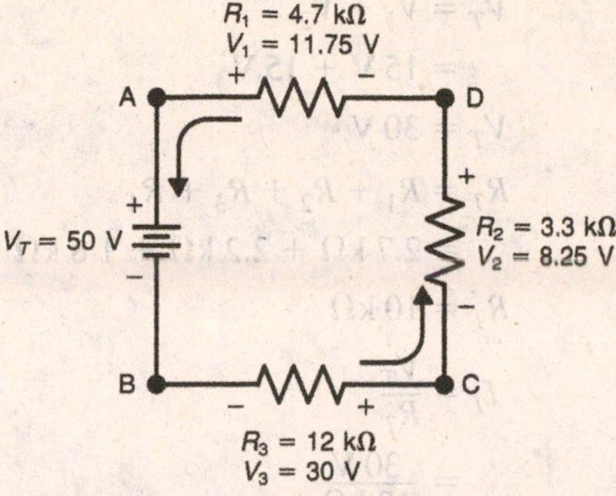

Fig. 9-13 Series circuit used to illustrate Kirchhoff's voltage law.

Voltages V_1, V_2, and V_3 are considered negative because for each of the voltage drops, the negative end of the potential difference is reached first. The source voltage is considered positive because its + terminal is reached first.

Going clockwise around the loop from point A, we have

$$V_1 + V_2 + V_3 - V_T = 0$$

or

$$11.75\text{ V} + 8.25\text{ V} + 30\text{ V} - 50\text{ V} = 0$$

Adding the V_T of 50 V to both sides of the equation results in

$$V_1 + V_2 + V_3 = V_T$$

or

$$11.75\text{ V} + 8.25\text{ V} + 30\text{ V} = 50\text{ V}$$

Notice that this is the basic rule for series circuits. The sum of the individual voltage drops must equal the applied voltage.

Solved Problem

Using Kirchhoff's voltage law, find voltages V_{AG}, V_{BG}, and V_{CG} in Fig. 9-14.

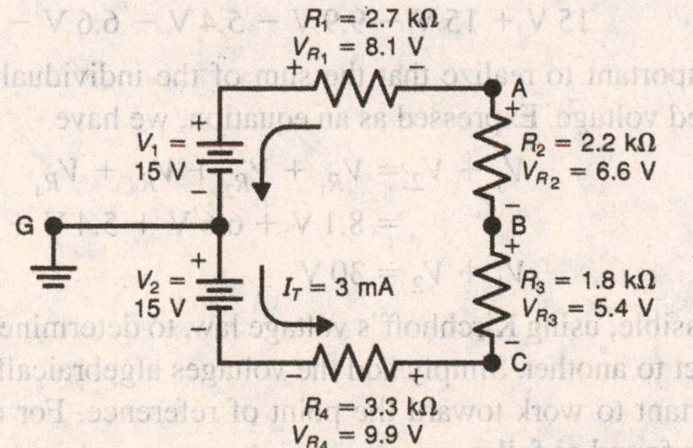

Fig. 9-14 Circuit used to illustrate Kirchhoff's voltage law.

Answer

In Fig. 9-14, we have two 15-V batteries connected in a series-aiding manner. Both V_1 and V_2 force current to flow through the circuit in the same direction. Notice the chassis ground connection at the V_1, V_2 junction. This is simply used for a point of reference, and no current will flow into or out of ground. The circuit can be solved as follows:

$$V_T = V_1 + V_2$$
$$= 15\text{ V} + 15\text{ V}$$
$$V_T = 30\text{ V}$$
$$R_T = R_1 + R_2 + R_3 + R_4$$
$$= 2.7\text{ k}\Omega + 2.2\text{ k}\Omega + 1.8\text{ k}\Omega + 3.3\text{ k}\Omega$$
$$R_T = 10\text{ k}\Omega$$
$$I_T = \frac{V_T}{R_T}$$
$$= \frac{30\text{ V}}{10\text{ k}\Omega}$$
$$I_T = 3\text{ mA}$$
$$V_{R_1} = I_T \times R_1$$
$$= 3\text{ mA} \times 2.7\text{ k}\Omega$$
$$V_{R_1} = 8.1\text{ V}$$
$$V_{R_2} = I_T \times R_2$$
$$= 3\text{ mA} \times 2.2\text{ k}\Omega$$
$$V_{R_2} = 6.6\text{ V}$$
$$V_{R_3} = I_T \times R_3$$
$$= 3\text{ mA} \times 1.8\text{ k}\Omega$$
$$V_{R_3} = 5.4\text{ V}$$
$$V_{R_4} = I_T \times R_4$$
$$= 3\text{ mA} \times 3.3\text{ k}\Omega$$
$$V_{R_4} = 9.9\text{ V}$$

Note the polarity of all voltage drops in Fig. 9-14.

In Fig. 9-14, we can use Kirchhoff's voltage law to determine if we have solved the circuit correctly. Going counterclockwise around the loop, we should obtain an algebraic sum of 0 V.

$$V_1 + V_2 - V_{R_4} - V_{R_3} - V_{R_2} - V_{R_1} = 0$$

Substituting the values from Fig. 9-14, we have

$$15\text{ V} + 15\text{ V} - 9.9\text{ V} - 5.4\text{ V} - 6.6\text{ V} - 8.1\text{ V} = 0$$

It is important to realize that the sum of the individual voltage drops must equal the applied voltage. Expressed as an equation, we have

$$V_1 + V_2 = V_{R_1} + V_{R_2} + V_{R_3} + V_{R_4}$$
$$= 8.1\text{ V} + 6.6\text{ V} + 5.4\text{ V} + 9.9\text{ V}$$
$$V_1 + V_2 = 30\text{ V}$$

It is possible, using Kirchhoff's voltage law, to determine the voltage at some point with respect to another. Simply add the voltages algebraically between the two points. It is important to work toward the point of reference. For example, in Fig. 9-14, voltage V_{AG} is found as follows:

$$V_{AG} = -V_{R_1} + V_1$$
$$= -8.1\text{ V} + 15\text{ V}$$
$$V_{AG} = 6.9\text{ V}$$

Going in the other direction from point *A* produces the same result.

$$V_{AG} = V_{R_2} + V_{R_3} + V_{R_4} - V_2$$
$$= 6.6\text{ V} + 5.4\text{ V} + 9.9\text{ V} - 15\text{ V}$$
$$V_{AG} = 6.9\text{ V}$$

Since there are less voltages to add going counterclockwise, it is the recommended solution for V_{AG}. Solving for V_{BG} and V_{CG}, we have

$$V_{BG} = -V_{R_2} - V_{R_1} + V_1$$
$$= -6.6\text{ V} - 8.1\text{ V} + 15\text{ V}$$
$$V_{BG} = 0.3\text{ V}$$
$$V_{CG} = V_{R_4} - V_2$$
$$= 9.9\text{ V} - 15\text{ V}$$
$$V_{CG} = -5.1\text{ V}$$

PRACTICE PROBLEMS

Sec. 9-2 For Figs. 9-15 through 9-24, solve for the unknowns listed.

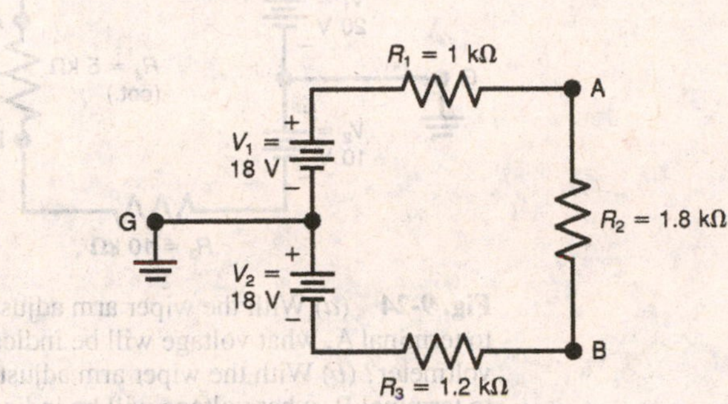

Fig. 9-15 Find V_{AG} and V_{BG}.

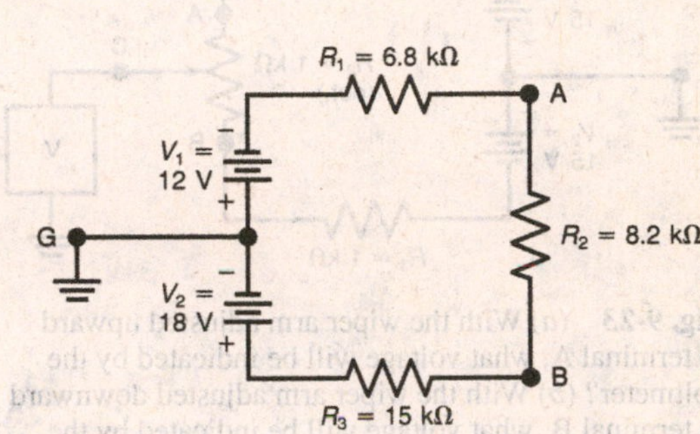

Fig. 9-16 Find V_{AG} and V_{BG}.

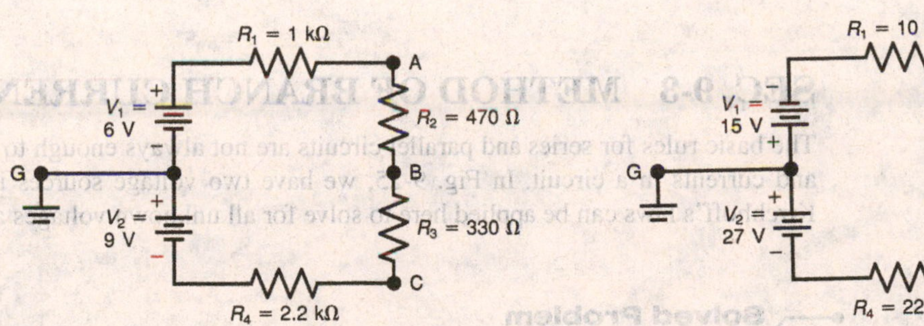

Fig. 9-17 Find V_{AG}, V_{BG}, and V_{CG}.

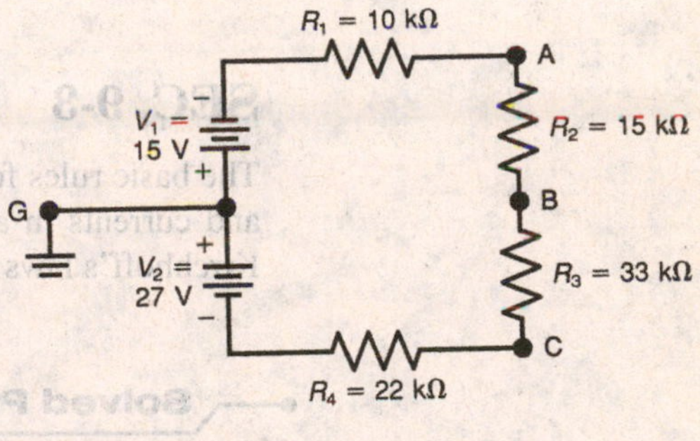

Fig. 9-18 Find V_{AG}, V_{BG}, and V_{CG}.

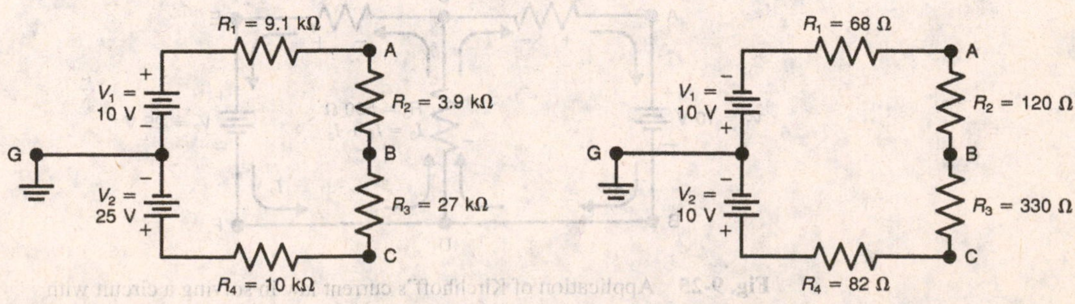

Fig. 9-19 Find V_{AG}, V_{BG}, and V_{CG}.

Fig. 9-20 Find V_{AG}, V_{BG}, and V_{CG}.

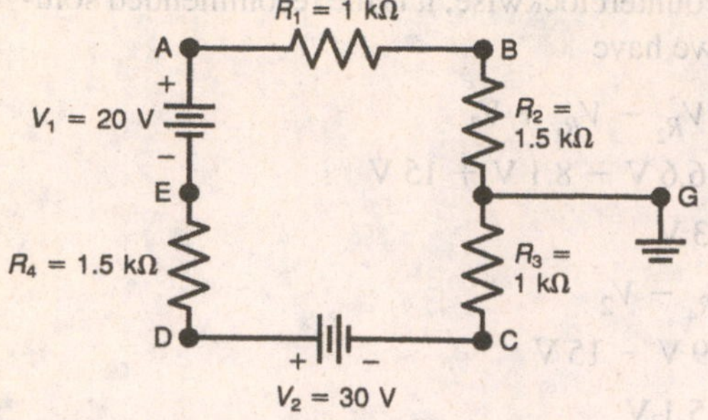

Fig. 9-21 Find V_{AG}, V_{BG}, V_{CG}, V_{DG}, and V_{EG}.

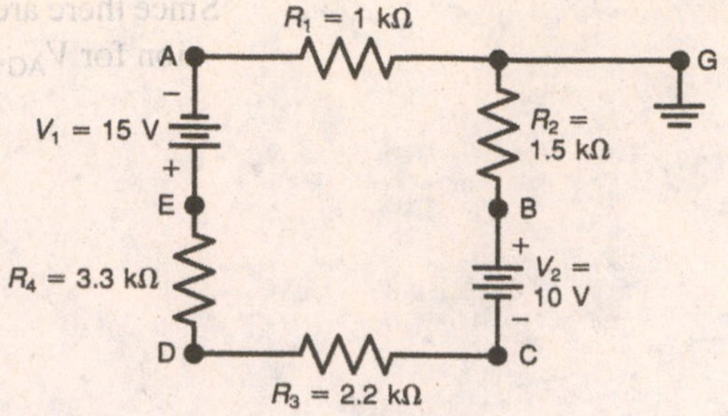

Fig. 9-22 Find V_{AG}, V_{BG}, V_{CG}, V_{DG}, and V_{EG}.

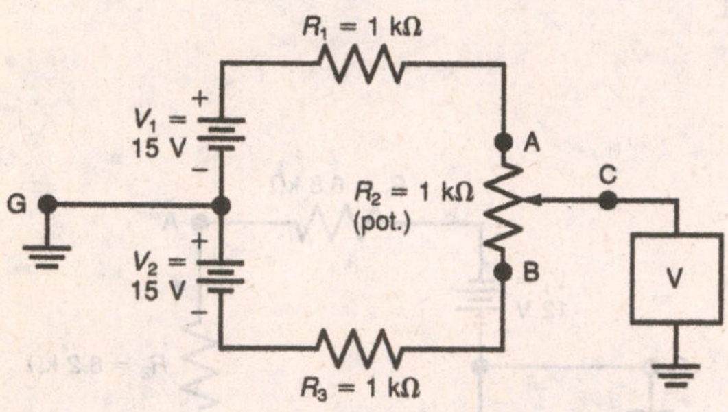

Fig. 9-23 (*a*) With the wiper arm adjusted upward to terminal A, what voltage will be indicated by the voltmeter? (*b*) With the wiper arm adjusted downward to terminal B, what voltage will be indicated by the voltmeter? (*c*) With the wiper arm adjusted midway between terminals A and B, what voltage will be indicated by the voltmeter?

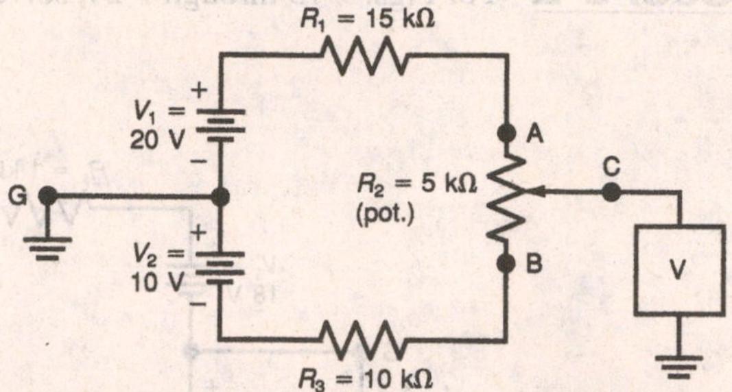

Fig. 9-24 (*a*) With the wiper arm adjusted upward to terminal A, what voltage will be indicated by the voltmeter? (*b*) With the wiper arm adjusted downward to terminal B, what voltage will be indicated by the voltmeter? (*c*) With the wiper arm adjusted midway between terminals A and B, what voltage will be indicated by the voltmeter?

SEC. 9-3 METHOD OF BRANCH CURRENTS

The basic rules for series and parallel circuits are not always enough to solve for all voltages and currents in a circuit. In Fig. 9-25, we have two voltage sources in different branches. Kirchhoff's laws can be applied here to solve for all unknown voltages and currents.

Solved Problem

Find all currents and voltages in Fig. 9-25.

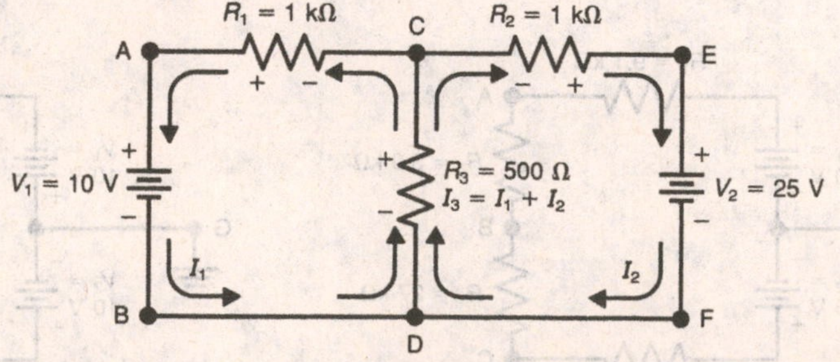

Fig. 9-25 Application of Kirchhoff's current law in solving a circuit with two voltage sources in different branches. Assumed directions of current flow are shown.

Answer

In the solution of currents in Fig. 9-25, we must first indicate the assumed direction of all currents. Remember that electrons entering a resistance produce a negative polarity, whereas currents leaving a resistance produce a positive polarity. In Fig. 9-25, we assume that the voltage source V_1 produces electron flow counterclockwise in the loop ABDCA. Also, we assume that the voltage source V_2 produces electron flow clockwise in the loop EFDCE.

The three currents in Fig. 9-25 are labeled I_1, I_2, and I_3. We can express I_3 in terms of I_1 and I_2 to reduce the number of equations used in the solution of this circuit. In Fig. 9-25, $I_3 = I_1 + I_2$. With only two unknowns, we need two equations, rather than three, to solve the circuit.

For Fig. 9-25, let us write two loop equations. For the loop with V_1 going clockwise from point A, we have

$$V_{R_1} + V_{R_3} - V_1 = 0$$

For the loop with V_2 going counterclockwise from point E, we have

$$V_{R_2} + V_{R_3} - V_2 = 0$$

Using the known values of all resistors, we can specify the IR voltage drops as follows:

$$V_{R_1} = I_1 R_1$$
$$V_{R_1} = I_1\,1\text{ k}\Omega$$

$$V_{R_2} = I_2 R_2$$
$$V_{R_2} = I_2\,1\text{ k}\Omega$$

$$V_{R_3} = I_3 R_3$$
$$V_{R_3} = I_3\,500\ \Omega$$

Since $I_3 = I_1 + I_2$, we have

$$V_{R_3} = (I_1 + I_2)500\ \Omega$$

Substituting these values into the loop equation with V_1, we have

$$1\text{ k}\Omega\, I_1 + 500\ \Omega(I_1 + I_2) - 10\text{ V} = 0$$

Substituting these values into the loop equation with V_2, we have

$$1\text{ k}\Omega\, I_2 + 500\ \Omega(I_1 + I_2) - 25\text{ V} = 0$$

These equations can be reduced to the following:

$$1.5\text{ k}\Omega\, I_1 + 500\ \Omega\, I_2 = 10\text{ V}$$
$$500\ \Omega\, I_1 + 1.5\text{ k}\Omega\, I_2 = 25\text{ V}$$

The solution of either I_1 or I_2 can be any one of the methods used to solve simultaneous linear equations. The method of elimination by addition or subtraction is used here. Multiply the top equation by 3, and then subtract the bottom equation from the top equation. This results in the following:

$$4\text{ k}\Omega\, I_1 = 5\text{ V}$$

Solving for I_1, we have

$$I_1 = \frac{5\text{ V}}{4\text{ k}\Omega}$$
$$I_1 = 1.25\text{ mA}$$

To find I_2, substitute 1.25 mA for I_1 in either of the two loop equations originally written. We use the bottom equation here.

Substituting for I_1, we have

$$500\ \Omega\ I_1 + 1.5\ \text{k}\Omega\ I_2 = 25\ \text{V}$$

$$500\ \Omega(1.25\ \text{mA}) + 1.5\ \text{k}\Omega\ I_2 = 25\ \text{V}$$

$$625\ \text{mV} + 1.5\ \text{k}\Omega\ I_2 = 25\ \text{V}$$

$$1.5\ \text{k}\Omega\ I_2 = 24.375\ \text{V}$$

$$I_2 = 16.25\ \text{mA}$$

To find I_3 through R_3, we add $I_1 + I_2$.

$$I_3 = I_1 + I_2$$

$$= 1.25\ \text{mA} + 16.25\ \text{mA}$$

$$I_3 = 17.5\ \text{mA}$$

If the solution of an unknown current is negative, the negative sign indicates that the assumed direction of current is wrong. The algebraic value of the solved current, however, must be used for substituting in the algebraic equations written for the assumed directions.

With all currents known, the voltage drops across R_1, R_2, and R_3 can be found.

$$V_{R_1} = I_1 \times R_1$$

$$= 1.25\ \text{mA} \times 1\ \text{k}\Omega$$

$$V_{R_1} = 1.25\ \text{V}$$

$$V_{R_2} = I_2 \times R_2$$

$$= 16.25\ \text{mA} \times 1\ \text{k}\Omega$$

$$V_{R_2} = 16.25\ \text{V}$$

$$V_{R_3} = I_3 \times R_3$$

$$= 17.5\ \text{mA} \times 500\ \Omega$$

$$V_{R_3} = 8.75\ \text{V}$$

All assumed current directions were correct for the circuit in Fig. 9-25. Figure 9-26 shows the network with all the correct current and voltage values.

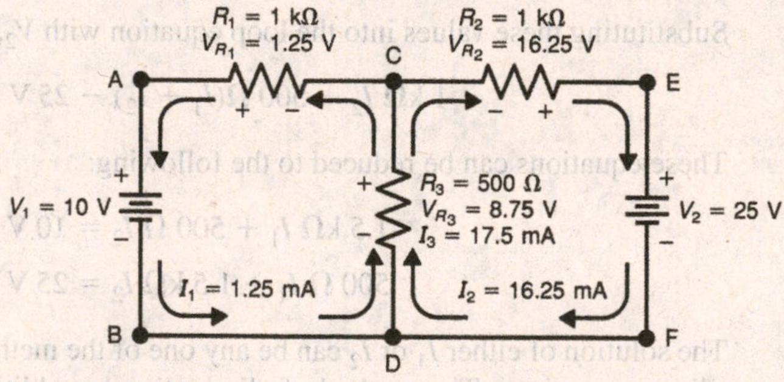

Fig. 9-26 Circuit showing all the correct current and voltage values for Fig. 9-25.

To check our work, let us use Kirchhoff's voltage and current laws.

At point C:

$$I_3 - I_2 - I_1 = 0$$

$$17.5\ \text{mA} - 16.25\ \text{mA} - 1.25\ \text{mA} = 0$$

At point D:

$$I_1 + I_2 - I_3 = 0$$

$$1.25\ \text{mA} + 16.25\ \text{mA} - 17.5\ \text{mA} = 0$$

For each inside loop, we have

$$1.25\text{ V} + 8.75\text{ V} - 10\text{ V} = 0 \text{ (clockwise from A)}$$
$$16.25\text{ V} + 8.75\text{ V} - 25\text{ V} = 0 \text{ (counterclockwise from E)}$$

For the outside loop, we have

$$+10\text{ V} - 25\text{ V} + 16.25\text{ V} - 1.25\text{ V} = 0 \text{ (counterclockwise from A)}$$

The solution for I_1, I_2, and I_3 can also be obtained by the method of mesh currents. However, this method is not covered here, since the solution method is nearly identical. It is recommended, however, that you refer to the text and review the method of mesh currents, and then try your hand at it when doing some of the practice problems that follow.

PRACTICE PROBLEMS

Sec. 9-3 For Figs. 9-27 through 9-36, find all unknown voltages and currents.

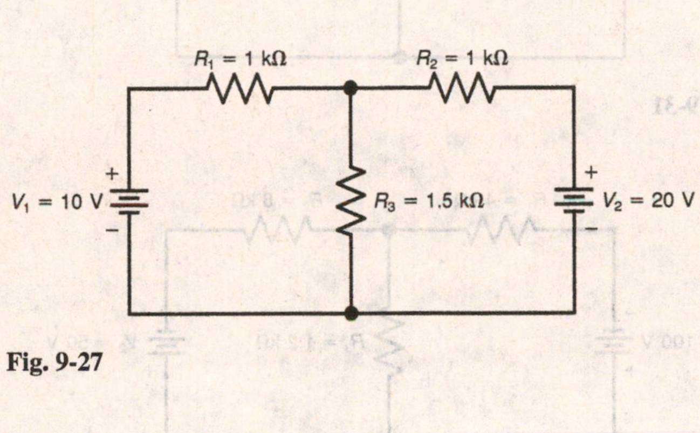

R1 = 1 kΩ
R2 = 1 kΩ
V1 = 10 V
R3 = 1.5 kΩ
V2 = 20 V

Fig. 9-27

R1 = 1.8 kΩ
R2 = 1.8 kΩ
V1 = 15 V
R3 = 100 Ω
V2 = 20 V

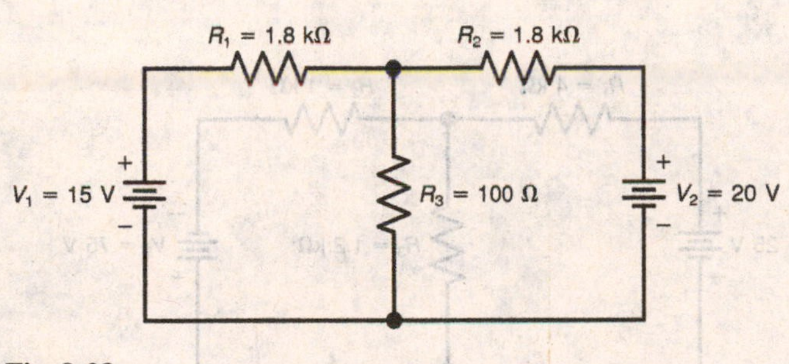

Fig. 9-28

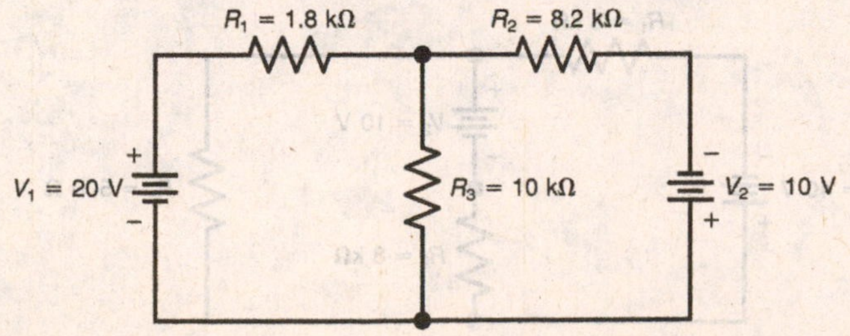

R1 = 1.8 kΩ
R2 = 8.2 kΩ
V1 = 20 V
R3 = 10 kΩ
V2 = 10 V

Fig. 9-29

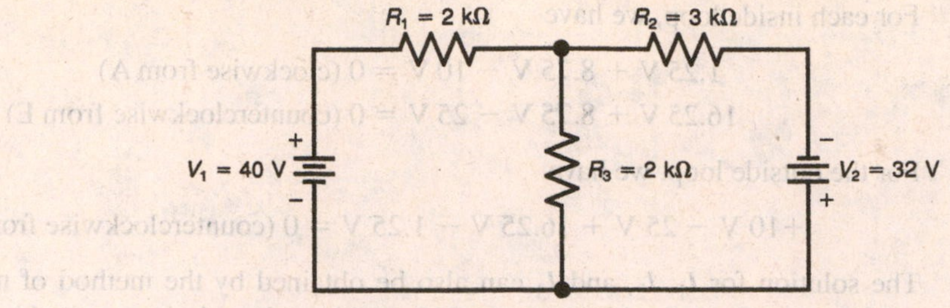

Fig. 9-30

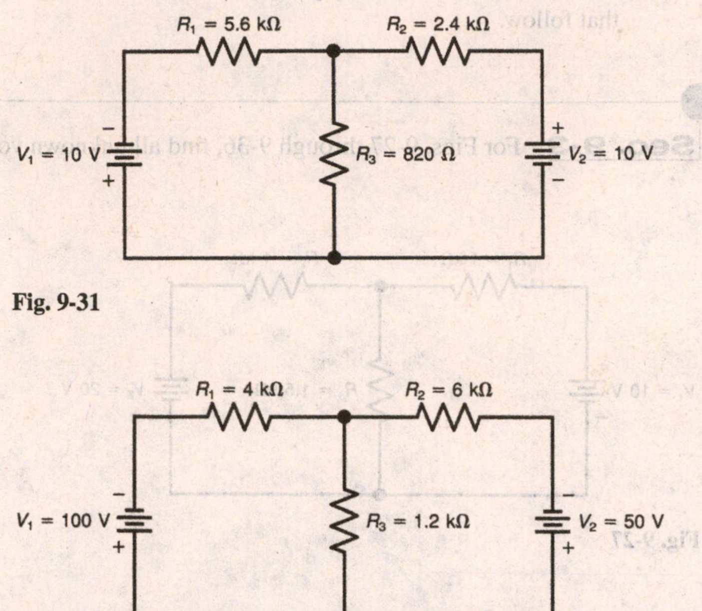

Fig. 9-31

Fig. 9-32

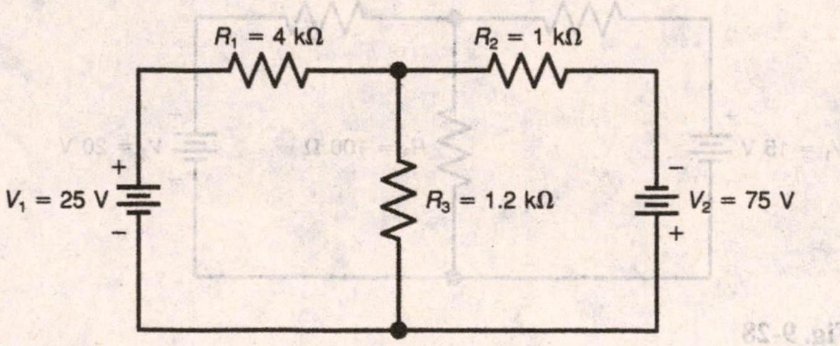

Fig. 9-33

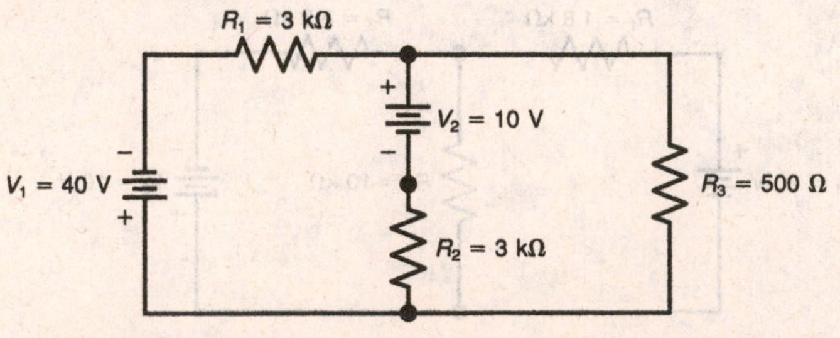

Fig. 9-34

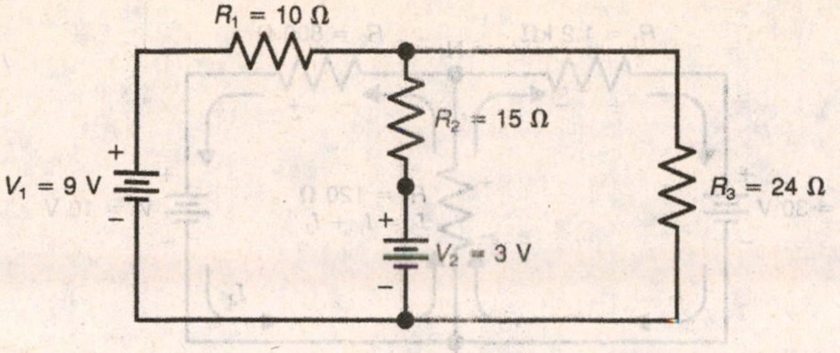

Fig. 9-35

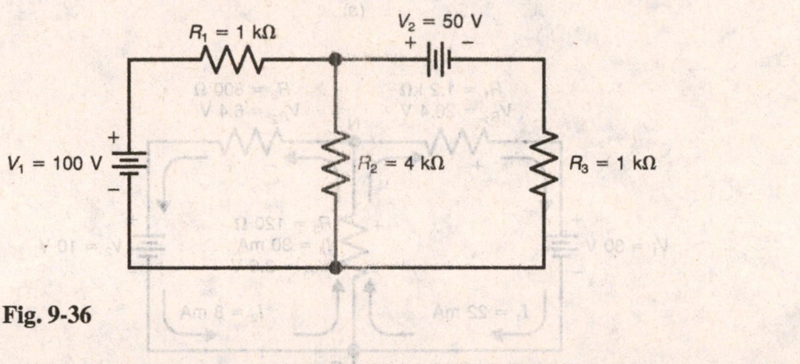

Fig. 9-36

SEC. 9-4 NODE-VOLTAGE ANALYSIS

Another method can be used to solve for all voltages and currents in circuits like those shown in Figs. 9-25 through 9-36. The method of node-voltage analysis allows us to specify the currents at a branch point using Kirchhoff's current law. A branch point is also called a node. A principal node has three or more connections. When using node-voltage analysis, node equations of current are written to satisfy Kirchhoff's current law. Solving the node equation allows us to calculate the unknown node voltage.

Solved Problem

Using node-voltage analysis, solve for all currents and voltages in Fig. 9-37*a*.

Answer

In Fig. 9-37*a*, point G is the reference node. The problem here is to find the node voltage, V_N, from N to G. Once this voltage is known, all other voltages and currents can be found. As before, we assume directions for currents I_1, I_2, and I_3. Note that V_N and V_{R_3} are the same. Current I_3 can be found using Ohm's law.

$$I_3 = \frac{V_N}{R_3}$$

The equations of current at point N in Fig. 9-37 are

$$I_3 - I_1 - I_2 = 0$$

or

$$\frac{V_N}{R_3} - \frac{V_{R_1}}{R_1} - \frac{V_{R_2}}{R_2} = 0$$

Here we have three unknowns. It is possible, however, to write V_{R1} and V_{R2} in terms of V_N and the known values of V_1 and V_2. For V_{R1}, we have

$$V_{R_1} = V_1 - V_N$$
$$V_{R_1} = 30\text{ V} - V_N$$

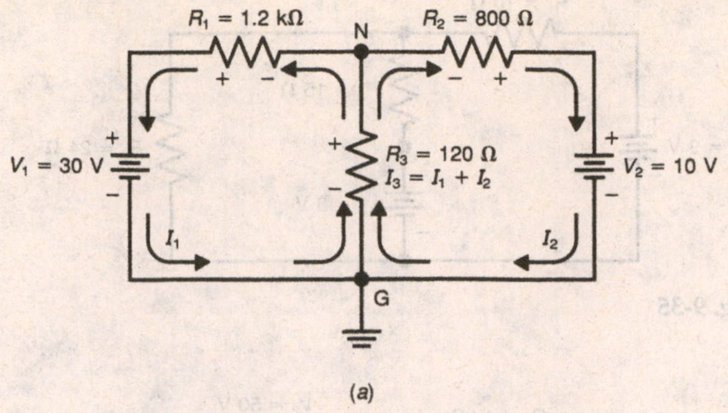

(a)

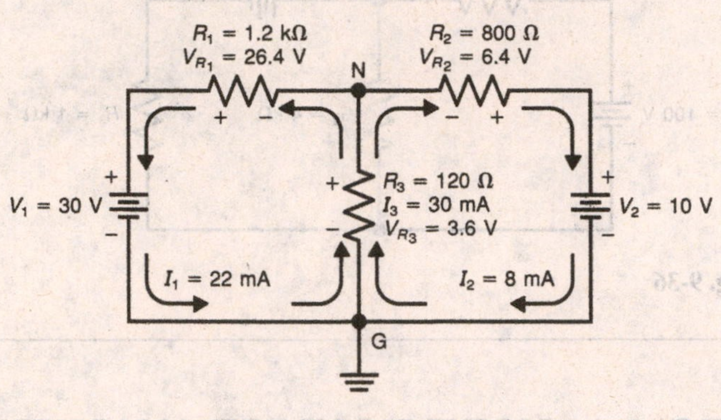

(b)

Fig. 9-37 Circuit used to apply node-voltage analysis. (*a*) Circuit showing assumed current directions. (*b*) Circuit showing all solved current and voltage values.

For V_{R_2}, we have

$$V_{R_2} = V_2 - V_N$$
$$V_{R_2} = 10\text{ V} - V_N$$

The equation for current at point N can be written in terms of V_N.

$$\frac{V_N}{R_3} - \frac{(V_1 - V_N)}{R_1} - \frac{(V_2 - V_N)}{R_2} = 0$$

Including all known values of resistance and voltage gives us the following:

$$\frac{V_N}{120\ \Omega} - \frac{30\text{ V} - V_N}{1.2\text{ k}\Omega} - \frac{10\text{ V} - V_N}{800\ \Omega} = 0$$

Clearing the fractions by multiplying by 2.4 kΩ gives us

$$20\ V_N - 2(30\text{ V} - V_N) - 3(10\text{ V} - V_N) = 0$$

Expanding to solve for V_N, we have

$$20\ V_N - 60\text{ V} + 2\ V_N - 30\text{ V} + 3\ V_N = 0$$
$$25V_N = 90\text{ V}$$
$$V_N = 3.6\text{ V}$$

Knowing this node voltage allows us to determine the other voltages and currents. Solving for V_{R_1}, V_{R_2}, I_1, and I_2, we proceed as follows:

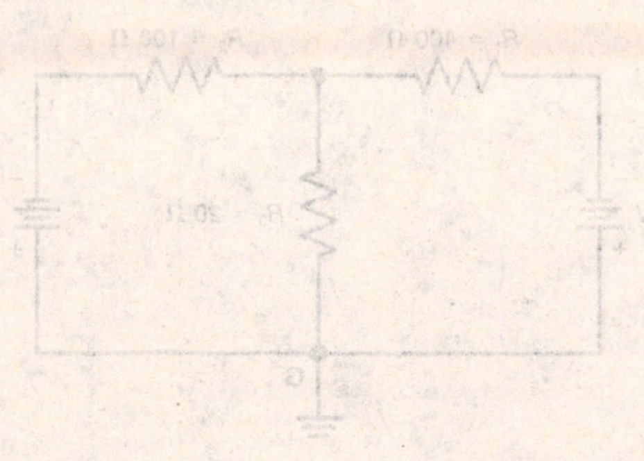

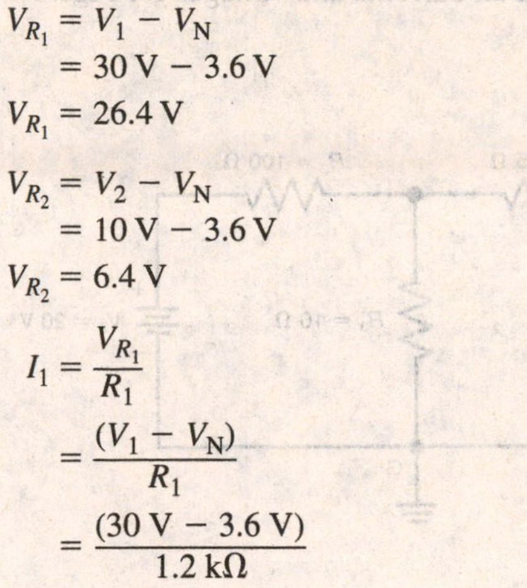

$$V_{R_1} = V_1 - V_N$$
$$= 30\text{ V} - 3.6\text{ V}$$
$$V_{R_1} = 26.4\text{ V}$$

$$V_{R_2} = V_2 - V_N$$
$$= 10\text{ V} - 3.6\text{ V}$$
$$V_{R_2} = 6.4\text{ V}$$

$$I_1 = \frac{V_{R_1}}{R_1}$$
$$= \frac{(V_1 - V_N)}{R_1}$$
$$= \frac{(30\text{ V} - 3.6\text{ V})}{1.2\text{ k}\Omega}$$
$$I_1 = 22\text{ mA}$$

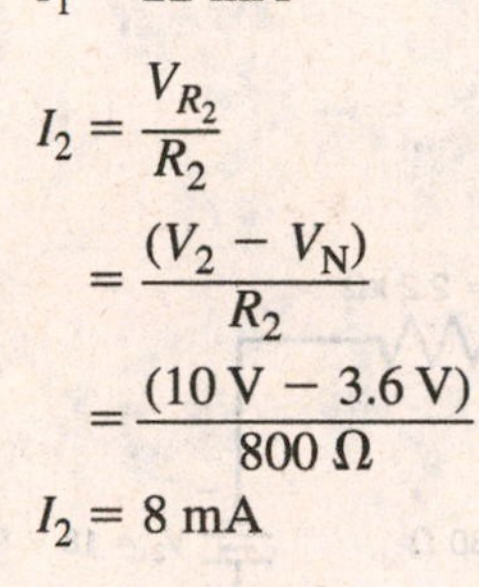

$$I_2 = \frac{V_{R_2}}{R_2}$$
$$= \frac{(V_2 - V_N)}{R_2}$$
$$= \frac{(10\text{ V} - 3.6\text{ V})}{800\ \Omega}$$
$$I_2 = 8\text{ mA}$$

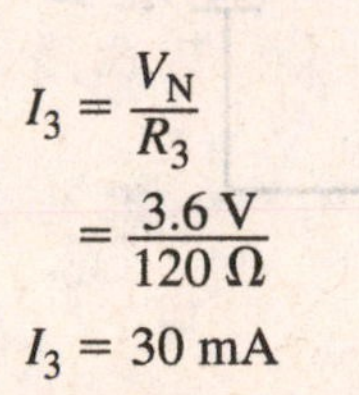

$$I_3 = \frac{V_N}{R_3}$$
$$= \frac{3.6\text{ V}}{120\ \Omega}$$
$$I_3 = 30\text{ mA}$$

Also, note that

$$I_3 = I_1 + I_2$$
$$= 22\text{ mA} + 8\text{ mA}$$
$$I_3 = 30\text{ mA}$$

If the solution for V_{R_1} or V_{R_2} is negative, it means that the assumed direction of current is wrong. Figure 9-37*b* shows the network with all the correct current and voltage values. The method of node-voltage analysis is actually shorter than the method of branch currents. Simpler methods of solution for complicated circuits are reviewed in Chap. 10, "Network Theorems."

Sec. 9-4 Find all currents and voltages for Figs. 9-38 through 9-42 using the method of node-voltage analysis.

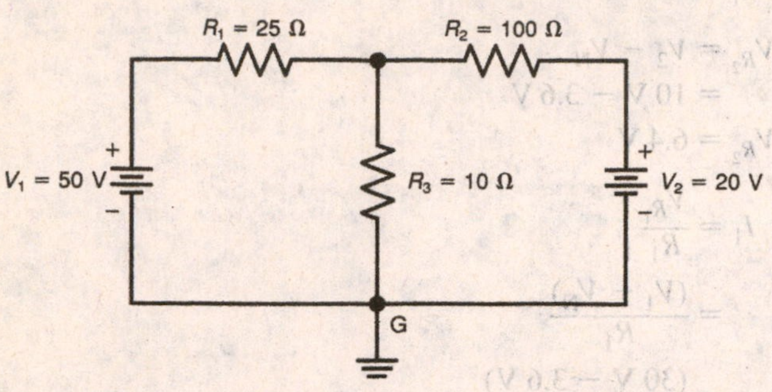

Fig. 9-38

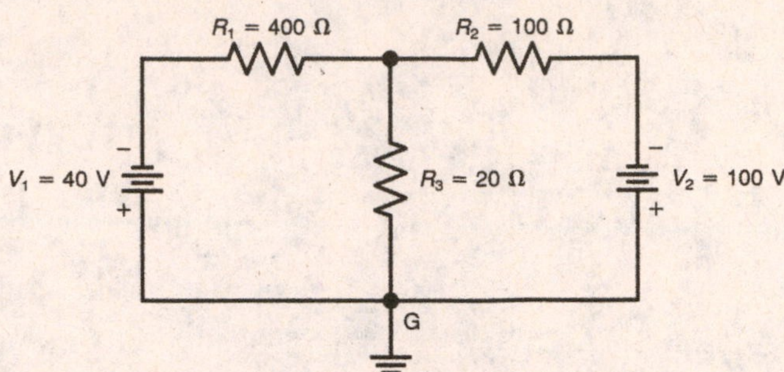

Fig. 9-39

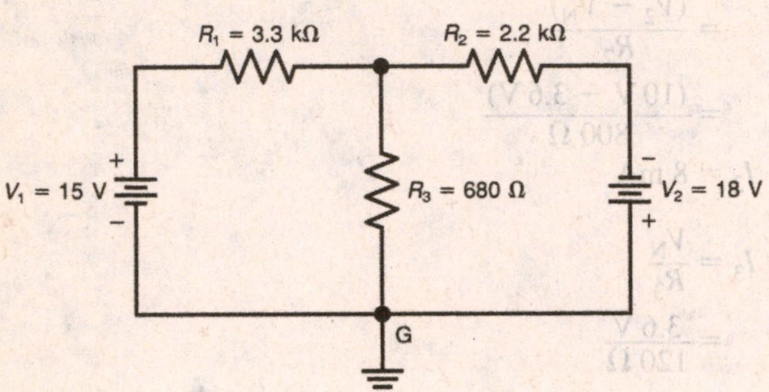

Fig. 9-40

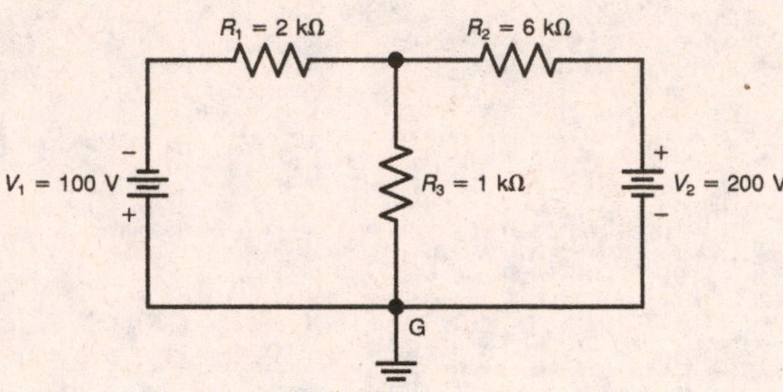

Fig. 9-41

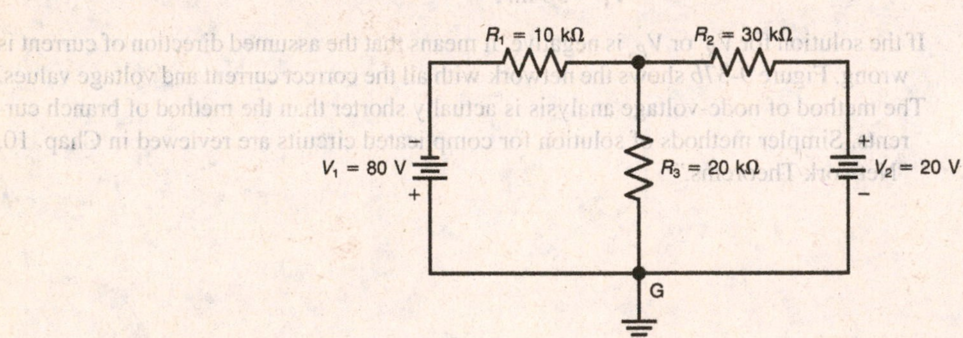

Fig. 9-42

Chapter 9: Kirchhoff's Laws

Answer True or False.

1. Kirchhoff's laws can be applied to all types of circuits because these laws do not depend on whether the components are connected in series, in parallel, or in series-parallel.
2. Kirchhoff's Current Law (KCL) states that the algebraic sum of the currents entering and leaving any point must equal the total current, I_T.
3. When writing a KCL equation, all currents flowing into a point are considered positive terms in the KCL equation.
4. When writing a KCL equation, all currents flowing away from a point are considered negative terms in the KCL equation.
5. Kirchhoff's Voltage Law (KVL) states that the algebraic sum of the voltage sources and voltage drops around any closed loop must equal zero.
6. When applying KVL, consider any voltage whose positive side is reached first as a positive term in the KVL equation and any voltage whose negative side is reached first as a negative term.
7. When solving for the voltage between two points in a circuit, such as V_{AG}, simply add the voltages algebraically between the two points, starting at point A and ending at point G.
8. When using the method of branch currents or node-voltage analysis to solve for the unknown values in a circuit, a negative solution indicates that the assumed direction of current was correct.
9. A branch point is called a node.
10. A principal node has three or more connections.

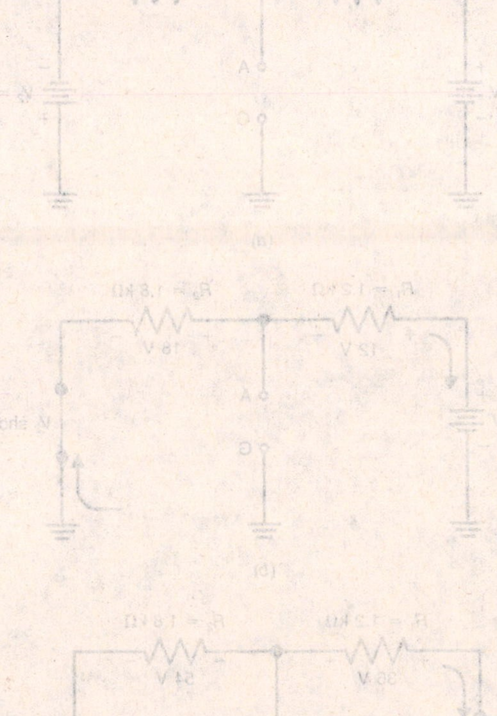

CHAPTER 10

Network Theorems

In this chapter, you will learn how to solve for the voltages and currents in a complex network using special theorems that simplify the method of solution. You will use the superposition theorem, Thevenin's theorem, Norton's theorem, and Millman's theorem to solve for the unknown values in a circuit. In addition, you will learn how to transform a wye (Y) network into a delta (Δ) network and vice versa.

SEC. 10-1 SUPERPOSITION THEOREM

As a definition, the superposition theorem states that in a network with two or more sources, the current or voltage for any component is the algebraic sum of the effects produced by each source acting separately. In order to apply the superposition theorem, all components must be both linear and bilateral. A component is linear if its V and I values are proportional. A component is bilateral if the magnitude of the current is the same for opposite polarities of voltage across its terminals.

Solved Problem

Using the superposition theorem, find the voltage V_{AG} in Fig. 10-1*a*.

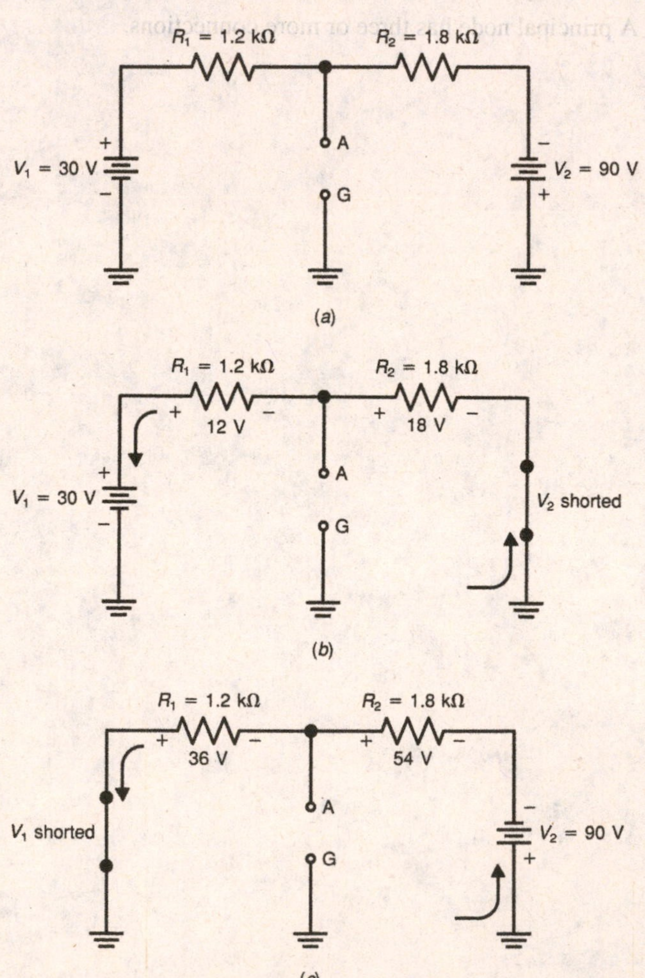

Fig. 10-1 Superposition theorem applied to a voltage divider with two voltage sources. (*a*) Original circuit. (*b*) Voltages with V_2 shorted. (*c*) Voltages with V_1 shorted.

Answer

To find the effect of V_1, short-circuit V_2 as shown in Fig. 10-1*b*. Notice that R_1 and R_2 form a series voltage divider for the source voltage V_1. The voltage across R_2 becomes the same as the voltage from point A to ground.

$$V_{R_2} = \frac{R_2}{R_1 + R_2} \times V_1$$
$$= \frac{1.8\ \text{k}\Omega}{1.2\ \text{k}\Omega + 1.8\ \text{k}\Omega} \times 30\ \text{V}$$
$$V_{R_2} = 18\ \text{V}$$

Next find the effect of V_2 alone with V_1 short-circuited, as shown in Fig. 10-1*c*. Notice that R_1 and R_2 form a series voltage divider for the source voltage V_2. The voltage across R_1 becomes the same as the voltage from point A to ground.

$$V_{R_1} = \frac{R_1}{R_1 + R_2} \times V_2$$
$$= \frac{1.2\ \text{k}\Omega}{1.2\ \text{k}\Omega + 1.8\ \text{k}\Omega} \times -90\ \text{V}$$
$$V_{R_1} = -36\ \text{V}$$

Finally the voltage at point A is

$$V_{AG} = \frac{R_2V_1}{R_1 + R_2} + \frac{R_1V_2}{R_1 + R_2}$$
$$= 18\ \text{V} - 36\ \text{V}$$
$$V_{AG} = -18\ \text{V}$$

Note: Since the two terms in the formula for V_{AG} have a common denominator, it is often shown as

$$V_{AG} = \frac{R_2V_1 + R_1V_2}{R_1 + R_2}$$

It is important to note the polarities of V_1 and V_2 when using this formula to find the voltage across points A and G.

PRACTICE PROBLEMS

Sec. 10-1 Using the superposition theorem, find the voltage V_{AG} in Figs. 10-2 through 10-9.

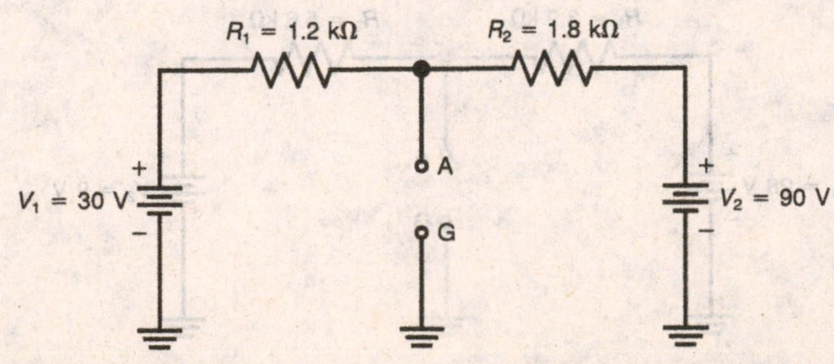

Fig. 10-2

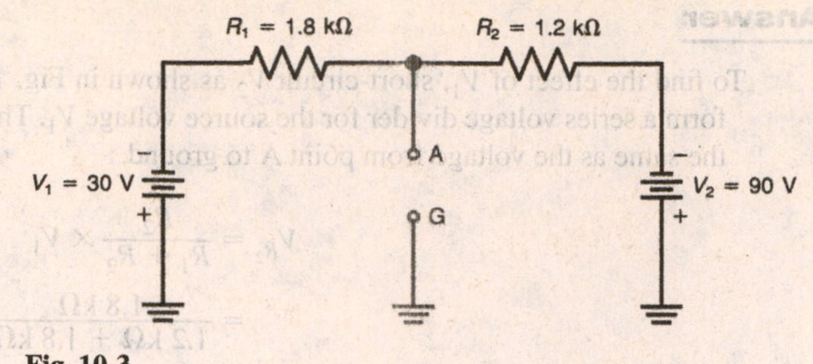

Fig. 10-3

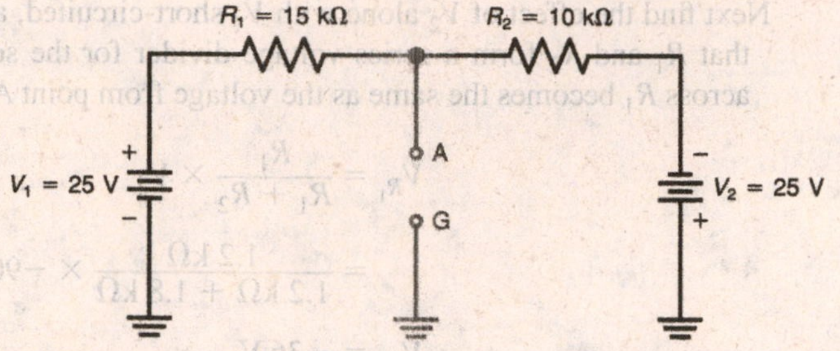

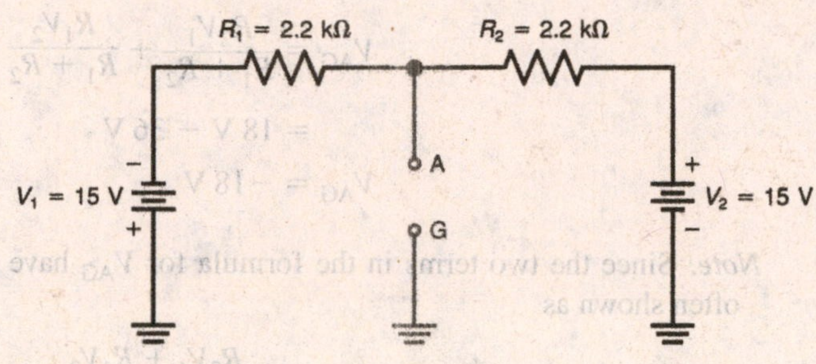

Fig. 10-4

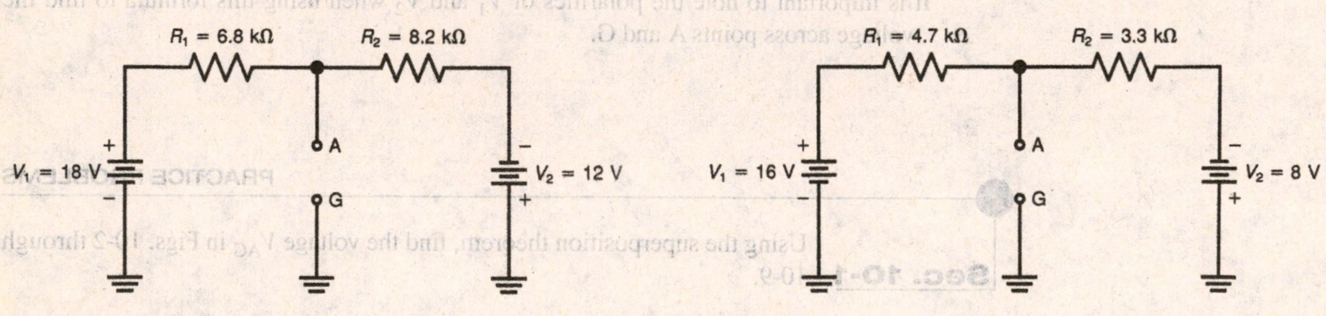

Fig. 10-5

Fig. 10-6

Fig. 10-7

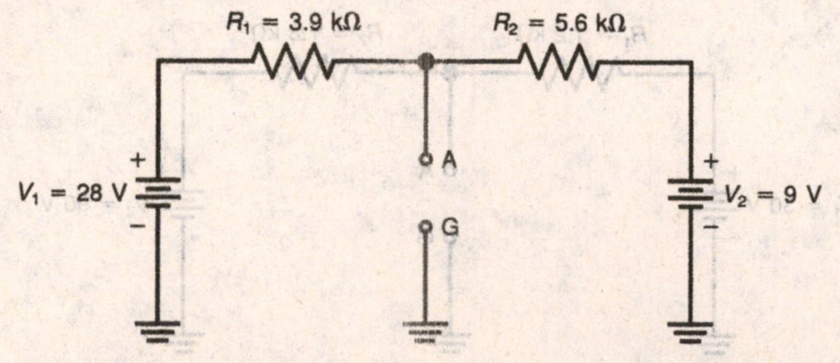

Fig. 10-8

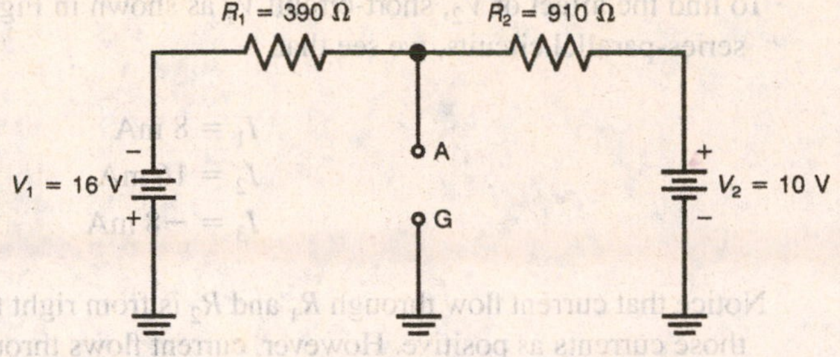

Fig. 10-9

Solved Problem

Using the superposition theorem, find all currents and voltages in Fig. 10-10*a*.

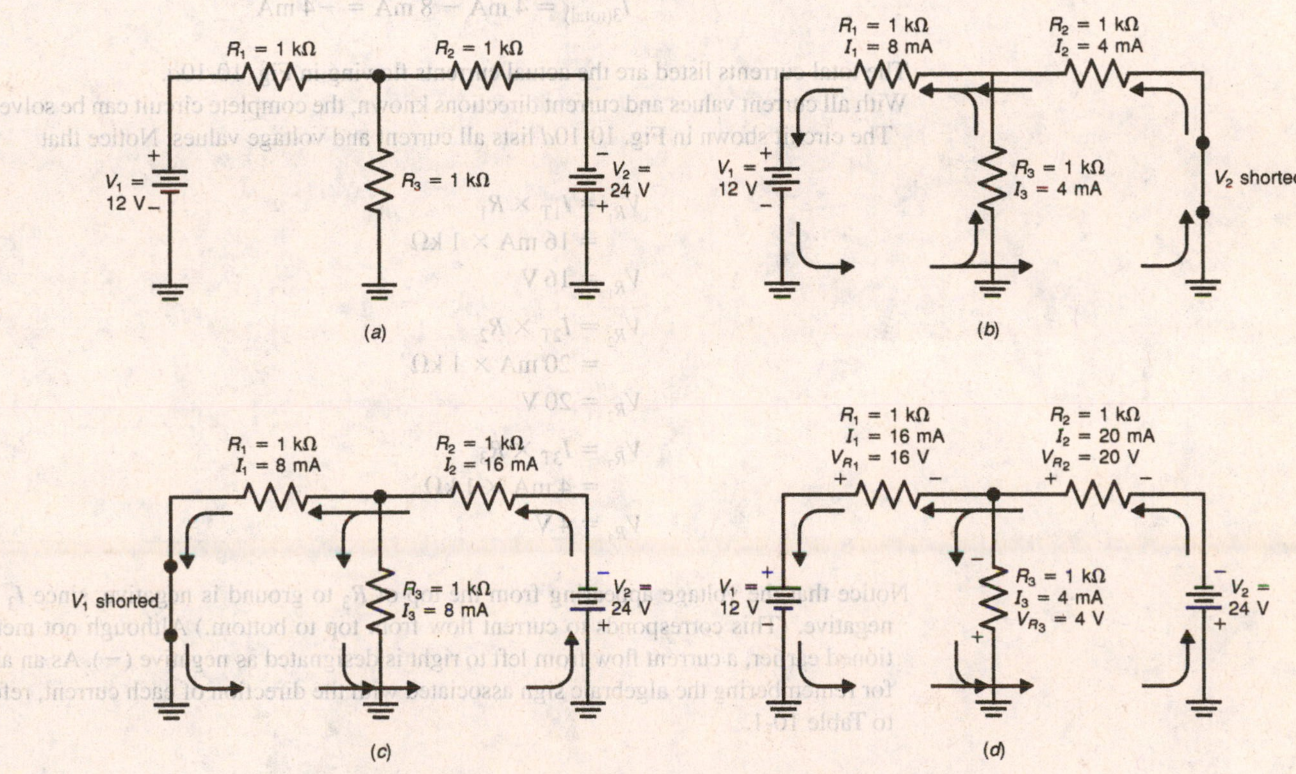

Fig. 10-10 Superposition theorem used to find all voltages and currents. (*a*) Original circuit. (*b*) Current values with V_2 shorted. (*c*) Current values with V_1 shorted. (*d*) Circuit showing all solved voltage and current values.

Answer

To find the effect of V_1, short-circuit V_2 as shown in Fig. 10-10*b*. Using Ohm's law for series-parallel circuits, we see that

$$I_1 = 8 \text{ mA}$$

$$I_2 = 4 \text{ mA}$$

$$I_3 = 4 \text{ mA}$$

Notice that the direction of current flow through R_3 is from bottom to top. Let us call this current direction positive (+). Also, notice that the direction of current flow through R_1 and R_2 is from right to left. Let us call this current direction positive (+) also.

To find the affect of V_2, short-circuit V_1 as shown in Fig. 10-10*c*. Using Ohm's law for series-parallel circuits, we see that

$$I_1 = 8 \text{ mA}$$
$$I_2 = 16 \text{ mA}$$
$$I_3 = -8 \text{ mA}$$

Notice that current flow through R_1 and R_2 is from right to left. Therefore, we will label those currents as positive. However, current flows through R_3 from top to bottom. We will designate this current direction as negative (−). Current $I_3 = -8$ mA for this condition.

The final step is to add algebraically the individual currents flowing when each source is acting separately.

$$I_{1(\text{total})} = 8 \text{ mA} + 8 \text{ mA} = 16 \text{ mA}$$
$$I_{2(\text{total})} = 4 \text{ mA} + 16 \text{ mA} = 20 \text{ mA}$$
$$I_{3(\text{total})} = 4 \text{ mA} - 8 \text{ mA} = -4 \text{ mA}$$

The total currents listed are the actual currents flowing in Fig. 10-10*a*.

With all current values and current directions known, the complete circuit can be solved. The circuit shown in Fig. 10-10*d* lists all current and voltage values. Notice that

$$V_{R_1} = I_{1T} \times R_1$$
$$= 16 \text{ mA} \times 1 \text{ k}\Omega$$
$$V_{R_1} = 16 \text{ V}$$
$$V_{R_2} = I_{2T} \times R_2$$
$$= 20 \text{ mA} \times 1 \text{ k}\Omega$$
$$V_{R_2} = 20 \text{ V}$$
$$V_{R_3} = I_{3T} \times R_3$$
$$= 4 \text{ mA} \times 1 \text{ k}\Omega$$
$$V_{R_3} = 4 \text{ V}$$

Notice that the voltage appearing from the top of R_3 to ground is negative, since I_3 is negative. (This corresponds to current flow from top to bottom.) Although not mentioned earlier, a current flow from left to right is designated as negative (−). As an aid for remembering the algebraic sign associated with the direction of each current, refer to Table 10-1.

TABLE 10-1 ALGEBRAIC SIGNS FOR CURRENTS

Current Direction	Algebraic Sign
→	−
←	+
↑	+
↓	−

As a final check of our answers, we can apply Kirchhoff's voltage and current laws. If the work is done correctly, the algebraic sum of the voltage sources and voltage drops around any closed loop should be zero. Likewise, the algebraic sum of the currents entering and leaving any point should be zero.

Sec. 10-1

For Figs. 10-11 through 10-20, solve for all currents and voltages using the superposition theorem.

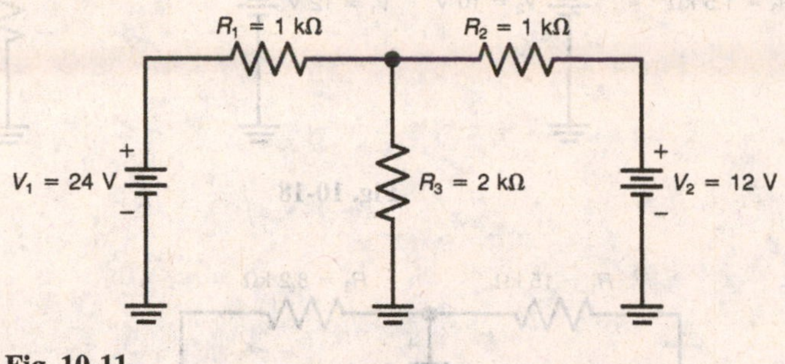

Fig. 10-11

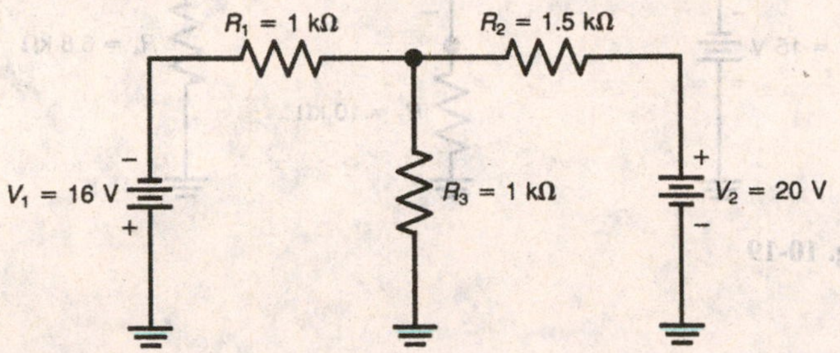

Fig. 10-12

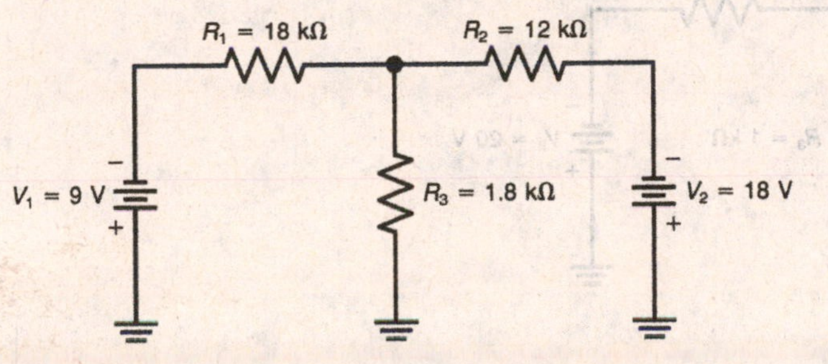

Fig. 10-13

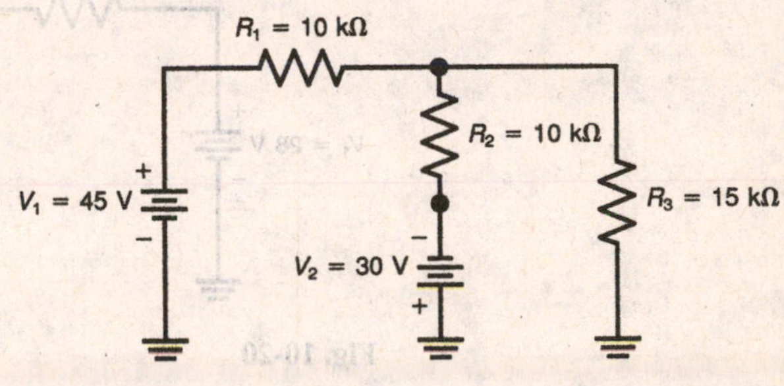

Fig. 10-14

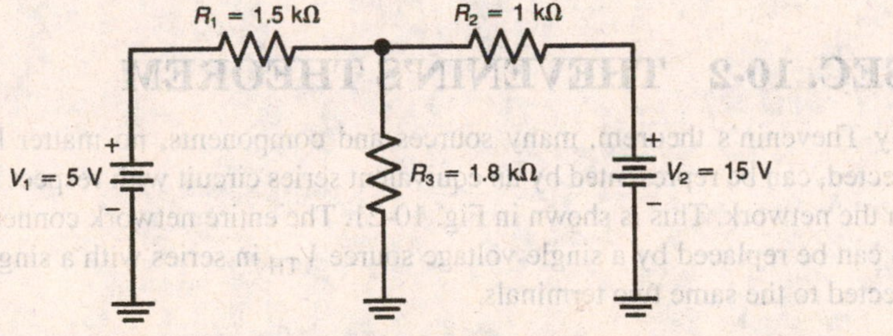

Fig. 10-15

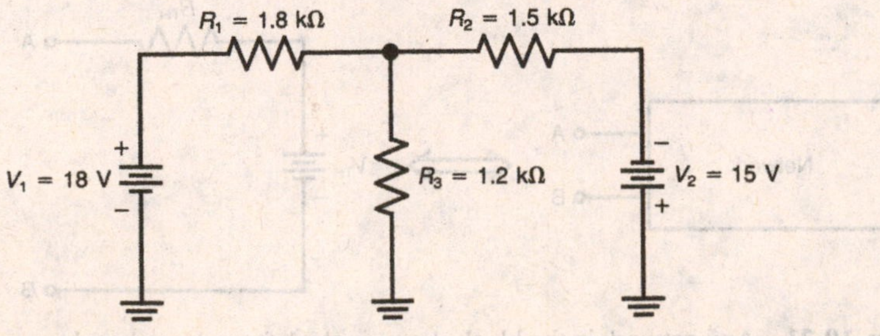

Fig. 10-16

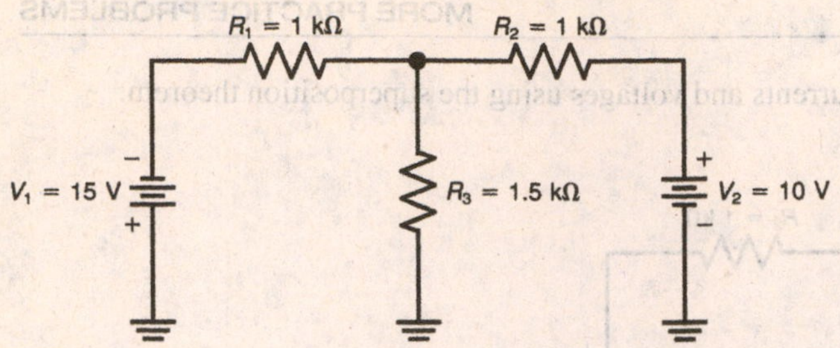

Fig. 10-17

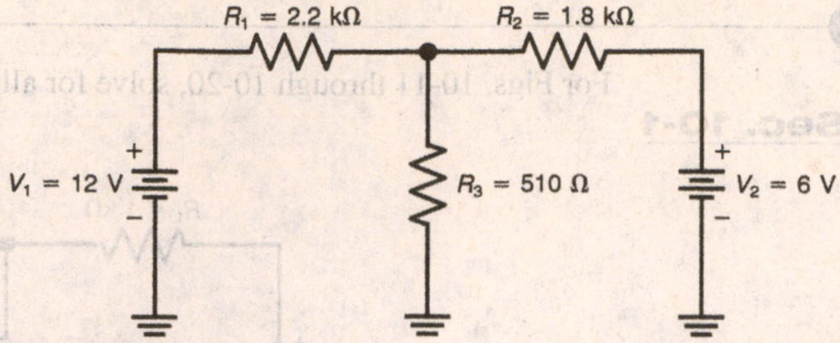

Fig. 10-18

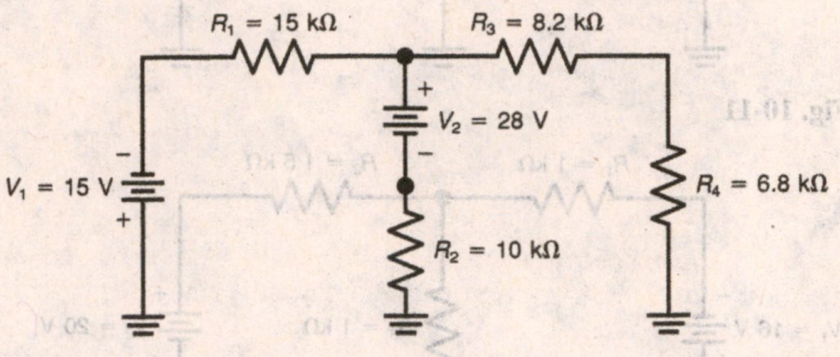

Fig. 10-19

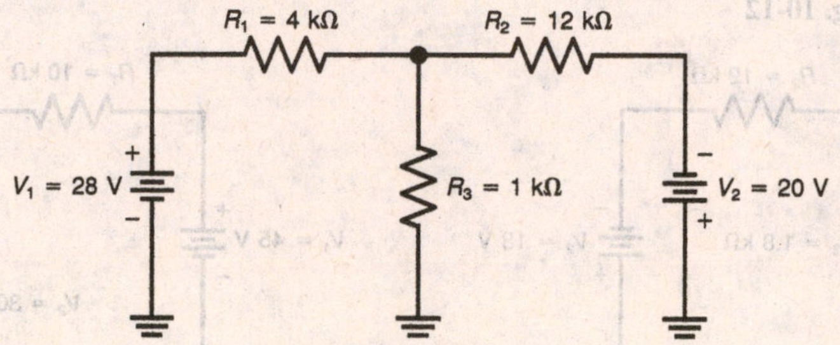

Fig. 10-20

SEC. 10-2 THEVENIN'S THEOREM

By Thevenin's theorem, many sources and components, no matter how they are interconnected, can be represented by an equivalent series circuit with respect to any pair of terminals in the network. This is shown in Fig. 10-21. The entire network connected to terminals A and B can be replaced by a single voltage source V_{TH} in series with a single resistance R_{TH}, connected to the same two terminals.

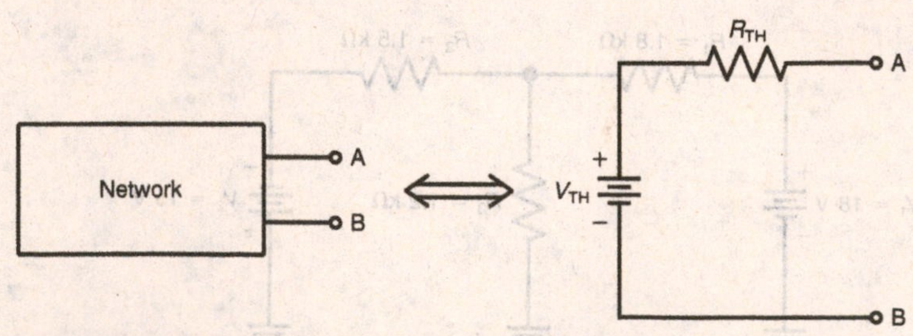

Fig. 10-21 Any network in the block shown at the left can be reduced to the Thevenin equivalent circuit shown at the right.

Solved Problem

For Fig. 10-22*a*, find the current I in R_L using Thevenin's theorem.

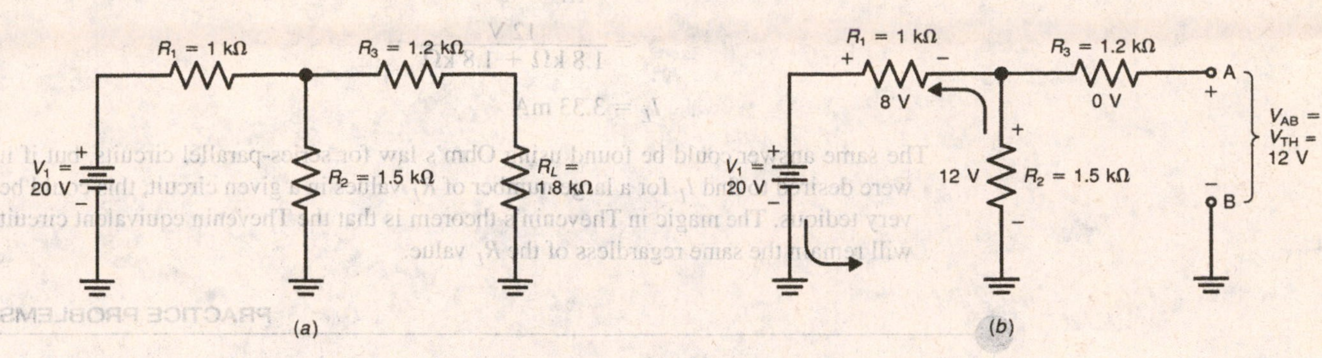

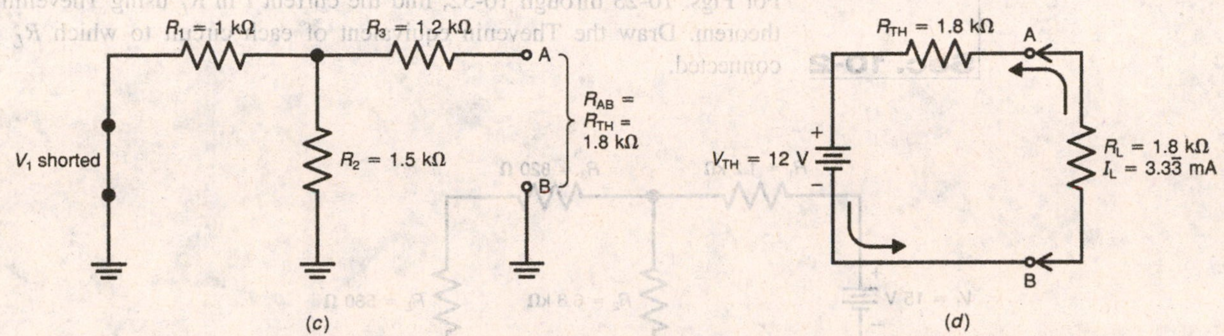

Fig. 10-22 Application of Thevenin's theorem. (*a*) Original circuit. (*b*) Determining the open-circuit voltage across terminals A and B after R_L has been removed. (*c*) Short-circuiting V_1 to find R_{AB}. (*d*) Thevenin equivalent circuit with R_L reconnected.

Answer

The first step required in thevenizing is to disconnect R_L from the circuit and label the points that were connected to R_L as A and B. This is shown in Fig. 10-22*b*.

The second step involves finding the open-circuit voltage across terminals A and B. This is the Thevenin equivalent voltage labeled V_{TH}. For Fig. 10-22*b*, $V_{AB} = V_{TH}$.

$$V_{AB} = V_{TH} = \frac{R_2}{R_1 + R_2} \times V_1$$

$$= \frac{1.5\ \text{k}\Omega}{1\ \text{k}\Omega + 1.5\ \text{k}\Omega} \times 20\ \text{V}$$

$$V_{TH} = 12\ \text{V}$$

Notice that R_3 does not affect V_{TH} for this condition, since no current flows through it. With $V_{R_3} = 0$ V, the voltage appearing from both the left side to ground and right side to ground will be the same (12 V).

The third step is to find the resistance, seen when looking back into terminals A and B, with the source voltage V_1 shorted. This is the Thevenin equivalent resistance R_{TH}. This condition is shown in Fig. 10-22*c*. For this circuit, $R_{AB} = R_{TH}$.

$$R_{TH} = R_3 + (R_1 \| R_2)$$

$$= 1.2\ \text{k}\Omega + (1\ \text{k}\Omega \| 1.5\ \text{k}\Omega)$$

$$R_{TH} = 1.8\ \text{k}\Omega$$

The final step is to replace the circuit connected to terminals A and B in Fig. 10-22*b* with its Thevenin equivalent. We can then reconnect R_L across terminals A and B and find I in R_L. This is shown in Fig. 10-22*d*.

$$I_L = \frac{V_{TH}}{R_{TH} + R_L}$$

$$= \frac{12\ \text{V}}{1.8\ \text{k}\Omega + 1.8\ \text{k}\Omega}$$

$$I_L = 3.33\ \text{mA}$$

The same answer could be found using Ohm's law for series-parallel circuits, but if it were desired to find I_L for a large number of R_L values in a given circuit, this could be very tedious. The magic in Thevenin's theorem is that the Thevenin equivalent circuit will remain the same regardless of the R_L value.

PRACTICE PROBLEMS

Sec. 10-2

For Figs. 10-23 through 10-32, find the current I in R_L using Thevenin's theorem. Draw the Thevenin equivalent of each circuit to which R_L is connected.

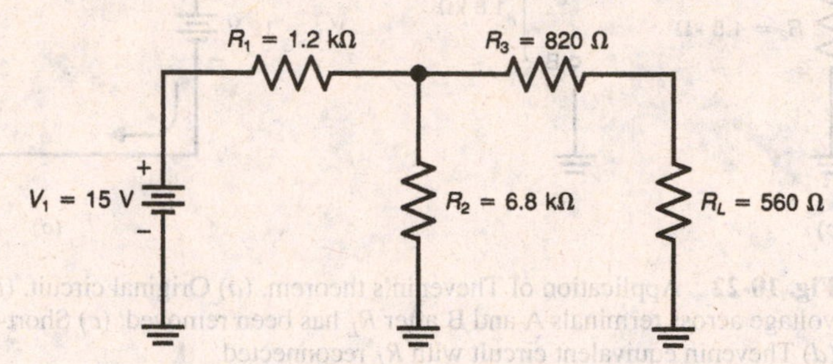

Fig. 10-23

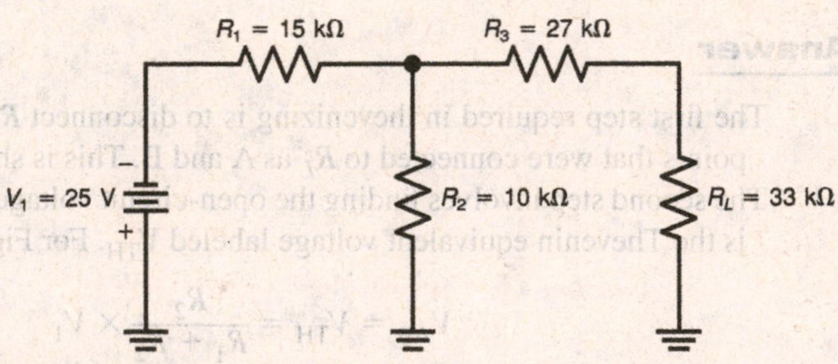

Fig. 10-24

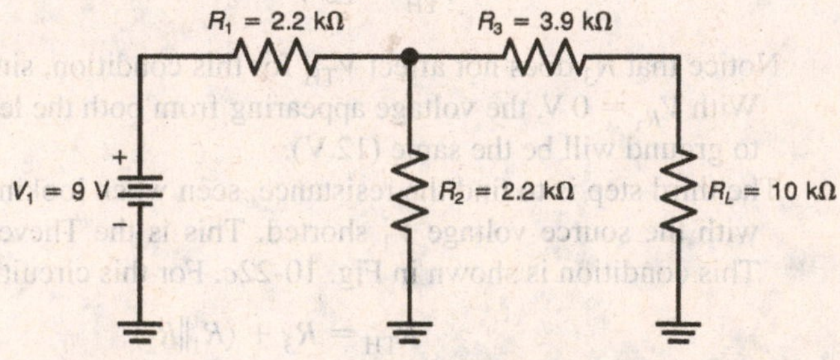

Fig. 10-25

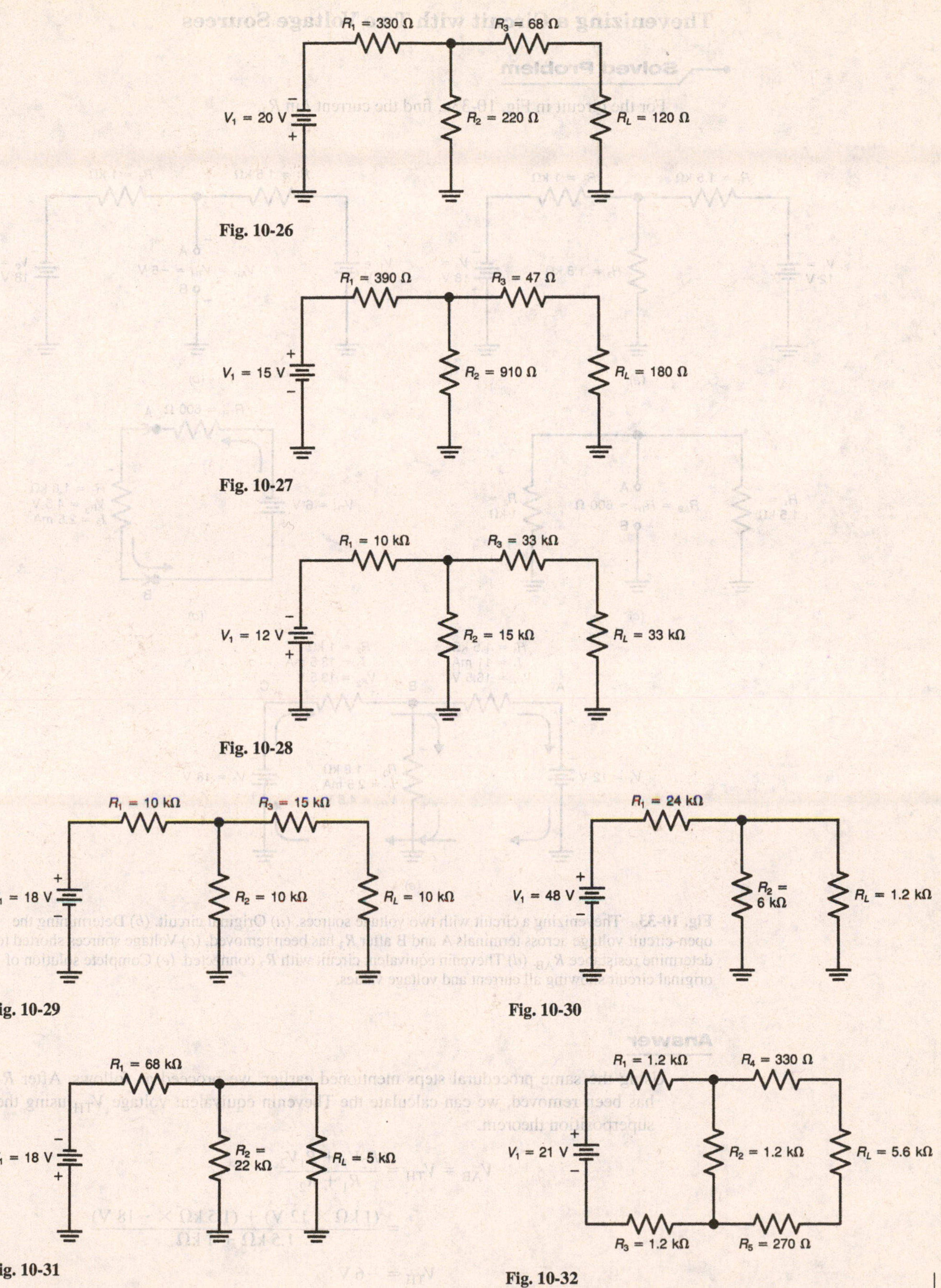

Fig. 10-26

Fig. 10-27

Fig. 10-28

Fig. 10-29

Fig. 10-30

Fig. 10-31

Fig. 10-32

Thevenizing a Circuit with Two Voltage Sources

Solved Problem

For the circuit in Fig. 10-33a, find the current I in R_3.

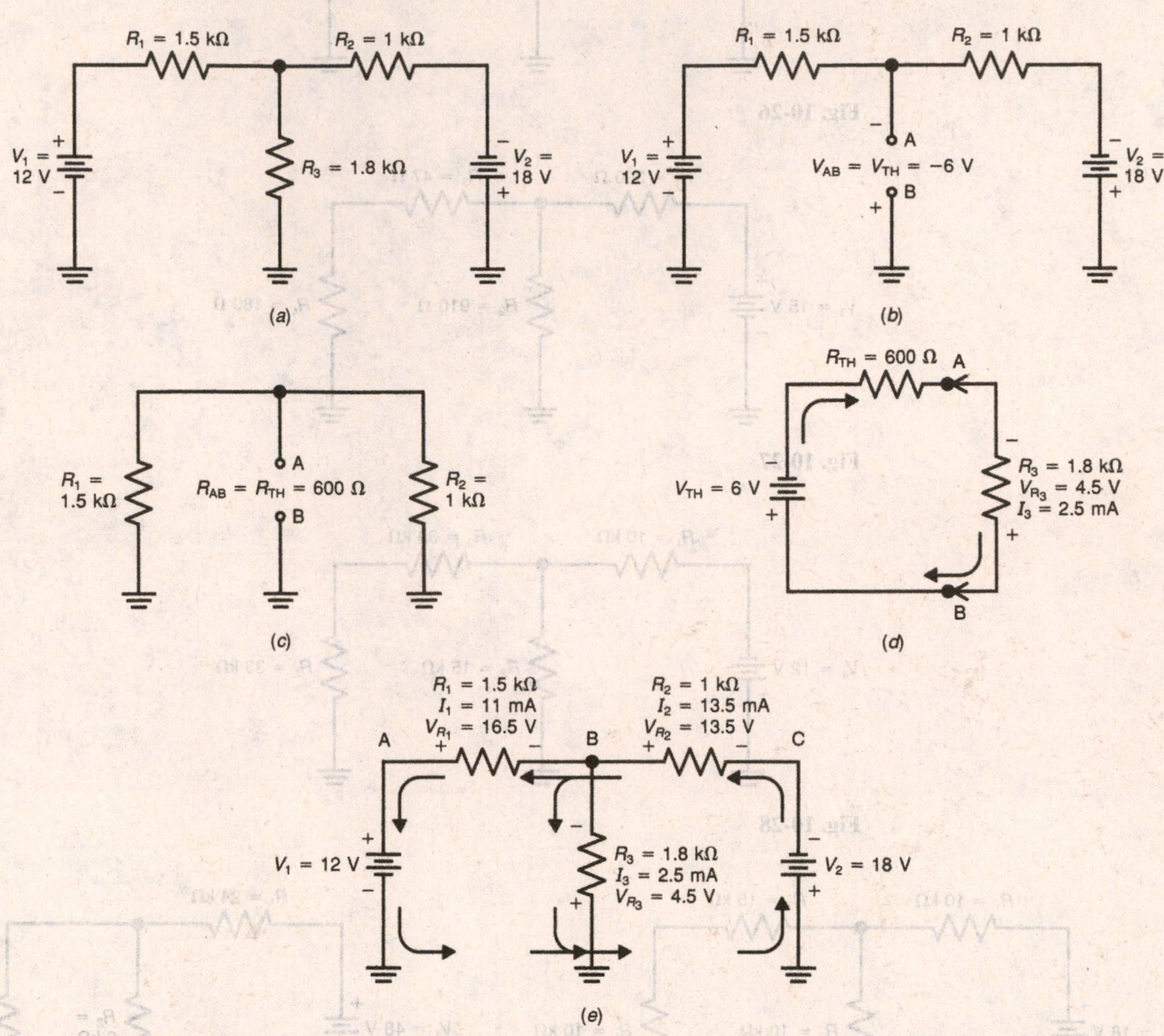

Fig. 10-33 Thevenizing a circuit with two voltage sources. (a) Original circuit. (b) Determining the open-circuit voltage across terminals A and B after R_3 has been removed. (c) Voltage sources shorted to determine resistance R_{AB}. (d) Thevenin equivalent circuit with R_3 connected. (e) Complete solution of original circuit showing all current and voltage values.

Answer

Using the same procedural steps mentioned earlier, we proceed as follows. After R_3 has been removed, we can calculate the Thevenin equivalent voltage V_{TH} using the superposition theorem.

$$V_{AB} = V_{TH} = \frac{R_2V_1 + R_1V_2}{R_1 + R_2}$$

$$= \frac{(1\text{ k}\Omega \times 12\text{ V}) + (1.5\text{ k}\Omega \times -18\text{ V})}{1.5\text{ k}\Omega + 1\text{ k}\Omega}$$

$$V_{TH} = -6\text{ V}$$

This is shown in Fig. 10-33b.

To calculate the Thevenin equivalent resistance R_{TH}, short-circuit V_1 and V_2 as shown in Fig. 10-33*c*. Notice that R_1 and R_2 are in parallel.

$$R_{TH} = \frac{R_1 \times R_2}{R_1 + R_2}$$

$$= \frac{1.5\ k\Omega \times 1\ k\Omega}{1.5\ k\Omega + 1\ k\Omega}$$

$$R_{TH} = 600\ \Omega$$

The final result is the Thevenin equivalent in Fig. 10-33*d* with a V_{TH} of 6 V and an R_{TH} of 600 Ω; V_{TH} produces a current through the total resistance of 600 Ω for R_{TH} and 1.8 kΩ for R_3.

$$I_3 = \frac{V_{TH}}{R_{TH} + R_3}$$

$$= \frac{6\ V}{600\ \Omega + 1.8\ k\Omega}$$

$$I_3 = 2.5\ mA$$

Notice that I_3 flows through R_3 from top to bottom.
It is important to realize that with I_3 known, V_{R_3} can be easily calculated.

$$V_{R_3} = I_3 \times R_3$$

$$= 2.5\ mA \times 1.8\ k\Omega$$

$$V_{R_3} = 4.5\ V$$

With V_{R_3} known, Kirchhoff's voltage law can be applied to solve for V_{R_1} and V_{R_2}. With V_{R_1} and V_{R_2} known, the solutions for I_1 and I_2 are also possible. Refer to Fig. 10-33*e*.

$$V_{R_1} = V_1 + V_{R_3}\ \text{(CCW from point A to B)}$$

$$= 12\ V + 4.5\ V$$

$$V_{R_1} = 16.5\ V$$

$$I_1 = \frac{V_{R1}}{R_1}$$

$$= \frac{16.5\ V}{1.5\ k\Omega}$$

$$I_1 = 11\ mA$$

$$V_{R_2} = V_2 + V_{R_3}\ \text{(CW from point C to B)}$$

$$= -18\ V + 4.5\ V$$

$$V_{R_2} = -13.5\ V$$

$$I_2 = \frac{V_{R_2}}{R_2}$$

$$= \frac{13.5\ V}{1\ k\Omega}$$

$$I_2 = 13.5\ mA$$

Notice that V_{R_1} was found by adding the voltages V_1 and V_{R_3} counterclockwise from points A to B, whereas V_{R_2} was found by adding the voltages V_2 and V_{R_3} clockwise from points C to B. The algebraic signs for V_{R_1} and V_{R_2} indicate the polarity of each voltage drop with respect to points A to B for V_{R_1} and points C to B for V_{R_2}.

Sec. 10-2

For Figs. 10-34 through 10-43, find all currents and voltages using Thevenin's theorem. Start by finding I_3 and V_{R_3} in each case. Also, draw the Thevenin equivalent of the circuit to which R_3 is connected.

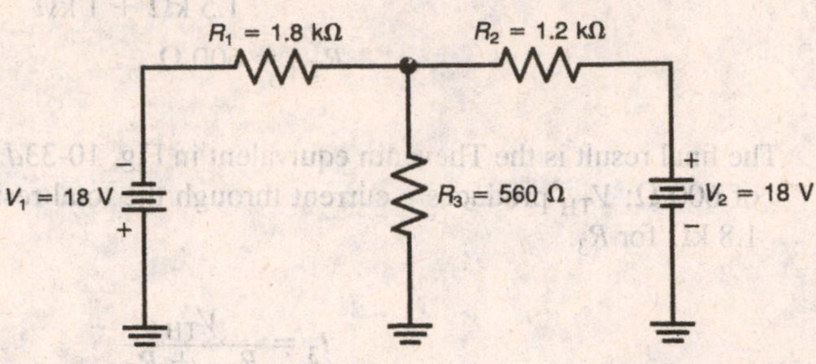

Fig. 10-34

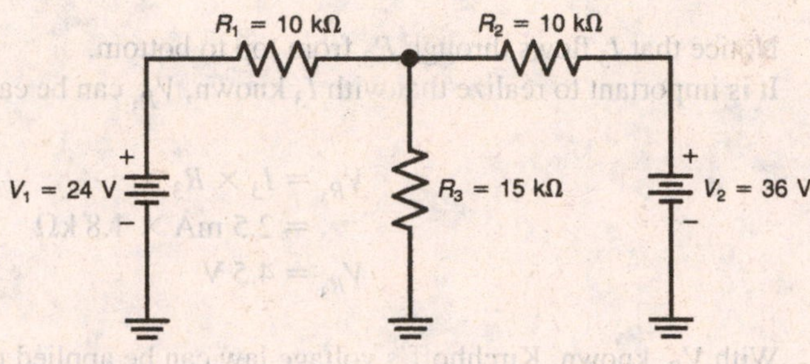

Fig. 10-35

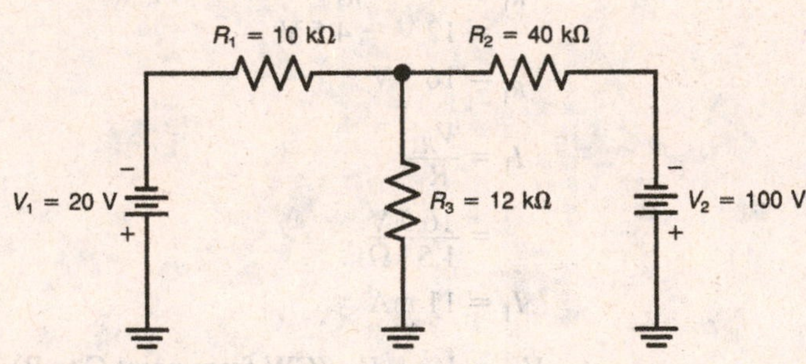

Fig. 10-36

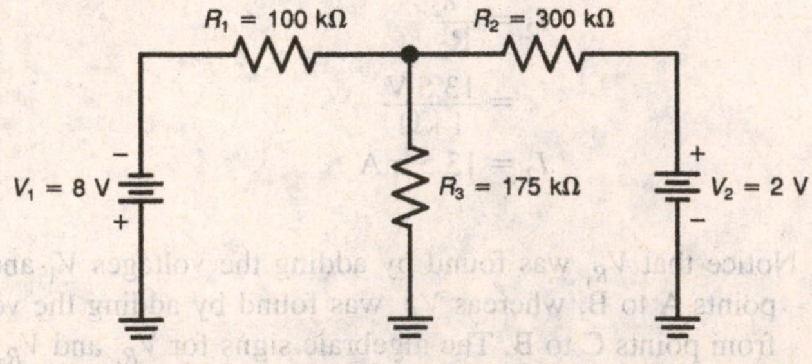

Fig. 10-37

R_1 = 50 Ω
R_2 = 150 Ω
V_1 = 10 V
R_3 = 25 Ω
V_2 = 25 V

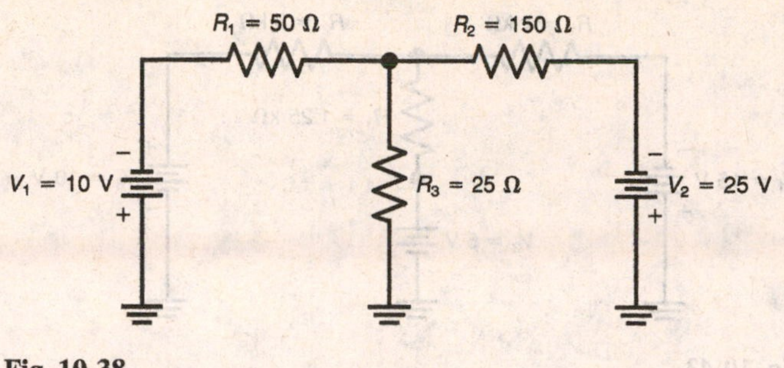

Fig. 10-38

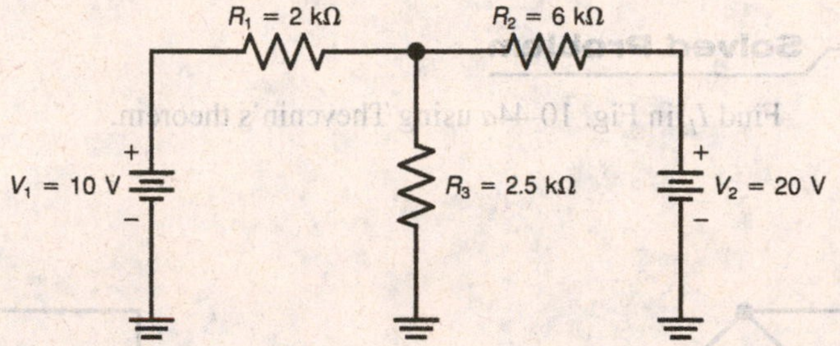

R_1 = 2 kΩ
R_2 = 6 kΩ
V_1 = 10 V
R_3 = 2.5 kΩ
V_2 = 20 V

Fig. 10-39

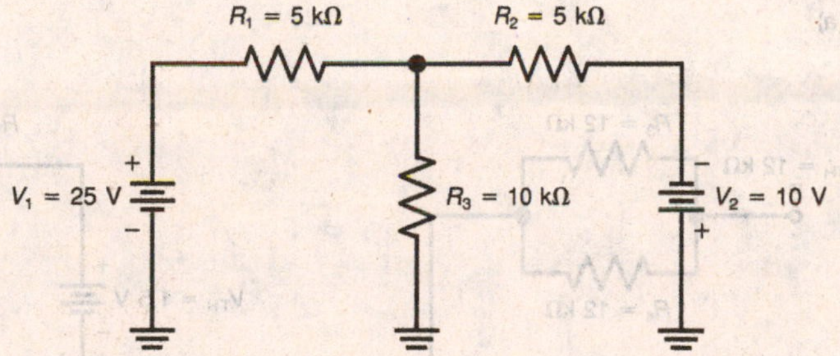

R_1 = 5 kΩ
R_2 = 5 kΩ
V_1 = 25 V
R_3 = 10 kΩ
V_2 = 10 V

Fig. 10-40

R_1 = 3.3 kΩ
R_2 = 4.7 kΩ
V_1 = 30 V
R_3 = 1.5 kΩ
V_2 = 60 V

Fig. 10-41

R_1 = 3.3 kΩ
R_2 = 2.2 kΩ
V_1 = 15 V
R_3 = 680 Ω
V_2 = 18 V

Fig. 10-42

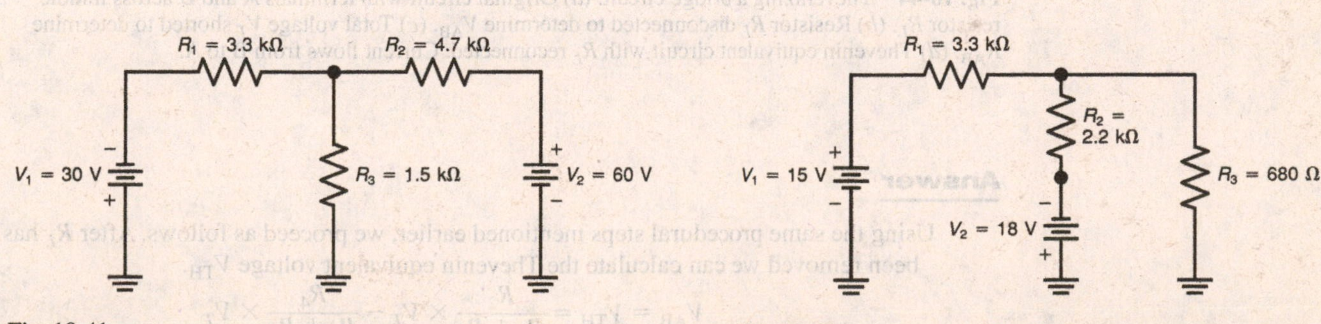

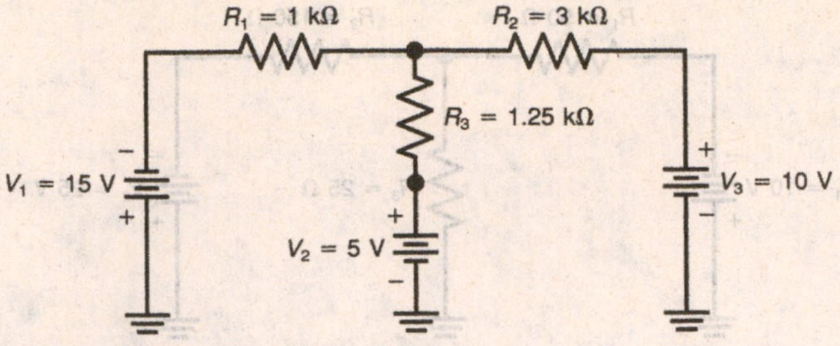

Fig. 10-43

Thevenizing a Bridge Circuit

Solved Problem

Find I_L in Fig. 10-44*a* using Thevenin's theorem.

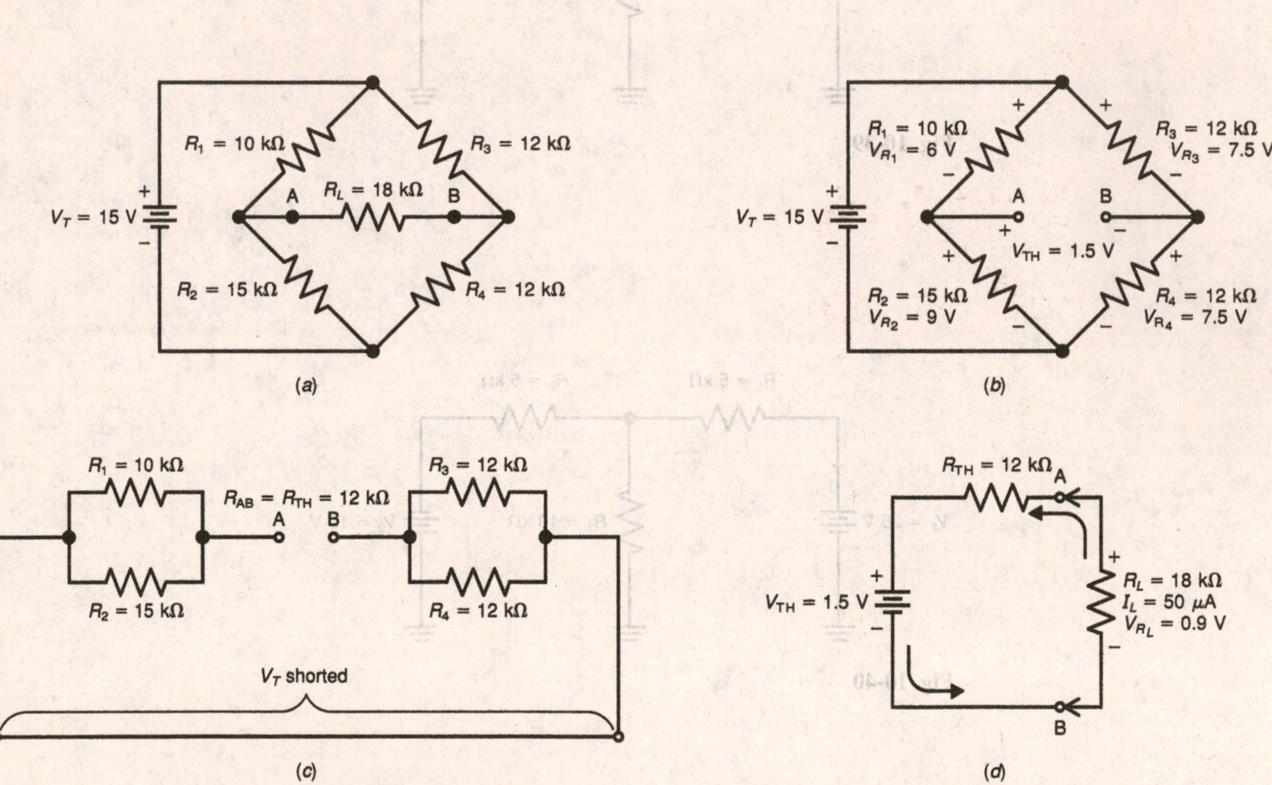

Fig. 10-44 Thevenizing a bridge circuit. (*a*) Original circuit with terminals A and B across middle resistor R_L. (*b*) Resistor R_L disconnected to determine V_{AB}. (*c*) Total voltage V_T shorted to determine R_{AB}. (*d*) Thevenin equivalent circuit with R_L reconnected. Current flows from B to A.

Answer

Using the same procedural steps mentioned earlier, we proceed as follows. After R_L has been removed we can calculate the Thevenin equivalent voltage V_{TH}.

$$V_{AB} = V_{TH} = \frac{R_2}{R_1 + R_2} \times V_T - \frac{R_4}{R_3 + R_4} \times V_T$$

Note that this is nothing more than $V_{R_2} - V_{R_4}$. Inserting the values from Fig. 10-44*a*, we have

$$V_{AB} = V_{TH} = \frac{15\text{ k}\Omega}{10\text{ k}\Omega + 15\text{ k}\Omega} \times 15\text{ V} - \frac{12\text{ k}\Omega}{12\text{ k}\Omega + 12\text{ k}\Omega} \times 15\text{ V}$$

$$= 9\text{ V} - 7.5\text{ V}$$

$$V_{TH} = 1.5\text{ V}$$

This is shown in Fig. 10-44*b*.

To calculate the Thevenin equivalent resistance R_{TH}, short-circuit V_T as shown in Fig. 10-44*c*. Notice that the parallel combination of R_1 and R_2 is in series with the parallel combination of R_3 and R_4.

$$R_{TH} = \frac{R_1 \times R_2}{R_1 + R_2} + \frac{R_3 \times R_4}{R_4 + R_4}$$

$$= \frac{10\text{ k}\Omega \times 15\text{ k}\Omega}{10\text{ k}\Omega + 15\text{ k}\Omega} + \frac{12\text{ k}\Omega \times 12\text{ k}\Omega}{12\text{ k}\Omega + 12\text{ k}\Omega}$$

$$R_{TH} = 12\text{ k}\Omega$$

The final result is shown in Fig. 10-44*d* with a V_{TH} of 1.5 V and an R_{TH} of 12 kΩ; V_{TH} produces a current through the total resistance of 12 kΩ for R_{TH} and 18 kΩ for R_L.

$$I_{R_L} = \frac{V_{TH}}{R_{TH} + R_L}$$

$$= \frac{1.5\text{ V}}{12\text{ k}\Omega + 18\text{ k}\Omega}$$

$$I_{R_L} = 50\ \mu\text{A}$$

$$V_{R_L} = I_L \times R_L$$

$$= 50\ \mu\text{A} \times 18\text{ k}\Omega$$

$$V_{R_L} = 0.9\text{ V}$$

Notice that current is flowing from B to A. This would be from the right to left in Fig. 10-44*a*, since point B is at the junction of R_3 and R_4.

EVEN MORE PRACTICE PROBLEMS

Sec. 10-2

For Figs. 10-45 through 10-56, find I_L and V_{R_L} using Thevenin's theorem. On your answer sheet, indicate with an arrow the direction of current flow through R_L, either ← or →.

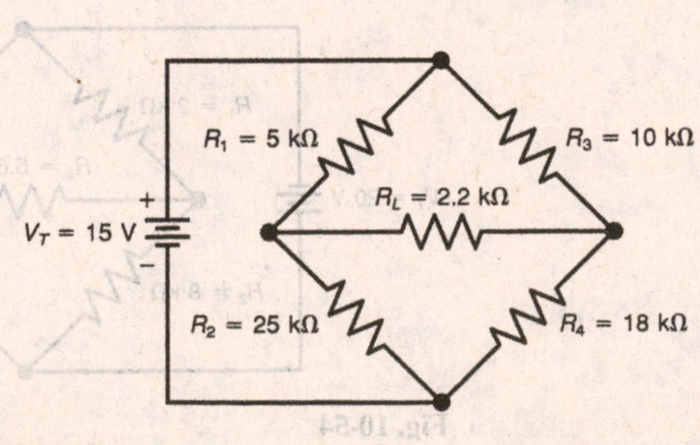

Fig. 10-45

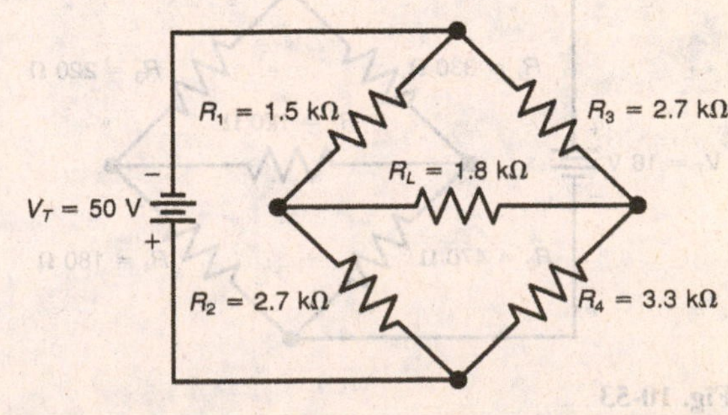

Fig. 10-46

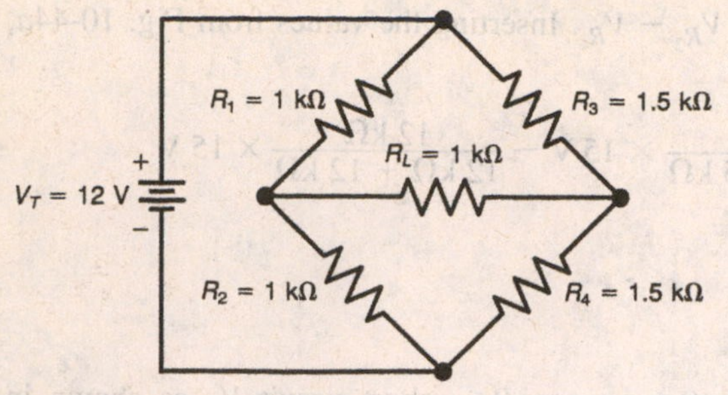

Fig. 10-47

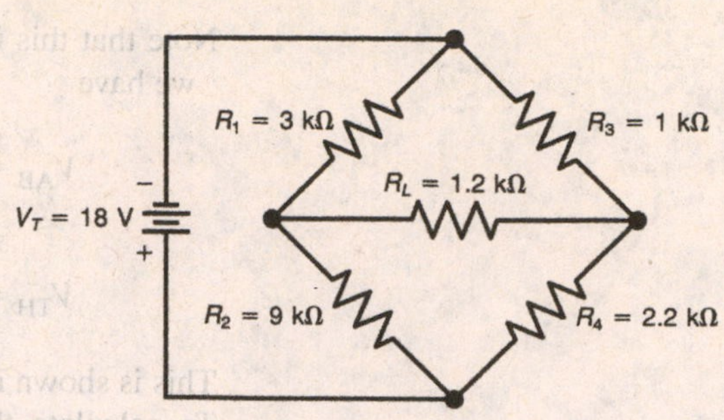

Fig. 10-48

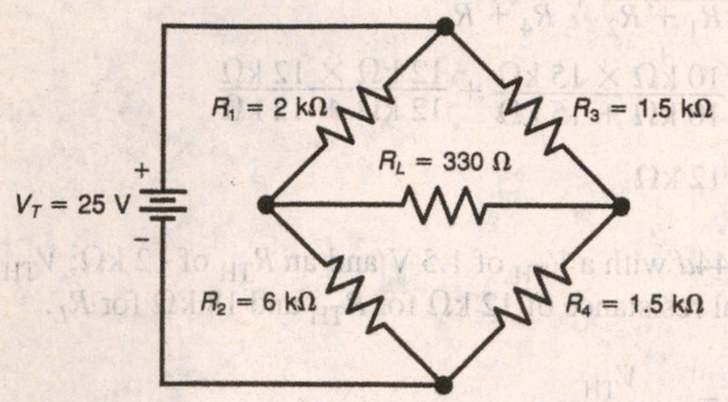

Fig. 10-49

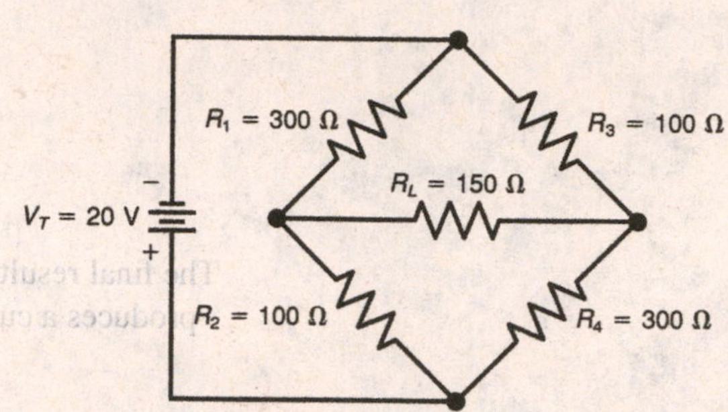

Fig. 10-50

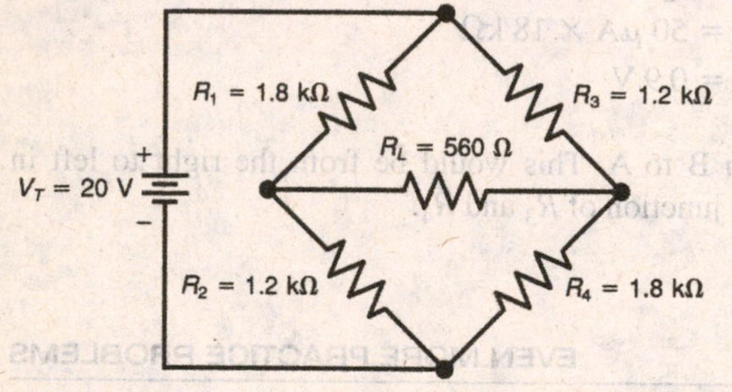

Fig. 10-51

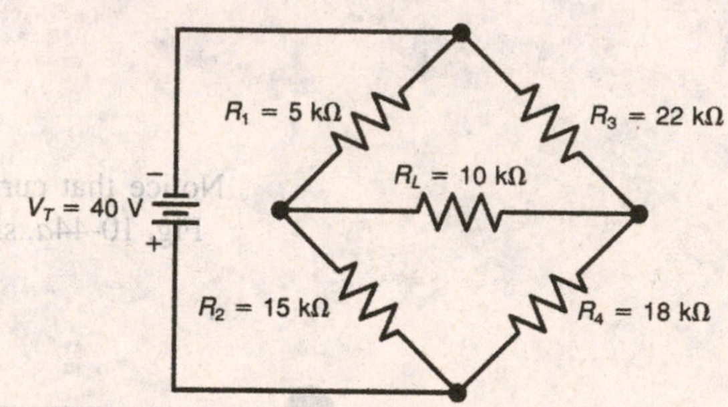

Fig. 10-52

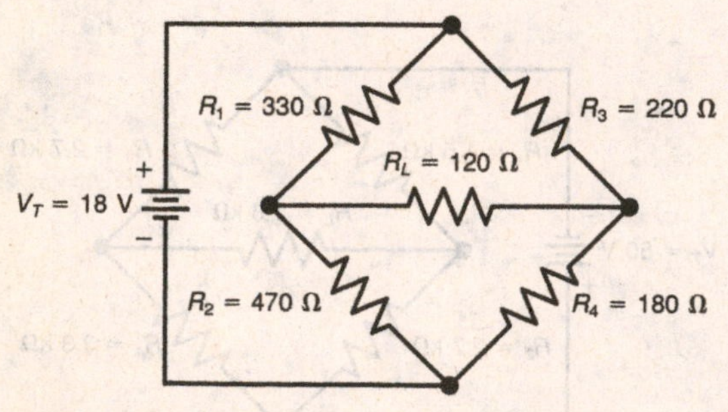

Fig. 10-53

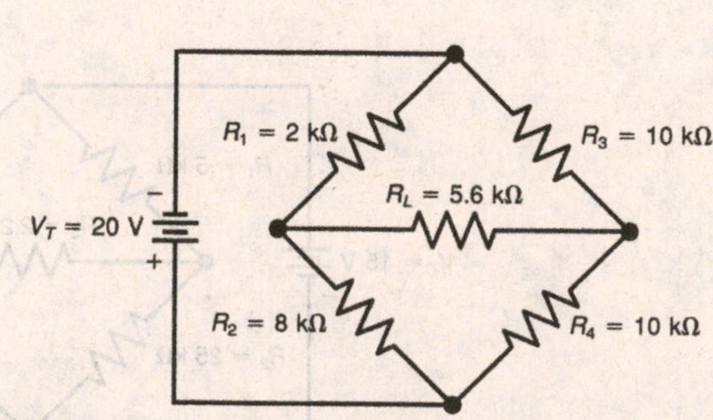

Fig. 10-54

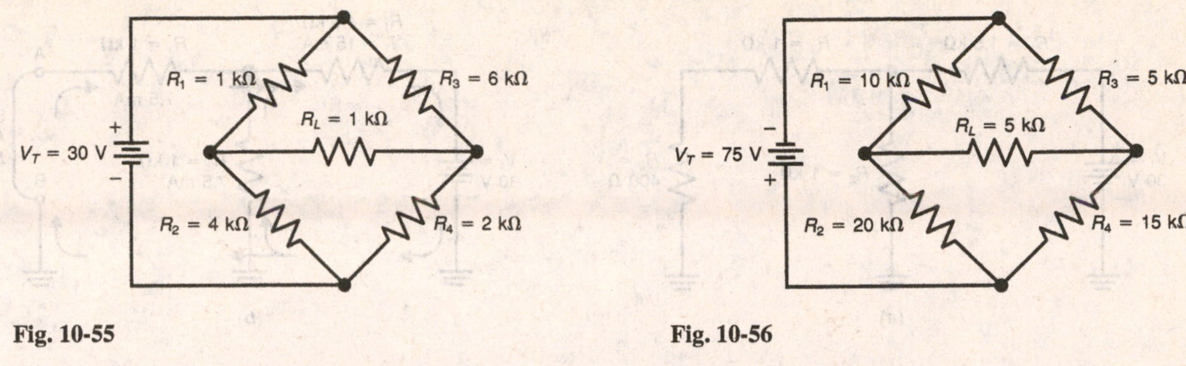

Fig. 10-55

Fig. 10-56

SEC. 10-3 NORTON'S THEOREM

By Norton's theorem, many sources and components, no matter how they are interconnected, can be represented by an equivalent circuit consisting of a constant-current generator, I_N, shunted by an internal resistance, R_N. Norton's theorem states that the entire network connected to terminals A and B in Fig. 10-57 can be replaced by a single current source in parallel with a single resistance, R_N, connected to the same two terminals. The dashed arrow in the current source points in the direction of electron flow.

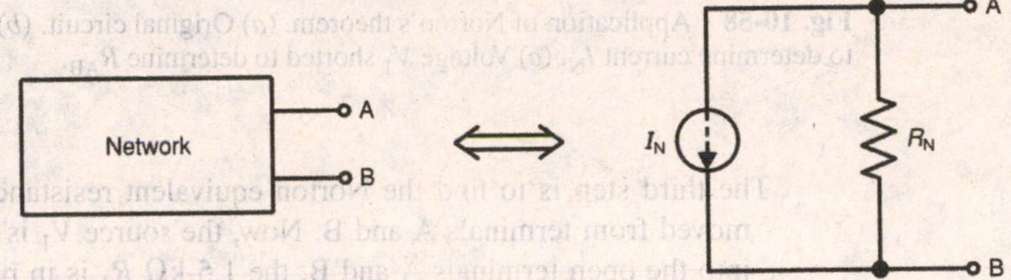

Fig. 10-57 Any network in the block at left can be reduced to the Norton equivalent circuit at right.

Solved Problem

For Fig. 10-58*a*, find the current I in R_L using Norton's theorem.

Answer

The first step required in nortonizing the circuit shown in Fig. 10-58*a* is to disconnect R_L from the circuit and label the points where R_L was connected as points A and B. This is shown in Fig. 10-58*b*.

The second step involves finding the short-circuit current through terminals A and B. The current flowing through the short-circuited terminals is the Norton equivalent current I_N. This current can be found using Ohm's law. See Fig. 10-58*b*.

$$I_T = \frac{V_1}{R_T}$$

$$= \frac{V_1}{R_1 + (R_2 \| R_3)}$$

$$= \frac{30\ \text{V}}{1.5\ \text{k}\Omega + 500\ \Omega}$$

$$I_T = 15\ \text{mA}$$

$$I_N = I_T - I_2$$

$$= 15\ \text{mA} - 7.5\ \text{mA}$$

$$I_N = 7.5\ \text{mA}$$

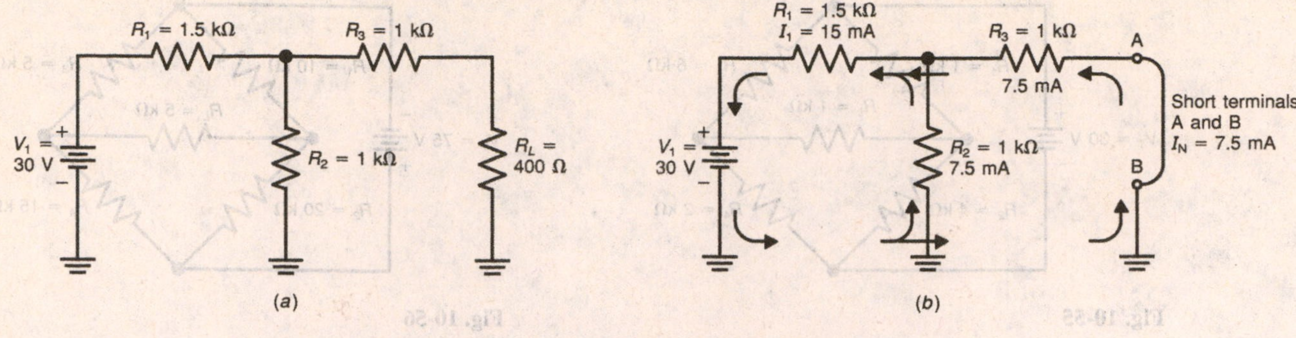

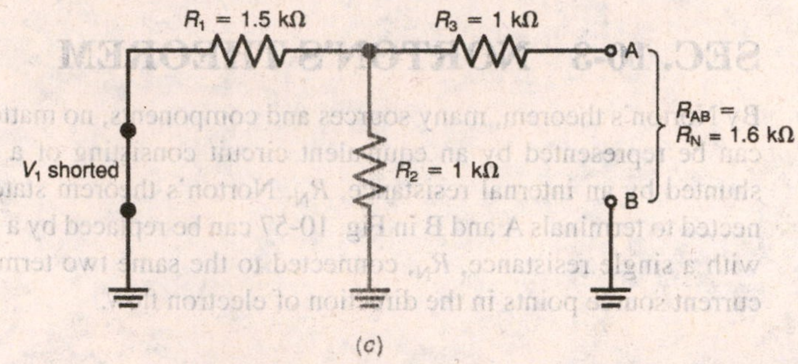

Fig. 10-58 Application of Norton's theorem. (*a*) Original circuit. (*b*) Short-circuit terminals A and B to determine current I_N. (*c*) Voltage V_1 shorted to determine R_{AB}.

The third step is to find the Norton equivalent resistance R_N. The short circuit is removed from terminals A and B. Now, the source V_1 is short-circuited. Looking back into the open terminals A and B, the 1.5-kΩ R_1 is in parallel with the 1-kΩ R_2. This combination has an equivalent resistance of

$$R_{EQ} = \frac{R_1 \times R_2}{R_1 + R_2}$$

$$= \frac{1.5\text{ k}\Omega \times 1\text{ k}\Omega}{1.5\text{ k}\Omega + 1\text{ k}\Omega}$$

$$R_{EQ} = 600\ \Omega$$

This 600 Ω is in series with the 1-kΩ R_3 to make $R_{AB} = R_N = 1.6$ kΩ. This is shown in Fig. 10-58*c*.

The fourth step is to replace the circuit connected to terminals A and B in Fig. 10-58*b* with its Norton equivalent. This is shown in Fig. 10-59*a*.

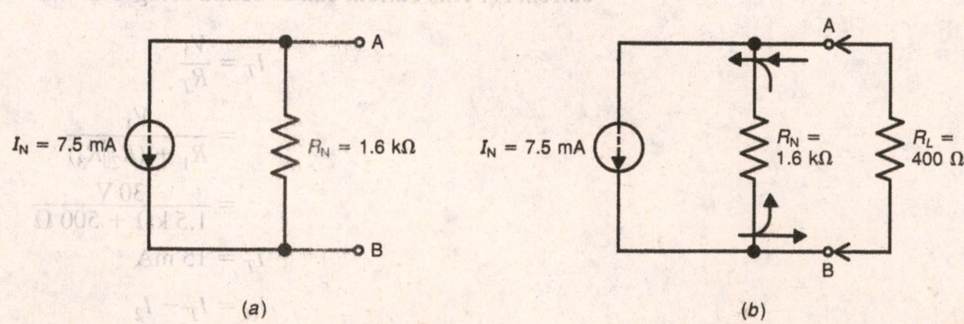

Fig. 10-59 (*a*) Norton equivalent circuit for the circuit connected to terminals A and B in Fig. 10-58*b*. (*b*) Norton equivalent circuit with R_L reconnected.

The final step is to reconnect R_L across terminals A and B as in Fig. 10-59*b* to solve for I_{R_L}.

$$I_{R_L} = \frac{R_N}{R_N + R_L} \times I_N$$
$$= \frac{1.6\ \text{k}\Omega}{1.6\ \text{k}\Omega + 400\ \Omega} \times 7.5\ \text{mA}$$
$$I_{R_L} = 6\ \text{mA}$$
$$V_{R_L} = I_{R_L} \times R_L$$
$$= 6\ \text{mA} \times 400\ \Omega$$
$$V_{R_L} = 2.4\ \text{V}$$

For current analysis, Norton's theorem can be used to reduce a network to a simple parallel circuit with a current source.

PRACTICE PROBLEMS

Sec. 10-3 For Figs. 10-60 through 10-67, find the current I in R_L using Norton's theorem. Draw the Norton equivalent of each circuit to which R_L is connected.

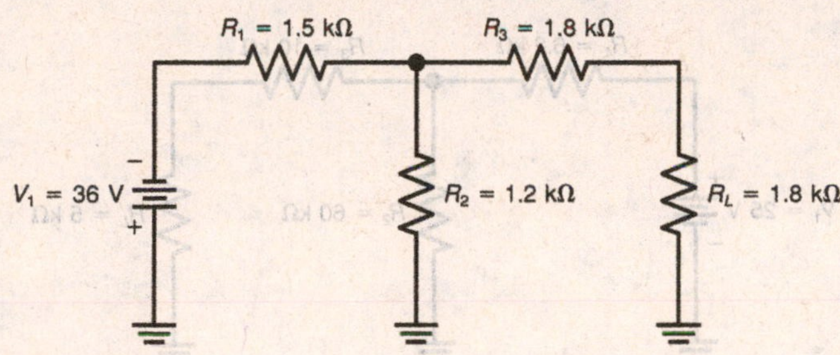

Fig. 10-60

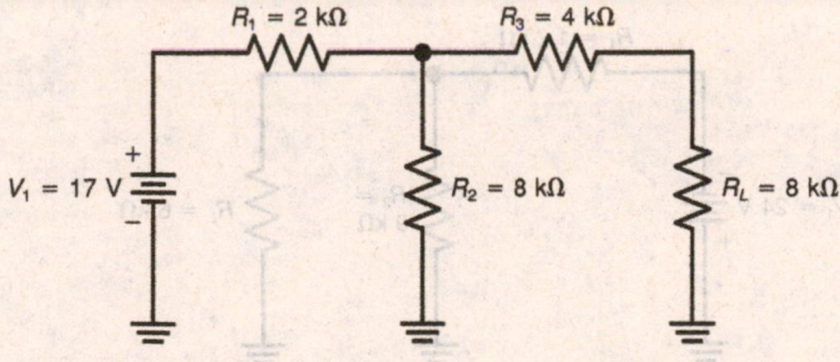

Fig. 10-61

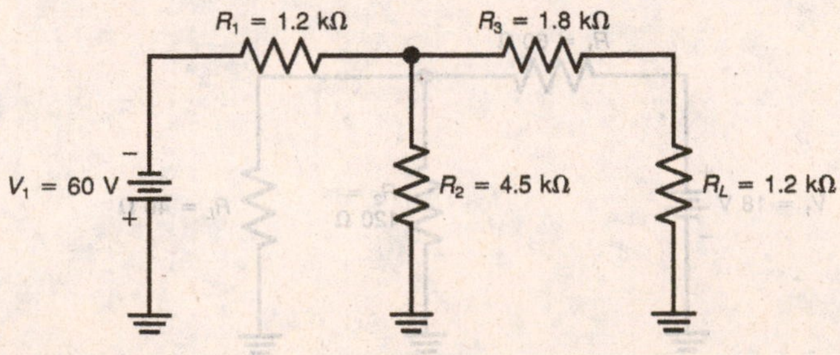

Fig. 10-62

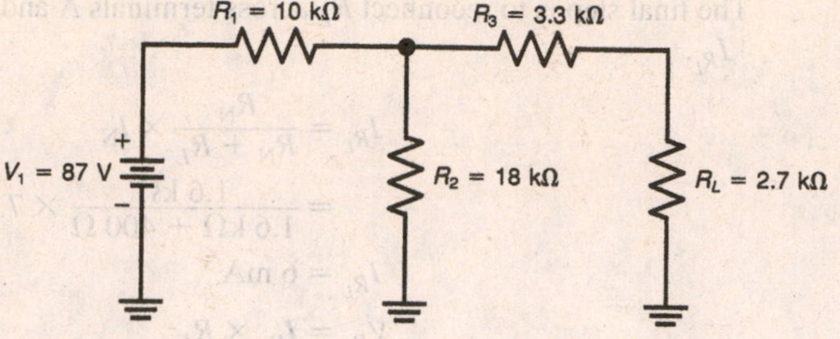

Fig. 10-63

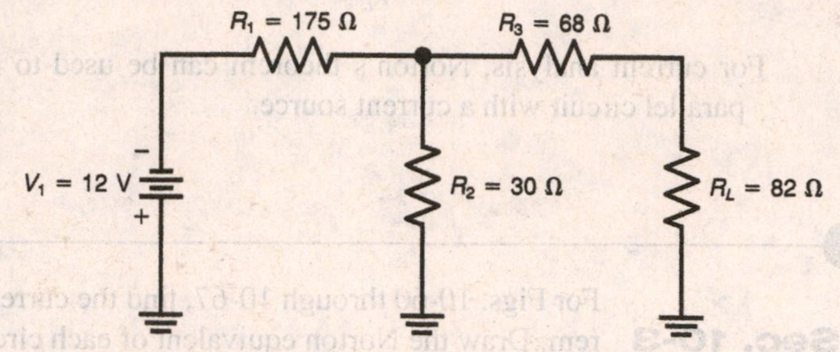

Fig. 10-64

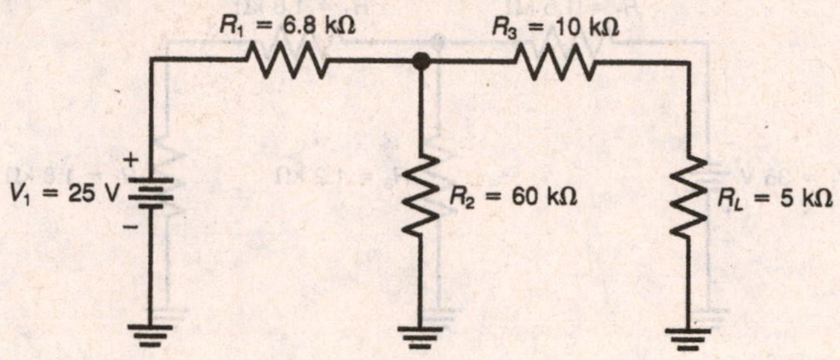

Fig. 10-65

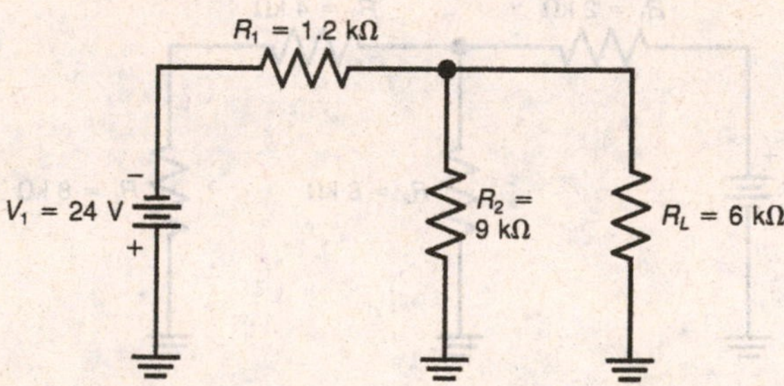

Fig. 10-66

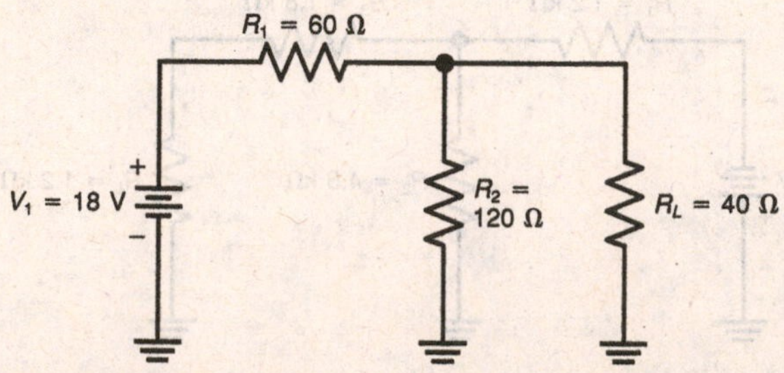

Fig. 10-67

Use Norton's theorem to solve for I_3 in Figs. 10-68 and 10-69. Draw the Norton equivalent of each circuit to which R_L is connected. (*Hint:* When A to B is shorted, the sources produce current flow independent of each other. The short-circuit current I_N is the algebraic sum of the source currents. Consider currents flowing from top to bottom through short-circuited terminals A and B as negative, and currents from bottom to top as positive.)

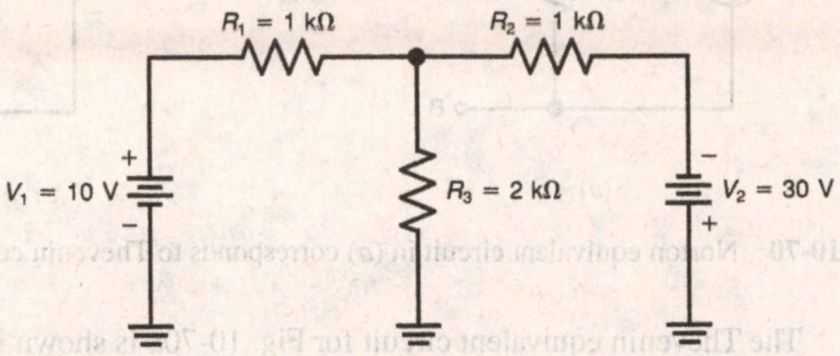

Fig. 10-68

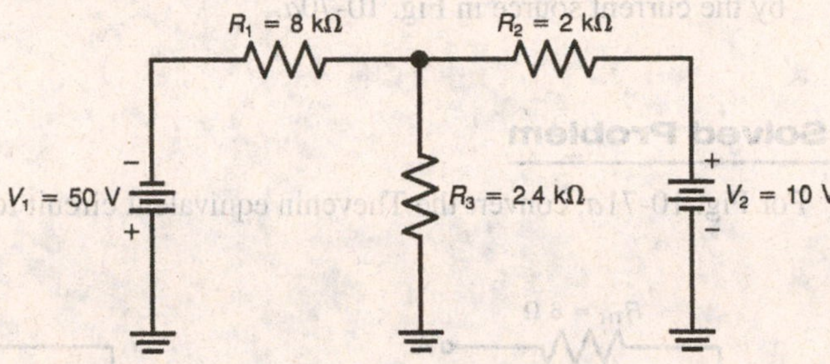

Fig. 10-69

SEC. 10-4 THEVENIN-NORTON CONVERSIONS

Thevenin's theorem says that any network can be represented by a voltage source and series resistance, whereas Norton's theorem says that the same network can be represented by a current source and shunt resistance. It is possible, therefore, to convert directly from a Thevenin form to a Norton form and vice versa. To convert from one form to another, use the following conversion formulas.

Thevenin from Norton:

$$R_{TH} = R_N$$
$$V_{TH} = I_N \times R_N$$

Norton from Thevenin:

$$R_N = R_{TH}$$
$$I_N = \frac{V_{TH}}{R_{TH}}$$

Solved Problem

For Fig. 10-70*a*, convert the Norton equivalent circuit to its Thevenin equivalent.

Answer

$$R_{TH} = R_N$$
$$R_{TH} = 200\ \Omega$$

$$V_{TH} = I_N \times R_N$$
$$= 0.5\text{ A} \times 200\ \Omega$$
$$V_{TH} = 100\text{ V}$$

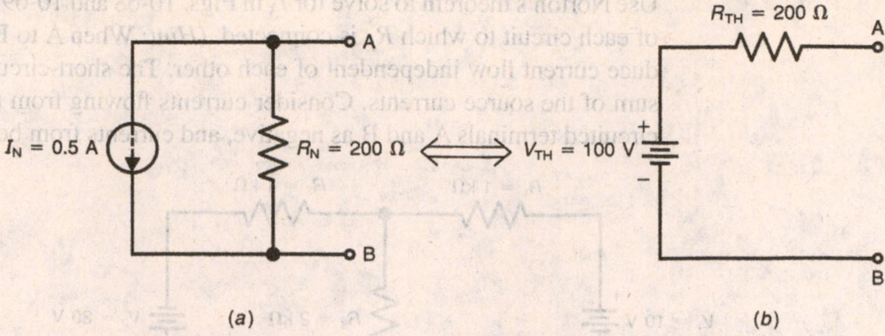

Fig. 10-70 Norton equivalent circuit in (*a*) corresponds to Thevenin equivalent circuit in (*b*).

The Thevenin equivalent circuit for Fig. 10-70*a* is shown in Fig. 10-70*b*. Notice that the polarity of V_{TH} in Fig. 10-70*b* corresponds to the direction of electron flow produced by the current source in Fig. 10-70*a*.

Solved Problem

For Fig. 10-71*a*, convert the Thevenin equivalent circuit to its Norton equivalent.

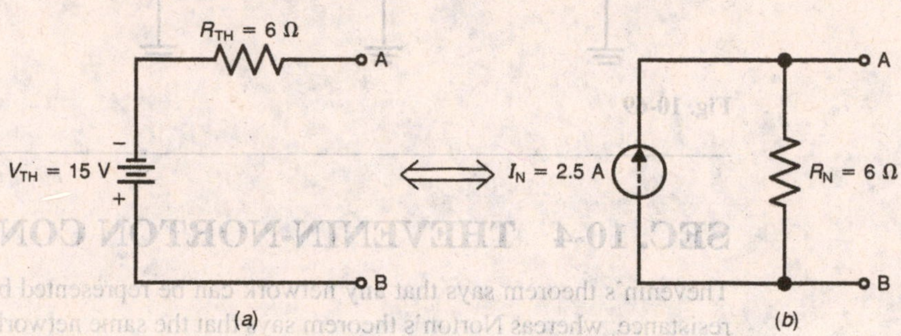

Fig. 10-71 Thevenin equivalent circuit in (*a*) corresponds to Norton equivalent circuit in (*b*).

Answer

$$R_N = R_{TH}$$
$$R_N = 6\ \Omega$$
$$I_N = \frac{V_{TH}}{R_{TH}}$$
$$= \frac{15\ \text{V}}{6\ \Omega}$$
$$I_N = 2.5\ \text{A}$$

The Norton equivalent circuit for Fig. 10-71*a* is shown in Fig. 10-71*b*. Notice the direction of electron flow in Fig. 10-71*b*. This produces current in the same direction as V_{TH} in Fig. 10-71*a*.

PRACTICE PROBLEMS

Sec. 10-4

Convert the Norton equivalent circuits shown in Figs. 10-72 through 10-75 to their Thevenin equivalents. Draw the equivalent circuits showing all values.

Convert the Thevenin equivalent circuits in Figs. 10-76 through 10-79 to their Norton equivalents. Draw the equivalent circuits showing all values.

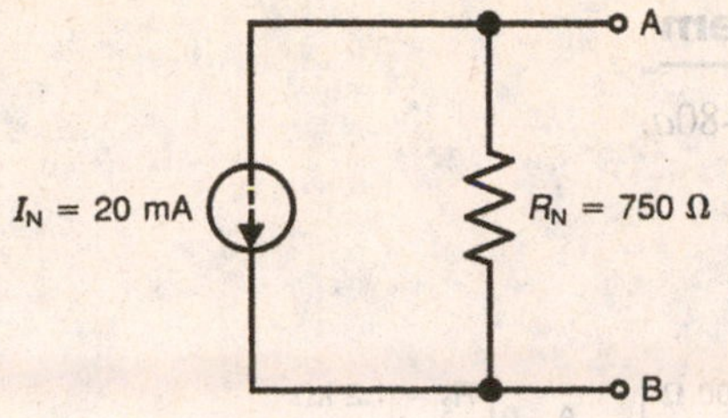

Fig. 10-72

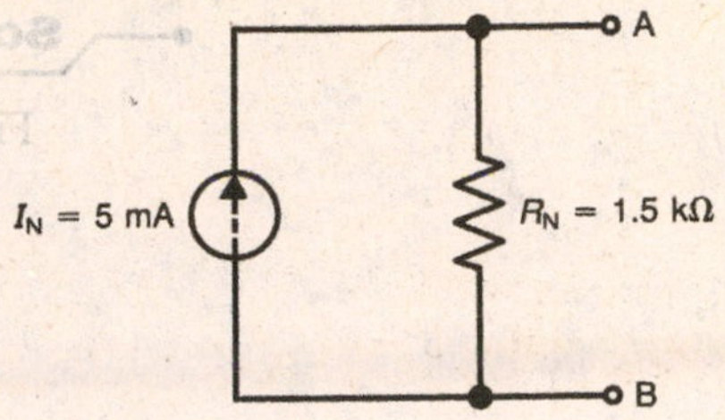

Fig. 10-73

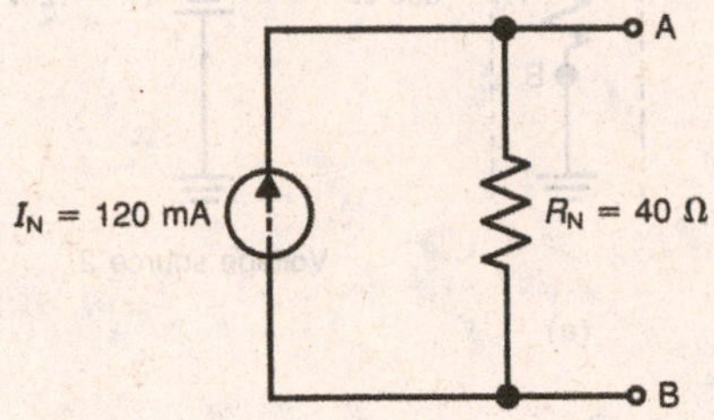

Fig. 10-74

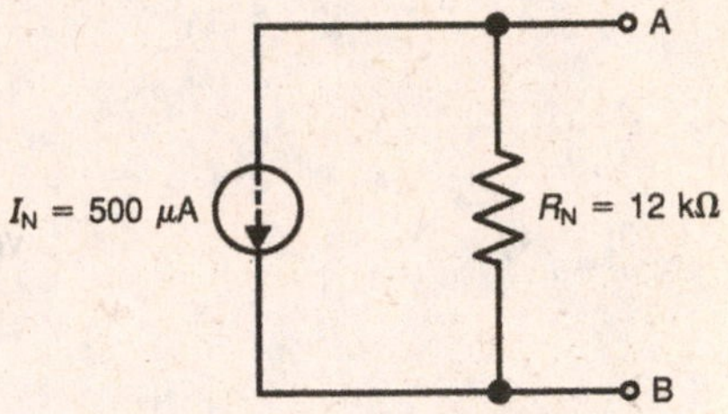

Fig. 10-75

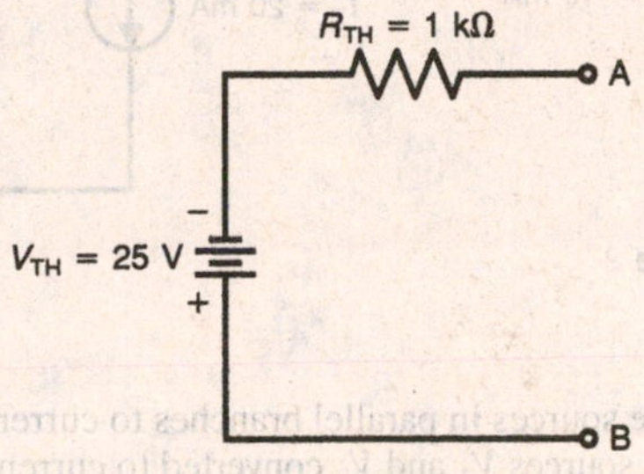

Fig. 10-76

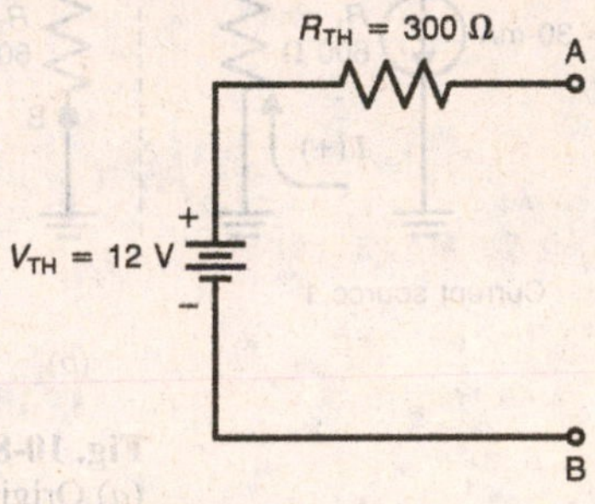

Fig. 10-77

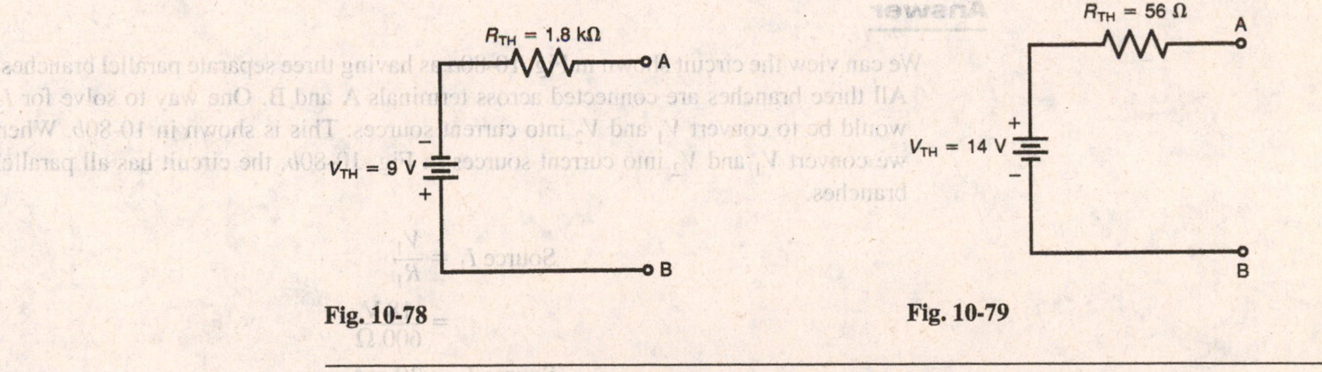

Fig. 10-78

Fig. 10-79

SEC. 10-5 CONVERSION OF VOLTAGE AND CURRENT SOURCES

Conversion of voltage and current sources can often simplify circuits, especially those with two or more sources. Current sources are easier for parallel connections where we can add or divide currents. Voltage sources are easier for series connections where we can add or divide voltages.

Solved Problem

Find I_3 in Fig. 10-80*a*.

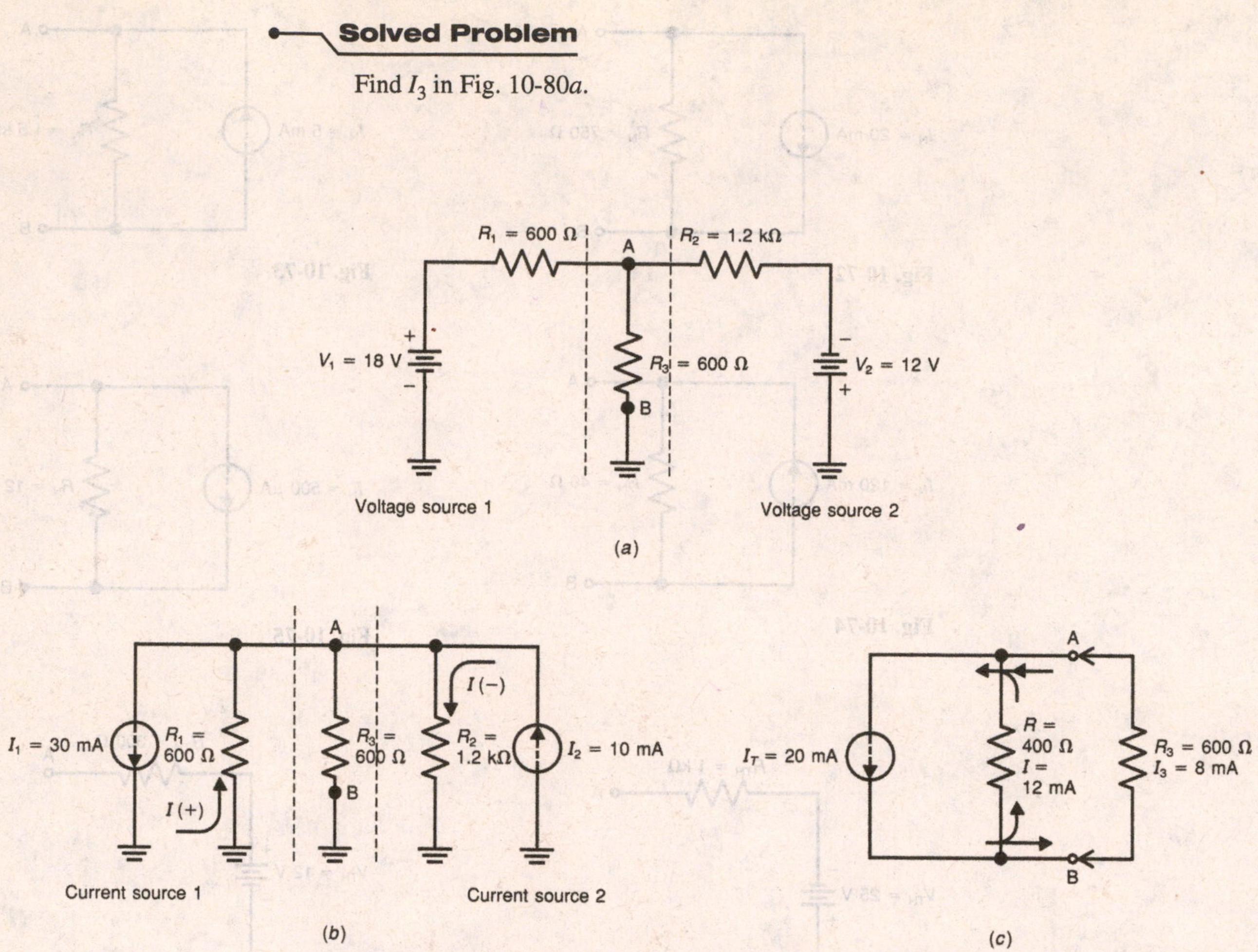

Fig. 10-80 Converting voltage sources in parallel branches to current sources that can be combined. (*a*) Original circuit. (*b*) Voltage sources V_1 and V_2 converted to current sources I_1 and I_2. (*c*) Combined equivalent current source with its shunt *R*. With R_L connected as shown, $I_3 = 8$ mA.

Answer

We can view the circuit shown in Fig. 10-80*a* as having three separate parallel branches. All three branches are connected across terminals A and B. One way to solve for I_3 would be to convert V_1 and V_2 into current sources. This is shown in 10-80*b*. When we convert V_1 and V_2 into current sources in Fig. 10-80*b*, the circuit has all parallel branches.

$$\text{Source } I_1 = \frac{V_1}{R_1}$$

$$= \frac{18\text{ V}}{600\ \Omega}$$

$$\text{Source } I_1 = 30\text{ mA}$$

$$\text{Source } I_2 = \frac{V_2}{R_2}$$

$$= \frac{12\text{ V}}{1.2\text{ k}\Omega}$$

$$\text{Source } I_2 = -10\text{ mA}$$

Since the source current I_1 produces current flow from bottom to top through its shunt *R*, we call it positive. Since I_2 produces current flow from top to bottom through its shunt R, we label it as negative. Current sources I_1 and I_2 can be combined for the one equivalent current source shown in Fig. 10-80*c*.

$$I_T = I_1 + I_2$$

For Fig. 10-80*b*, we have

$$I_T = 30 \text{ mA} - 10 \text{ mA}$$
$$I_T = 20 \text{ mA}$$

The shunt R of each current source shown in Fig. 10-80*b* can also be combined as one resultant parallel path.

$$\frac{R_1 \times R_2}{R_1 + R_2} = \frac{600\ \Omega \times 1200\ \Omega}{600\ \Omega + 1200\ \Omega}$$
$$\frac{R_1 \times R_2}{R_1 + R_2} = 400\ \Omega$$

The equivalent circuit for Fig. 10-80*a* and 10-80*b* is shown in Fig. 10-80*c*.
For Fig. 10-80*c*, we have

$$I_3 = \frac{R}{R + R_3} \times I_T$$
$$= \frac{400\ \Omega}{400\ \Omega + 600\ \Omega} \times 20 \text{ mA}$$
$$I_3 = 8 \text{ mA}$$

A negative value for I_T would indicate that the current source is producing current flow downward through R and R_3.

PRACTICE PROBLEMS

Sec. 10-5

Solve for I_3 in Figs. 10-81 through 10-84. Indicate the value of current source I_1 and current source I_2, as well as the combined equivalent current source I_T. Indicate also the combined shunt resistance the two will have.

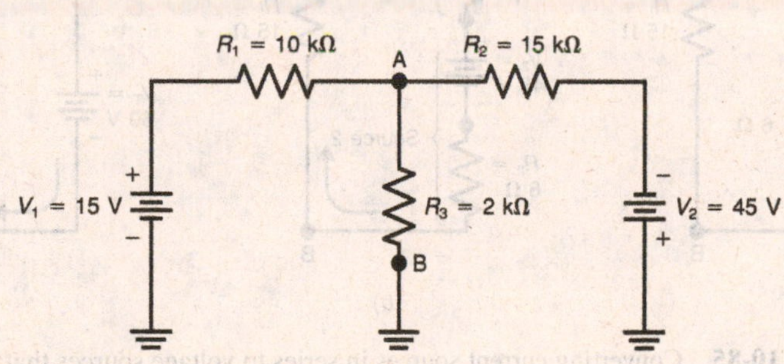

Fig. 10-81

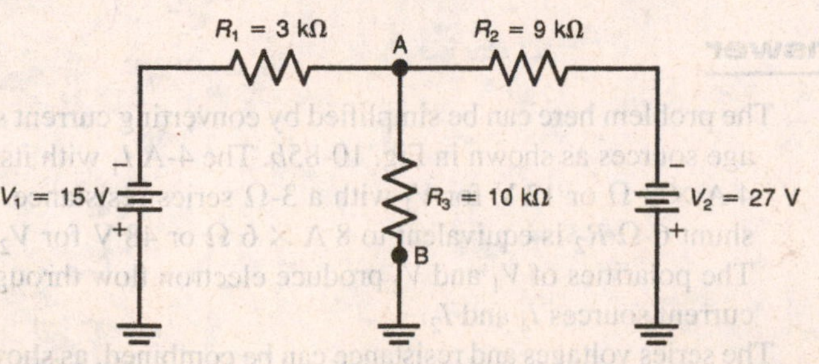

Fig. 10-82

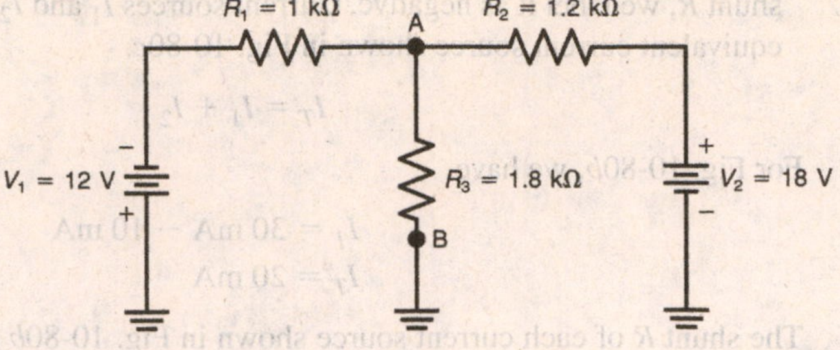

Fig. 10-83

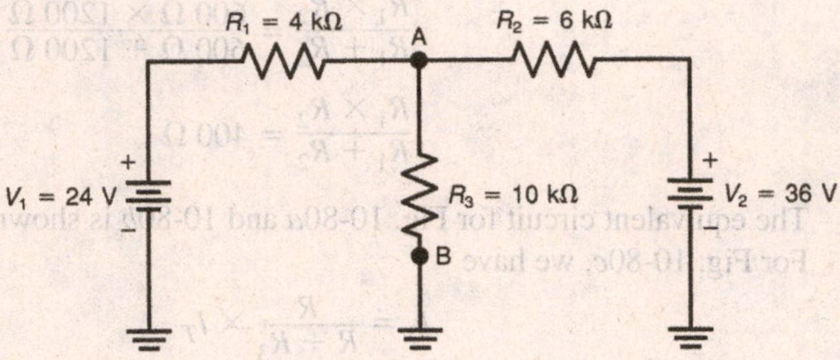

Fig. 10-84

Solved Problem

Find the current I in R_L for Fig. 10-85*a*.

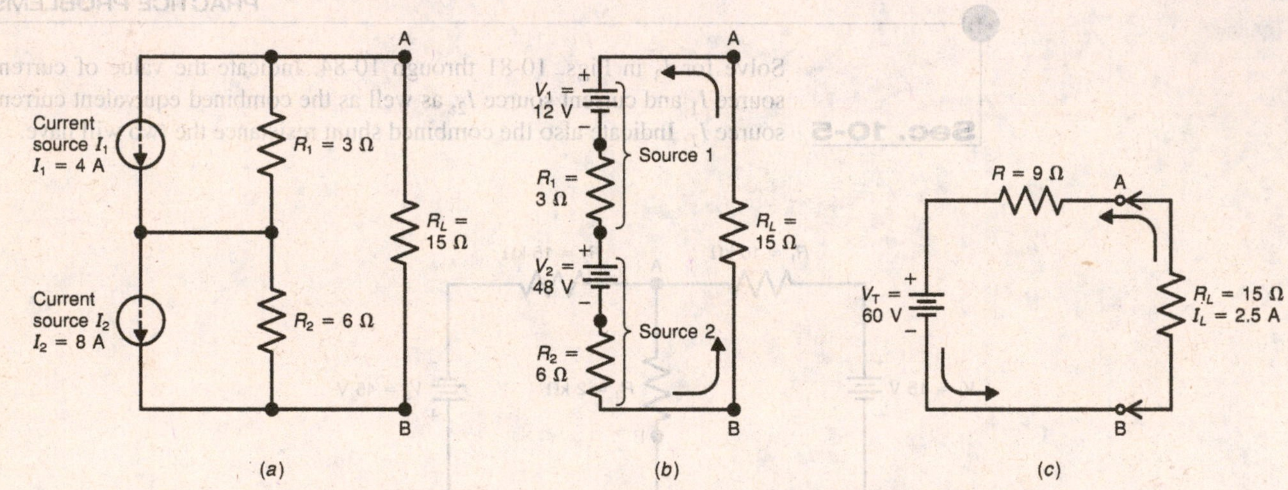

Fig. 10-85 Converting current sources in series to voltage sources that can be combined. (*a*) Original circuit. (*b*) Current sources I_1 and I_2 converted to series V_1 and V_2. (*c*) Circuit with combined voltage and resistance values. With R_L connected as shown, I_L = 2.5 A.

Answer

The problem here can be simplified by converting current sources I_1 and I_2 to series voltage sources as shown in Fig. 10-85*b*. The 4-A I_1 with its shunt 3-Ω R_1 is equivalent to 4 A × 3 Ω or 12 V for V_1 with a 3-Ω series resistance. Similarly, the 8-A I_2 with its shunt 6-Ω R_2 is equivalent to 8 A × 6 Ω or 48 V for V_2 with a 6-Ω series resistance. The polarities of V_1 and V_2 produce electron flow through R_L in the same direction as current sources I_1 and I_2.

The series voltages and resistance can be combined, as shown in Fig. 10-85*c*. The 12 V of V_1 and 48 V of V_2 are added because they are series aiding, resulting in the total voltage

V_T of 60 V. The resistances R_1 and R_2 are added for a total resistance of 9 Ω. To find I_L, use Ohm's law.

$$I_L = \frac{V_T}{R_T}$$

$$= \frac{60\text{ V}}{9\ \Omega + 15\ \Omega}$$

$$I_L = 2.5\text{ A}$$

If V_1 and V_2 were series opposing (as would be determined by the current sources), the total voltage would be the difference between the two source values with polarity being the same as that of the larger source.

MORE PRACTICE PROBLEMS

Sec. 10-5

Solve for I in R_L in Figs. 10-86 through 10-89. Indicate the value of voltage source V_1 and voltage source V_2, as well as the combined equivalent voltage source V_T. Also, indicate the combined resistance of the two voltages sources.

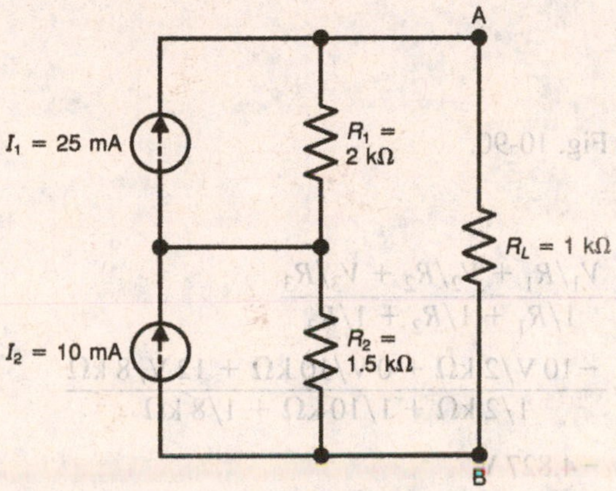

Fig. 10-86

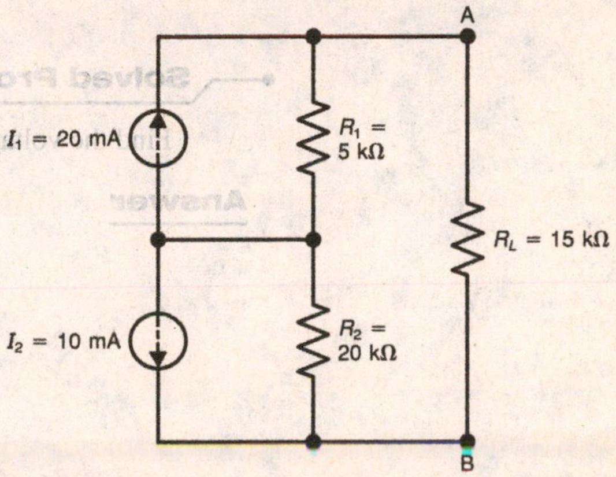

Fig. 10-87

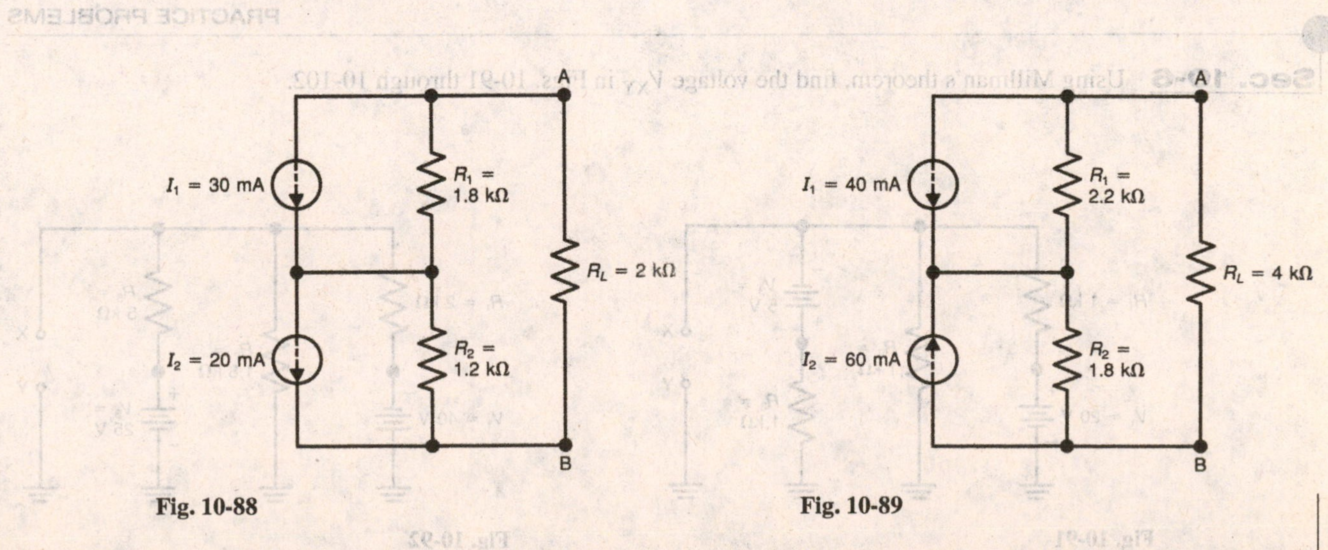

Fig. 10-88

Fig. 10-89

SEC. 10-6 MILLMAN'S THEOREM

Millman's theorem provides a shortcut for finding the common voltage across any number of parallel branches with different voltages. A typical example is shown in Fig. 10-90. A general equation used to determine the voltage across points X and Y is shown here.

$$V_{XY} = \frac{V_1/R_1 + V_2/R_2 + V_3/R_3 + \cdots + V_n/R_n}{1/R_1 + 1/R_2 + 1/R_3 + \cdots + 1/R_n}$$

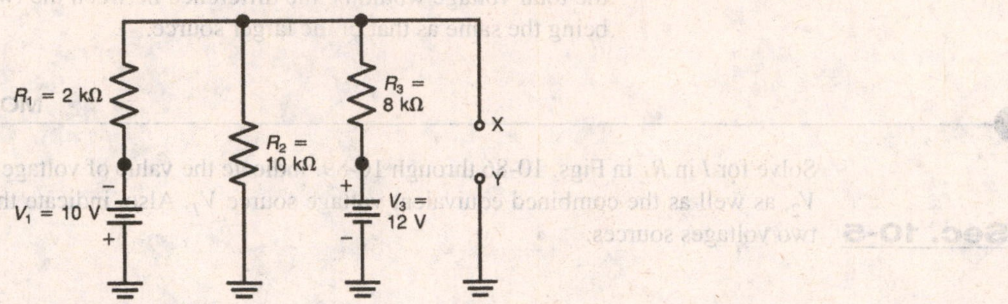

Fig. 10-90 Millman's theorem applied to find the voltage V_{XY}.

Solved Problem

Find the voltage V_{XY} in Fig. 10-90.

Answer

$$V_{XY} = \frac{V_1/R_1 + V_2/R_2 + V_3/R_3}{1/R_1 + 1/R_2 + 1/R_3}$$

$$= \frac{-10\text{ V}/2\text{ k}\Omega + 0\text{ V}/10\text{ k}\Omega + 12\text{ V}/8\text{ k}\Omega}{1/2\text{ k}\Omega + 1/10\text{ k}\Omega + 1/8\text{ k}\Omega}$$

$$V_{XY} = -4.827\text{ V}$$

Finding the value of V_{XY} gives the net effect of all the sources in determining the voltage at X with respect to Y.

PRACTICE PROBLEMS

Sec. 10-6 Using Millman's theorem, find the voltage V_{XY} in Figs. 10-91 through 10-102.

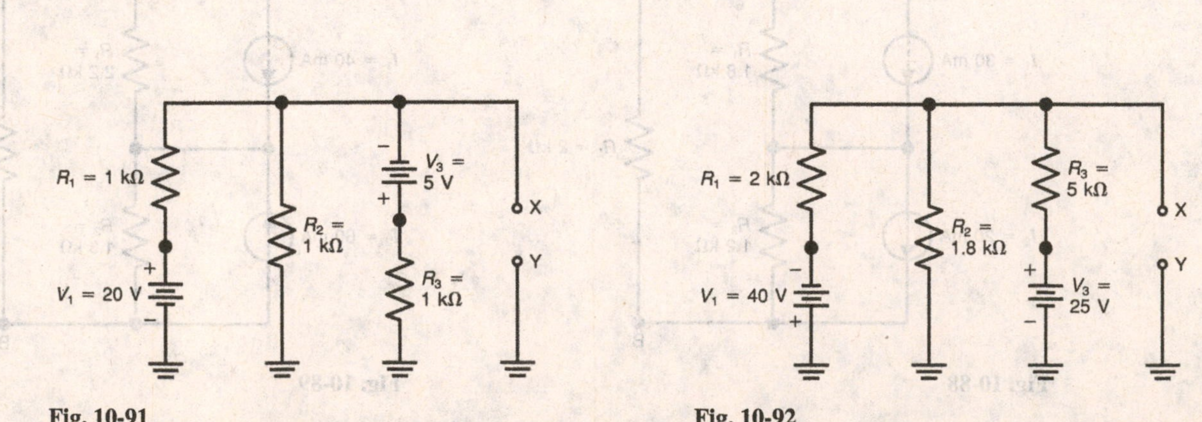

Fig. 10-91

Fig. 10-92

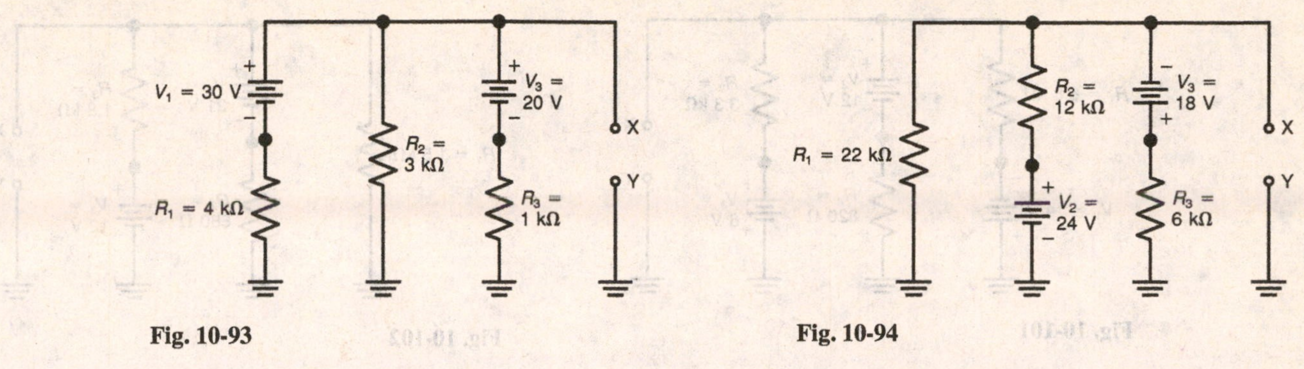

Fig. 10-93

Fig. 10-94

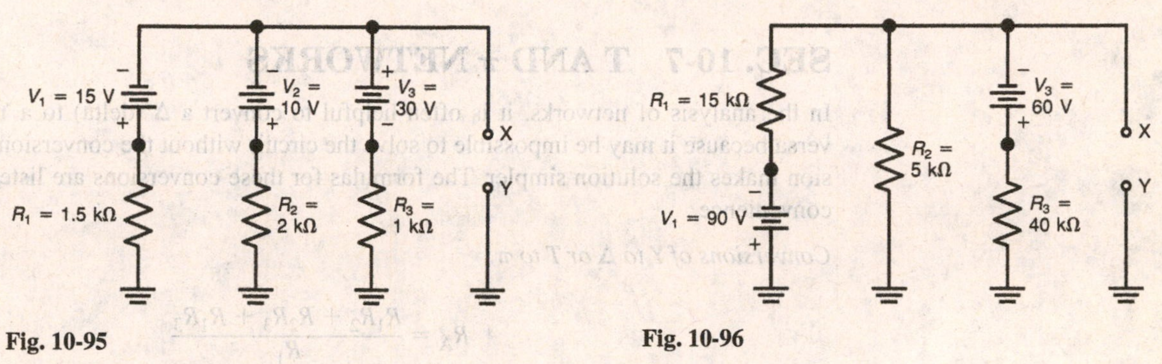

Fig. 10-95

Fig. 10-96

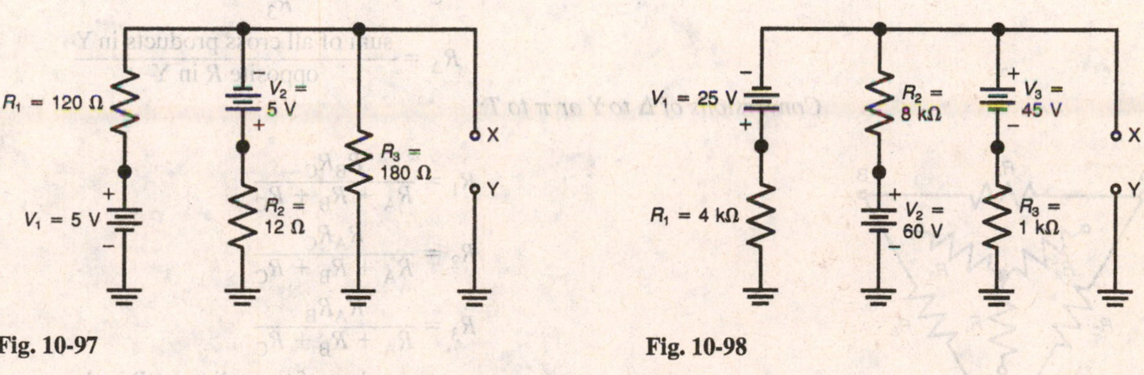

Fig. 10-97

Fig. 10-98

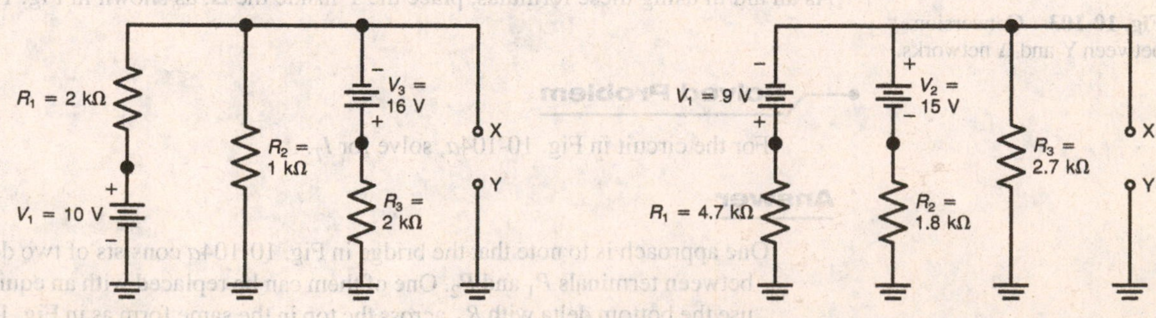

Fig. 10-99

Fig. 10-100

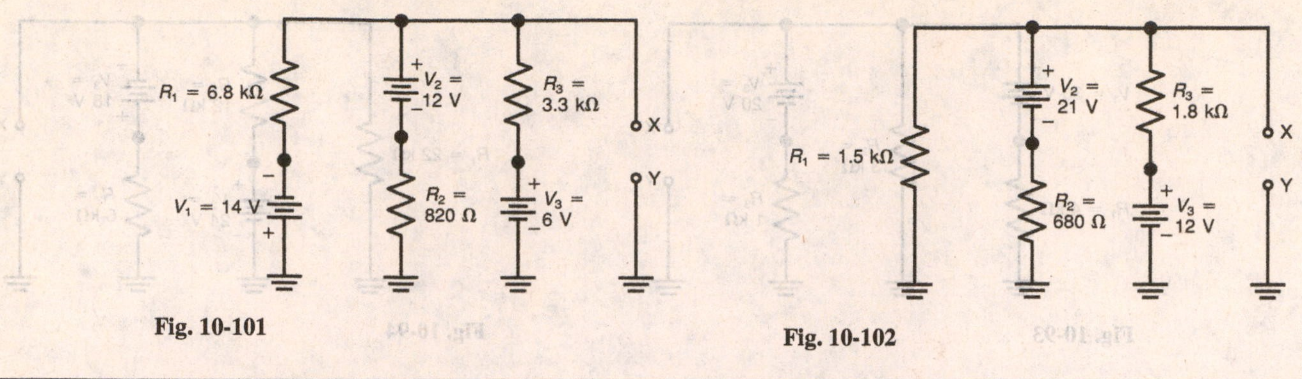

Fig. 10-101

Fig. 10-102

SEC. 10-7 T AND π NETWORKS

In the analysis of networks, it is often helpful to convert a Δ (delta) to a Y (wye) or vice versa because it may be impossible to solve the circuit without the conversion, or the conversion makes the solution simpler. The formulas for these conversions are listed here for your convenience.

Conversions of Y to Δ or T to π:

$$R_A = \frac{R_1R_2 + R_2R_3 + R_1R_3}{R_1}$$

$$R_B = \frac{R_1R_2 + R_2R_3 + R_1R_3}{R_2}$$

$$R_C = \frac{R_1R_2 + R_2R_3 + R_1R_3}{R_3}$$

$$R_\Delta = \frac{\text{sum of all cross products in Y}}{\text{opposite } R \text{ in Y}}$$

Conversions of Δ to Y or π to T:

$$R_1 = \frac{R_BR_C}{R_A + R_B + R_C}$$

$$R_2 = \frac{R_AR_C}{R_A + R_B + R_C}$$

$$R_3 = \frac{R_AR_B}{R_A + R_B + R_C}$$

$$R_Y = \frac{\text{product of two adjacent } R \text{ in } \Delta}{\text{sum of all } R \text{ in } \Delta}$$

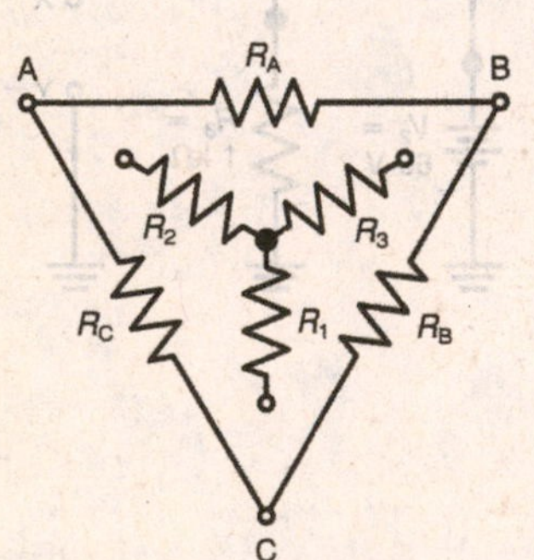

Fig. 10-103 Conversion between Y and Δ networks.

As an aid in using these formulas, place the Y inside the Δ, as shown in Fig. 10-103.

Solved Problem

For the circuit in Fig. 10-104*a*, solve for I_T.

Answer

One approach is to note that the bridge in Fig. 10-104*a* consists of two deltas connected between terminals P_1 and P_2. One of them can be replaced with an equivalent wye. We use the bottom delta with R_A across the top in the same form as in Fig. 10-103. We then replace this delta $R_AR_BR_C$ by an equivalent wye $R_1R_2R_3$, as shown in Fig. 10-104*b*.

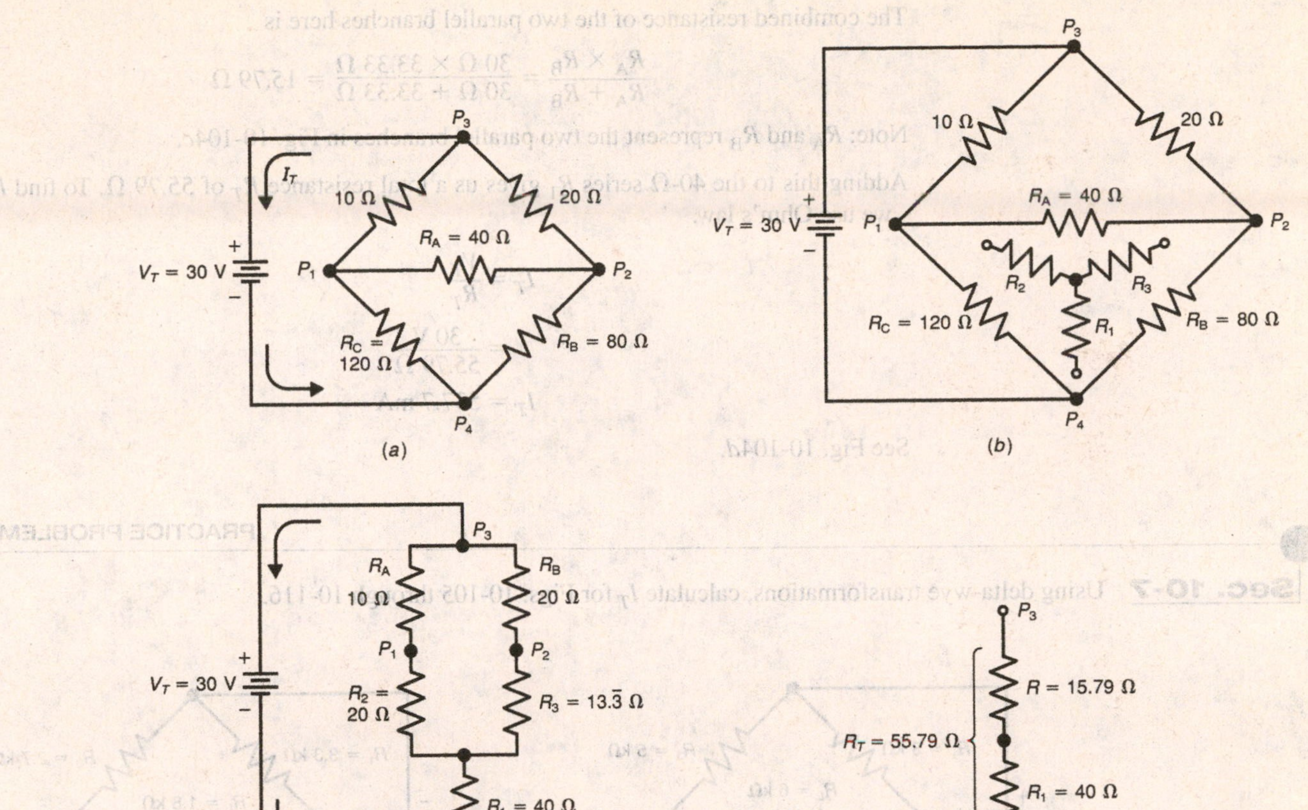

Fig. 10-104 Solving a bridge circuit by Δ-to-Y conversion. (*a*) Original circuit. (*b*) How Y of R_1, R_2, and R_3 corresponds to Δ of R_A, R_B, and R_C. (*c*) Series-parallel circuit resulting from Δ-to-Y conversions. (*d*) Resistance R_T between points P_3 and P_4.

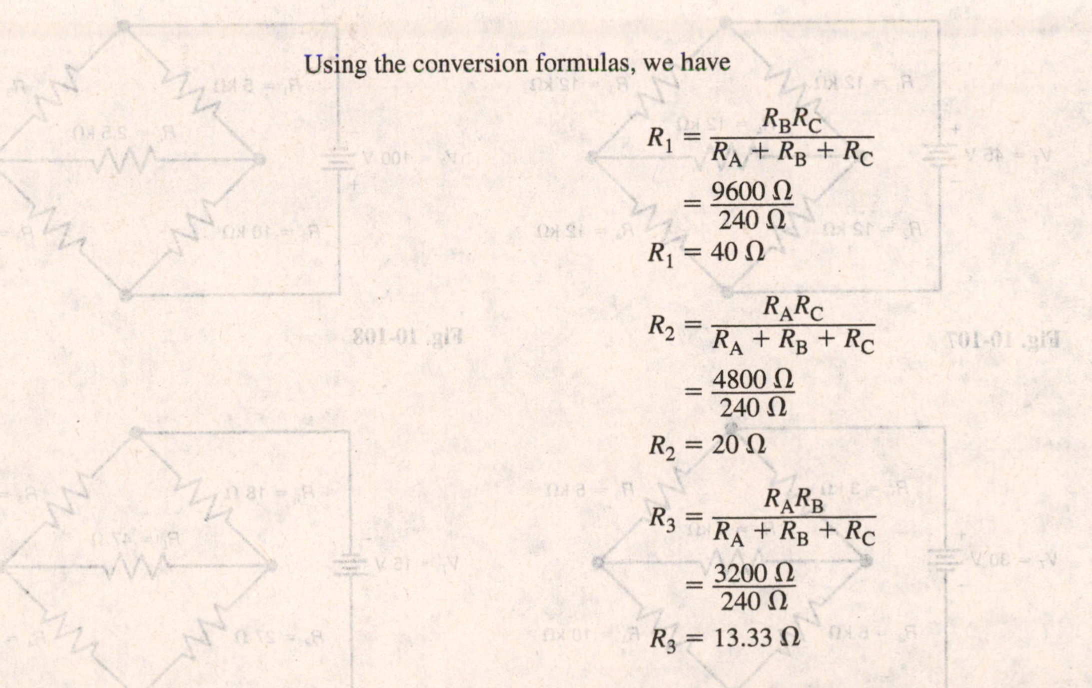

Using the conversion formulas, we have

$$R_1 = \frac{R_B R_C}{R_A + R_B + R_C} = \frac{9600\ \Omega}{240\ \Omega}$$

$$R_1 = 40\ \Omega$$

$$R_2 = \frac{R_A R_C}{R_A + R_B + R_C} = \frac{4800\ \Omega}{240\ \Omega}$$

$$R_2 = 20\ \Omega$$

$$R_3 = \frac{R_A R_B}{R_A + R_B + R_C} = \frac{3200\ \Omega}{240\ \Omega}$$

$$R_3 = 13.33\ \Omega$$

We next use these values for R_1, R_2, and R_3 in an equivalent wye to replace the original delta. Then the resistances form the series-parallel circuit in Fig. 10-104*c*.

The combined resistance of the two parallel branches here is

$$\frac{R_A \times R_B}{R_A + R_B} = \frac{30\ \Omega \times 33.33\ \Omega}{30\ \Omega + 33.33\ \Omega} = 15.79\ \Omega$$

Note: R_A and R_B represent the two parallel branches in Fig. 10-104*c*.

Adding this to the 40-Ω series R_1 gives us a total resistance R_T of 55.79 Ω. To find I_T, we use Ohm's law.

$$I_T = \frac{V_T}{R_T}$$

$$= \frac{30\ \text{V}}{55.79\ \Omega}$$

$$I_T = 537.7\ \text{mA}$$

See Fig. 10-104*d*.

PRACTICE PROBLEMS

Sec. 10-7 Using delta-wye transformations, calculate I_T for Figs. 10-105 through 10-116.

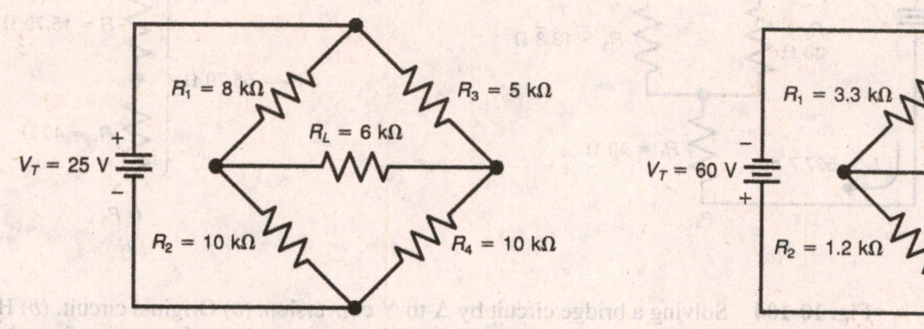

Fig. 10-105

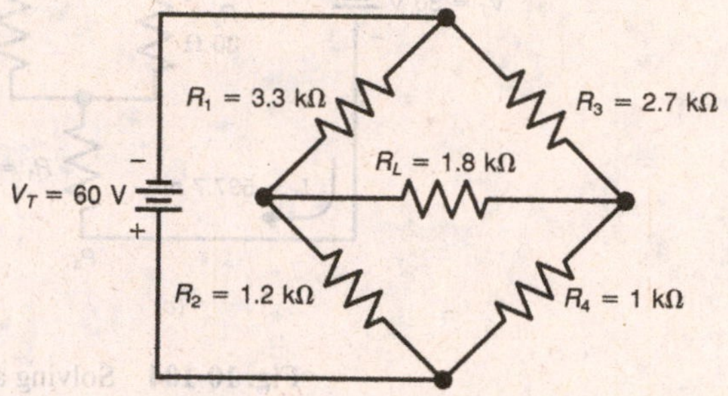

Fig. 10-106

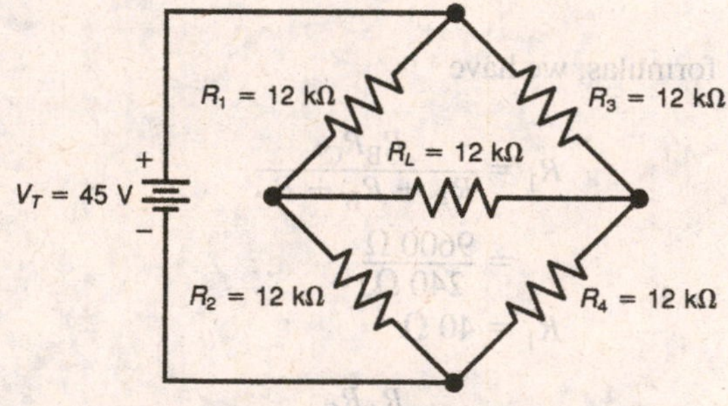

Fig. 10-107

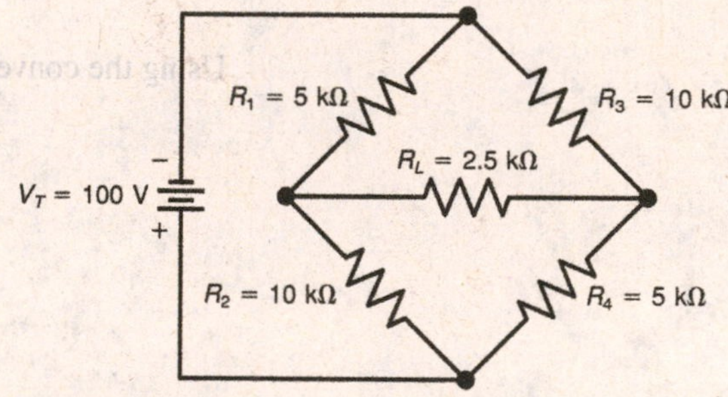

Fig. 10-108

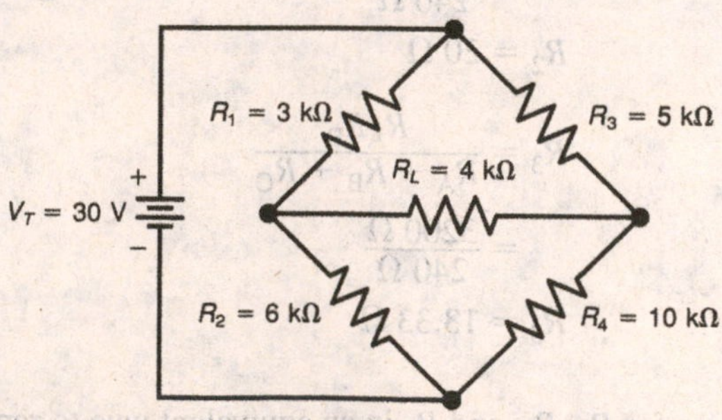

Fig. 10-109

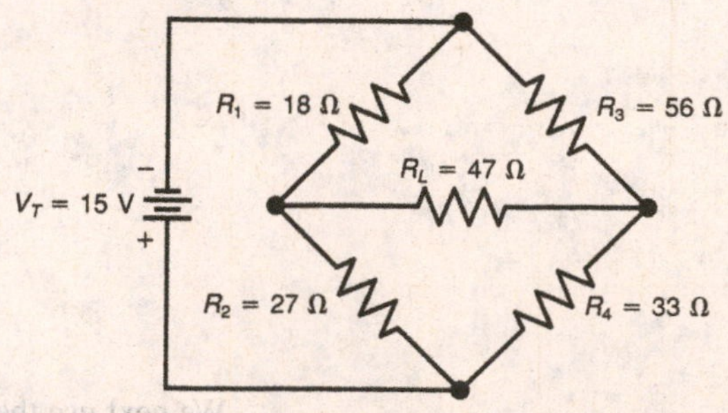

Fig. 10-110

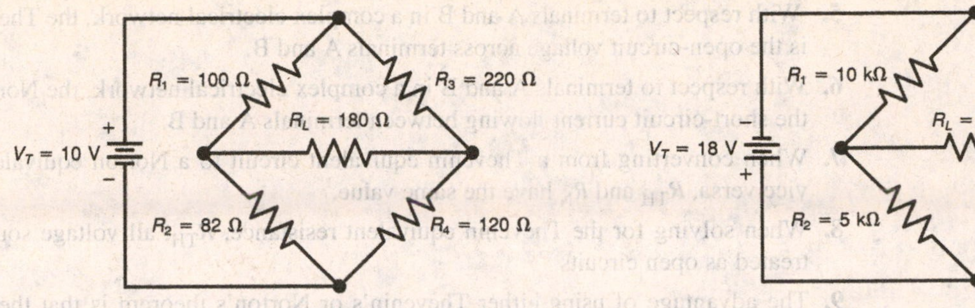

Fig. 10-111

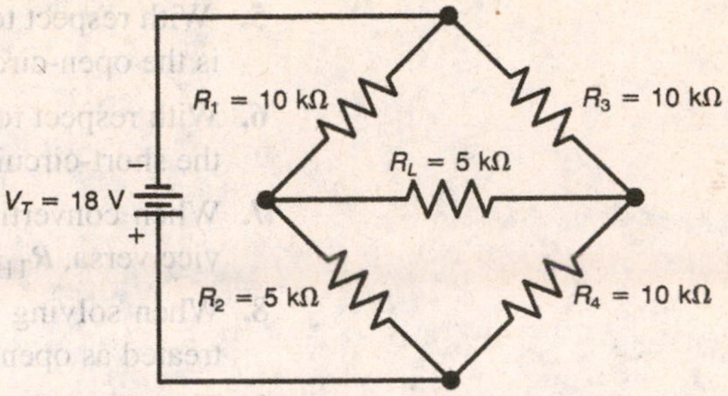

Fig. 10-112

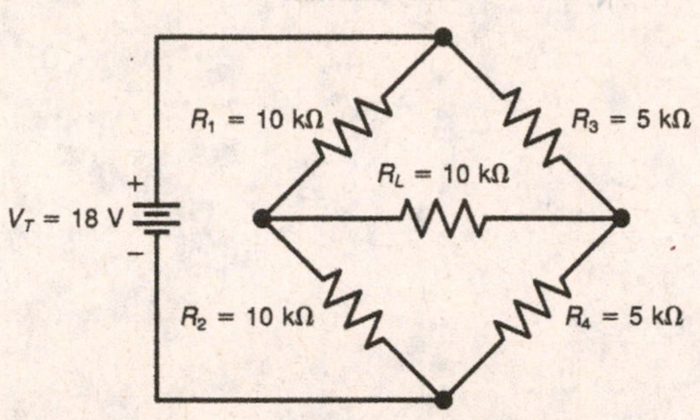

Fig. 10-113

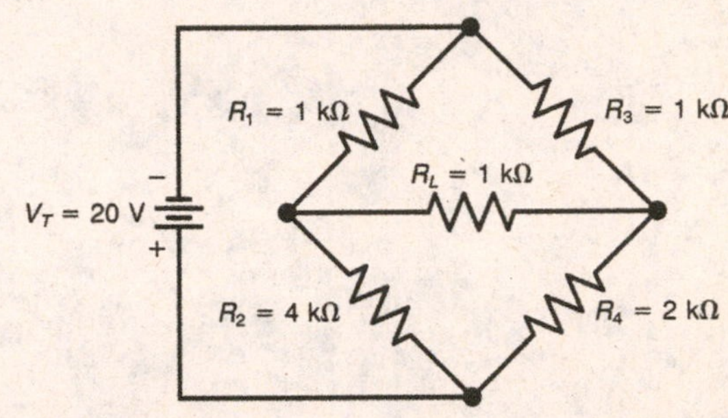

Fig. 10-114

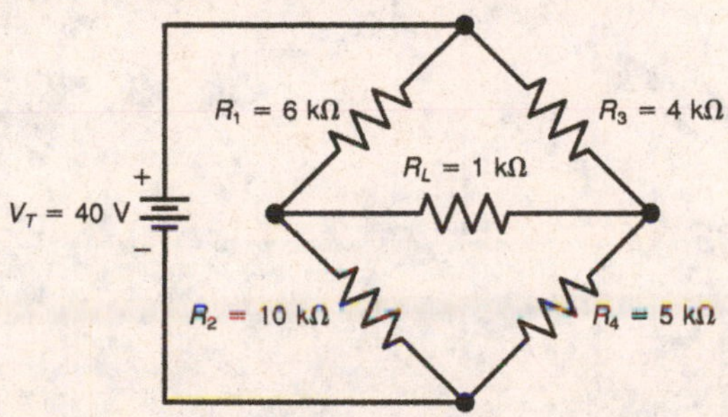

Fig. 10-115

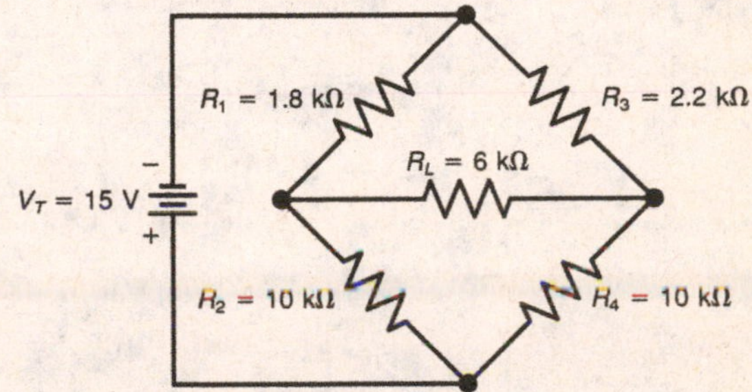

Fig. 10-116

END OF CHAPTER TEST

Chapter 10: Network Theorems

Answer True or False.

1. The superposition theorem states that in a network with two or more sources, the current or voltage for any component is the algebraic sum of the effects produced by each source acting separately.
2. In order to apply the superposition theorem, all components must be both linear and bilateral.
3. Thevenin's theorem states that with respect to any pair of terminals in a complex electrical network, the entire network can be replaced by a single voltage source in series with a single resistance.
4. Norton's theorem states that with respect to any pair of terminals in a complex electrical network, the entire network can be replaced by a single current source in series with a single resistance.

5. With respect to terminals A and B in a complex electrical network, the Thevenin voltage is the open-circuit voltage across terminals A and B.

6. With respect to terminals A and B in a complex electrical network, the Norton current is the short-circuit current flowing between terminals A and B.

7. When converting from a Thevenin equivalent circuit to a Norton equivalent circuit, or vice versa, R_{TH} and R_N have the same value.

8. When solving for the Thevenin equivalent resistance, R_{TH}, all voltage sources must be treated as open circuits.

9. The advantage of using either Thevenin's or Norton's theorem is that these equivalent circuits remain the same even if the load resistance changes.

10. When determining the Thevenin equivalent voltage, V_{TH}, or the Norton current, I_N, the polarity of V_{TH} or direction of I_N doesn't really matter.

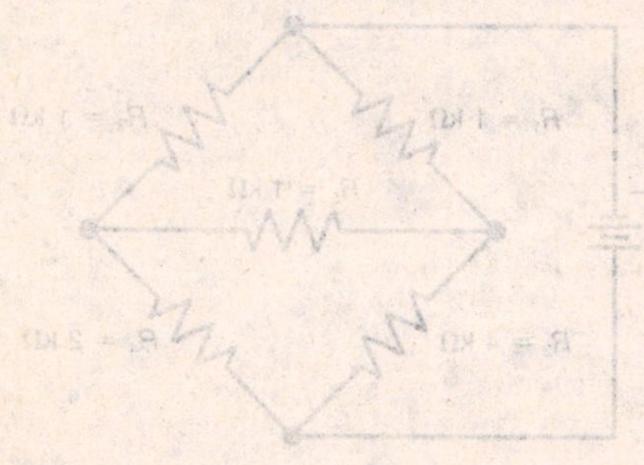

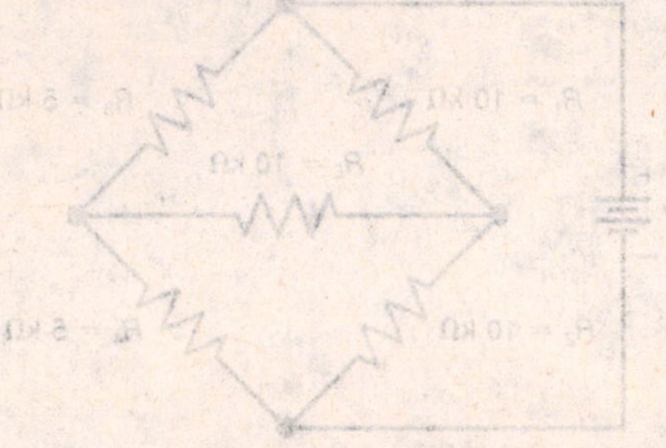

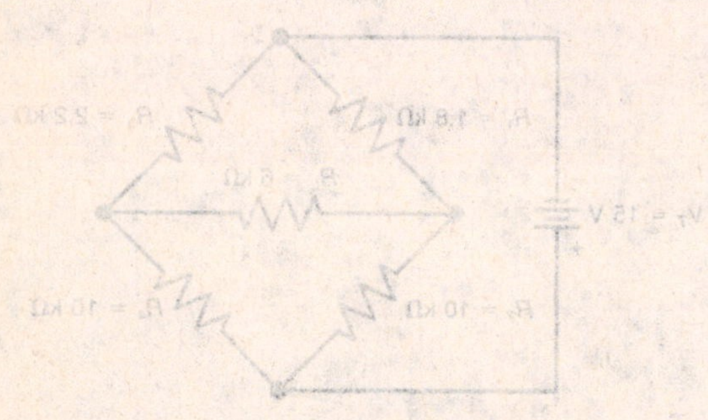

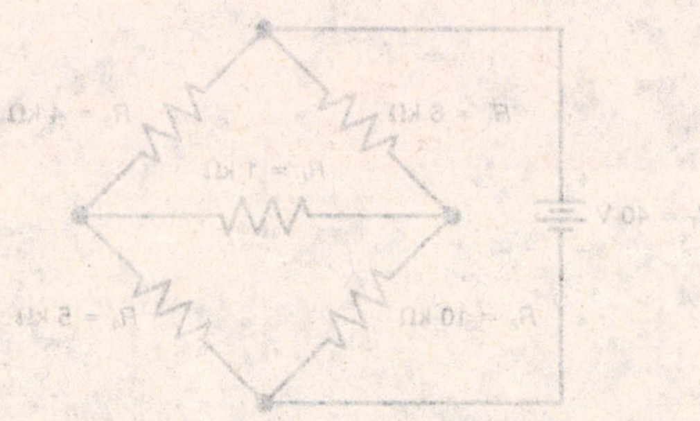

CHAPTER

11

Conductors and Insulators

Materials that have very low resistance are used as conductors, whereas materials having very high resistance are used as insulators. An ideal conductor would have zero resistance, and an ideal insulator would have infinite resistance. Unfortunately, these ideal values of resistances for conductors and insulators are not easily achievable. For most cases, the resistance of the conductors is small enough to be ignored. However, in some cases, the resistance may need to be considered. Likewise, the high resistance of an insulator can be considered infinite for most cases, but in some cases the finite value of resistance may need to be considered.

SEC. 11-1 DETERMINING THE RESISTANCE OF WIRE CONDUCTORS

For conductors that carry large amounts of current, the resistance value of the conductors is of great importance. The high current values can produce a significant IR voltage drop across the conductors, causing the I^2R power loss to become significant. This I^2R power loss is in the form of heat, which can cause the wire's insulation to burn if the heat becomes excessive.

For any given metal, a larger diameter wire with a greater circular area has less resistance than a smaller diameter wire with less circular area.

Table 11-1 lists the standard wire sizes in the system known as the American Wire Gage (AWG), or Brown and Sharpe (B&S) Gage. It can be seen from this table that as the gage number increases, the diameter and circular area of the wire decrease. The higher the gage number, the thinner the wire and the greater will be its resistance. The cross-sectional area of round wire is measured in circular mils, abbreviated *cmil.* The cmil area of round wire is found by squaring the diameter: cmil = d^2. To find the resistance of a conductor of any size or length, we use the following formula.

$$R = \rho \times \frac{l}{A}$$

where R = total resistance in ohms
l = length of wire in feet
A = cross-sectional area in cmil
ρ = specific resistance in cmil · Ω/ft

The factor ρ enables the resistance of different materials to be compared according to their nature, without regard to different lengths or areas. Table 11-2 lists the specific resistance values of the different metals having the standard wire size of a 1-ft length with a cross-sectional area of 1 cmil. The specific resistance ρ is measured in cmil · Ω/ft.

The temperature coefficient of a conductor tells us how its resistance changes with temperature. The symbol used to represent a conductor's temperature coefficient is the symbol alpha (α). A positive value for α indicates that R increases when temperature increases. A negative value for α indicates that R decreases when temperature increases.

TABLE 11-1 COPPER-WIRE TABLE

Gage No.	Diameter, Mil	Circular-Mil Area	Ohms per 1,000 Ft of Copper Wire at Room Temperature*	Gage No.	Diameter, Mil	Circular-Mil Area	Ohms per 1,000 Ft of Copper Wire at Room Temperature*
1	289.3	83,690	0.1264	21	28.46	810.1	13.05
2	257.6	66,370	0.1593	22	25.35	642.4	16.46
3	229.4	52,640	0.2009	23	22.57	509.5	20.76
4	204.3	41,740	0.2533	24	20.10	404.0	26.17
5	181.9	33,100	0.3195	25	17.90	320.4	33.00
6	162.0	26,250	0.4028	26	15.94	254.1	41.62
7	144.3	20,820	0.5080	27	14.20	201.5	52.48
8	128.5	16,510	0.6405	28	12.64	159.8	66.17
9	114.4	13,090	0.8077	29	11.26	126.7	83.44
10	101.9	10,380	1.018	30	10.03	100.5	105.2
11	90.74	8234	1.284	31	8.928	79.70	132.7
12	80.81	6530	1.619	32	7.950	63.21	167.3
13	71.96	5178	2.042	33	7.080	50.13	211.0
14	64.08	4107	2.575	34	6.305	39.75	266.0
15	57.07	3257	3.247	35	5.615	31.52	335.0
16	50.82	2583	4.094	36	5.000	25.00	423.0
17	45.26	2048	5.163	37	4.453	19.83	533.4
18	40.30	1624	6.510	38	3.965	15.72	672.6
19	35.89	1288	8.210	39	3.531	12.47	848.1
20	31.96	1022	10.35	40	3.145	9.88	1069

* 20 to 25°C or 68 to 77°F is considered average room temperature.

TABLE 11-2 PROPERTIES OF CONDUCTING MATERIALS*

Material	Description and Symbol	ρ = Specific Resistance at 20°C, cmil · Ω/ft	Temperature Coefficient per °C, α	Melting Point, °C
Aluminum	Element (Al)	17	0.004	660
Carbon	Element (C)	†	−0.0003	3000
Constantan	Alloy, 55% Cu, 45% Ni	295	0 (average)	1210
Copper	Element (Cu)	10.4	0.004	1083
Gold	Element (Au)	14	0.004	1063
Iron	Element (Fe)	58	0.006	1535
Manganin	Alloy, 84% Cu, 12% Mn, 4% Ni	270	0 (average)	910
Nichrome	Alloy, 65% Ni, 23% Fe, 12% Cr	676	0.0002	1350
Nickel	Element (Ni)	52	0.005	1452
Silver	Element (Ag)	9.8	0.004	961
Steel	Alloy, 99.5% Fe, 0.5% C	100	0.003	1480
Tungsten	Element (W)	33.8	0.005	3370

* Listings are approximate only, since precise values depend on exact composition of material.
† Carbon has about 2,500 to 7,500 times the resistance of copper. Graphite is a form of carbon.

The resistance of a conductor at any temperature can be found using the formula

$$R_T = R_0 + R_0\,(\alpha \times \Delta t)$$

where R_T = the resistance at any temperature
R_0 = the resistance of the conductor at 20°C
α = temperature coefficient of material
Δt = change in temperature from 20°C

Solved Problem

Use Tables 11-1 and 11-2 to find the resistance of 50 ft of No. 28 copper wire.

Answer

Note from Table 11-1 that the cross-sectional area is 159.8 cmil for No. 28 gage copper wire. The specific resistance for copper wire obtained from Table 11-2 is 10.4 cmil · Ω/ft. Using this information, we can find the total resistance.

$$R = \rho \times \frac{l}{A}$$

$$= 10.4 \text{ cmil} \cdot \Omega/\text{ft} \times \frac{50 \text{ ft}}{159.8 \text{ cmil}}$$

$$R = 3.25\ \Omega$$

Solved Problem

An aluminum wire has 20 Ω of resistance at 20°C. Calculate its resistance at 100°C.

Answer

The temperature rise Δt is 80°C. From Table 11-2, we see that the temperature coefficient (α) for aluminum is 0.004. Using this information, we find the new resistance as follows:

$$R_T = R_0 + R_0(\alpha \times \Delta t)$$

$$= 20\ \Omega + 20\ \Omega(0.004 \times 80°\text{C})$$

$$R_T = 26.4\ \Omega$$

PRACTICE PROBLEMS

Sec. 11-1

1. How much is the resistance of 1,000 ft of No. 14 gage copper wire?
2. How much is the resistance of 1,000 ft of No. 14 gage aluminum wire?
3. How much is the resistance of 1,000 ft of No. 14 gage steel wire?
4. Compare the resistances of two conductors: **(a)** 1,000 ft of No. 20 gage copper wire and **(b)** 1,000 ft of No. 23 gage copper wire.
5. Compare the resistances of two conductors: **(a)** 500 ft of No. 12 gage silver wire and **(b)** 500 ft of No. 12 gage steel wire.
6. Compare the resistances of two conductors: **(a)** 2,500 ft of No. 10 gage copper wire and **(b)** 2,500 ft of No. 16 gage copper wire.
7. A 14-gage copper wire has 20 Ω of resistance at 20°C. Calculate its length.
8. A 16-gage steel wire has 50 Ω of resistance at 20°C. Calculate its length.
9. An 8-gage aluminum wire has 10 Ω of resistance at 20°C. Calculate its length.
10. A tungsten wire has 25 Ω of resistance at 20°C. Calculate its resistance at 100°C.

11. A steel wire has 15 Ω of resistance at 20°C. Calculate its resistance at 75°C.
12. An aluminum wire has 40 Ω of resistance at 20°C. Calculate its resistance at 150°C.
13. The resistance of a copper wire is 10 Ω at 100°C. Calculate its resistance at 20°C.
14. The resistance of a steel wire is 5 Ω at 150°C. Calculate its resistance at 20°C.
15. The resistance of an aluminum wire is 100 Ω at 200°C. Calculate its resistance at 0°C.
16. The resistance of a silver wire is 20 Ω at 250°C. Calculate its resistance at −10°C.
17. The resistance of a nichrome wire is 1.5 kΩ at 75°C. Calculate its resistance at 20°C.
18. The resistance of a No. 16 gage copper wire is 20 Ω at 50°C. Calculate its length.
19. The resistance of a No. 10 gage aluminum wire is 100 Ω at 80°C. Calculate its length.
20. The resistance of a No. 28 gage steel wire is 200 Ω at 60°C. Calculate its length.
21. What is the resistance for each conductor of a 100-ft extension cord that uses No. 14 gage copper wire?
22. **(a)** If the extension cord in Prob. 21 carries 12 A of current, calculate the voltage drop across each conductor. **(b)** How much power is dissipated by each conductor for this current?
23. What is the resistance for each conductor of a 200-ft extension cord that uses No. 10 gage copper wire?
24. **(a)** If the extension cord in Prob. 23 carries 10 A of current, calculate the voltage drop across each conductor. **(b)** How much power is dissipated by each conductor for this current?
25. A length of copper wire has a resistance of 30 Ω at 20°C. Calculate its resistance at −5°C.
26. **(a)** An electric clothes dryer is connected to the 220-V power line through a 50-ft power cord that uses No. 10 gage copper wire. Calculate the voltage drop across each conductor in the power cord if the dryer draws 30 A of current when drying. **(b)** How much power is dissipated by each conductor in the power cord?
27. A motor receives its power through No. 10 gage copper wire from a generator located 250 ft away. The voltage at the motor is 210 V when 25 A of current is drawn. Calculate the voltage across the generator output terminals.
28. A 1-hp motor having an efficiency of 60% is connected to the 120-V power line using a 50-ft extension cord. The extension cord uses No. 12 gage copper wire for its conductors. Calculate the voltage available at the motor.
29. A toaster uses 180 in. of No. 26 tungsten wire for its heating elements. If the temperature of the wire rises to 1,800°C when toasting, calculate the current drawn from the 120-V power line.
30. A 125-ft extension cord using No. 16 gage copper wire connects a 10-A load to the 120-V power line. Calculate the voltage available at the load.
31. A stranded wire consists of 207 individual strands of No. 37 gage copper wire. **(a)** Calculate the equivalent gage size in solid wire. **(b)** If the total length is 2,000 ft, calculate its resistance. **(c)** What is the resistance of each 2,000-ft strand?
32. A stranded wire used in the distribution of electrical power consists of four strands of No. 12 gage aluminum wire. **(a)** Calculate the equivalent gage size in solid wire. **(b)** If the total length is 500 ft, calculate its resistance. **(c)** What is the resistance of each 500-ft strand?
33. What is the smallest gage size of copper wire that will limit the line drop to 5 V with 240 V AC applied with a 20-A load? The total line length is 200 ft.
34. What is the smallest gage size of aluminum wire that will limit the line drop to 10 V with 120 V AC applied and a 10-A load? The total line length is 250 ft.
35. A 100-ft extension cord uses 20 strands of No. 23 gage copper wire for each conductor. If the extension cord connects the 120-V AC power line to a 15-A load, calculate the voltage available at the load.

Chapter 11: Conductors and Insulators

Answer True or False.

1. In most cases, the resistance of an insulator is so high that it can be considered infinite.
2. In most cases, the resistance of a wire conductor is small enough to be ignored.
3. The cross-sectional area of round wire is measured in circular mils (cmil).
4. For a given length, a larger diameter wire has more resistance than a smaller diameter wire.
5. For a given diameter wire, a longer length corresponds to a higher resistance.
6. The higher the gage number, the larger the circular mil (cmil) area of the wire.
7. A metal wire with a positive temperature coefficient increases in resistance as its temperature increases.
8. A 100-ft length of No. 14 gage copper wire has less resistance than a 100-ft length of No. 14 gage aluminum wire.
9. The specific resistance, ρ, measured in cmil · Ω/ft, is used to compare the resistance of different materials according to their nature, without regard to their length or areas.
10. A 50-ft length of No. 18 gage copper wire has less resistance than a 250-ft length of No. 18 gage copper wire.

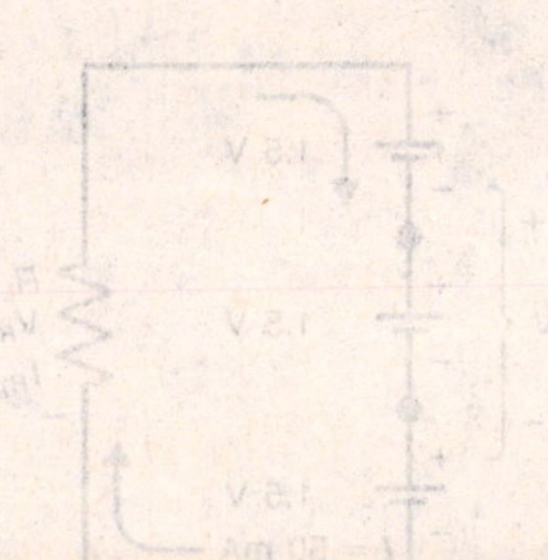

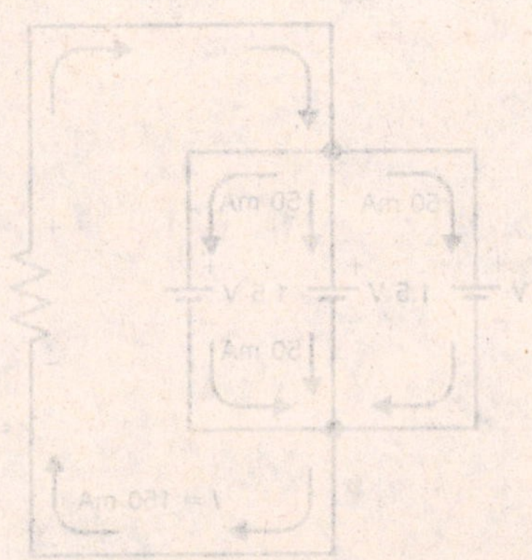

CHAPTER

12

Batteries

A battery is a device that converts chemical energy into electrical energy. In most cases, a battery consists of a combination of individual voltaic cells connected either in series or in parallel and, in some cases, a series-parallel arrangement. The purpose of a battery is to provide a steady DC voltage of fixed polarity.

SEC. 12-1 CONNECTION OF CELLS

The battery voltage can be increased by connecting cells in series. The total battery voltage equals the sum of the individual cell voltages. The current capacity of the battery is the same as for one cell, because the current flows through all series cells. Figure 12-1 shows three 1.5-V cells, each with a current capacity of 100 mA. Connected in series, the total battery voltage is 4.5 V. With this battery voltage, the 90-Ω load draws 50 mA of current. This current flow of 50 mA is the same through the load R_L and each series connected cell.

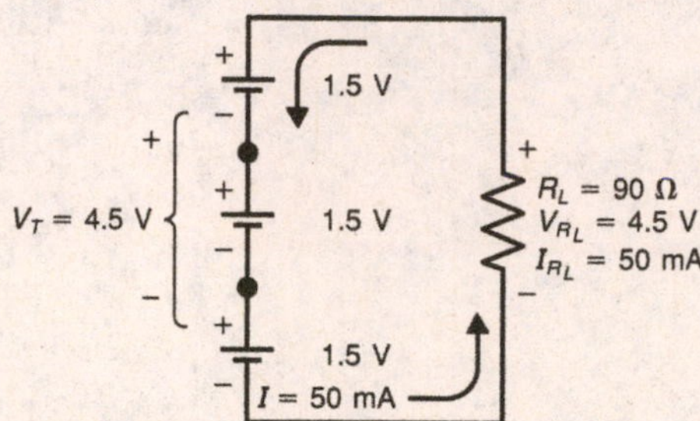

Fig. 12-1 Cells in series to add voltages.

For more current capacity, cells can be connected in parallel. All the negative terminals are connected together, and all the positive terminals are connected together. The total battery voltage is the same as for one cell. However, the battery can deliver more current than can one cell alone. Figure 12-2 shows three cells connected in parallel. Each cell of the battery

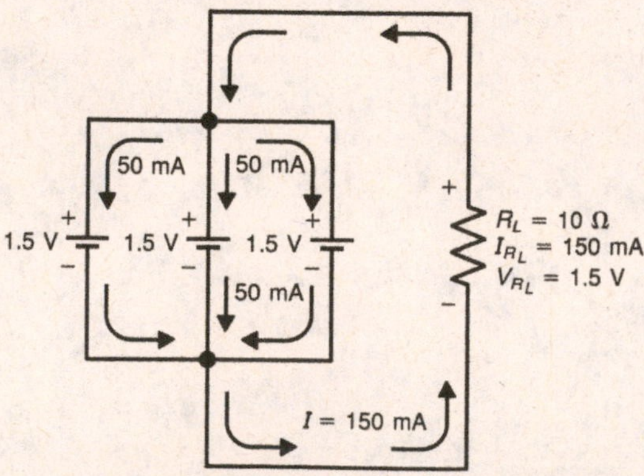

Fig. 12-2 Cells in parallel to increase current capacity.

supplies one-third of the total load current. The parallel connection of the cells is equivalent to increasing the size of the electrodes and electrolyte.

In order to obtain a higher battery voltage with increased current capacity, we can connect cells in a series-parallel configuration. Figure 12-3 shows four 1.5-V cells connected in a series-parallel arrangement to obtain a total battery voltage of 3 V and a current capacity equal to twice that of one cell.

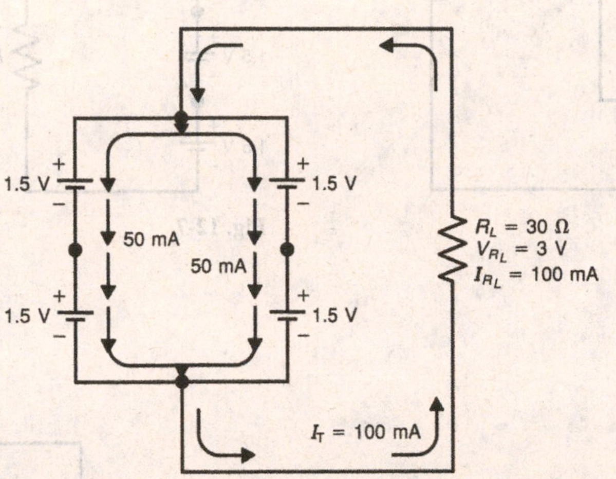

Fig. 12-3 Cells connected in series-parallel to provide higher output voltage with more current capacity.

For each combination shown, it is assumed that the condition of each cell is identical. One bad cell for any arrangement of cells shown will cause the total battery voltage or current capacity or both to differ from the ideal values shown.

PRACTICE PROBLEMS

Sec. 12-1

For Figs. 12-4 through 12-12, find the load voltage V_L, load current I_L, and the current supplied by each cell in the battery. Assume that all cells are identical and that the current capacity for each cell is not being exceeded for the load conditions represented.

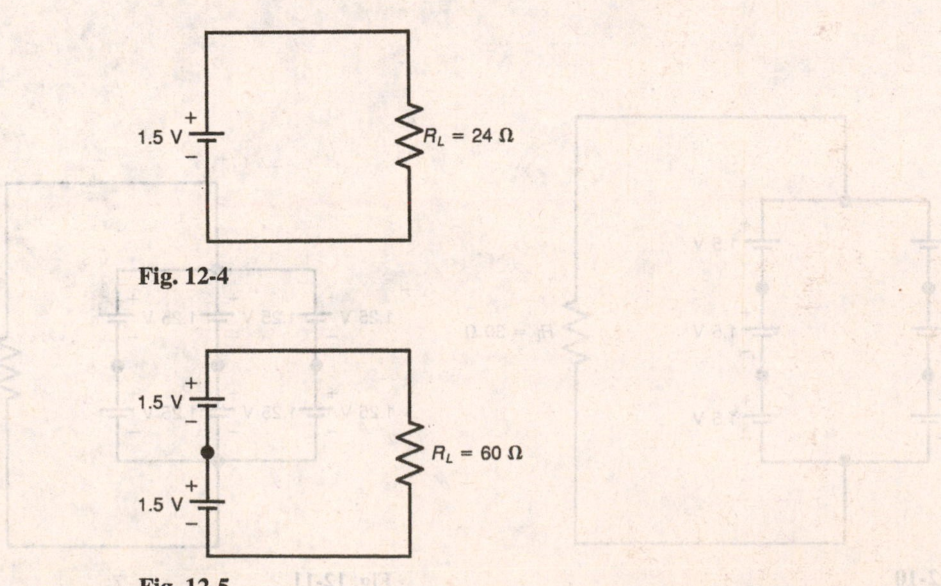

Fig. 12-4

Fig. 12-5

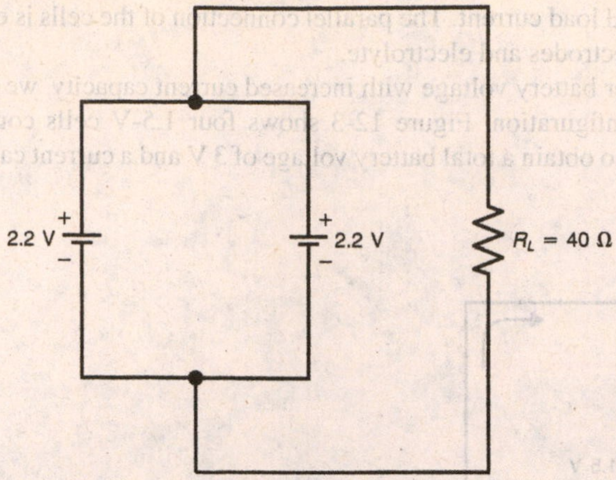

Fig. 12-6

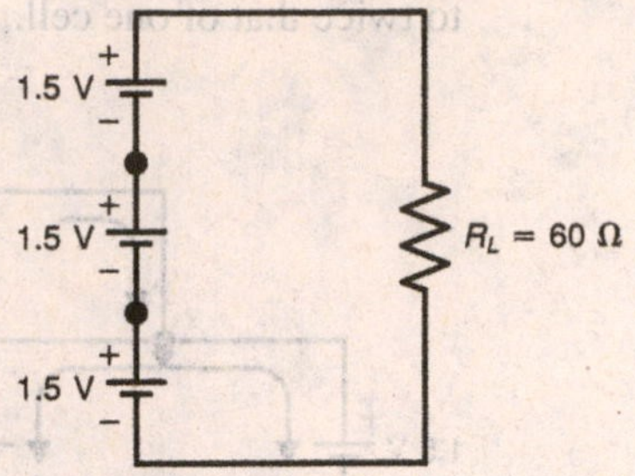

Fig. 12-7

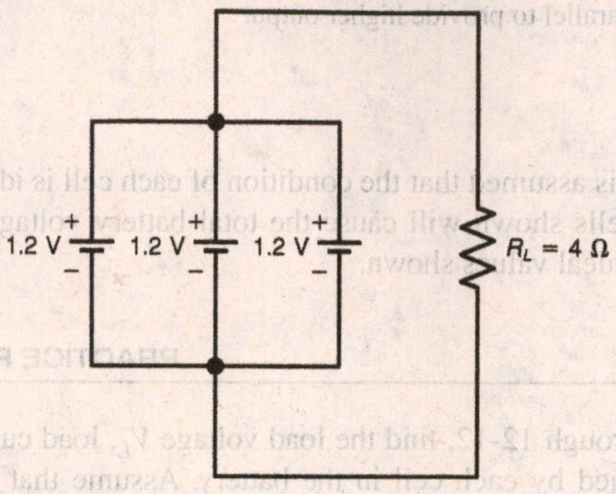

Fig. 12-8

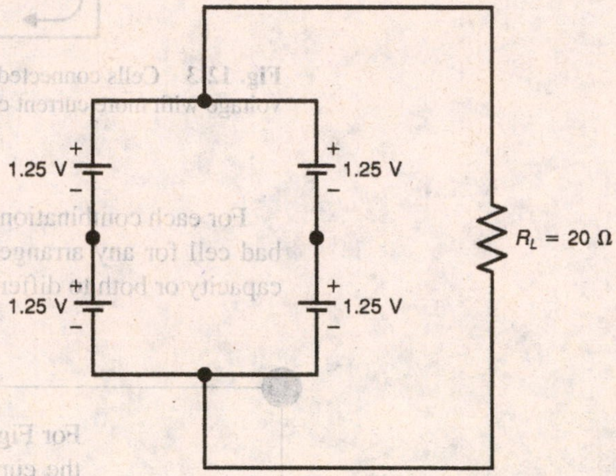

Fig. 12-9

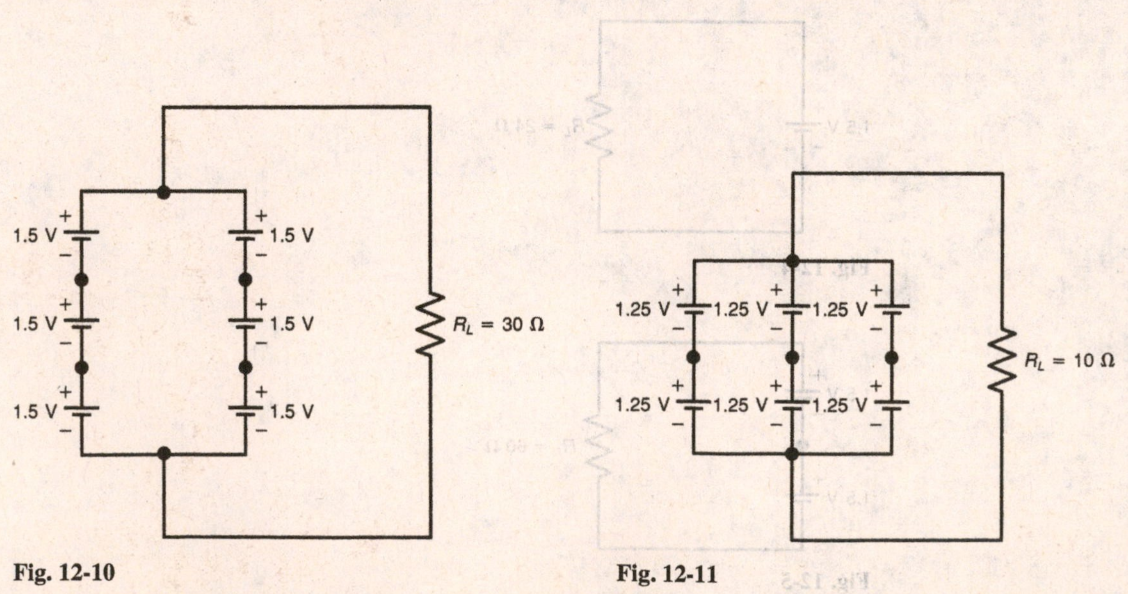

Fig. 12-10

Fig. 12-11

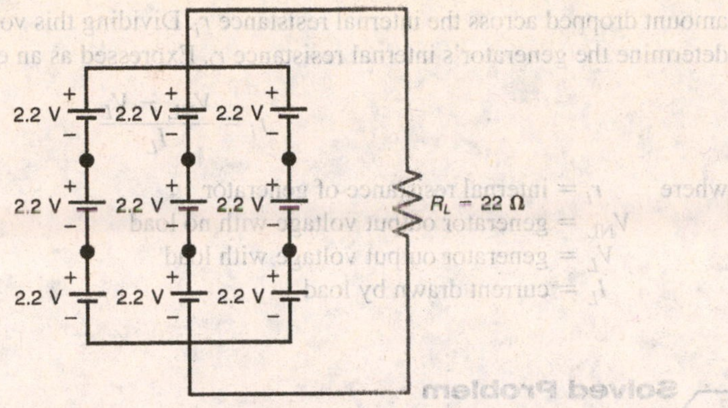

Fig. 12-12

SEC. 12-2 INTERNAL RESISTANCE OF A GENERATOR

All generators have internal resistance. In Fig. 12-13, we label the internal resistance r_i. As can be seen, the internal resistance r_i is in series with the cell that produces the potential difference of 1.6 V. The internal resistance is important when the generator must supply current to a load connected to its output terminals. The load current produced by the potential difference of the generator causes a voltage to be dropped across its own internal resistance r_i. This $I_L r_i$ voltage drop subtracts from the generated potential difference, resulting in less voltage available at the output terminals of the generator. The internal r_i is, therefore, the opposition to load current inside the generator. If, in Fig. 12-13, the output terminals become short-circuited, the only resistance limiting current flow would be that of r_i. For Fig. 12-13, the short-circuit current would be

$$I_{\text{short circuit}} = \frac{V_{NL}}{r_i}$$

$$= \frac{1.6\text{ V}}{0.1\ \Omega}$$

$$I_{\text{short circuit}} = 16\text{ A}$$

The internal resistance of any generator can be measured indirectly by determining the difference between the no-load and full-load output voltage when a specific value of load current is drawn. The difference between the no-load and full-load output voltage is the

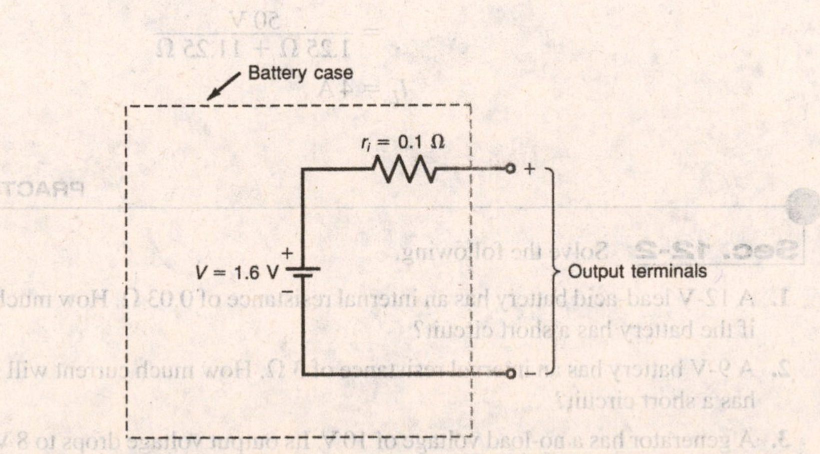

Fig. 12-13 Internal resistance r_i is in series with the cell voltage.

amount dropped across the internal resistance r_i. Dividing this voltage drop by I_L allows us to determine the generator's internal resistance r_i. Expressed as an equation, we have

$$r_i = \frac{V_{NL} - V_L}{I_L}$$

where r_i = internal resistance of generator
V_{NL} = generator output voltage with no load
V_L = generator output voltage with load
I_L = current drawn by load

Solved Problem

Calculate r_i for Fig. 12-14 if the output of the generator drops from 50 V with zero load current to 45 V with 4 A of load current.

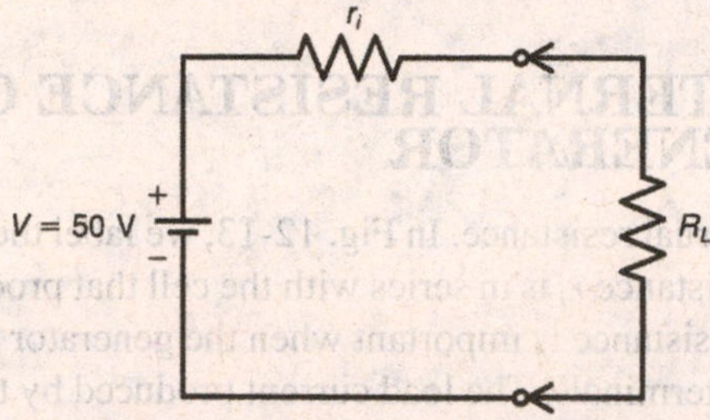

Fig. 12-14 Circuit used to illustrate the effects of internal resistance r_i.

Answer

The solution is found using the equation shown for r_i

$$r_i = \frac{V_{NL} - V_L}{I_L}$$

$$= \frac{50\text{ V} - 45\text{ V}}{4\text{ A}}$$

$$r_i = 1.25\ \Omega$$

Again, it must be realized that r_i is the opposition to load current inside the generator. The load R_L and resistance r_i act together to limit current flow. With r_i known, we have the following equation. (*Note:* $R_L = V_L/I_L = 11.25\ \Omega$ in Fig. 12-14.)

$$I_L = \frac{V_{NL}}{r_i + R_L}$$

$$= \frac{50\text{ V}}{1.25\ \Omega + 11.25\ \Omega}$$

$$I_L = 4\text{ A}$$

PRACTICE PROBLEMS

Sec. 12-2 Solve the following.

1. A 12-V lead-acid battery has an internal resistance of 0.03 Ω. How much current will flow if the battery has a short circuit?
2. A 9-V battery has an internal resistance of 3 Ω. How much current will flow if the battery has a short circuit?
3. A generator has a no-load voltage of 10 V. Its output voltage drops to 8 V with an R_L value of 0.4 Ω. Calculate r_i.

4. A generator has a no-load voltage of 40 V. Its output voltage drops to 36 V with an R_L value of 2.25 Ω. Calculate r_i.
5. A 6-V battery delivers a current of 960 mA to a 6-Ω load. What is the internal resistance of the battery?
6. A 24-V generator delivers a current of 4.8 A to a 4.5-Ω load. What is the internal resistance of the battery?
7. Three identical cells, each having a no-load voltage of 1.5 V, are connected in series to form a battery. With a 21-Ω load connected, the voltage available from the battery is 4.2 V. Calculate the r_i of each cell in the battery.
8. Six identical cells, each having a no-load voltage of 2.1 V, are connected in series to form a battery. When a load current of 375 mA is drawn from the battery, the voltage available at its output terminals drops to 12 V. Calculate the r_i of each cell in the battery.
9. Two identical batteries, each having a no-load voltage of 6 V, are connected in parallel. When a load current of 3 A is drawn from the battery, the load voltage equals 5.4 V. Calculate the r_i of each battery.
10. Four identical batteries, each having a no-load voltage of 12.6 V, are connected in parallel. When a load current of 300 A is drawn from the battery, the load voltage equals 12 V. Calculate the r_i of each battery.
11. Four identical cells, each having a no-load voltage of 1.25 V, are connected in a series-parallel arrangement to obtain a battery voltage of 2.5 V. If the load voltage equals 2.2 V when I_L = 1.2 A, what is the r_i for each cell in the battery?
12. Nine identical cells, each having a no-load voltage of 2.2 V, are connected in a series-parallel arrangement to obtain a battery voltage of 6.6 V. If the load voltage equals 6 V when R_L = 0.04 Ω, what is the r_i for each cell in the battery?

SEC. 12-3 CONSTANT VOLTAGE AND CONSTANT CURRENT SOURCES

A generator with very low internal resistance is considered to be a constant voltage source. For this type of generator, the output voltage remains constant even though the load current may vary over a very wide range. This idea can best be explained by referring to Fig. 12-15. Notice that the internal resistance of the voltage source is 0.01 Ω. The output voltage for several different R_L values ranging from 0.1 Ω to 1 MΩ is shown in Table 12-1.

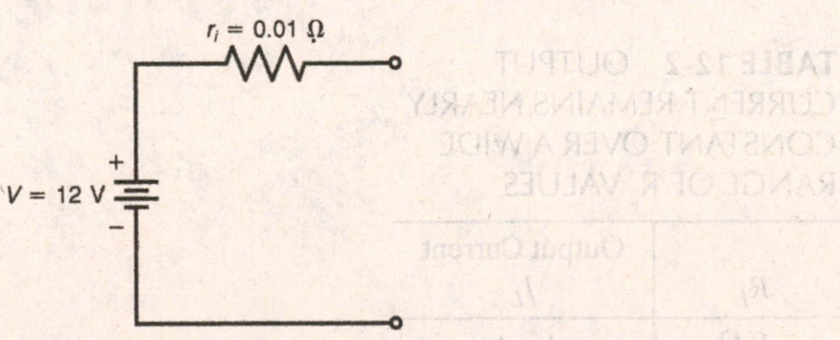

Fig. 12-15 Constant voltage generator with low r_i.

The load current that exists for any R_L value is found using the equation shown earlier.

$$I_L = \frac{V_{\mathrm{NL}}}{r_i + R_L}$$

TABLE 12-1 OUTPUT VOLTAGE REMAINS NEARLY CONSTANT OVER A WIDE RANGE OF R_L VALUES

R_L	Output Voltage V_L	Load Current I_L
0.1 Ω	10.9 V	109.09 A
1 Ω	11.88 V	11.88 A
10 Ω	11.98 V	1.198 A
100 Ω	11.99 V	119.98 mA
1 kΩ	11.99 V	11.99 mA
10 kΩ	12 V	1.2 mA
100 kΩ	12 V	120 μA
1 MΩ	12 V	12 μA

The load current value for each R_L value in Table 12-1 is found using this formula. The load voltage can be found one of two ways:

$$V_L = I_L \times R_L$$

or

$$V_L = \frac{R_L}{r_i + R_L} \times V_{NL}$$

In either case, V_L is the same. From Table 12-1, it can be seen that V_L remains relatively constant as I_L varies from 12 μA to 109 A. For this reason, V_L is referred to as a constant voltage source.

A generator with very high internal resistance is considered to be a constant current source. For this type of generator, the output current remains constant even though the load resistance may vary over a wide range. This idea can be best explained by referring to Fig. 12-16. Notice that the internal resistance of the generator is 10 MΩ. The output current for several different R_L values ranging from 0 Ω to 1 MΩ is shown in Table 12-2.

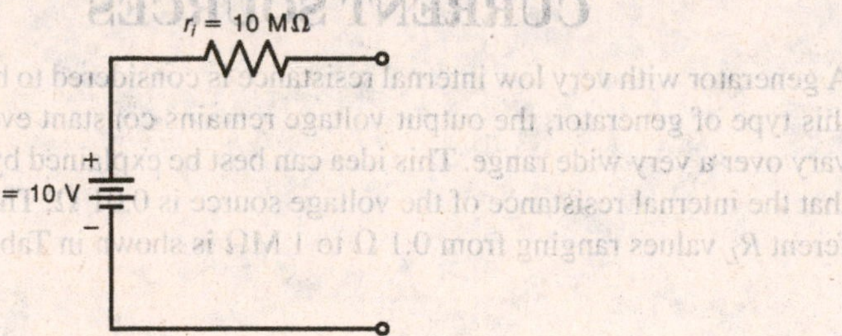

Fig. 12-16 Constant current source with high r_i.

TABLE 12-2 OUTPUT CURRENT REMAINS NEARLY CONSTANT OVER A WIDE RANGE OF R_L VALUES

R_L	Output Current I_L
0 Ω	1 μA
1 Ω	1 μA
10 Ω	1 μA
100 Ω	1 μA
1 kΩ	1 μA
10 kΩ	0.999 μA
100 kΩ	0.990 μA
1 MΩ	0.909 μA

The load current for any value of R_L is found as follows:

$$I_L = \frac{V_{NL}}{r_i + R_L}$$

From Table 12-2, it can be seen that I_L remains relatively constant as R_L varies from 0 Ω to 1 MΩ. For this reason I_L is considered to be a constant current source.

The current source in Fig. 12-16 often appears as shown in Fig. 12-17. The arrow symbol inside the circle represents a current source.

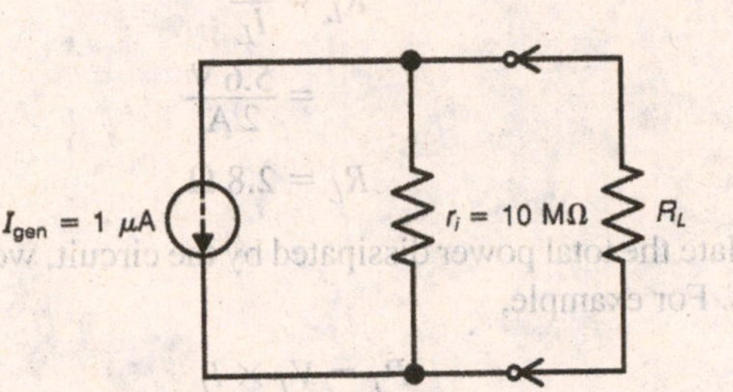

Fig. 12-17 Equivalent current source for Fig. 12-16.

The current source is connected across the parallel combination of r_i and R_L. The load current I_L can be found for any value of R_L by using the laws governing parallel circuits. The load current I_L can be found using the following formula:

$$I_L = \frac{r_i}{r_i + R_L} \times I_{gen}$$

For an R_L value of 10 kΩ, I_L is found as follows:

$$I_L = \frac{10\text{ M}\Omega}{10\text{ M}\Omega + 10\text{ k}\Omega} \times 1\ \mu\text{A}$$

$$I_L = 0.999\ \mu\text{A}$$

When the load resistance R_L connected to a generator equals the internal resistance r_i of the generator, we say that the load and generator are matched. The matching is important because it is this condition that delivers maximum power to the load R_L. If it is desirable to obtain maximum voltage across R_L, the load should have as high a resistance as possible. This condition also produces maximum efficiency because the low current value causes a very small percentage of the total power to be dissipated in the generator's own internal resistance.

Solved Problem

A 6-V battery develops a terminal voltage of 5.6 V when delivering 2 A to a load resistor R_L.

a. What is the internal resistance r_i of the battery?
b. What is the load resistance R_L?
c. How much total power is being dissipated?
d. How much power is being absorbed by R_L?
e. What is the efficiency of the power transfer?
f. Would the battery be considered a constant voltage source?

Answer

a. To calculate r_i,

$$r_i = \frac{V_{NL} - V_L}{I_L}$$
$$= \frac{6\ \text{V} - 5.6\ \text{V}}{2\ \text{A}}$$
$$r_i = 0.2\ \Omega$$

b. To calculate the load resistance R_L, we proceed as follows:

$$R_L = \frac{V_L}{I_L}$$
$$= \frac{5.6\ \text{V}}{2\ \text{A}}$$
$$R_L = 2.8\ \Omega$$

c. To calculate the total power dissipated by the circuit, we can use any one of the power formulas. For example,

$$P_T = V_T \times I_L$$
$$= 6\ \text{V} \times 2\ \text{A}$$
$$P_T = 12\ \text{W}$$

d. The power dissipated by R_L is found as follows:

$$P_L = V_L \times I_L$$
$$= 5.6\ \text{V} \times 2\ \text{A}$$
$$P_L = 11.2\ \text{W}$$

e. To determine the efficiency of power transfer, we determine how much of the total power, P_T, is delivered to the load, R_L. Expressed as an efficiency, we have

$$\%\ \text{efficiency} = \frac{P_L}{P_T} \times 100$$
$$= \frac{11.2\ \text{W}}{12\ \text{W}} \times 100$$
$$\%\ \text{efficiency} = 93.3\%$$

f. The battery used in this problem is considered a constant voltage source because the internal resistance r_i of 0.2 Ω will drop only a small portion of the battery voltage for R_L values 2 Ω or higher.

PRACTICE PROBLEMS

Sec. 12-3 Solve the following.

1. A 1.5-V cell develops a terminal voltage of 1.25 V when delivering 250 mA to a load resistor R_L.
   a. What is the internal resistance r_i of the battery?
   b. What is the load resistance R_L?
   c. How much total power is being dissipated?
   d. How much power is being absorbed by R_L?
   e. What is the efficiency of the power transfer?
   f. What R_L value will cause the power transfer from generator to load to be maximum?
   g. How much power will be dissipated by R_L when the generator and load are matched?
   h. What is the efficiency of power transfer for the matched load condition?

2. A 9-V battery develops a terminal voltage of 8.4 V when delivering 1.2 A to a load resistance R_L.
   a. What is the internal resistance r_i of the battery?
   b. What is the load resistance R_L?
   c. How much total power is being dissipated?
   d. How much power is being absorbed by R_L?
   e. What is the efficiency of the power transfer?
   f. How much power will be dissipated by R_L when the generator and load are matched?
   g. What is the efficiency of power transfer for the matched load condition?
3. A 10-V source has an internal resistance, r_i, of 50 Ω. Calculate I_L, V_L, P_L, P_T, and the efficiency of power transfer for the following R_L values: 10 Ω, 25 Ω, 50 Ω, 100 Ω, 500 Ω, and 1 kΩ.
4. A 25-V source has an internal resistance, r_i, of 100 Ω. Calculate I_L, V_L, P_L, P_T, and the efficiency of power transfer for the following R_L values: 20 Ω, 50 Ω, 100 Ω, 200 Ω, 500 Ω, and 1 kΩ.
5. Three identical 1.5-V cells are connected in series to form a battery. Each cell has an internal resistance of 0.1 Ω.
   a. What is the internal resistance, r_i, of the battery?
   b. What is the terminal voltage when a 1.5-Ω load is connected?
   c. How much total power is dissipated for the R_L value of 1.5 Ω in **b**?
   d. How much power is absorbed by the 1.5-Ω load in **b**?
   e. What is the efficiency of the power transfer when the 1.5-Ω load is connected across this battery?
   f. How much power is lost in each cell of the battery when a 1.2-Ω load is connected?
6. Three identical 6-V batteries are connected in parallel. Each cell has an internal resistance of 0.15 Ω.
   a. What is the terminal voltage when a load current of 15 A exists?
   b. How much total power is dissipated for the 15-A load in **a**?
   c. How much power is being absorbed by the load in **a**?
   d. What is the efficiency of the power transfer when the 15-A load is connected?
   e. How much power is supplied to the load by each battery?
7. Three 1.5-V cells are connected in series to form a battery. Two of the three cells are identical, having an internal resistance of 0.1 Ω each. One weak cell has an internal resistance of 2 Ω.
   a. What is the terminal voltage when a 2.8-Ω load is connected?
   b. What is the efficiency of the power transfer for the 2.8-Ω load in **a**?
8. How does the condition of an ohmmeter's battery affect the setting of the 0-Ω adjust control on the $R \times 1$ range?
9. On which range should the condition of an ohmmeter's internal battery be checked? Why?
10. Can the condition of a battery be checked by connecting a DMM, having an input resistance of 10 MΩ, across the output terminals? Why?
11. For the constant current source shown in Fig. 12-18, find I_L for the R_L values listed.

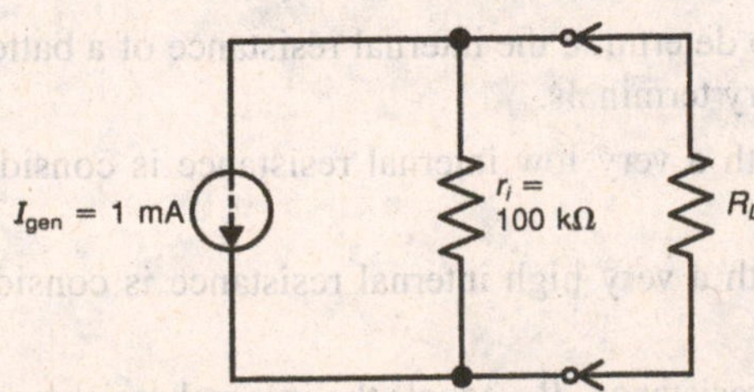

Fig. 12-18 Find I_L for R_L = 0 Ω, 1 Ω, 10 Ω, 100 Ω, 1 kΩ, and 10 kΩ.

12. For the constant current source shown in Fig. 12-19, find I_L for the R_L values listed.

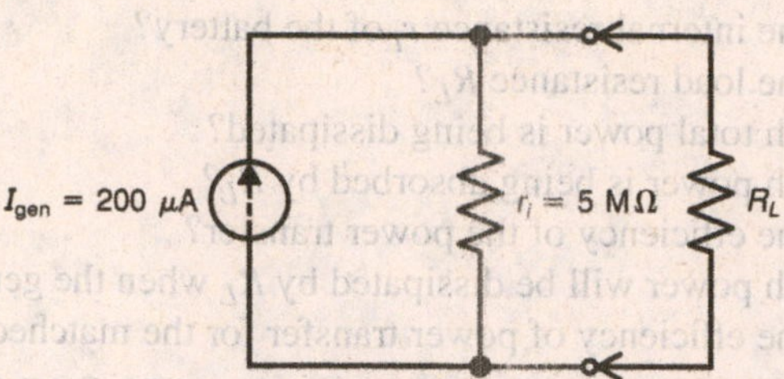

Fig. 12-19 Find I_L for R_L = 0 Ω, 1 Ω, 10 Ω, 500 Ω, 10 kΩ, 50 kΩ, 200 kΩ, and 500 kΩ.

13. For the constant voltage source shown in Fig. 12-20, find V_L for the R_L values listed.

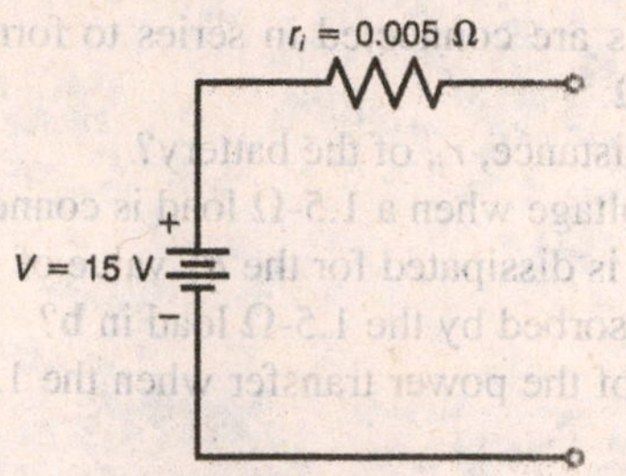

Fig. 12-20 Find V_L for R_L = 0.05 Ω, 0.1 Ω, 1 Ω, 5 Ω, 10 Ω, 100 Ω, 500 Ω, 1 kΩ, 5 kΩ, and 10 kΩ.

14. **(a)** For Fig. 12-18, what R_L value will produce a maximum transfer of power from generator to load? **(b)** How much power will be delivered to R_L for this matched-load condition?

END OF CHAPTER TEST

Chapter 12: Batteries

Answer True or False.

1. A battery consists of a combination of individual voltaic cells connected in series or in parallel and in some cases in series-parallel.
2. A battery provides a steady DC voltage of fixed polarity.
3. To increase the current capacity, voltaic cells are connected in parallel.
4. To increase the voltage output, voltaic cells are connected in series.
5. The voltage output of a cell or battery decreases as the load current increases because some voltage is dropped across its own internal resistance, r_i.
6. The no-load voltage of a generator is always less than its full-load voltage.
7. The best way to determine the internal resistance of a battery is to connect an ohmmeter across the battery terminals.
8. A generator with a very low internal resistance is considered to be a constant voltage source.
9. A generator with a very high internal resistance is considered to be a constant current source.
10. When the load resistance, R_L, equals the internal resistance, r_i, of a generator, maximum power is transferred from the generator to the load.

CHAPTER 13

Magnetism

In this chapter, you will be introduced to magnetism and magnetic fields. Not only will you become familiar with the units associated with magnetic fields, but you will also learn how to convert between the units associated with each quantity.

SEC. 13-1 THE MAGNETIC FIELD

Permanent magnets and electromagnets possess a magnetic field of force that can do the work of attraction or repulsion. Figure 13-1 shows each type of magnet with its surrounding magnetic field. Either type of magnet can be considered a generator for the external magnetic field. The group of magnetic field lines that exist for any magnet is called *magnetic flux.* A strong magnetic field will have more lines of force than a weak magnetic field. The units of magnetic flux are the line, the maxwell (Mx), and the weber (Wb), and they are related as follows.

$$1\ \text{Mx} = 1\ \text{magnetic field line}$$
$$1\ \text{Wb} = 1 \times 10^8\ \text{Mx}$$
$$1\ \mu\text{Wb} = 1 \times 10^{-6}\ \text{Wb} \times 10^8\ \text{Mx/Wb} = 100\ \text{Mx}$$

The μWb unit is commonly used since the Wb unit is too large for typical fields. The symbol used to represent magnetic flux is the Greek letter phi (ϕ). For the units of magnetic flux ϕ, the maxwell is a cgs unit, whereas the weber is an mks or SI unit. All the units of magnetic flux apply to the magnetic field produced by either the permanent magnet in Fig. 13-1*a* or the electromagnet in Fig. 13-1*b*.

In Fig. 13-1, it is evident that the concentration of the magnetic field is greatest at the magnet's north and south poles. The magnetic field's flux-line density is the number of magnetic-field lines per unit area of a section perpendicular to the direction of flux. The symbol used to represent flux density is B. The formula used to determine flux density is

$$B = \frac{\phi}{A}$$

where ϕ = number of flux lines
A = area in which flux lines are concentrated
B = flux density

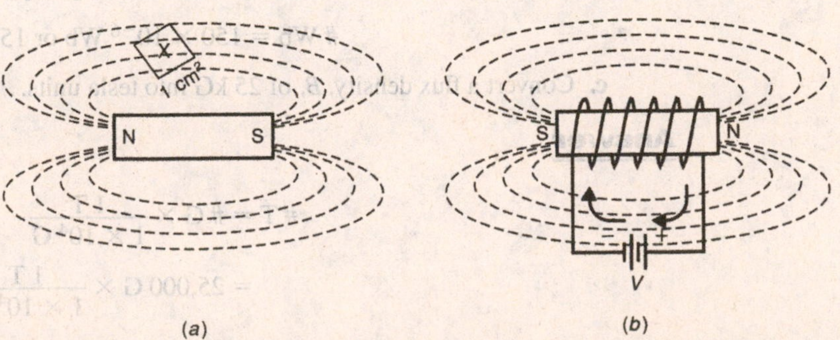

Fig. 13-1 Magnetic field of force. (*a*) Magnetic field surrounding a permanent magnet. (*b*) Magnetic field surrounding an electromagnet.

The gauss (G) is a cgs unit of flux density. This unit is one magnetic field line per square centimeter or $1G = \frac{1\text{ Mx}}{\text{cm}^2}$. In Fig. 13-1*a*, for point X, the flux-line density is three lines per square centimeter, or 3 G.

A larger unit of flux density is the tesla (T), an SI unit. The tesla unit of flux density is one weber of magnetic flux per square meter, or $1\text{ T} = \frac{1\text{ Wb}}{\text{m}^2}$. Since the gauss unit is so small, it is often used in kilogauss units, where 1 kg = 10^3 G.

Conversion between Units

It is often necessary to convert between the different units when dealing with magnetic flux, ϕ, and flux density, B. To convert from maxwells (Mx) to webers (Wb) or vice versa, use the following conversions:

$$\#\text{ Wb} = \#\text{ Mx} \times \frac{1\text{ Wb}}{1 \times 10^8\text{ Mx}}$$

$$\#\text{ Mx} = \#\text{ Wb} \times \frac{1 \times 10^8\text{ Mx}}{1\text{ Wb}}$$

To convert from teslas (T) to gauss (G) units or vice versa, use the following conversions:

$$\#\text{ G} = \#\text{ T} \times \frac{1 \times 10^4\text{ G}}{1\text{ T}}$$

$$\#\text{ T} = \#\text{ G} \times \frac{1\text{ T}}{1 \times 10^4\text{ G}}$$

Solved Problems

a. Convert 4.5 Wb of magnetic flux, ϕ, into maxwell units.

Answer

$$\#\text{ Mx} = \#\text{ Wb} \times \frac{1 \times 10^8\text{ Mx}}{1\text{ Wb}}$$

$$= 4.5\text{ Wb} \times \frac{1 \times 10^8\text{ Mx}}{1\text{ Wb}}$$

$$\#\text{ Mx} = 4.5 \times 10^8\text{ Mx}$$

b. Convert a magnetic flux, ϕ, of 15,000 Mx into weber units.

Answer

$$\#\text{ Wb} = \#\text{ Mx} \times \frac{1\text{ Wb}}{1 \times 10^8\text{ Mx}}$$

$$= 15{,}000\text{ Mx} \times \frac{1\text{ Wb}}{1 \times 10^8\text{ Mx}}$$

$$\#\text{ Wb} = 150 \times 10^{-6}\text{ Wb or } 150\ \mu\text{Wb}$$

c. Convert a flux density, B, of 25 kG into tesla units.

Answer

$$\#\text{ T} = \#\text{ G} \times \frac{1\text{ T}}{1 \times 10^4\text{ G}}$$

$$= 25{,}000\text{ G} \times \frac{1\text{ T}}{1 \times 10^4\text{ G}}$$

$$\#\text{ T} = 2.5\text{ T}$$

d. Convert a flux density, B, of 0.025 T into gauss units.

Answer

$$\#G = \#T \times \frac{1 \times 10^4 \text{ G}}{1 \text{ T}}$$

$$= 0.025 \text{ T} \times \frac{1 \times 10^4 \text{ G}}{1 \text{ T}}$$

$$\#G = 250 \text{ G}$$

Permeability

The ability of a material to concentrate magnetic flux is called its permeability. The symbol is μ. The symbol for relative permeability is μ_r, where the subscript r indicates relative. There are no units for relative permeability because μ_r is a comparison of two flux densities, and the units cancel in the equation. Numerical values of μ_r for different materials are compared with air or vacuum. As an equation for relative permeability we have

$$\mu_r = \frac{\text{flux density in specified material}}{\text{flux density in air}}$$

Again, there are not units for μ_r because they cancel in the equation.

Solved Problem

The flux density, B, is 100 G in the air core of an electromagnet. When an iron core is inserted the flux density, B, increases to 15 T. What is the relative permeability of the iron core?

Answer

First, convert 100 G to tesla units.

$$\#T = \#G \times \frac{1 \text{ T}}{1 \times 10^4 \text{ G}}$$

$$= 100 \text{ G} \times \frac{1 \text{ T}}{1 \times 10^4 \text{ G}}$$

$$\#T = 0.01 \text{ T or } 1 \times 10^{-2} \text{ T}$$

Next, determine μ_r.

$$\mu_r = \frac{\text{flux density in iron core}}{\text{flux density in air}}$$

$$= \frac{15 \text{ T}}{0.01 \text{ T}}$$

$$\mu_r = 1{,}500$$

PRACTICE PROBLEMS

Sec. 13-1 Solve the following.

1. What symbol is used to represent (**a**) magnetic flux, (**b**) flux density?
2. What are the units of (**a**) magnetic flux, (**b**) flux density?
3. Define (**a**) 1 maxwell, (**b**) 1 weber.
4. Define (**a**) 1 gauss, (**b**) 1 tesla.

5. Define the term *relative permeability.*
6. What is the unit for relative permeability?

For Probs. 7–26, make the conversions listed.

7. 20,000 Mx to Wb
8. 100 T to G
9. 2,500 G to T
10. 45,000 μWb to Mx
11. 1,500 Mx to Wb
12. 0.05 T to kG
13. 150 kG to T
14. 40 kG to T
15. 0.25 Wb to Mx
16. 480 μWb to Mx
17. 0.002 T to G
18. 0.0015 Wb to Mx
19. 0.005 Wb to Mx
20. 5 T to kG
21. 15 kG to T
22. 500 μWb to Mx
23. 100 G to T
24. 7.5 T to G
25. 1,000 μWb to Mx
26. 900 Mx to μWb
27. A permanent magnet has a flux ϕ of 2,000 μWb. How many lines of force does this flux represent?
28. An electromagnet has a flux ϕ of 30,000 μWb. How many lines of force does this flux represent?
29. A magnet produces 40,000 Mx of magnetic flux. Convert this flux value to weber units.
30. With a flux of 25 μWb through an area of 0.004 m^2, what is the flux density in tesla units?
31. The flux density at a south pole of a magnet is 2,800 G. What is the flux density in tesla units?
32. The ϕ is 5,000 Mx through 2 cm^2. How much is B in gauss units?
33. How much is B in tesla units for a flux ϕ of 400 μWb through an area of 0.025 m^2?
34. With a flux of 150 μWb through an area of 0.25 m^2, what is the flux density in gauss units?
35. The flux density is 500 G in the air core of an electromagnet. When an iron core is inserted, the flux density in the core is 24 T. How much is the relative permeability of the core?
36. The flux density is 1.2 T in the iron core of an electromagnet. When the iron core is removed, the flux density in the air core is 240 G. How much is the relative permeability of the core?
37. The flux density in an iron core is 2 T. If the area of the core is 50 cm^2, calculate the total number of flux lines in the core.
38. The flux density in an iron core is 450 G. If the area of the core is 1×10^{-2} m^2, calculate the total number of flux lines in the core.

Chapter 13: Magnetism Answer True or False.

1. The term magnetic flux is used to describe the group of magnetic field lines that exist around a magnet.
2. The Greek letter phi (ϕ) is the symbol for magnetic flux.
3. One maxwell (Mx) equals one magnetic field line.
4. One weber (Wb) equals 100 maxwells.
5. A magnetic field's flux density is the number of magnetic field lines per unit area.
6. The symbol for flux density is B.
7. $1 \text{ gauss (G)} = \dfrac{1 \text{ Mx}}{\text{cm}^2}$.
8. $1 \text{ tesla (T)} = \dfrac{1 \text{ Wb}}{\text{m}^2}$.
9. The ability of a material to concentrate magnetic flux is called its permeability.
10. The relative permeability (μ_r) is a ratio of two flux densities.

CHAPTER

14

Electromagnetism

In this chapter, you will be introduced to the units and unit conversions associated with electromagnetism. You will also be introduced to Ohm's law for magnetic circuits and to Faraday's law of induced voltage.

SEC. 14-1 MAGNETOMOTIVE FORCE

A magnetic field is always associated with current flow. Therefore, the magnetic units can be derived from the current that produces the field. With an electromagnet, the strength of the magnetic field will depend on two factors: (1) the amount of current flow in the turns of the coil and (2) the number of turns in a specific length. The electromagnet serves as a bar magnet, with opposite poles at its ends, providing a magnetic field proportional to the ampere-turns (A · t). As a formula,

$$\text{A} \cdot \text{t} = N \times I$$

where N = number of turns of the coil
I = current flowing through the coil in amperes

The quantity NI specifies the amount of magnetizing force, magnetic potential, or magnetomotive force (mmf). In general, mmf = NI. The practical unit for mmf is the ampere-turn, which is an SI unit. The cgs unit of mmf is the gilbert, abbreviated Gb. Furthermore,

$$1.26 \text{ Gb} = 1 \text{ A} \cdot \text{t}$$

Conversion between Units

To convert from A · t to Gb or vice versa, use the following conversion formulas.

$$\# \text{ Gb} = \# \text{ A} \cdot \text{t} \times \frac{1.26 \text{ Gb}}{1 \text{ A} \cdot \text{t}}$$

$$\# \text{ A} \cdot \text{t} = \# \text{ Gb} \times \frac{1 \text{ A} \cdot \text{t}}{1.26 \text{ Gb}}$$

Solved Problems

a. Calculate the A · t of mmf for a coil with 500 turns and a 20 mA current.

Answer

$$\begin{aligned} \text{mmf} &= NI \\ &= 500 \text{ t} \times 20 \text{ mA} \\ \text{mmf} &= 10 \text{ A} \cdot \text{t} \end{aligned}$$

b. A coil with 900 turns must produce 360 A · t of mmf. How much current is necessary?

Answer

$$\begin{aligned} \text{I} &= \frac{\text{mmf}}{N} \\ &= \frac{360 \text{ A} \cdot \text{t}}{900 \text{ t}} \\ \text{I} &= 400 \text{ mA} \end{aligned}$$

c. Convert 18.9 Gb to A · t.

Answer

$$\#\,A\cdot t = \#\,Gb \times \frac{1\,A\cdot t}{1.26\,Gb}$$

$$= 18.9\,Gb \times \frac{1\,A\cdot t}{1.26\,Gb}$$

$$\#\,A\cdot t = 15\,A\cdot t$$

d. Convert 2,850 A · t to Gb.

Answer

$$\#\,Gb = \#\,A\cdot t \times \frac{1.26\,Gb}{1\,A\cdot t}$$

$$= 2{,}850\,A\cdot t \times \frac{1.26\,Gb}{1\,A\cdot t}$$

$$\#\,Gb = 3{,}591\,Gb$$

PRACTICE PROBLEMS

Sec. 14-1 For Probs. 1–10, make the conversions listed.

1. 1,000 A · t to Gb
2. 500 Gb to A · t
3. 6,300 Gb to A · t
4. 250 A · t to Gb
5. 3,780 Gb to A · t
6. 12 A · t to Gb
7. 50,400 Gb to A · t
8. 440 A · t to Gb
9. 8 Gb to A · t
10. 200 A · t to Gb
11. Calculate the ampere-turns for a coil with 20,000 turns and a 2.5-mA current.
12. Calculate the ampere-turns for a coil with 120 turns and a 50-mA current.
13. A coil with 5,000 turns must produce 120 A · t of mmf. How much current is necessary?
14. A coil with 40,000 turns must produce 1,000 A · t of mmf. How much current is necessary?
15. **(a)** A coil having 15,000 turns has a resistance of 120 Ω. Calculate the mmf in ampere-turns if the coil is connected across a 9-V battery. **(b)** How much is mmf in gilberts?
16. A coil with 4,000 turns has an mmf of 240 Gb. How much current is flowing through the coil?
17. A coil with 150 turns must produce 630 Gb of mmf. How much current is necessary?
18. **(a)** A coil having 125 turns has a resistance of 4 Ω. Calculate the mmf in gilberts if the coil is connected across a 3-V battery. **(b)** How much is mmf in ampere-turns?
19. A coil with 12,000 turns has an mmf of 360 A · t. How much current is flowing through the coil?
20. Calculate the mmf in gilberts for a coil with 3,000 turns and a current of 9 mA.

SEC. 14-2 INTENSITY OF MAGNETIC FIELD

The ampere-turns or gilberts of mmf specify the magnetizing force, but the intensity of a magnetic field depends on the length of the coil. The mks unit of field intensity H is the ampere-turn per meter or

$$H = \frac{\text{A} \cdot \text{t}}{\text{m}}$$

The H is the intensity at the center of an air core coil. With an iron core, H is the field intensity through the entire iron core.

The H is basically mmf per unit of length. In SI units, H is in A · t/m. The cgs unit for H is the oersted, abbreviated Oe.

$$1 \text{ Oe} = \frac{1 \text{ Gb}}{\text{cm}}$$

Conversion between Units

It is often necessary to convert from $\frac{\text{A} \cdot \text{t}}{\text{m}}$ to Oe or vice versa. When this is necessary, use the following conversion formulas:

$$\# \frac{\text{A} \cdot \text{t}}{\text{m}} = \# \text{ Oe} \times \frac{\frac{1 \text{ A} \cdot \text{t}}{\text{m}}}{0.0126 \text{ Oe}}$$

$$\# \text{ Oe} = \# \frac{\text{A} \cdot \text{t}}{\text{m}} \times \frac{0.0126 \text{ Oe}}{\frac{1 \text{ A} \cdot \text{t}}{\text{m}}}$$

Solved Problems

a. Convert a field intensity, H, of 200 $\frac{\text{A} \cdot \text{t}}{\text{m}}$ to Oe.

Answer

$$\# \text{ Oe} = \# \frac{\text{A} \cdot \text{t}}{\text{m}} \times \frac{0.0126 \text{ Oe}}{\frac{1 \text{ A} \cdot \text{t}}{\text{m}}}$$

$$= 200 \frac{\text{A} \cdot \text{t}}{\text{m}} \times \frac{0.0126 \text{ Oe}}{\frac{1 \text{ A} \cdot \text{t}}{\text{m}}}$$

$$\# \text{ Oe} = 2.52 \text{ Oe}$$

Notice that the $\frac{\text{A} \cdot \text{t}}{\text{m}}$ units cancel, leaving Oe.

b. Convert 2.52 Oe back into $\frac{\text{A} \cdot \text{t}}{\text{m}}$ units.

Answer

$$\# \frac{\text{A} \cdot \text{t}}{\text{m}} = \# \text{ Oe} \times \frac{\frac{1 \text{ A} \cdot \text{t}}{\text{m}}}{0.0126 \text{ Oe}}$$

$$= 2.52 \text{ Oe} \times \frac{\frac{1 \text{ A} \cdot \text{t}}{\text{m}}}{0.0126 \text{ Oe}}$$

$$\# \frac{\text{A} \cdot \text{t}}{\text{m}} = 200 \frac{\text{A} \cdot \text{t}}{\text{m}}$$

Sec. 14-2 For Probs. 1–12, make the following conversions.

1. 400 A · t/m to Oe
2. 3.15 Oe to A · t/m
3. 1,250 A · t/m to Oe
4. 0.126 Oe to A · t/m
5. 900 A·t/m to Oe
6. 1,260 Oe to A · t/m
7. 252 Oe to A · t/m
8. 0.54 A · t/m to Oe
9. 693 Oe to A · t/m
10. 6,250 A · t/m to Oe
11. 12.6 Oe to A · t/m
12. 40 A · t/m to Oe

SEC. 14-3 PERMEABILITY

The permeability (μ) of a material describes its ability to concentrate magnetic flux. A good magnetic material with high relative permeability (μ_r) can concentrate magnetic flux and produce a large value of flux density B for a specified H. These factors are related as follows:

$$B = \mu \times H$$

or

$$\mu = \frac{B}{H}$$

For SI units, B is in teslas, and H is in A · t/m.

In SI units, the permeability of air or vacuum is 1.26×10^{-6}. The symbol is μ_0. Therefore, to find μ when μ_r is known, we must multiply μ_r by 1.26×10^{-6} T/(A · t/m).

$$\mu = \mu_r \times 1.26 \times 10^{-6} \text{ T/(A} \cdot \text{t/m)}$$

If $\mu_r = 100$, then

$$\mu = 100 \times 1.26 \times 10^{-6} \text{ T/(A} \cdot \text{t/m)}$$
$$\mu = 126 \times 10^{-6} \text{ T/(A} \cdot \text{t/m)}$$

Solved Problem

A given coil with an iron core has a field intensity H of 250 A · t/m. The relative permeability, μ_r of the core is 200. Calculate B in teslas.

Answer

$$B = \mu \times H$$

Converting μ_r to μ, we have

$$\mu = \mu_r \times 1.26 \times 10^{-6} \text{ T/(A} \cdot \text{t/m)}$$
$$= 200 \times 1.26 \times 10^{-6} \text{ T/(A} \cdot \text{t/m)}$$
$$\mu = 252 \times 10^{-6} \text{ T/(A} \cdot \text{t/m)}$$

To find B, we multiply μ by H as follows:

$$\begin{aligned} B &= \mu \times H \\ &= 252 \times 10^{-6}\ \text{T/(A}\cdot\text{t/m)} \times 250\ \text{A}\cdot\text{t/m} \\ &= 63 \times 10^{-3}\ \text{T} \\ B &= 0.063\ \text{T} \end{aligned}$$

Notice that the A · t and m units cancel, leaving B in teslas.

PRACTICE PROBLEMS

Sec. 14-3 Solve the following

1. Find μ in SI units when $\mu_r = 2{,}500$.
2. Find μ in SI units when $\mu_r = 6{,}000$.
3. Find μ_r when $\mu = 37.8 \times 10^{-3}$ T/(A · t/m).
4. Find B in teslas when $\mu_r = 400$ and $H = 200$ A · t/m.
5. Find B in teslas when $\mu_r = 50$ and $H = 150$ A · t/m.
6. Find H in A · t/m when $B = 12.6$ T and $\mu_r = 100$.
7. A coil with an iron core has a field intensity H of 4,000 A · t/m. The relative permeability μ_r of the core is 125. Calculate B in teslas.
8. A coil with an iron core has a field intensity of 25 A · t/m. The relative permeability μ_r of the core is 600. Calculate B in teslas.
9. A coil with an iron core has a field intensity H of 150 A · t/m. The relative permeability μ_r of the core is 400. Calculate B in teslas.
10. The flux density in an iron-core coil is 0.02 T. If the relative permeability of the core is 500, what is H in A · t/m?
11. Calculate the relative permeability, μ_r, of an iron core when a field intensity H of 50 A · t/m produces a flux density B of 0.063 T.
12. Calculate the relative permeability, μ_r, of an iron core when a field intensity H of 1,500 A·t/m produces a flux density B of 0.252 T.

SEC. 14-4 OHM'S LAW FOR MAGNETIC CIRCUITS

In comparison to electric circuits, the magnetic flux ϕ corresponds to current. The flux ϕ is produced by ampere-turns NI of mmf. The opposition to the production of magnetic flux is called its *reluctance,* which is comparable to resistance. The symbol for reluctance is $\mathcal{R}$. Reluctance is inversely proportional to permeability.

The three factors—flux, ampere-turns, and reluctance—are related as follows:

$$\phi = \frac{\text{mmf}}{\mathcal{R}}$$

The mmf is considered to produce flux ϕ in a magnetic material against the opposition of its reluctance. There are no specific units for reluctance, but it can be considered an mmf/ϕ ratio in ampere-turns per weber or A · t/Wb.

Solved Problems

a. A coil has an mmf of 250 A· t and a reluctance of 1.25×10^6 A · t/Wb. Calculate the total flux in μWb.

Answer

Since $\phi = \text{mmf}/\mathcal{R}$, we have

$$\phi = \frac{250 \text{ A·t}}{1.25 \times 10^6 \text{ A·t/Wb}}$$
$$= 200 \times 10^{-6} \text{ Wb}$$
$$\phi = 200\ \mu\text{Wb}$$

b. A magnetic material has a total flux of 32 μWb with an mmf of 40 A · t. Calculate the reluctance in A · t/Wb.

Answer

Since $\phi = \text{mmf}/\mathcal{R}$, $\mathcal{R} = \text{mmf}/\phi$, and we have

$$\mathcal{R} = \frac{40 \text{ A·t}}{32 \times 10^{-6} \text{ Wb}}$$
$$\mathcal{R} = 1.25 \times 10^6 \text{ A·t/Wb}$$

c. If the reluctance of a material is 100 × 10^6 A · t/Wb, how much mmf would be required to produce a flux ϕ value of 250 μWb?

Answer

Since $\phi = \text{mmf}/\mathcal{R}$, $\text{mmf} = \phi \times \mathcal{R}$, and we have

$$\text{mmf} = 100 \times 10^6 \text{ A·t/Wb} \times 250 \times 10^{-6} \text{ Wb}$$
$$\text{mmf} = 25{,}000 \text{ A·t}$$

PRACTICE PROBLEMS

Sec. 14-4 Solve the following

1. A coil has an mmf of 75 A · t and a reluctance of 50 × 10^6 A · t/Wb. Calculate the total flux in microwebers.
2. A magnetic material has a total flux of 1,200 μWb with an mmf of 20 A · t. Calculate the reluctance in A · t/Wb.
3. If the reluctance of a material is 25 × 10^6 A · t/Wb, how much mmf would be required to produce a flux ϕ value of 100 μWb?
4. A magnetic material has a total flux of 1 μWb with an mmf of 15 A · t. Calculate the reluctance in A · t/Wb.
5. A coil has an mmf of 500 A · t and a reluctance of 0.05 × 10^6 A · t/Wb. Calculate the total flux in webers.
6. If the reluctance of a material is 0.15 × 10^6 A · t/Wb, how much mmf would be required to produce a flux ϕ of 2,000 μWb?
7. A coil has an mmf of 40,000 A · t and a reluctance of 15 × 10^6 A · t/Wb. Calculate the total flux in microwebers.
8. If the reluctance of a material is 5.0 × 10^6 A · t/Wb, how much mmf is required to produce a flux ϕ of 300 μWb?
9. What is the reluctance of a material that produces 1,000 μWb of flux when the applied mmf is 6,000 A · t?
10. What is the reluctance of a material that produces 600 μWb of flux when the applied mmf is 200 A · t?

SEC. 14-5 FARADAY'S LAW OF INDUCED VOLTAGE

When magnetic flux cuts across a conductor, a voltage will be induced across the ends of the conductor. Either the magnetic flux or the conductor can move. Specifically, the amount of the induced voltage is determined by the following equation:

$$V_{\text{ind}} = N \times \frac{d\phi}{dt}$$

where N = number of turns in coil

$\frac{d\phi}{dt}$ = rate of magnetic flux cutting across the coil (webers per second)

This is Faraday's law. The equation tells us that the amount of induced voltage depends on three factors.

1. The number of flux lines in webers that cut across the conductor.
2. The number of turns in the coil.
3. The rate at which the magnetic flux cuts across the conductor.

It is very important to realize here that the flux must be moving (i.e., expanding or collapsing) in order for there to be an induced voltage across the ends of a coil. A stationary magnetic field cannot induce a voltage in a stationary coil!

Solved Problem

A magnetic field cuts across a coil having 1,000 turns at the rate of 2,000 μWb/s. Calculate the induced voltage.

Answer

Since $V_{\text{ind}} = N \times d\phi/dt$,

$$V_{\text{ind}} = 1{,}000 \times \frac{2{,}000\ \mu\text{Wb}}{\text{s}}$$

$$V_{\text{ind}} = 2\ \text{V}$$

PRACTICE PROBLEMS

Sec. 14-5 Solve the following

1. A magnetic field cuts across a coil having 2,000 turns at the rate of 1,000 μWb/s. Calculate the induced voltage.
2. A magnetic field cuts across a coil having 250 turns at the rate of 0.04 Wb/s. Calculate the induced voltage.
3. A magnetic flux of 150 Mx cuts across a coil of 12,000 turns in 1 ms. Calculate the induced voltage.
4. A coil of 1,800 turns has an induced voltage of 150 V. Calculate the rate of flux change $d\phi/dt$ in Wb/s.
5. The magnetic flux of a coil changes from 10,000 μWb to 4,000 μWb in 2.5 ms. How much is $d\phi/dt$?
6. The magnetic flux of a coil changes from 1,000 Mx to 16,000 Mx in 4 μs. If the coil has 200 turns, what is the amount of induced voltage?
7. The magnetic flux around a coil is 80 μWb. If the coil has 50,000 turns, what is the amount of induced voltage when the flux remains stationary?
8. A coil has 120-V induced voltage when the rate of flux change is 0.4 Wb/s. How many turns are in the coil?

9. A coil has 50-V induced voltage when the rate of flux change is 10,000 μWb/s. How many turns are in the coil?
10. A magnetic flux of 0.05 Wb cuts across a coil of 10,000 turns in 250 ms. Calculate the amount of induced voltage for each turn in the coil.
11. How does the polarity of an induced voltage relate to Lenz's law?

END OF CHAPTER TEST

Chapter 14: Electromagnetism

Answer True or False.

1. A magnetic field is always associated with current flow.
2. For an electromagnet, the strength of the magnetic field is determined by the amount of current in the coil and the number of coil turns.
3. For an electromagnet, the larger the current and the greater the number of coil turns, the greater are the ampere-turns (A·t) of magnetomotive force.
4. The units of magnetomotive force (mmf) are the ampere-turn (A·t) and the gilbert (Gb).
5. The permeability of a material describes its ability to concentrate magnetic flux.
6. The units of field intensity are the tesla and the weber.
7. The opposition to the production of magnetic flux is called reluctance.
8. In comparison to electric circuits, magnetic flux, ϕ, corresponds to current.
9. The rate at which magnetic field lines cut across a conductor has little or no effect on the amount of voltage induced in a conductor.
10. A stationary magnetic field cannot produce an induced voltage in a stationary coil.

CHAPTER

Alternating Voltage and Current

An alternating voltage is a voltage that continuously varies in magnitude and periodically reverses in polarity. An alternating voltage is shown in Fig. 15-1. The variations up and down on the waveform show the changes in magnitude. It can also be seen in Fig. 15-1 that the magnitude of the waveform varies with respect to time. Because of its characteristics, the waveform in Fig. 15-1 is often referred to as a time-varying voltage.

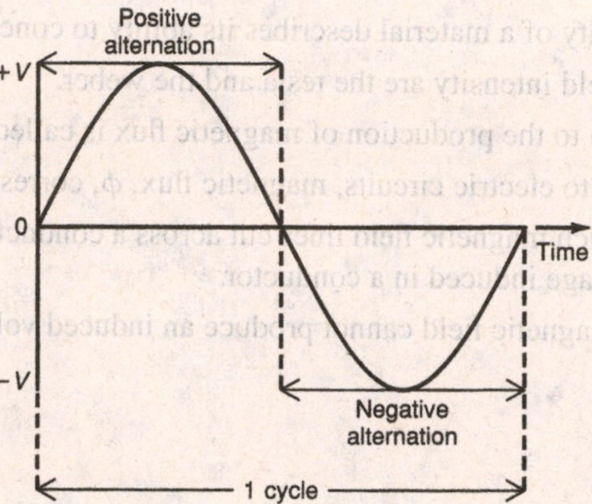

Fig. 15-1 One cycle of alternating voltage showing positive and negative alternations.

A conductor loop rotating through a magnetic field can generate an AC voltage. Figure 15-2 illustrates how the waveform of Fig. 15-1 can be generated. The conductor loop rotates through the magnetic field to generate the induced AC voltage across its open terminals. The polarity of the induced voltage across the ends of the conductor loop is determined by the polarity of the magnetic field and the direction in which the conductors cut across the flux lines. The instantaneous value of voltage is determined by Faraday's law of induced voltage.

SEC. 15-1 ANGULAR MEASURE

One complete revolution of the conductor loop in Fig. 15-2 will produce the waveform in Fig. 15-1. One complete revolution around the circle is called a *cycle.* A cycle is technically defined as including the variations between two successive points having the same value and varying in the same direction. A half-cycle of revolution for the conductor loop of Fig. 15-2 is called an *alternation.* It is very common to refer to either the positive or the negative alternation of a waveform. In Fig. 15-1, we see that the half-cycle above zero is referred to as the positive alternation, whereas the half-cycle below zero is referred to as the negative alternation. Since the cycle of voltage in Fig. 15-1 corresponds to the rotation of a conductor loop around a circle, it is convenient to consider parts of the cycle in angles. One complete cycle

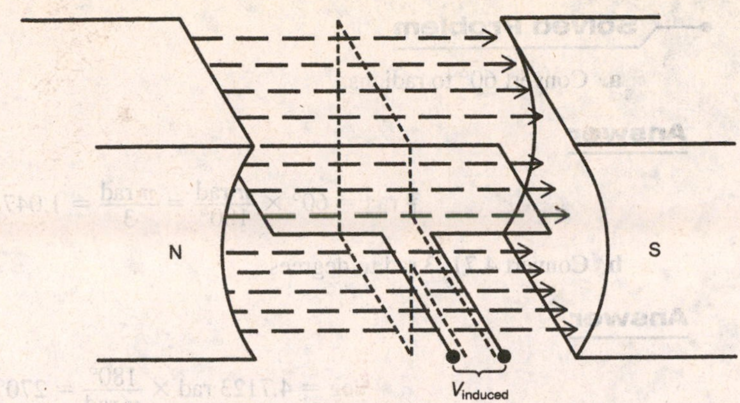

Fig. 15-2 Conductor loop rotating through a magnetic field generates a sinusoidal waveform, as shown in Fig. 15-1.

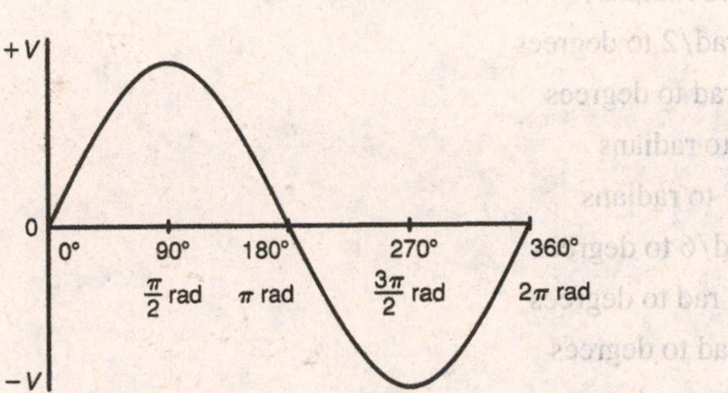

Fig. 15-3 One complete cycle corresponds to 360° or 2π rad.

corresponds to 360°. One-half cycle corresponds to 180°, and one-quarter cycle corresponds to 90°. This is shown in Fig. 15-3.

It is often common to see angles specified in radians. This is because a radian is the angular part of a circle that includes an arc equal to the radius r of the circle. Specifically, 1 radian (abbreviated rad) corresponds to an angle of 57.3°. This is depicted in Fig. 15-4.

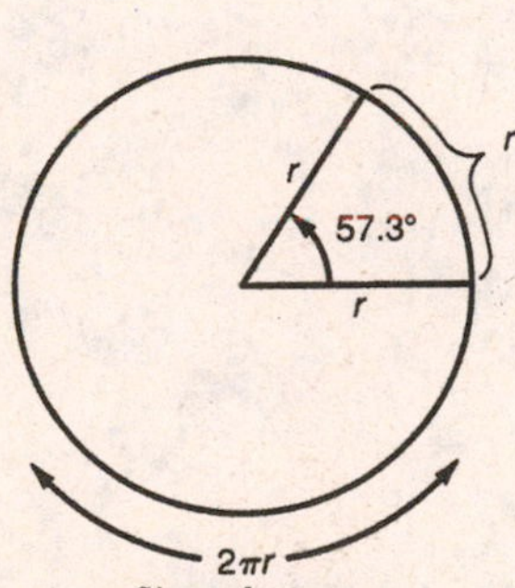

Fig. 15-4 One radian is the angle equal to 57.3°. The complete circle with 360° has 2π rad.

The circumference (C) around the circle in Fig. 15-4 equals $2\pi r$ in length, where r is the radius. The greek letter π, whose value is approximately 3.142, is used to designate the ratio of the circumference to the diameter for any circle. Since $C = 2\pi r$ and 1 rad uses one length of r, the complete circle includes 2π rad. A circle also includes 360°. Therefore, 2π rad = 360°. Knowing the relationships that exist for the circle in Fig. 15-4 allows us to state the following:

$$2\pi \text{ rad} = 360°$$
$$\pi \text{ rad} = 180°$$
$$1 \text{ rad} \cong 57.3°$$

It is often necessary to know how to convert from radians to degrees and vice versa. To convert from degrees to radians, use the following equation:

$$\# \text{ rad} = \# \text{ deg} \times \frac{\pi \text{ rad}}{180°}$$

To convert from radians to degrees, use the following equation:

$$\# \text{ deg} = \# \text{ rad} \times \frac{180°}{\pi \text{ rad}}$$

Solved Problem

a. Convert 60° to radians.

Answer

$$\# \text{ rad} = 60° \times \frac{\pi \text{ rad}}{180°} = \frac{\pi \text{ rad}}{3} = 1.047 \text{ rad}$$

b. Convert 4.7123 rad to degrees.

Answer

$$\# \text{ deg} = 4.7123 \text{ rad} \times \frac{180°}{\pi \text{ rad}} = 270°$$

PRACTICE PROBLEMS

Sec. 15-1 Make the following conversions.

1. 15° to radians
2. 3π rad/2 to degrees
3. 1.5 rad to degrees
4. 75° to radians
5. 445° to radians
6. π rad/6 to degrees
7. 1.45 rad to degrees
8. 3π rad to degrees
9. 210° to radians
10. 30° to radians
11. 0.5 rad to degrees
12. 0.2 rad to degrees
13. 720° to radians
14. 2.5 rad to degrees
15. 120° to radians
16. 310° to radians
17. π rad to degrees
18. 330° to radians
19. π rad/2 to degrees
20. 150° to radians
21. How many degrees are included in one cycle?
22. How many radians are included in one cycle?
23. List the individual factors that affect the amount of induced voltage for the rotating conductor loop of Fig. 15-2.
24. How many radians are included for one-quarter cycle?
25. How many degrees are included for three-quarters of a cycle?

SEC. 15-2 CALCULATING THE INSTANTANEOUS VALUE OF A SINUSOIDAL WAVEFORM

The voltage waveforms in Figs. 15-1 and 15-3 are called *sine waves.* This is because the amount of AC voltage is proportional to the sine of the angle of rotation for the rotating conductor loop in the magnetic field. The sine is a trigonometric function of an angle equal to the ratio of the opposite side to the hypotenuse. The instantaneous value of a sine-wave voltage for any angle of rotation is expressed by the following formula:

$$v_{inst} = V_{max} \times \sin \theta$$

where v_{inst} = instantaneous voltage
V_{max} = maximum voltage value
θ = the angle of rotation
sin = abbreviation for sine

A sine wave has its minimum values at 0° and 180°, and its maximum values at 90° and 270°. Since the sine of 270° is −1, v_{inst} for this angle corresponds to the sine wave's maximum negative value. Waveform A in Fig. 15-5 is a sine wave.

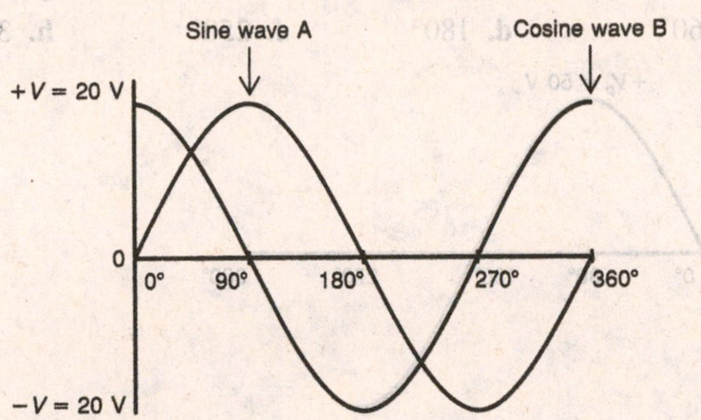

Fig. 15-5 Relationship between a sine wave and a cosine wave.

A cosine wave is represented as waveform B in Fig. 15-5. The *cosine* is a trigonometric function of an angle equal to the ratio of the adjacent side to the hypotenuse. The instantaneous value of voltage for a cosine wave for any angle of rotation is expressed by the following formula:

$$v_{inst} = V_{max} \times \cos \theta$$

where v_{inst} = instantaneous voltage
V_{max} = maximum voltage value
θ = the angle of rotation
cos = abbreviation for cosine

A cosine wave has its maximum values at 0° and 180°, and its minimum values at 90° and 270°. Since the cosine of 180° is −1, v_{inst} for this angle corresponds to the cosine waveform's maximum negative value.

Figure 15-5 allows us to compare a sine wave to a cosine wave. Both waveforms have a peak or maximum value of 20 V.

Solved Problems

a. For waveform A in Fig. 15-5, calculate the instantaneous voltage at 120°.

Answer

Since $v_{inst} = V_{max} \times \sin \theta$ for a sine wave, we have

$$\begin{aligned} v_{inst} &= 20 \text{ V} \times \sin 120° \\ &= 20 \text{ V} \times 0.866 \\ v_{inst} &= 17.32 \text{ V} \end{aligned}$$

b. For waveform B in Fig. 15-5, calculate the instantaneous voltage at 120°.

Answer

Since $v_{inst} = V_{max} \times \cos \theta$ for a cosine wave, we have

$$\begin{aligned} v_{inst} &= 20\text{ V} \times \cos 120° \\ &= 20\text{ V} \times (-0.5) \\ v_{inst} &= -10\text{ V} \end{aligned}$$

PRACTICE PROBLEMS

Sec. 15-2

1. For the waveform shown in Fig. 15-6, calculate the value of instantaneous voltage for the angles listed.

   a. 30° c. 135° e. 210° g. 300°
   b. 60° d. 180° f. 250° h. 330°

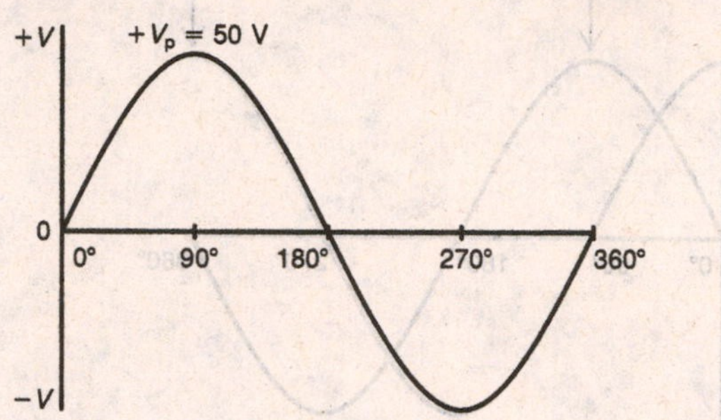

Fig. 15-6

2. For the waveform shown in Fig. 15-7, calculate the value of instantaneous voltage for the angles listed.

   a. 0° c. 60° e. 120° g. 315°
   b. 45° d. 90° f. 225° h. 330°

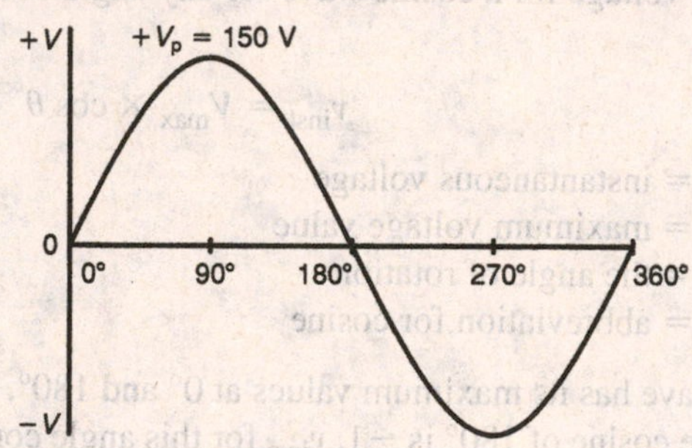

Fig. 15-7

3. For the waveform shown in Fig. 15-8, calculate the value of instantaneous voltage for the angles listed.

   a. 0° c. 90° e. 180° g. 225°
   b. 60° d. 150° f. 210° h. 315°

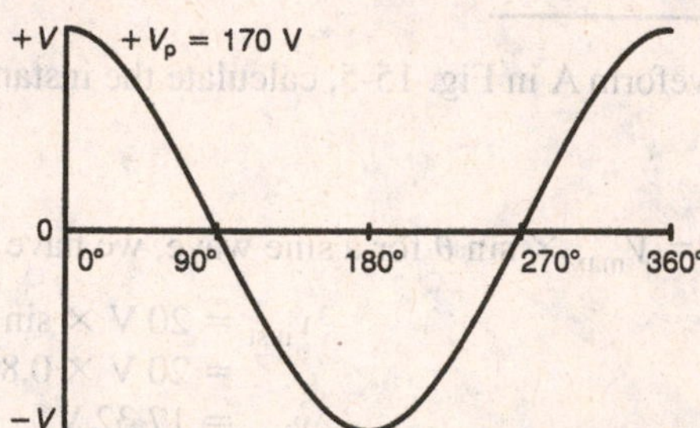

Fig. 15-8

4. For the waveform shown in Fig. 15-9, calculate the value of instantaneous voltage for the angles listed.

a. 0°	**c.** 45°	**e.** 120°	**g.** 210°
b. 30°	**d.** 90°	**f.** 160°	**h.** 270°

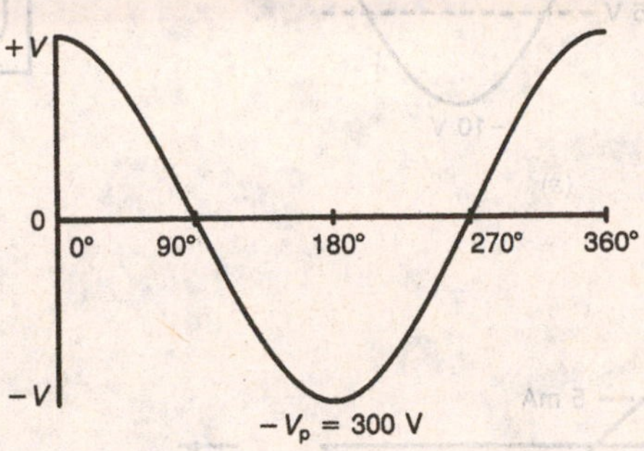

Fig. 15-9

5. A sine wave of voltage has an instantaneous value of 125 V when $\theta = 30°$. Calculate the waveform's maximum voltage at 90°.
6. A cosine wave of current has an instantaneous value of 50 mA when $\theta = 300°$. Calculate the waveform's maximum current value at 0°.
7. A cosine wave of voltage has an instantaneous value of −51.96 V when $\theta = 210°$. Calculate the waveform's maximum voltage value at 0°.
8. A sine wave of voltage has an instantaneous value of 17.5 V when $\theta = 45°$. Calculate the waveform's voltage at 150°.
9. A sine wave of voltage has an instantaneous value of 14.14 V at 30°. Calculate the waveform's voltage at 145°.
10. A cosine wave of voltage has an instantaneous value of 10 V at 30°. Calculate the voltage at 150°.
11. A sine wave of voltage has an instantaneous value of −100 V at $3\pi/2$ rad. Calculate the voltage at $\pi/4$ rad.
12. A cosine wave of voltage has an instantaneous value of 259.8 V at $\pi/3$ rad. Calculate the voltage at π rad.

SEC. 15-3 VOLTAGE AND CURRENT VALUES FOR SINUSOIDAL WAVEFORMS

When an alternating voltage is connected across a resistance, alternating current will flow. The sine wave of voltage in Fig. 15-10*a* represents the applied voltage for the generator in Fig. 15-10*b*. The current flow that is produced by the generator of Fig. 15-10*b* is shown in Fig. 15-10*c*.

In Fig. 15-10*b*, arrow A indicates the direction of current flow for the positive alternation of the voltage waveform. Arrow B indicates the direction of current flow for the negative alternation of the voltage waveform. The current flow for any instant in time can be found by knowing the instantaneous value of voltage. Then

$$i = v/R$$

where i = instantaneous value of current
v = instantaneous value of voltage

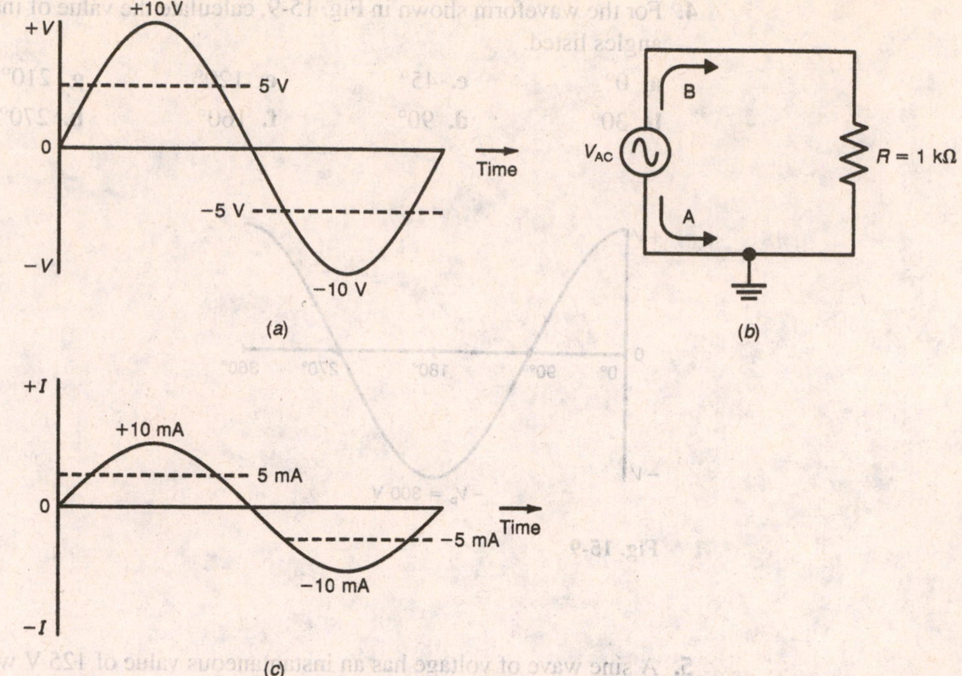

Fig. 15-10 Sine wave alternating voltage across *R* produces a sine wave of alternating current in *R*. (*a*) Waveform of applied voltage. (*b*) Circuit. (*c*) Waveform of circuit current.

The amplitude for the voltage and current waveforms of Fig. 15-10 are specified using several different terms. They are instantaneous, peak, peak-to-peak, root mean square (rms), and average.

The peak value of an AC waveform is its maximum voltage, or current value. The peak value applies to either the positive or negative peak. In Fig. 15-10*a*, the peak value of voltage is 10 V. In Fig. 15-10*c*, the peak value of current is 10 mA because $i = v/R$, where $v = 10$ V and $R = 1$ kΩ.

The peak-to-peak value of an AC waveform is the amplitude difference that exists between the waveform's maximum positive and negative excursions. For Fig. 15-10*a*, the peak-to-peak voltage is $+10\text{ V} - (-10\text{ V}) = 20$ V p-p. The current waveform of Fig. 15-10*c*, has a peak-to-peak value of 20 mA.

The average value of a sine wave of voltage or current is stated for one alternation. The average value of a sine wave is determined by adding several instantaneous values of voltage or current for one alternation and then finding the average of all these values. The average value of voltage or current can be found by multiplying the waveform's peak value by 0.637. In Fig. 15-10*a*, the voltage waveform has an average value of

$$
\begin{aligned}
V_{av} &= 0.637 \times V_p \\
&= 0.637 \times 10\text{ V} \\
V_{av} &= 6.37\text{ V}
\end{aligned}
$$

where V_{av} = average voltage for one alternation
V_p = peak, or maximum voltage

The average value of the current waveform in Fig. 15-10*c* has a value of

$$
\begin{aligned}
I_{av} &= 0.637 \times I_p \\
&= 0.637 \times 10\text{ mA} \\
I_{av} &= 6.37\text{ mA}
\end{aligned}
$$

where I_{av} = average current for one alternation
I_p = peak, or maximum current

The rms value of a sine wave specifies its root-mean-square value, or effective value. It is the most common way of specifying a sine wave's amplitude. The rms value of voltage or current is determined using one alternation of the waveform. It is found by squaring several instantaneous values, determining the average of those values for one alternation, and finally taking the square root of this average value. The rms value of voltage or current for any sine wave can be found by multiplying the waveform's peak value by 0.707. In Fig. 15-10*a*, the voltage waveform has an rms value of

$$\begin{aligned} V_{rms} &= 0.707 \times V_p \\ &= 0.707 \times 10\text{ V} \\ V_{rms} &= 7.07\text{ V} \end{aligned}$$

In Fig. 15-10*c*, the current waveform has an rms value of

$$\begin{aligned} I_{rms} &= 0.707 \times I_p \\ &= 0.707 \times 10\text{ mA} \\ I_{rms} &= 7.07\text{ mA} \end{aligned}$$

A comparison of the different amplitude measurements is shown in Fig. 15-11.

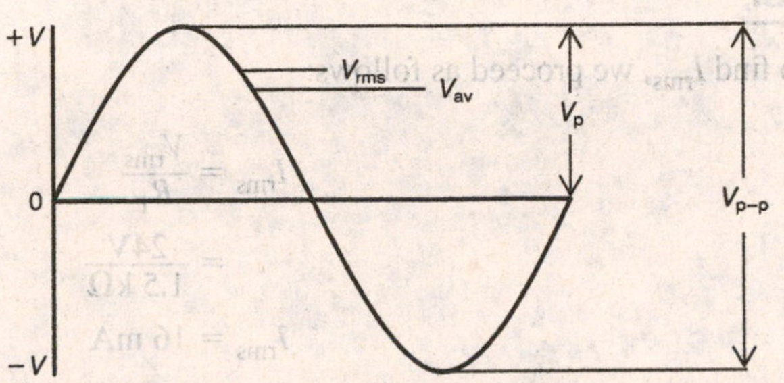

Fig. 15-11 Waveform showing relationship between the different amplitude values used to measure AC voltage and current.

It is often necessary to convert from one amplitude form to another. For example, it may be necessary to know the rms value of a waveform when the peak value is known. Table 15-1 shows the conversion formulas that make it possible to convert from one amplitude form to another.

TABLE 15-1 CONVERSION FORMULAS BETWEEN DIFFERENT AMPLITUDE MEASUREMENTS

Stated Conversion	Formula
V_p to V_{rms}	$V_{rms} = V_p \times 0.707$
V_{rms} to V_p	$V_p = V_{rms} \times 1.414$
V_p to V_{av}	$V_{av} = V_p \times 0.637$
V_{av} to V_p	$V_p = V_{av} \times 1.57$
V_{rms} to V_{p-p}	$V_{p-p} = V_{rms} \times 2.828$
V_{p-p} to V_{rms}	$V_{rms} = V_{p-p} \times 0.3535$
V_{av} to V_{rms}	$V_{rms} = V_{av} \times 1.11$
V_p to V_{p-p}	$V_{p-p} = 2\text{ V peak}$
V_{p-p} to V_p	$V_p = V_{p-p}/2$
V_{rms} to V_{av}	$V_{av} = V_{rms} \times 0.9$
V_{p-p} to V_{av}	$V_{av} = V_{p-p} \times 0.318$
V_{av} to V_{p-p}	$V_{p-p} = V_{av} \times 3.14$

Solved Problem

For the AC circuit shown in Fig. 15-12, find I_{rms}, I_p, I_{av}, $I_{p\text{-}p}$, and P_1.

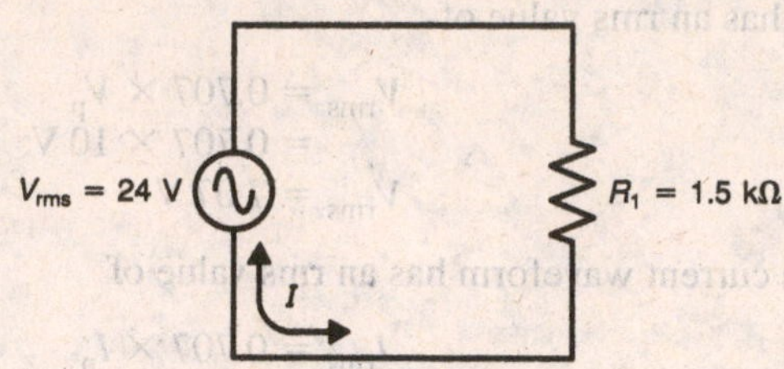

Fig. 15-12 An AC circuit.

Answer

To find I_{rms}, we proceed as follows:

$$I_{rms} = \frac{V_{rms}}{R_1}$$
$$= \frac{24\text{V}}{1.5\ \text{k}\Omega}$$
$$I_{rms} = 16\ \text{mA}$$

To find I_p, $I_{p\text{-}p}$, and I_{av}, use Table 15-1. Find I_p first.

$$I_p = I_{rms} \times 1.414$$
$$= 16\ \text{mA} \times 1.414$$
$$I_p = 22.62\ \text{mA}$$

Then $I_{p\text{-}p}$ can be found by multiplying $I_p \times 2$.

$$I_{p\text{-}p} = I_p \times 2$$
$$= 22.62\ \text{mA} \times 2$$
$$I_{p\text{-}p} = 45.25\ \text{mA}$$

Finally, I_{av} is found as follows:

$$I_{av} = I_p \times 0.637$$
$$= 22.62\ \text{mA} \times 0.637$$
$$I_{av} = 14.4\ \text{mA}$$

It should be pointed out that the current values can be found by specifying the voltage in peak, rms, or average and then determining I using V/R. Either method is appropriate.

The power dissipation in R_1 is calculated using rms values of current and/or voltage. For the resistor R_1, the power dissipation is found as follows:

$$P = V_{rms} \times I_{rms}$$
$$= 24\ \text{V} \times 16\ \text{mA}$$
$$P = 384\ \text{mW}$$

The rms value of current or voltage is often referred to as the effective value. An rms voltage of 24 V is just as effective in producing heat in a resistance as 24 V DC. Hence, the term *effective*.

Sec. 15-3 For Figs. 15-13 through 15-18, solve for the unknowns listed.

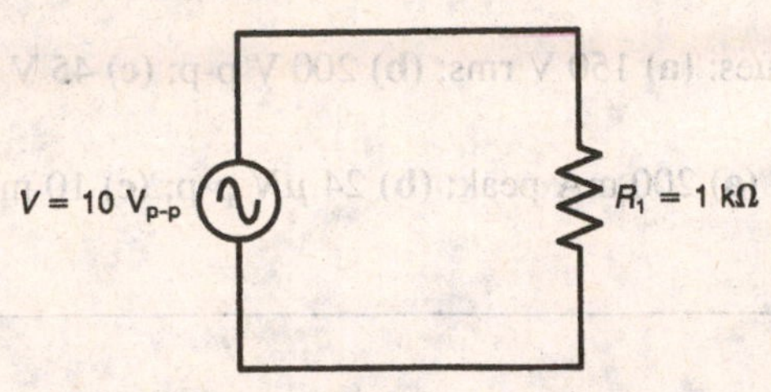

Fig. 15-13 Solve for I_{p-p}, I_p, I_{rms}, I_{av}, V_p, V_{rms}, V_{av}, and P_1.

Fig. 15-14 Solve for I_{p-p}, I_p, I_{rms}, I_{av}, V_p, V_{p-p}, V_{av}, and P_1.

Fig. 15-15 Solve for I_{p-p}, I_p, I_{rms}, I_{av}, V_{p-p}, V_{rms}, V_{av}, and P_1.

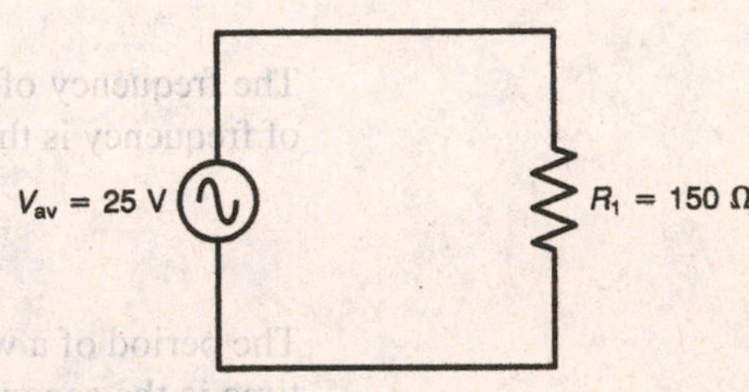

Fig. 15-16 Solve for I_{p-p}, I_p, I_{rms}, I_{av}, V_{p-p}, V_p, V_{rms}, and P_1.

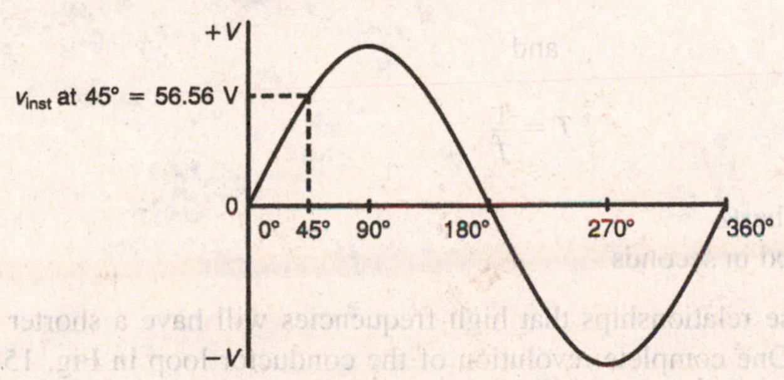

Fig. 15-17 Solve for V_p, V_{rms}, V_{av}, and V_{p-p}.

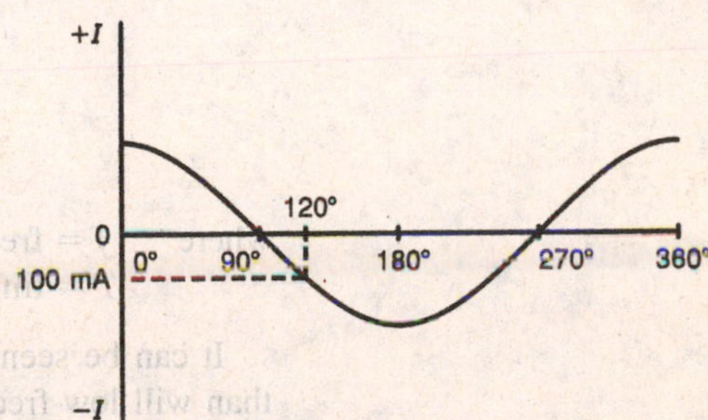

Fig. 15-18 Solve for I_{p-p}, I_p, I_{rms}, and I_{av}.

WORD PROBLEMS

Sec. 15-3 For Probs. 1–10, make the conversions listed.

1. Convert the following into rms values: **(a)** 50 V p-p; **(b)** 1 A peak; **(c)** 40 V av; **(d)** 20 V p-p.
2. Convert the following into average values: **(a)** 25 μA rms; **(b)** 115 V rms; **(c)** 300 mA p-p; **(d)** 120 V peak.
3. Convert the following into peak values: **(a)** 15 V rms; **(b)** 200 mV av; **(c)** 2 kV p-p; **(d)** 120 V rms.
4. Convert the following into peak-to-peak values: **(a)** 36 V rms; **(b)** 50 V av; **(c)** 5 μV peak; **(d)** 18 V rms.
5. Convert the following into average values: **(a)** 2 V peak; **(b)** 80 μV rms; **(c)** 55 V p-p; **(d)** 30 V rms.

6. Convert the following into rms values: **(a)** 350 mA p-p; **(b)** 60 V av; **(c)** 10 V peak; **(d)** 75 mA p-p.
7. Convert the following into peak-to-peak values: **(a)** 100 μA rms; **(b)** 45 V av; **(c)** 22 V peak; **(d)** 2 mA av.
8. Convert the following into peak values: **(a)** 210 V av; **(b)** 66 V rms; **(c)** 15 A rms; **(d)** 1210 V p-p.
9. Convert the following into average values: **(a)** 150 V rms; **(b)** 200 V p-p; **(c)** 45 V peak; **(d)** 88 V peak.
10. Convert the following into rms values: **(a)** 200 mA peak; **(b)** 24 μV p-p; **(c)** 10 mA av; **(d)** 180 μV p-p.

SEC. 15-4 RELATIONSHIP BETWEEN FREQUENCY, TIME, AND WAVELENGTH

The frequency of a waveform is the number of cycles per second that it completes. The unit of frequency is the hertz (Hz).

$$1 \text{ Hz} = 1 \text{ cycle per second}$$

The period of a waveform is the length of time required to complete one cycle. The unit of time is the second. The frequency and period are reciprocals of each other. This is shown as

$$f = \frac{1}{T}$$

and

$$T = \frac{1}{f}$$

where f = frequency in hertz
T = time for period in seconds

It can be seen from these relationships that high frequencies will have a shorter period than will low frequencies. One complete revolution of the conductor loop in Fig. 15-2 corresponds to one cycle of alternating voltage. If the conductor loop makes 60 complete revolutions through the magnetic field in 1 s, the frequency is 60 Hz. The period for one cycle at this frequency is

$$T = \frac{1}{f}$$

$$= \frac{1}{60 \text{ Hz}}$$

$$T = 16.66 \text{ ms}$$

If the speed of rotation for the conductor loop is increased to 120 complete revolutions per 1 s, the period decreases. The period can be found as follows:

$$T = \frac{1}{f}$$

$$= \frac{1}{120 \text{ Hz}}$$

$$T = 8.33 \text{ ms}$$

Notice that for the higher frequency of 120 Hz, less time is required for one cycle.

Figure 15-19 shows a sine wave having a period of 500 μs. Its frequency is found as follows:

$$f = \frac{1}{T}$$

$$= \frac{1}{500\ \mu\text{s}}$$

$$f = 2\ \text{kHz}$$

This frequency of 2 kHz tells us that the waveform completes 2,000 cycles per second. It is also important to realize that one-quarter cycle takes 125 μs and one-half cycle takes 250 μs.

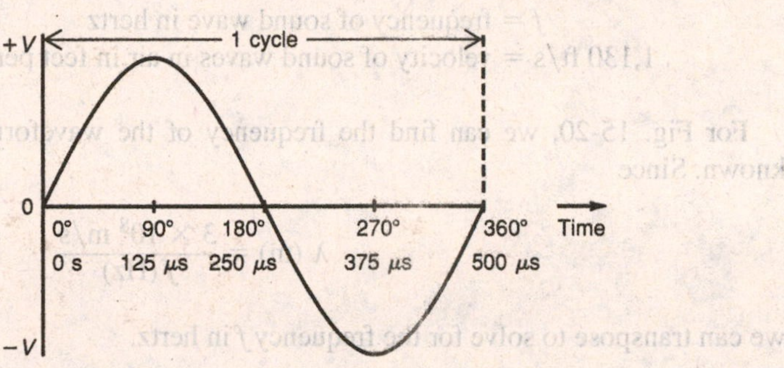

Fig. 15-19 The period of a waveform is the length of time it takes to complete one cycle.

When a periodic waveform is considered with respect to distance, one cycle includes the wavelength. One wavelength is the distance traveled by the waveform in one cycle. Figure 15-20 shows the distance traveled by a radio wave in one cycle. The wavelength is dependent on the frequency of the waveform and the velocity of transmission.

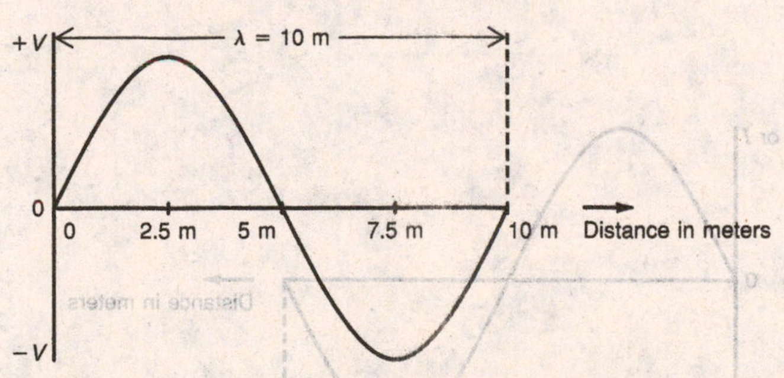

Fig. 15-20 Wavelength λ is the distance traveled by a wave in one cycle.

This, expressed as an equation, is

$$\lambda = \frac{\text{velocity}}{\text{frequency}}$$

where λ = symbol for wavelength

Radio waves traveling through space have a velocity approaching that of the speed of light. Sound waves caused by variations in air pressure travel with a much lower velocity.

The wavelength of radio waves is found using this formula:

$$\lambda\ (\text{m}) = \frac{3 \times 10^8\ \text{m/s}}{f\,(\text{Hz})}$$

where λ (m) = the distance in meters the radio wave travels to complete one cycle
f = frequency of radio wave in hertz
3×10^8 m/s = velocity of radio wave in space in meters per second (m/s)

The wavelength of sound waves is found using the following formula:

$$\lambda\ (\text{ft}) = \frac{1{,}130\ \text{ft/s}}{f\,(\text{Hz})}$$

where λ (ft) = the distance in feet the sound wave travels to complete one cycle
f = frequency of sound wave in hertz
1,130 ft/s = velocity of sound waves in air in feet per second (ft/s)

For Fig. 15-20, we can find the frequency of the waveform because its wavelength is known. Since

$$\lambda\ (\text{m}) = \frac{3 \times 10^8\ \text{m/s}}{f\,(\text{Hz})}$$

we can transpose to solve for the frequency f in hertz.

$$f\,(\text{Hz}) = \frac{3 \times 10^8\ \text{m/s}}{\lambda\ (\text{m})}$$

$$= \frac{3 \times 10^8\ \text{m/s}}{10\ \text{m}}$$

$$f\,(\text{Hz}) = 30\ \text{MHz}$$

Solved Problem

A radio frequency wave is shown in Fig. 15-21. Find its frequency f, and its period T.

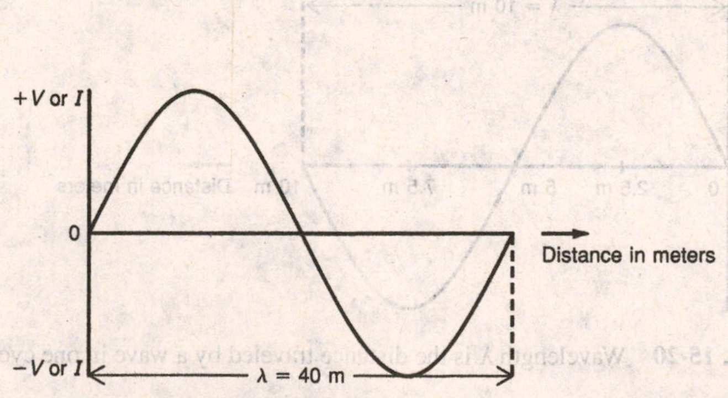

Fig. 15-21 Radio frequency wave.

Answer

Since

$$\lambda\ (\text{m}) = \frac{3 \times 10^8\ \text{m/s}}{f\,(\text{Hz})}$$

we have

$$f\,(\text{Hz}) = \frac{3 \times 10^8 \text{ m/s}}{\lambda \text{ (m)}}$$

$$= \frac{3 \times 10^8 \text{ m/s}}{40 \text{ m}}$$

$$f\,(\text{Hz}) = 7.5 \text{ MHz}$$

To find the waveform's period, proceed as follows:

$$T = \frac{1}{f}$$

$$= \frac{1}{7.5 \text{ MHz}}$$

$$T = 0.133\ \mu\text{s}$$

WORD PROBLEMS

Sec. 15-4 Solve the following.

1. Calculate the period T for the following frequencies: **(a)** 1 kHz; **(b)** 2 kHz; **(c)** 10 kHz; **(d)** 250 kHz.
2. Calculate the frequency for the following periods: **(a)** 25 ms; **(b)** 10 μs; **(c)** 40 μs; **(d)** 5 ms.
3. **(a)** Calculate λ in meters for a radio wave with a frequency of 4 MHz. **(b)** Calculate the waveform's period.
4. **(a)** Calculate λ in meters for a radio wave with a frequency of 50 MHz. **(b)** Calculate the waveform's period.
5. **(a)** Calculate λ in feet for a sound wave with a frequency of 1 kHz. **(b)** Calculate the waveform's period.
6. **(a)** Calculate the frequency of a radio wave that has a wavelength of 20 m. **(b)** Calculate the waveform's period.
7. **(a)** Calculate the frequency of a sound wave that has a wavelength of 0.452 ft. **(b)** Calculate the waveform's period.
8. **(a)** Calculate the frequency of a radio wave that has a wavelength of 2 m. **(b)** Calculate the waveform's period.
9. A sound wave completes one cycle in 4 ms. **(a)** Calculate wavelength in ft. **(b)** Calculate the waveform's frequency.
10. A radio wave completes one cycle in 0.025 μs. **(a)** Calculate wavelength in meters. **(b)** Calculate the waveform's frequency.

PRACTICE PROBLEMS

Sec. 15-4 For the waveforms shown in Figs. 15-22 through 15-25, solve for the unknowns listed.

SEC. 15-5 RELATIONSHIP BETWEEN TIME, FREQUENCY, AND PHASE

Referring back to Fig. 15-5, we see that sine wave A reaches its maximum value 90° after waveform B. The 90° phase angle between the two is maintained throughout the cycle.

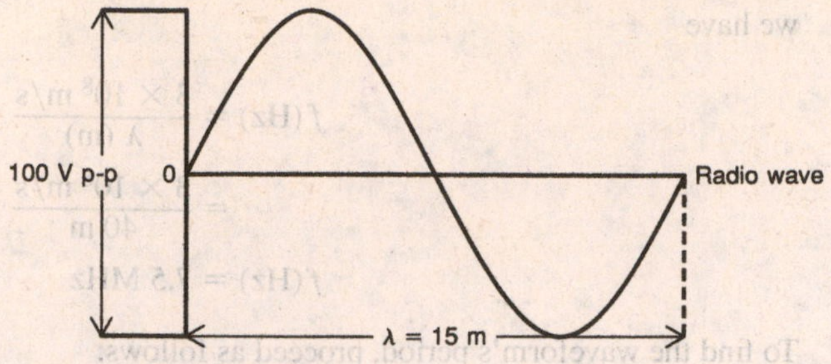

Fig. 15-22 Solve for period, frequency, V_{rms}, and v_{inst} at 30°.

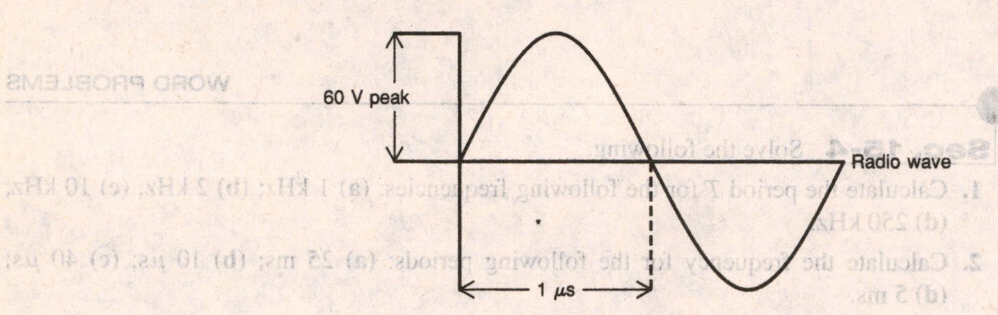

Fig. 15-23 Solve for period, frequency, λ, V_{av}, and v_{inst} at 60°.

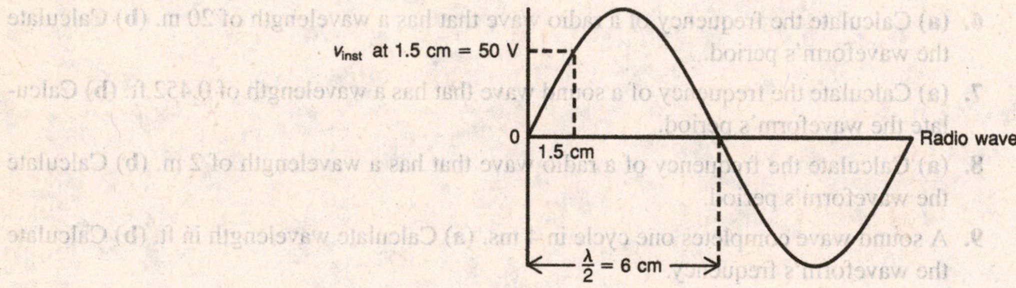

Fig. 15-24 Solve for period, frequency, λ, and V_p.

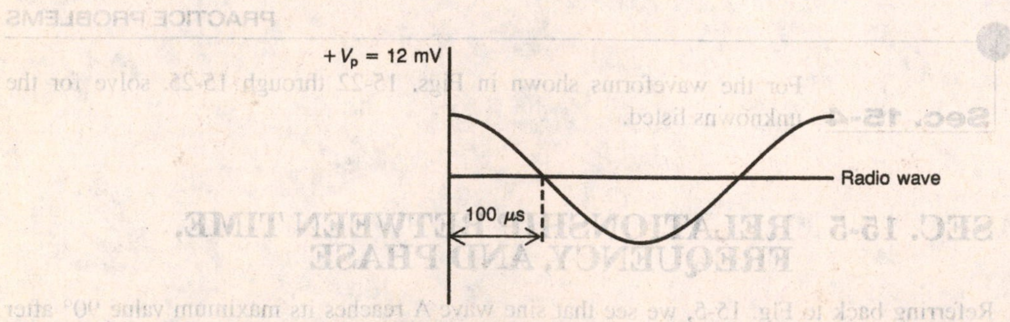

Fig. 15-25 Solve for period, frequency, λ, V_{rms}, and v_{inst} at 240°.

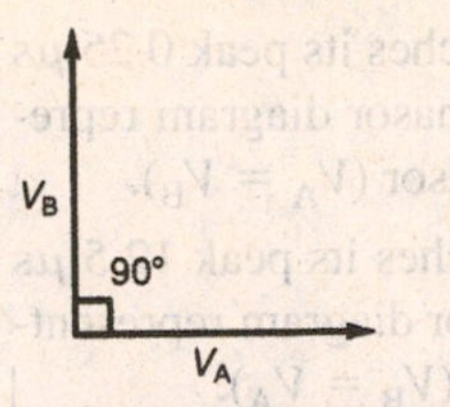

Fig. 15-26 Phasor diagram used to illustrate the angular difference of 90° for the waveforms shown in Fig. 15-5.

When comparing the phase relationship of two waveforms, it is much more convenient to use phasor diagrams. The phase angle of 90° in Fig. 15-5 can be shown as in Fig. 15-26.

For the waveforms in Fig. 15-5, the 90° phase relationship indicates a difference in time. Waveform A reaches its maximum value later in time than does waveform B. To find the time for a phase angle θ, the following formula can be used:

$$t = \frac{\theta}{360^\circ} \times \text{period}$$

where t = time difference in seconds
θ = angular difference between the waveforms
period = length of time to complete one cycle

When comparing the phase angle between two waveforms, the amplitudes can be different, but the frequencies must be the same. The phase angle corresponds to the entire cycle of voltage or current but is often shown at one angle, usually the starting point. The reference phasor is horizontal, corresponding to 0°.

Solved Problem

Two waveforms, A and B, have the same frequency of 1 kHz. Waveform B lags waveform A by 30°. **(a)** Find the time difference to which this angle corresponds. **(b)** Draw the phasor diagram that shows this phase relationship (assume the amplitudes to be identical: $V_A = V_B$).

Answer

a.

$$t = \frac{\theta}{360^\circ} \times \text{period}$$

$$= \frac{30^\circ}{360^\circ} \times 1 \text{ ms}$$

$$t = 83.3\ \mu\text{s}$$

b. Both phasor diagrams shown in Fig. 15-27 represent the angular difference of 30°. Either one can be used.

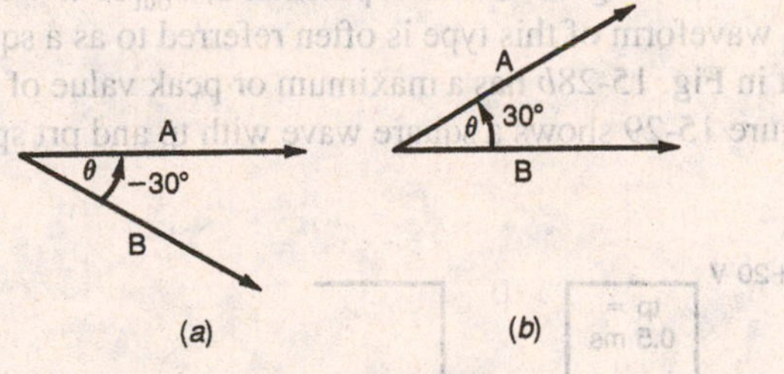

Fig. 15-27 Phasor diagram illustrating the angular difference of 30°.

PRACTICE PROBLEMS

Sec. 15-5 Solve the following.

1. For two waves with a frequency of 50 kHz, how much time is a phase angle of **(a)** 30°, **(b)** 45°, **(c)** 90°, **(d)** 150°, and **(e)** 210°?
2. For two waves with a frequency of 20 MHz, how much time is a phase angle of **(a)** 15°, **(b)** 40°, **(c)** 180°, **(d)** 240°, and **(e)** 330°?
3. For two waves with a frequency of 100 Hz, how much time is a phase angle of **(a)** 10°, **(b)** 60°, **(c)** 120°, **(d)** 200°, and **(e)** 300°?

4. Waveforms A and B have a frequency of 400 kHz. Waveform B reaches its peak 0.25 μs before waveform A. **(a)** Calculate the phase angle. **(b)** Draw the phasor diagram representing the phase difference, using waveform B as the reference phasor ($V_A = V_B$).
5. Waveforms A and B have a frequency of 10 kHz. Waveform A reaches its peak 12.5 μs after waveform B. **(a)** Calculate the phase angle. **(b)** Draw the phasor diagram representing the phase difference, using waveform A as the reference phasor ($V_B = V_A$).

SEC. 15-6 ANALYZING NONSINUSOIDAL WAVEFORMS

Any waveform that is not a sine wave or a cosine wave is a nonsinusoidal waveform. The term *pulse* is often used to describe a voltage or current waveform that is not sinusoidal. A pulse usually has sharp rising and falling edges. A positive pulse is one that has a positive excursion from a 0-V reference level; whereas a negative pulse is one that has a negative excursion from a 0-V reference level. A series of successive pulses is called a *pulse train.* The duration of the pulse is called the *pulse width,* or *pulse time,* abbreviated tp. The period of a rectangular waveform is often referred to as the pulse repetition time, abbreviated prt. The circuit shown in Fig. 15-28*a* can be used to generate the rectangular-type waveform in Fig. 15-28*b*.

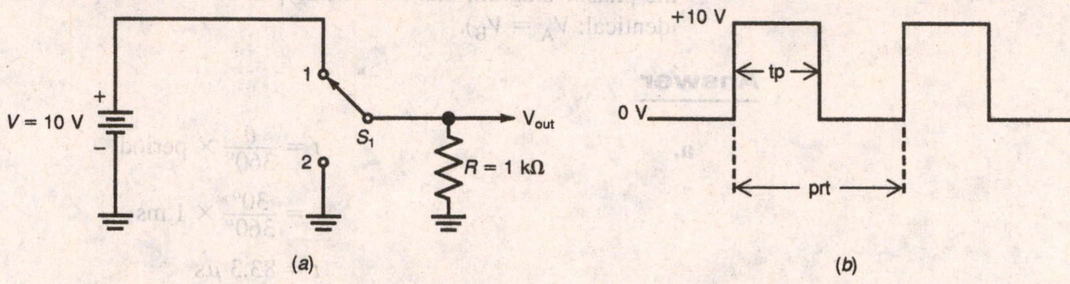

Fig. 15-28 (*a*) Circuit used to generate a rectangular-type waveform. (*b*) Waveform generated by moving switch S_1 between positions 1 and 2.

When S_1 in Fig. 15-28*a* is in position 1, V_{out} is at +10 V. When S_1 is in position 2, V_{out} is at 0 V. A waveform of this type is often referred to as a square wave. The amplitude of the pulse shown in Fig. 15-28*b* has a maximum or peak value of +10 V.

Figure 15-29 shows a square wave with tp and prt specified.

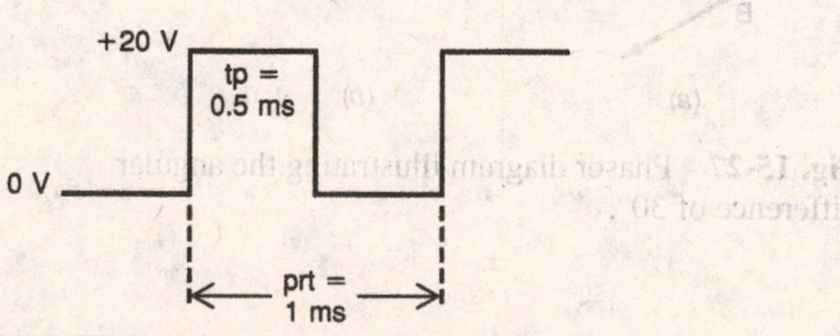

Fig. 15-29 Square wave showing pulse time tp and pulse repetition time prt.

The number of pulses per second is called the *pulse repetition frequency,* abbreviated prf. The prf and prt are related as follows:

$$\text{prf} = \frac{1}{\text{prt}} \quad \text{and} \quad \text{prt} = \frac{1}{\text{prf}}$$

For the waveform of Fig. 15-29, prf is found as shown.

$$\text{prf} = \frac{1}{\text{prt}}$$
$$= \frac{1}{1\text{ ms}}$$
$$\text{prf} = 1\text{ kHz}$$

The duty cycle of a rectangular waveform is the ratio of the pulse time tp to the total period prt. This is shown as

$$\%\text{ duty cycle} = \frac{\text{tp}}{\text{prt}} \times 100$$

For the waveform in Fig. 15-29, the duty cycle is

$$\%\text{ duty cycle} = \frac{0.5\text{ ms}}{1\text{ ms}} \times 100$$
$$\%\text{ duty cycle} = 50\%$$

As mentioned earlier, the pulse can be positive or negative. When a waveform contains both positive and negative values, the duty cycle can be specified for either the positive or the negative pulse.

Another very important characteristic of a nonsinusoidal waveform is its average value. The average value for any waveform is its DC value, either voltage or current. The average value of a sine wave is specified for a half-cycle for comparison purposes. The average value for a sine wave through one full cycle is zero. To determine the average value for the waveform in Fig. 15-29, we need to find its average height or amplitude level over one full cycle. First, we must determine the area of the pulse A_p. This is found as follows:

$$A_p = \text{tp} \times \pm V_\text{p}\ (\text{V-s})$$

where A_p = area of the pulse
tp = pulse time
$\pm V_\text{p}$ = waveform's peak value
V-s = volt-second units

Substituting the values from Fig. 15-29 gives us the following:

$$A_p = 0.5\text{ ms} \times 20\text{ V}$$
$$A_p = +10\text{ V-ms}$$

Notice that the area is given in volt-second units.

The next step in determining the waveform's average value is to divide the area of the pulse A_p by the pulse repetition time prt. This is shown as

$$V_\text{av} = \frac{A_p}{\text{prt}}$$

or more commonly as

$$V_\text{av} = \frac{\text{tp}}{\text{prt}} \times \pm V_\text{p}$$

Notice that this just tells us the waveform's average height or amplitude level over one full cycle. Substituting the values from Fig. 15-29 gives us

$$V_\text{av} = \frac{0.5\text{ ms}}{1\text{ ms}} \times 20\text{ V}$$
$$V_\text{av} = 10\text{ V}$$

The average value of voltage is the value indicated by a DC voltmeter. Notice that in the equation for V_{av} that the second unit of time cancels.

It is important to note that if a waveform should contain both positive and negative values, the numerical value of A_p would be the algebraic sum of the two, $+A_p$ and $-A_p$.

Solved Problem

For the waveform shown in Fig. 15-30, determine prf, % duty cycle, and V_{av}.

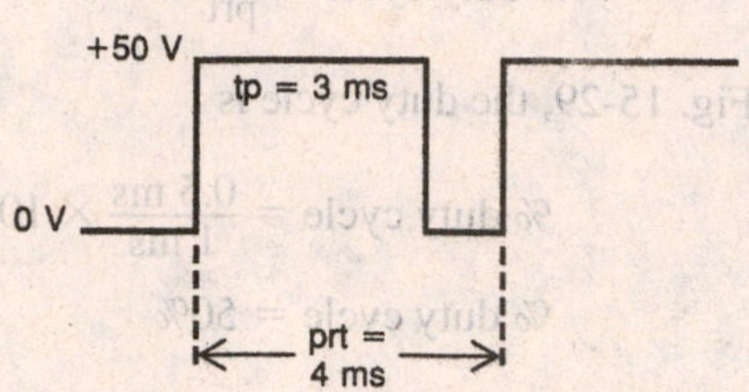

Fig. 15-30

Answer

Since prt is known, we have

$$\text{prf} = \frac{1}{\text{prt}}$$
$$= \frac{1}{4\text{ ms}}$$
$$\text{prf} = 250\text{ Hz}$$

The % duty cycle is found as follows:

$$\%\text{ duty cycle} = \frac{\text{tp}}{\text{prt}} \times 100$$
$$= \frac{3\text{ ms}}{4\text{ ms}} \times 100$$
$$\%\text{ duty cycle} = 75\%$$

To find V_{av}, we find A_p and then divide by prt. First,

$$A_p = \text{tp} \times \pm V_p$$
$$= 3\text{ ms} \times 50\text{ V}$$
$$A_p = 0.15\text{ V-s}$$

and then,

$$V_{av} = \frac{A_p}{\text{prt}}$$
$$= \frac{0.15\text{ V-s}}{4\text{ ms}}$$
$$V_{av} = 37.5\text{ V}$$

A DC voltmeter placed across the terminals of the generator producing this waveform would measure 37.5 V.

Sec. 15-6 For Figs. 15-31 through 15-38, solve for the unknowns listed.

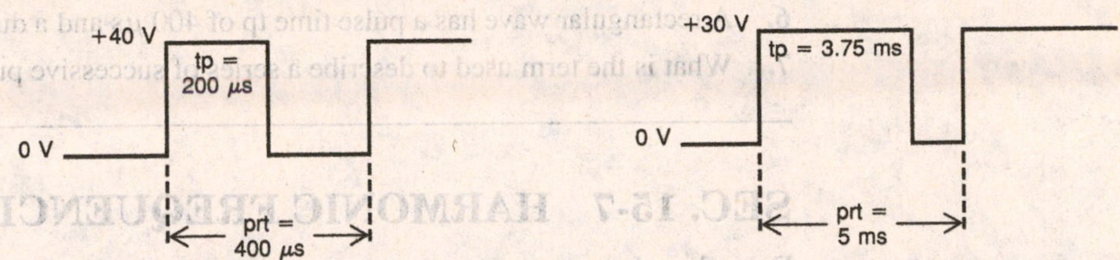

Fig. 15-31 Solve for prf, % duty cycle, and V_{av}.

Fig. 15-32 Solve for prf, % duty cycle, and V_{av}.

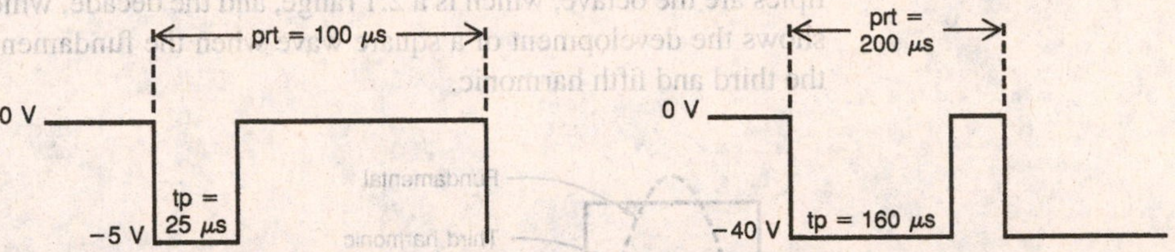

Fig. 15-33 Solve for prf, % duty cycle, and V_{av}.

Fig. 15-34 Solve for prf, % duty cycle, and V_{av}.

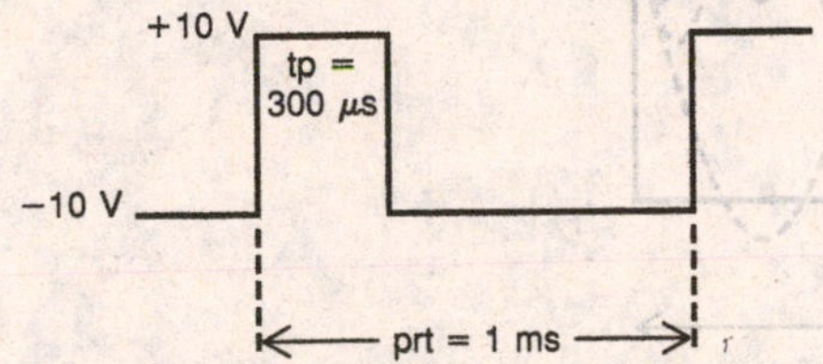

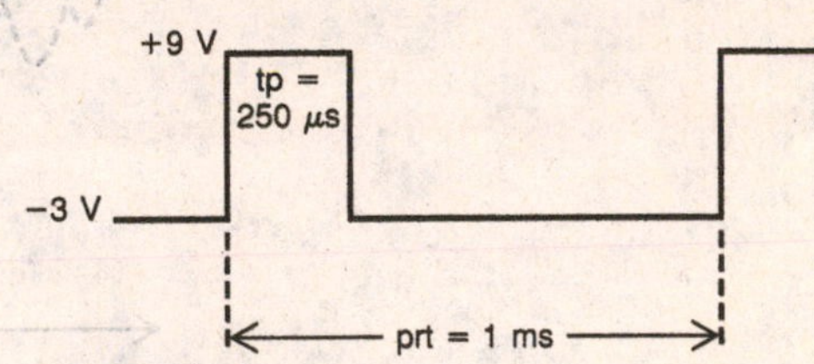

Fig. 15-35 Solve for prf, % duty cycle, and V_{av}.

Fig. 15-36 Solve for prf, % duty cycle, and V_{av}.

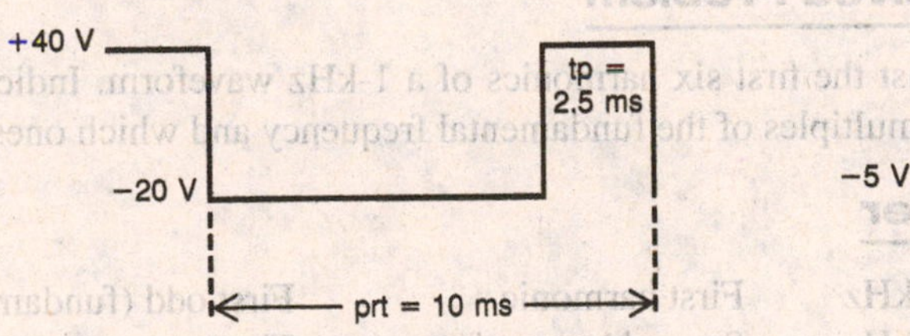

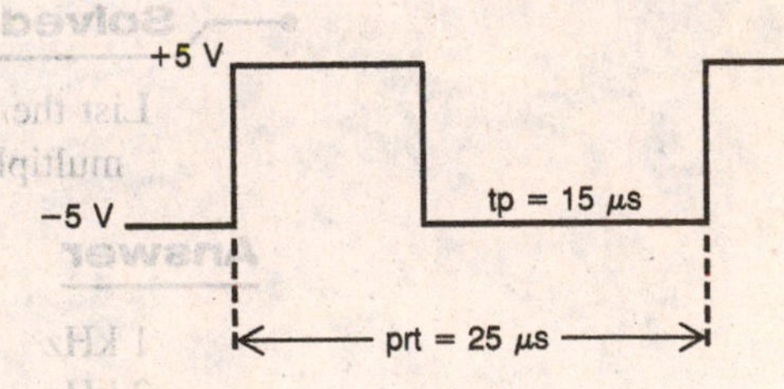

Fig. 15-37 Solve for prf, % duty cycle, and V_{av}.

Fig. 15-38 Solve for prf, % duty cycle, and V_{av}.

WORD PROBLEMS

Sec. 15-6 Solve the following.

1. A rectangular wave has a pulse time tp of 40 μs and a duty cycle of 80%. Calculate prf and prt.
2. A 4-kHz rectangular wave has a pulse time tp of 50 μs. Calculate the duty cycle.
3. A rectangular wave has a pulse time tp of 4 ms and a duty cycle of 33.3%. Calculate prf and prt.

4. What is the pulse time tp of a rectangular wave that has a frequency of 5 kHz and a 20% duty cycle?
5. What is the pulse time tp of a rectangular wave that has a frequency of 20 kHz and a 60% duty cycle?
6. A rectangular wave has a pulse time tp of 400 μs and a duty cycle of 25%. Calculate prf.
7. What is the term used to describe a series of successive pulses?

SEC. 15-7 HARMONIC FREQUENCIES

Even though square waves appear as being very basic, they are actually quite complex. A square wave consists of a pure sine wave at the fundamental frequency plus several odd-order harmonics. Exact multiples of the fundamental frequency are called *harmonics.* Even multiples are even harmonics, and odd multiples are odd harmonics. Two other units for frequency multiples are the octave, which is a 2:1 range, and the decade, which is a 10:1 range. Figure 15-39 shows the development of a square wave when the fundamental frequency is combined with the third and fifth harmonic.

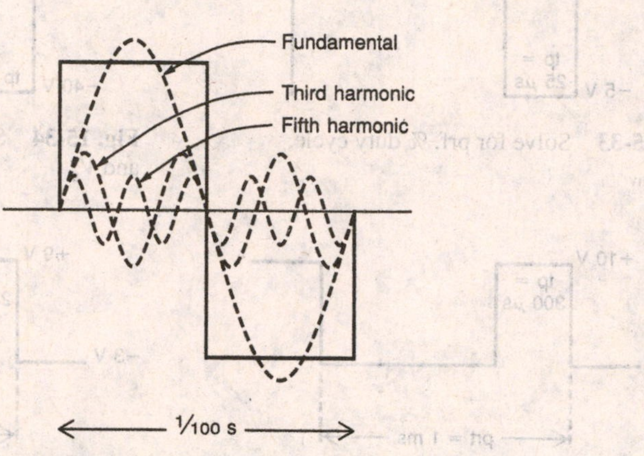

Fig. 15-39 Fundamental and harmonic frequencies for a square wave.

Solved Problem

List the first six harmonics of a 1-kHz waveform. Indicate which harmonics are even multiples of the fundamental frequency and which ones are odd multiples.

Answer

1 kHz	First harmonic	First odd (fundamental frequency)
2 kHz	Second harmonic	First even
3 kHz	Third harmonic	Second odd
4 kHz	Fourth harmonic	Second even
5 kHz	Fifth harmonic	Third odd
6 kHz	Sixth harmonic	Third even

PRACTICE PROBLEMS

Sec. 15-7

1. What is the fifth harmonic of 5 kHz?
2. What is the third harmonic of 800 Hz?
3. What is the second even harmonic of 10 kHz?
4. What is the third odd harmonic of 150 Hz?

5. What is the sixth odd harmonic of 40 Hz?
6. What is the fourth even harmonic of 60 Hz?
7. Calculate the wavelength for the fifth harmonic of a 540-kHz radio wave.
8. Calculate the wavelength for the second even harmonic of a 3.5-MHz radio wave.
9. What is the first harmonic of 25 kHz?
10. What is the fourth octave above 5 kHz?
11. What is the frequency 2 decades above 200 Hz?
12. Lowering the frequency of 1 kHz by 2 octaves corresponds to what frequency?
13. The fifth harmonic of a frequency has a period of 50 μs. What is the fundamental frequency?
14. The period of a waveform is 100 μs. Calculate the period of the waveform 2 octaves below.
15. Raising the frequency of 250 Hz by 3 decades corresponds to what frequency?

SEC. 15-8 OSCILLOSCOPE MEASUREMENTS

An oscilloscope (scope) is a test instrument that is used to measure the amplitude, period, and frequency of a repetitive waveform such as a sine wave or square wave. The oscilloscope graphs or draws a picture of the signal that is applied to its input. Most oscilloscopes available today have the ability to measure two input signals at the same time and are, therefore, referred to as dual-trace oscilloscopes. The input signal is brought into the oscilloscope through a shielded cable, which has a probe on one end for connecting to the circuit under test. A dual trace oscilloscope is shown in Fig. 15-40.

Voltage and Time Measurements

An oscilloscope is normally used to make two basic measurements: amplitude and time. After making these two measurements, other values can be determined. Figure 15-41 shows the screen of a typical oscilloscope. This screen is also known as the scope's graticule. The

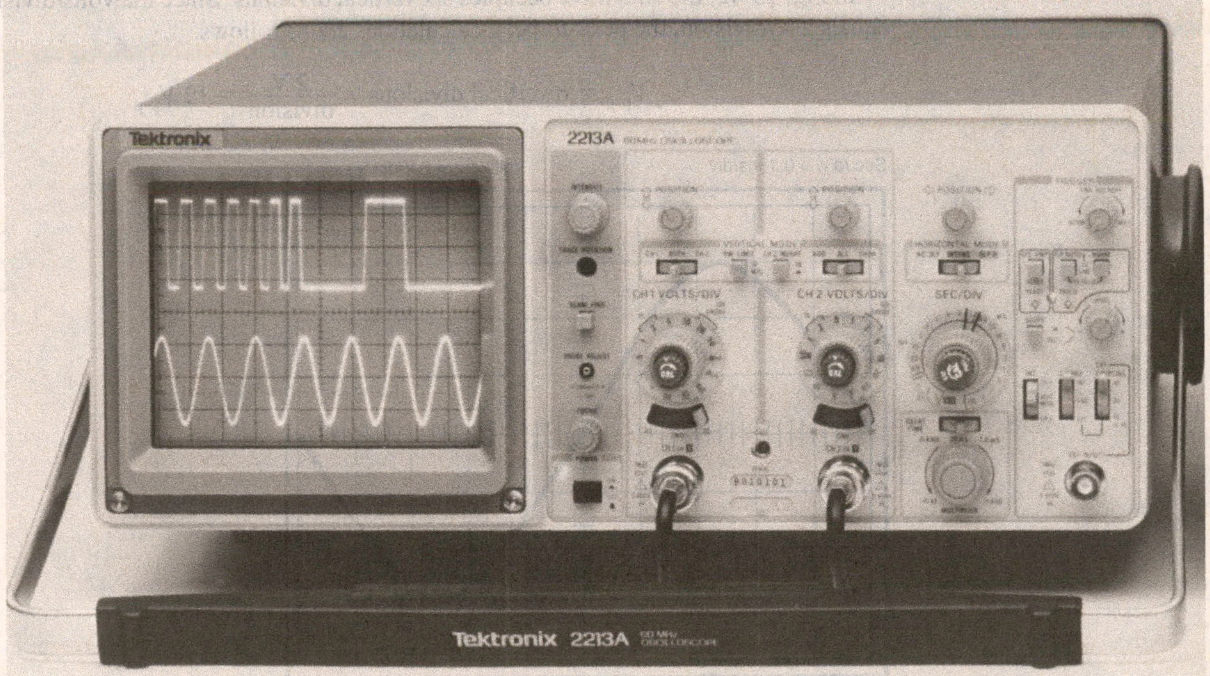

Fig. 15-40 Dual-trace oscilloscope.
Reproduced by permission of Tektronix, Inc.

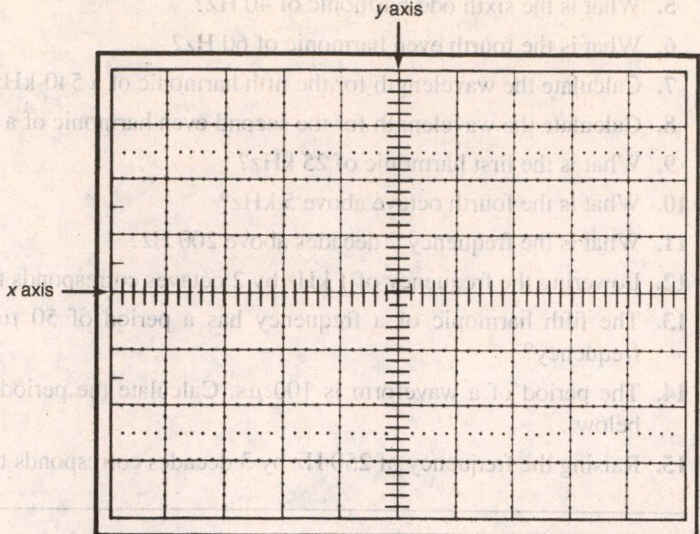

Fig. 15-41 Oscilloscope screen (graticule).

vertical or *Y* axis represents values of voltage amplitude, whereas the horizontal or *X* axis represents values of time. The volts/division control on the oscilloscope determines the amount of voltage needed at the scope input to deflect the electron beam one division vertically on the *Y* axis. The seconds/division control on the oscilloscope determines the time it takes for the scanning electron beam to scan one horizontal division. In Fig. 15-41, note that there are eight vertical divisions and ten horizontal divisions.

Refer to the sine wave being displayed on the oscilloscope graticule in Fig. 15-42. To calculate the peak-to-peak value of the waveform, simply count the number of vertical divisions occupied by the waveform and then multiply this number by the volts/division setting. Expressed as a formula,

$$V_{\text{p-p}} = \#\text{ vertical divisions} \times \frac{\text{volts}}{\text{division}}\text{ setting}$$

In Fig. 15-42, the sine wave occupies six vertical divisions. Since the volts/division setting equals 2 V/division, the peak-to-peak calculations are as follows.

$$V_{\text{p-p}} = 6\text{ vertical divisions} \times \frac{2\text{ V}}{\text{division}} = 12\ V_{\text{p-p}}$$

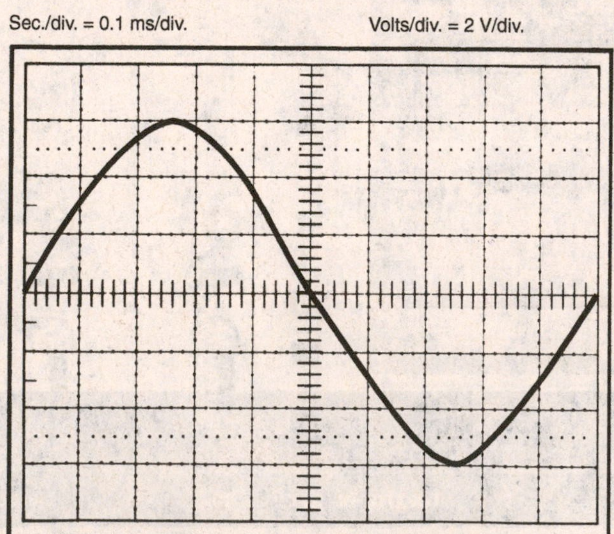

Fig. 15-42 Determining $V_{\text{p-p}}$, *T*, and *f* from the sine wave displayed on the scope graticule.

To calculate the period, T, of the waveform, all you do is count the number of horizontal divisions occupied by one cycle. Then, simply multiply the number of horizontal divisions by the sec./division setting. Expressed as a formula,

$$T = \# \text{ horizontal divisions} \times \frac{\text{sec.}}{\text{division}} \text{ setting}$$

In Fig. 15-42, one cycle of the sine wave occupies exactly ten horizontal divisions. Since the sec./division setting is set to 0.1 ms/div., the calculations for T are as follows:

$$T = 10 \text{ horizontal divisions} \times \frac{0.1 \text{ ms}}{\text{div.}} = 1 \text{ ms}$$

With the period, T, known, the frequency, f, can be found as follows:

$$f = \frac{1}{T} = \frac{1}{1 \text{ ms}} = 1 \text{ kHz}$$

Solved Problem

In Fig. 15-43, determine the peak-to-peak voltage, the period, T, and the frequency, f, of the displayed waveform.

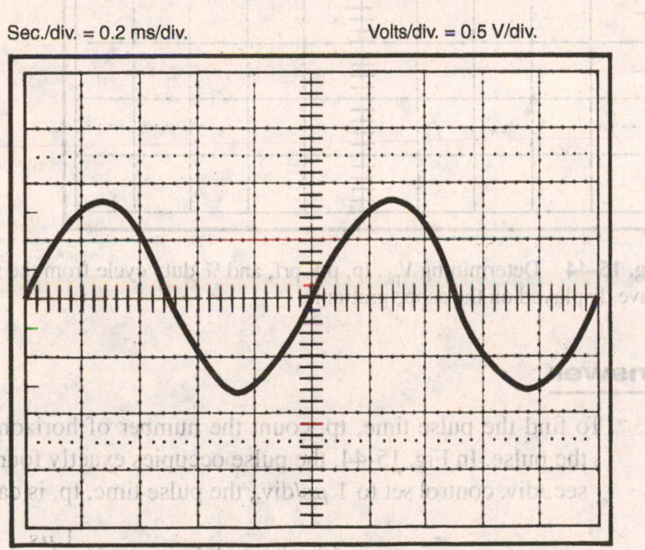

Fig. 15-43 Determining $V_{\text{p-p}}$, T, and f from the sine wave displayed on the scope graticule.

Answer

Careful study of the scopes graticule reveals that the height of the waveform occupies 3.4 vertical divisions. With the volts/div. setting at 0.5 V/div., the peak-to-peak voltage is calculated as follows:

$$V_{\text{p-p}} = 3.4 \text{ vertical divisions} \times \frac{0.5 \text{ V}}{\text{div.}} = 1.7\ V_{\text{p-p}}$$

To find the period, T, of the displayed waveform, count the number of horizontal divisions occupied by just one cycle. By viewing the scopes graticule, we see that one cycle occupies five horizontal divisions. Since the sec./div. control is set to 0.2 ms/div., the period, T, is calculated as:

$$T = 5 \text{ horizontal divisions} \times \frac{0.2 \text{ ms}}{\text{div.}} = 1 \text{ ms}$$

To calculate the frequency, f, take the reciprocal of the period, T.

$$f = \frac{1}{T} = \frac{1}{1 \text{ ms}} = 1 \text{ kHz}$$

Solved Problem

In Fig. 15-44, determine the pulse time, tp, pulse repetition time, prt, and the peak value, V_{pk}, of the displayed waveform. Also, calculate the waveforms % duty cycle and the pulse repetition frequency, prf.

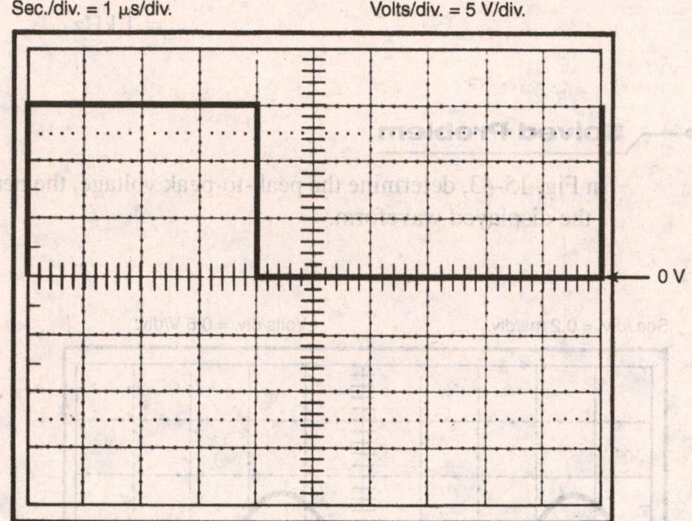

Fig. 15-44 Determining V_{pk}, tp, prt, prf, and % duty cycle from the rectangular wave displayed on the scope graticule.

Answer

To find the pulse time, tp, count the number of horizontal divisions occupied by just the pulse. In Fig. 15-44, the pulse occupies exactly four horizontal divisions. With the sec./div. control set to 1 μs/div., the pulse time, tp, is calculated as

$$\text{tp} = 4 \text{ horizontal divisions} \times \frac{1\ \mu\text{s}}{\text{div.}} = 4\ \mu\text{s}$$

The pulse repetition time, prt, is found by actually counting the number of horizontal divisions occupied by one cycle of the waveform. Since one cycle occupies 10 horizontal divisions, the pulse repetition time, prt, is calculated as follows:

$$\text{prt} = 10 \text{ horizontal divisions} \times \frac{1\ \mu\text{s}}{\text{div.}} = 10\ \mu\text{s}$$

With tp and prt known, the % duty cycle is calculated as follows:

$$\begin{aligned}\% \text{ duty cycle} &= \frac{\text{tp}}{\text{prt}} \times 100 \\ &= \frac{4\ \mu\text{s}}{10\ \mu\text{s}} \times 100 \\ &= 40\%\end{aligned}$$

The pulse repetition frequency, prf, is calculated by taking the reciprocal of prt.

$$\text{prf} = \frac{1}{\text{prt}}$$

$$= \frac{1}{10\ \mu\text{s}}$$

$$= 100\ \text{kHz}$$

The peak value of the waveform is based on the fact that the baseline value of the waveform is 0 V as shown. The positive peak of the waveform is shown to be three vertical divisions above zero. Since the volts/div. setting of the scope is 5 V/div. the peak value of the waveform is

$$V_{pk} = 3\ \text{vertical divisions} \times \frac{5\ \text{V}}{\text{div.}} = 15\ \text{V}$$

Notice that the waveform shown in Fig. 15-44 is entirely positive because the waveform's pulse makes a positive excursion from the zero-volt reference.

Input Coupling Switch

Every oscilloscope has a three-position switch that selects the method of coupling (connecting) the input signal to the oscilloscope's vertical amplifier. This switch is called an input coupling switch, and it has three different settings: AC, GND, and DC. When the input coupling switch is set to *AC*, the input signal is capacitively coupled to the vertical amplifier inside the oscilloscope. This means that any DC component in the signal is blocked and only the AC part of the signal is passed. When the switch is set to *Ground (GND),* the scopes vertical amplifier input is grounded to provide a zero-volt reference point. Setting the switch to *Ground,* however, does not ground the input signal being measured. With the switch set to *DC,* all the signal components, DC and AC, will be directly coupled to the scope's vertical amplifier. In Fig. 15-43, the input coupling switch could be set to either DC or AC to display the waveform shown. In Fig. 15-44, however, the input coupling switch must be set to *DC* to display the positive nature of V_{pk}. For a more detailed look at oscilloscopes, see Appendix E, in *Grob's Basic Electronics* text.

PRACTICE PROBLEMS

Sec. 15-8 In Figs. 15-45 through 15-54 determine the following values: $V_{p\text{-}p}$, T, and f.

In Figs. 15-55 through 15-64, determine the following values: t_p, prt, % duty cycle, prf, and V_{pk}.

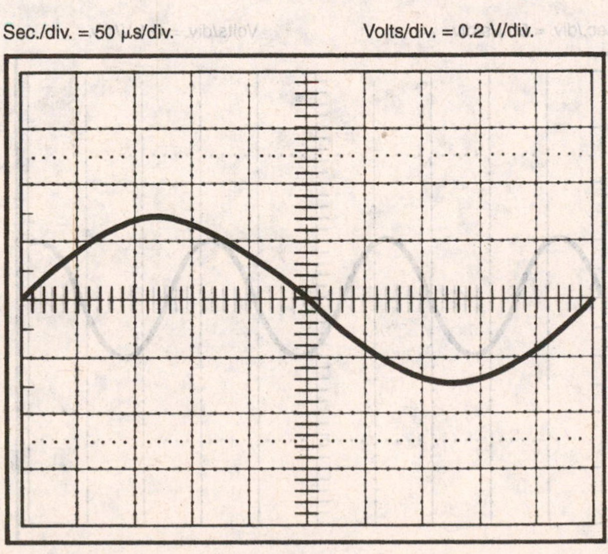

Fig. 15-45

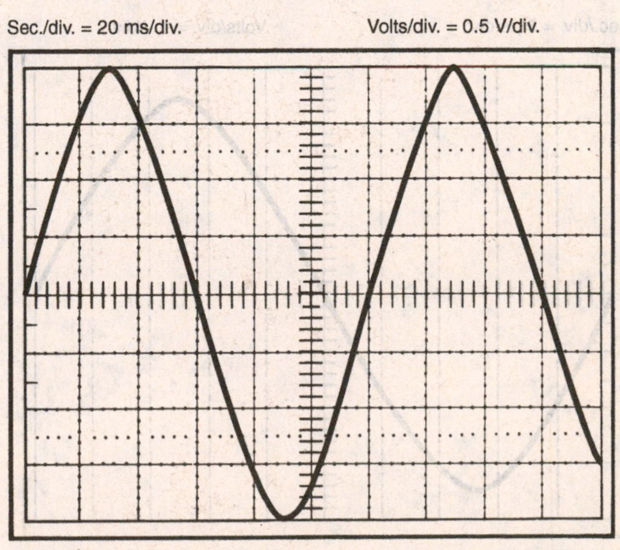

Fig. 15-46

Sec./div. = 20 μs/div. Volts/div. = 1 V/div.

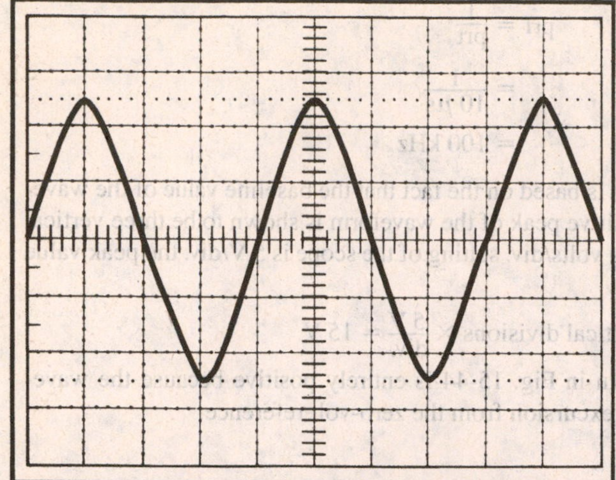

Fig. 15-47

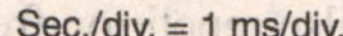

Sec./div. = 1 ms/div. Volts/div. = 0.2 V/div.

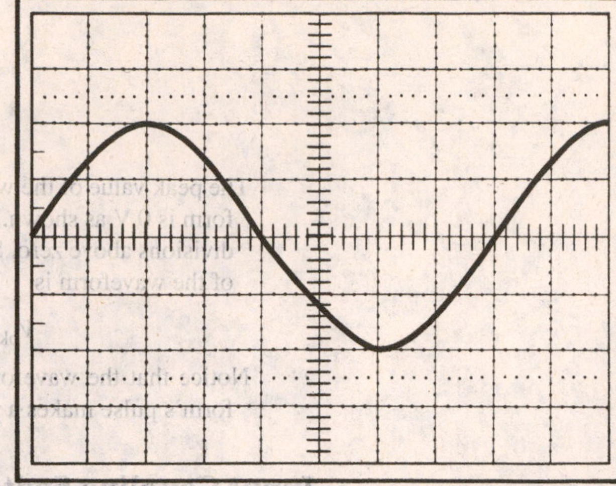

Fig. 15-48

Sec./div. = 0.1 ms/div. Volts/div. = 2 V/div.

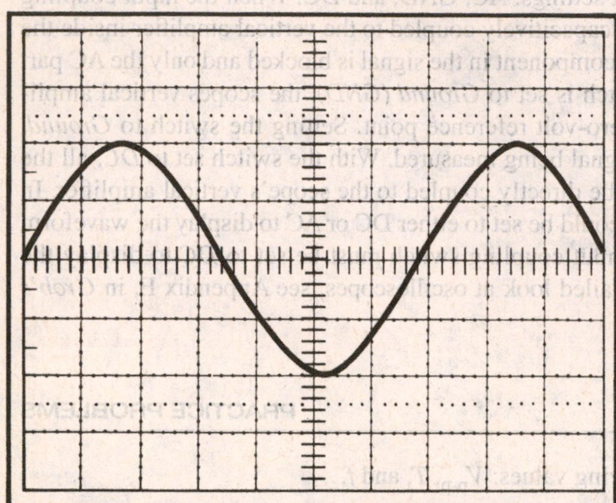

Fig. 15-49

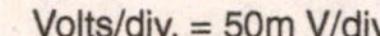

Sec./div. = 2 ms/div. Volts/div. = 50m V/div.

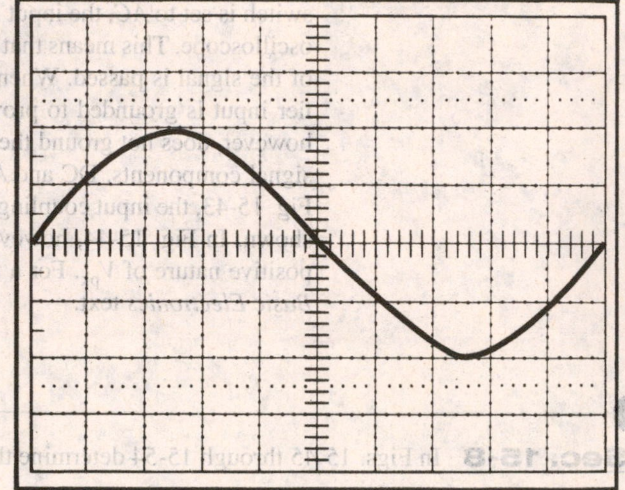

Fig. 15-50

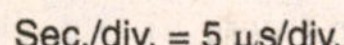

Sec./div. = 5 μs/div. Volts/div. = 1 V/div.

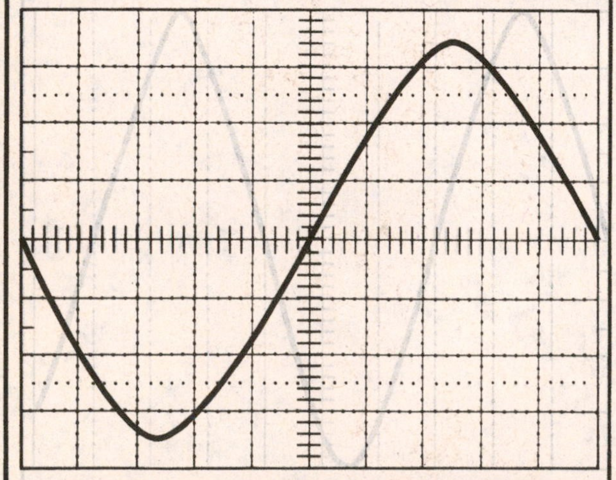

Fig. 15-51

Sec./div. = 50 μs/div. Volts/div. = 0.1 V/div.

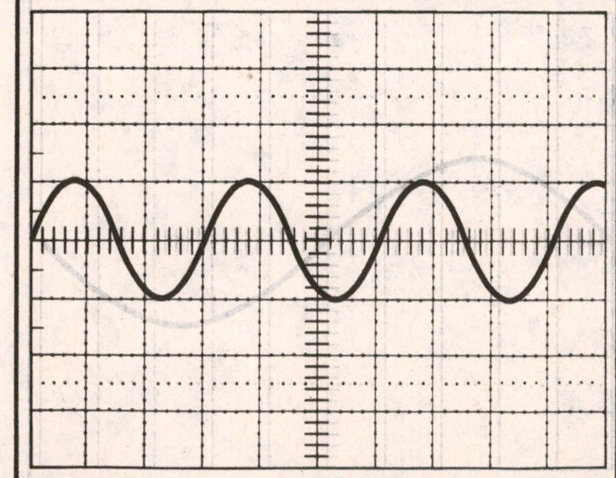

Fig. 15-52

Sec./div. = 0.5 µs/div. Volts/div. = 0.2 V/div.

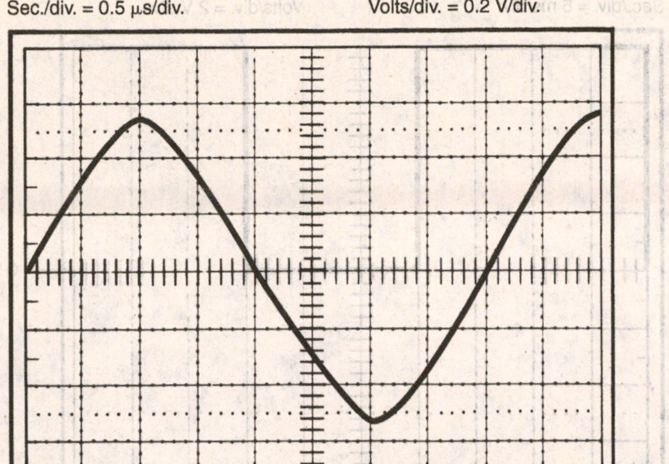
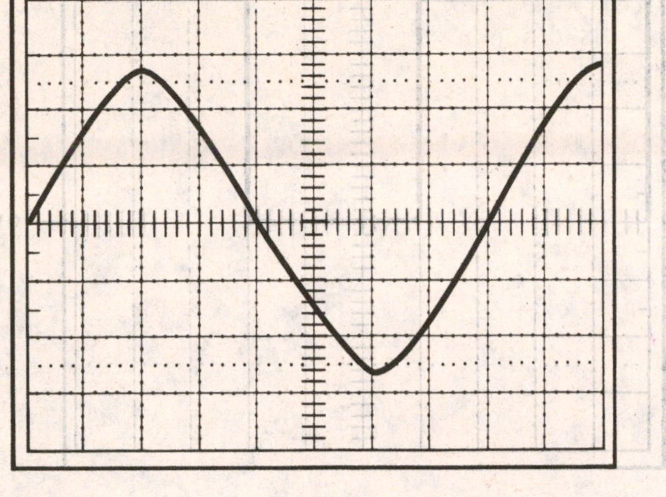

Fig. 15-53

Sec./div. = .1 ms/div. Volts/div. = 5 V/div.

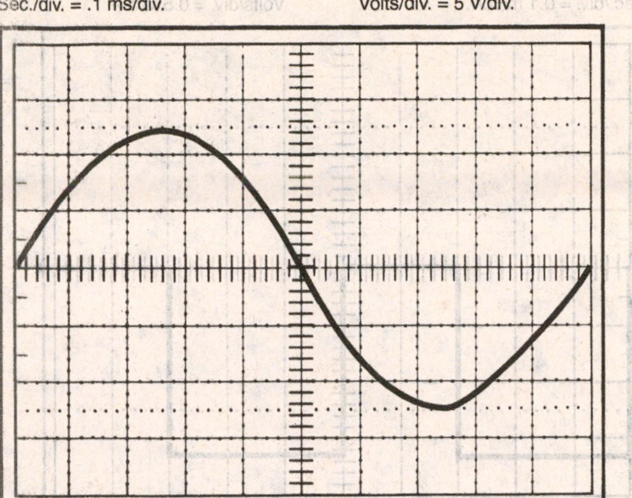

Fig. 15-54

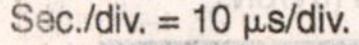

Sec./div. = 0.2 ms/div. Volts/div. = 1 V/div.

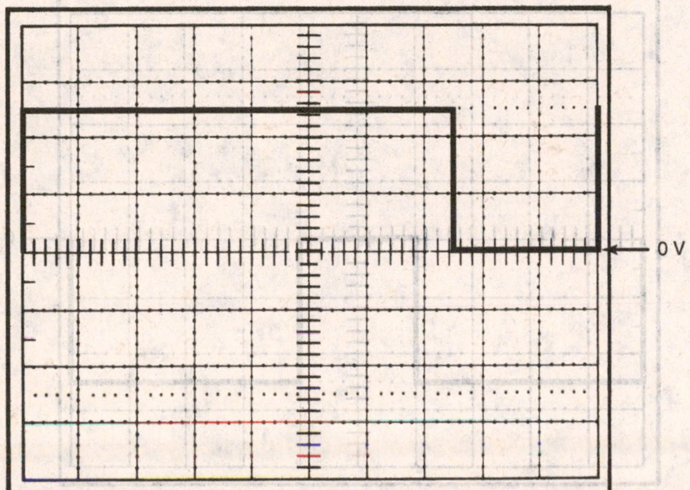

Fig. 15-55

Sec./div. = 10 µs/div. Volts/div. = 2 V/div.

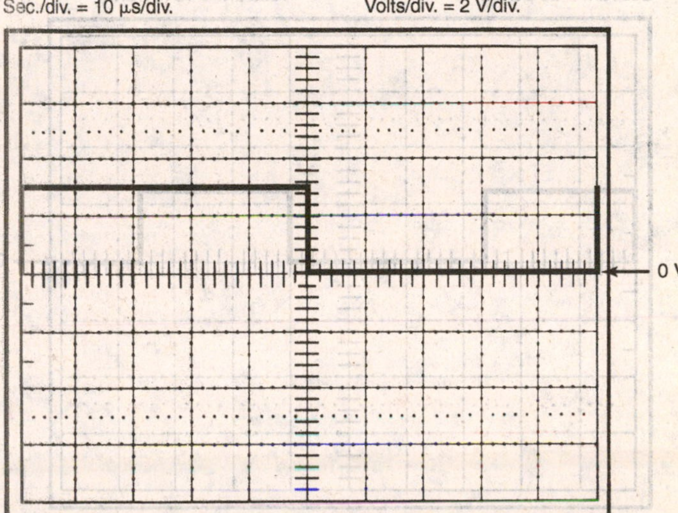

Fig. 15-56

Sec./div. = 0.2 ms/div. Volts/div. = 0.5 V/div.

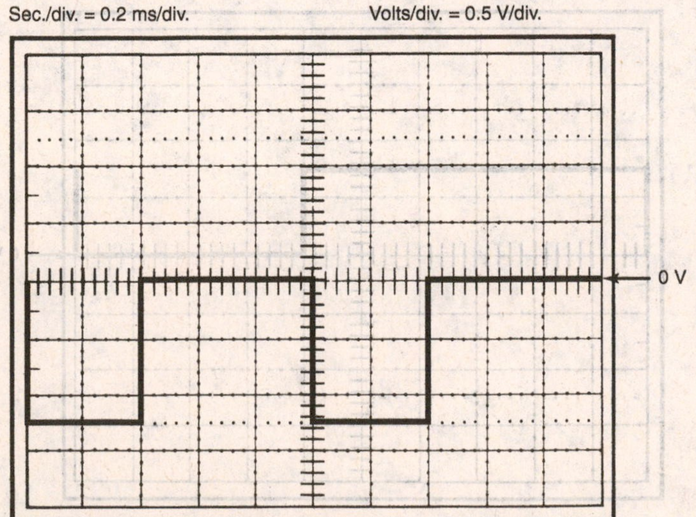

Fig. 15-57

Sec./div. = 2 µs/div. Volts/div. = 0.2 V/div.

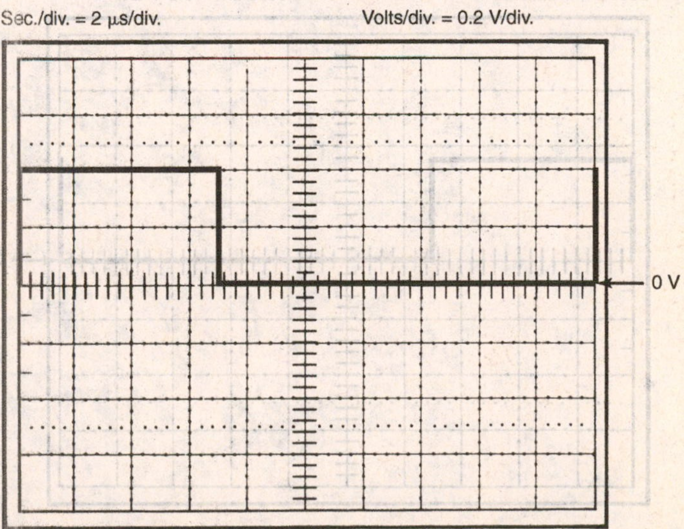

Fig. 15-58

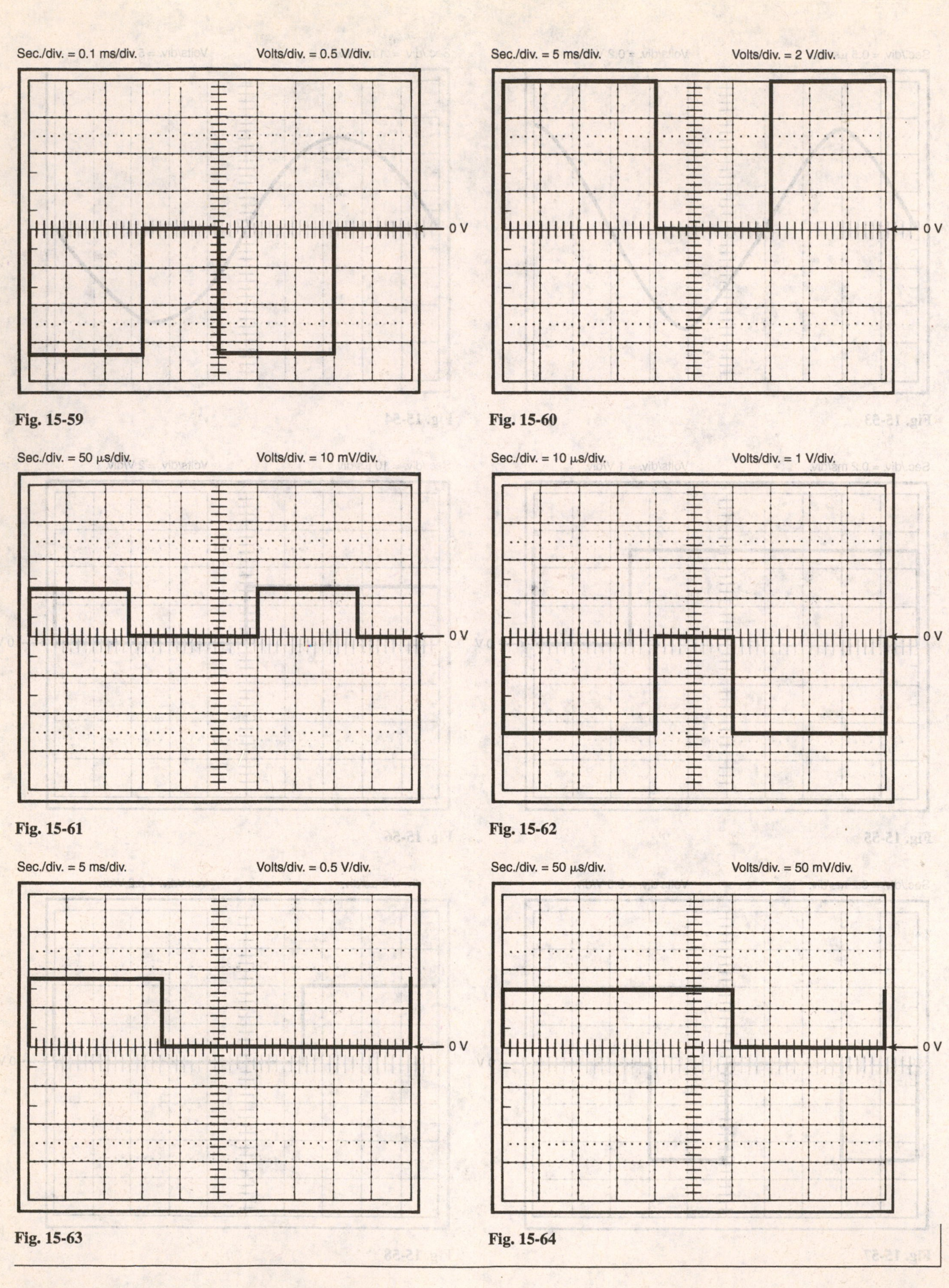

Fig. 15-59

Fig. 15-60

Fig. 15-61

Fig. 15-62

Fig. 15-63

Fig. 15-64

Chapter 15: Alternating Voltage and Current

Answer True or False.

1. An alternating voltage is a voltage that continuously varies in magnitude and periodically reverses in polarity.
2. In Fig. 15-2, one complete revolution of the conductor loop through the magnetic field is called a cycle.
3. One cycle corresponds to 360°.
4. 1 radian corresponds to an angle of approximately 57.3°.
5. One cycle includes 2π radians.
6. One half-cycle of sine wave alternating voltage is called an alternation.
7. A cosine wave has its maximum values at 90 and 270°.
8. At 30°, the instantaneous value of a sine wave is one-half its peak value.
9. The rms value of an AC waveform is less than its average value.
10. A sine wave and cosine wave are 90° out of phase.
11. The rms value of a sine wave can be found by multiplying the waveform's peak value by 0.707.
12. The peak-to-peak value of a sine wave is twice its peak value.
13. When calculating power in AC circuits, always use rms values of voltage and current.
14. The frequency of an AC waveform is the number of complete cycles it goes through in 1 second.
15. The higher the frequency, the shorter the period, T.
16. Sound waves and radio waves both have the same velocity.
17. When comparing the phase relationship between two waveforms, they must both have the same amplitude and frequency.
18. Any waveform that is not a sine or a cosine wave is a nonsinusoidal waveform.
19. The percent duty cycle of a rectangular waveform is a ratio of its pulse time, t_p, to the pulse repetition time, prt.
20. The DC value of any waveform is its average height or amplitude level over one full cycle.
21. Exact multiples of the fundamental frequency are called harmonics.
22. An octave corresponds to a 2:1 range in frequency.
23. The second odd harmonic of a 5 kHz waveform is 10 kHz.
24. A decade corresponds to a 10:1 range in frequencies.
25. An oscilloscope is a test instrument that is capable of measuring the amplitude and period of an AC waveform.

CHAPTER

16

Capacitance

Capacitance is the ability of a dielectric to hold or store an electric charge. Specifically, a capacitor consists of two metal plates separated by an insulator. The symbol for capacitance is *C*. The basic unit of capacitance is the farad (F). Practical capacitors have capacitance values in millionths of a farad. The most common units of capacitance include the microfarad, abbreviated μF, and the picofarad, abbreviated pF. Although traditionally not used, the nanofarad (nF) unit of capacitance is gaining acceptance in the electronics industry.

SEC. 16-1 DETERMINING THE CAPACITANCE VALUE

Capacitance is a physical constant, indicating the capacitance in terms of how much charge can be stored for a given amount of applied voltage. In terms of physical construction, the capacitance *C* can be calculated using the formula

$$C = K_\epsilon \times \frac{A}{d} \times 8.85 \times 10^{-12}\text{ F}$$

where C = capacitance in farads
K_ϵ = dielectric constant
A = area of either plate in square meters
d = distance in meters between plates

The constant factor 8.85×10^{-12} is the absolute permittivity of air or vacuum.

One farad is the amount of capacitance that will store 1 C of charge when a charging voltage of 1 V is applied. Expressed as an equation, we have

$$C = \frac{Q}{V}$$

where C = capacitance in farads
Q = charge stored in dielectric in coulombs
V = voltage applied across capacitor plates

To determine the stored charge, Q, when the capacitance and voltage are known, the formula is rearranged as follows:

$$Q = CV$$

Solved Problems

a. How much charge is stored in a 10-μF capacitor with 50 V across its plates?

Answer

$$Q = CV$$
$$= 10\ \mu\text{F} \times 50\text{ V} = 10 \times 10^{-6}\text{ F} \times 50\text{ V}$$
$$Q = 500 \times 10^{-6}\text{ C} = 500\ \mu\text{C}$$

b. Calculate the capacitance C for a capacitor that has the following characteristics: air dielectric, $A = 0.5\text{ m}^2$ for each plate, and $d = 0.01$ cm.

Answer

$$C = K_\epsilon \times \frac{A}{d} \times 8.85 \times 10^{-12}\text{ F}$$

$$= 1 \times \frac{0.5\text{ m}^2}{0.0001\text{ m}} \times 8.85 \times 10^{-12}\text{ F}$$

$$C = 0.04425\ \mu\text{F } or \text{ } 44.25\text{ nF}$$

Note that the relative permittivity of air is 1.

PRACTICE PROBLEMS

Sec. 16-1 Solve the following.

1. How much charge is stored in a 25-μF capacitor with 80 V across its plates?
2. How much charge is stored in a 6,800-μF capacitor with 50 V across its plates?
3. A capacitor stores 10,000 μC of charge with 20 V across its plates. Calculate C.
4. A capacitor stores 250 μC of charge with 100 V across its plates. Calculate C.
5. What voltage will be across the plates of a 2-μF capacitor if it stores 100 μC of charge?
6. What voltage will be across the plates of a 0.05-μF capacitor if it stores 20 μC of charge?
7. How much charge is stored in a 1,000-μF capacitor with 15 V across its plates?
8. A capacitor stores 1 μC of charge with 100 V across its plates. Calculate C.
9. What voltage will be across the plates of a 0.01-μF capacitor if it stores 0.4 μC of charge?
10. Calculate the capacitance C for a capacitor with the physical characteristics listed: air dielectric, $A = 1\text{ cm}^2$ for each plate, and $d = 0.05$ cm.
11. Calculate the capacitance C for a capacitor with the physical characteristics listed: dielectric constant = 200, $A = 5\text{ cm}^2$, and $d = 0.02$ cm.
12. Calculate the capacitance C for a capacitor with the physical characteristics listed: dielectric constant = 600, $A = 40\text{ cm}^2$, and $d = 2$ cm.
13. Calculate the capacitance C for a capacitor with the physical characteristics listed: dielectric constant = 80, $A = 20\text{ cm}^2$, and $d = 1$ cm.
14. The distance separating two plates of a capacitor is reduced by a factor of 4, and the area for each plate is cut in half. How is the capacitance C affected?
15. What capacitor rating is reduced as the two plates of a capacitor are brought closer together?
16. A constant current of 5 mA charges a capacitor for 100 ms. How much charge is stored by the capacitor? (*Hint*: $Q = I \times T$)
17. A constant current of 20 mA charges a 25-μF capacitor for 0.2 s. What voltage is across the capacitor?
18. A constant current of 100 mA charges a 0.1-μF capacitor for 50 μs. What voltage is across the capacitor?

SEC. 16-2 CALCULATING TOTAL CAPACITANCE

Connecting capacitors in parallel is equivalent to increasing the plate area. Therefore, the total capacitance for parallel-connected capacitors is the sum of the individual capacitances. That is,

$$C_T = C_1 + C_2 + C_3 + \cdots + C_n$$

where C_T = total capacitance in farads

Connecting capacitors in series is equivalent to increasing the distance between the capacitor plates. When capacitors are connected in series, the total capacitance, C_T, is found using the reciprocal formula.

$$C_T = \frac{1}{1/C_1 + 1/C_2 + 1/C_3 + \cdots + 1/C_n}$$

Solved Problems

a. Three capacitors of 10, 20, and 50 μF are shown connected in parallel in Fig. 16-1. Calculate C_T.

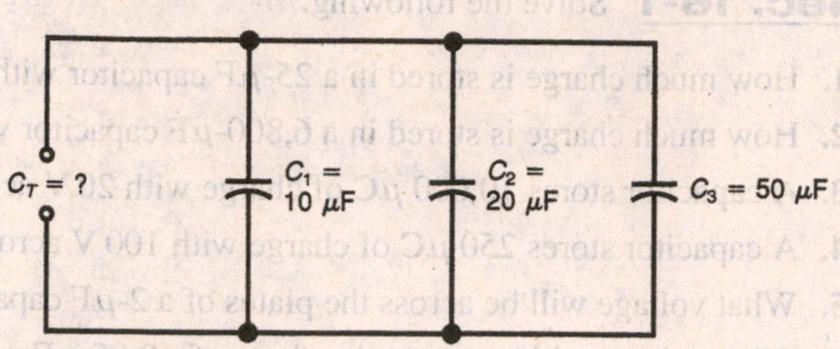

Fig. 16-1 Circuit with three capacitors connected in parallel.

Answer

The total capacitance for Fig. 16-1 is

$$\begin{aligned} C_T &= C_1 + C_2 + C_3 \\ &= 10\ \mu\text{F} + 20\ \mu\text{F} + 50\ \mu\text{F} \\ C_T &= 80\ \mu\text{F} \end{aligned}$$

b. Three capacitors of 8, 10, and 40 μF are shown connected in series in Fig. 16-2. Calculate C_T.

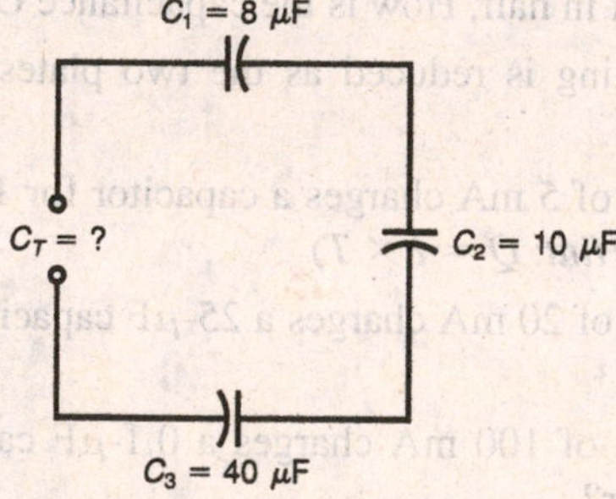

Fig. 16-2 Circuit with three capacitors connected in series.

Answer

The total capacitance for Fig. 16-2 is

$$\begin{aligned} C_T &= \frac{1}{1/C_1 + 1/C_2 + 1/C_3} \\ &= \frac{1}{1/8\ \mu\text{F} + 1/10\ \mu\text{F} + 1/40\ \mu\text{F}} \\ C_T &= 4\ \mu\text{F} \end{aligned}$$

Sec. 16-2 Solve the following.

1. Four 10-μF capacitors are connected in parallel. Calculate C_T.
2. Three 20-μF capacitors are connected in parallel. Calculate C_T.
3. Three capacitors of 0.1, 0.05, and 0.01 μF are connected in parallel. Calculate C_T.
4. Three capacitors—C_1, C_2, and C_3—are in parallel. If $C_2 = 3\ C_1$, and $C_3 = 4\ C_2$, calculate C_1, C_2, and C_3 for $C_T = 0.072\ \mu$F.
5. Four 20-μF capacitors are connected in series. Calculate C_T.
6. Five 0.02-μF capacitors are connected in series. Calculate C_T.
7. Three capacitors of 0.04, 0.01, and 0.008 μF are connected in series. Calculate C_T.
8. Two series-connected capacitors have a total capacitance of 50 pF. If the ratio $C_1/C_2 = \frac{1}{4}$, calculate C_1 and C_2.
9. Two series-connected capacitors have a total capacitance of 40 nF. If the ratio $C_1/C_2 = \frac{1}{3}$, calculate C_1 and C_2.
10. Three 0.06-μF capacitors, having a breakdown voltage rating of 500 V each, are connected in series. What is the breakdown voltage rating for this combination? What is C_T?
11. A 100-μF capacitor is charged to 100 V. If this charged capacitor is connected across an uncharged 400-μF capacitor, what voltage will exist across the two capacitors?
12. A 20-μF capacitor is charged to 50 V. If this charged capacitor is connected across an uncharged 80-μF capacitor, what voltage will exist across the two capacitors?

SEC. 16-3 VOLTAGE DIVISION FOR SERIES-CONNECTED CAPACITORS

When capacitors are connected in series, the voltage across each C is inversely proportional to its capacitance. The smaller capacitance has a larger portion of the applied voltage. The reason is that capacitors connected in series all receive the same charging current. The charging current $I = Q/T$, or $Q = I \times T$. Then, in order for the smaller capacitors to store the same amount of charge as the larger capacitors, the smaller capacitors must have a larger portion of the applied voltage. This can be seen by examining the formula $V = Q/C$. With Q being the same for all capacitors in series, V must be larger across the smaller capacitors.

Solved Problem

Refer to Fig. 16-3. Calculate V_{C_1}, V_{C_2}, and V_{C_3}.

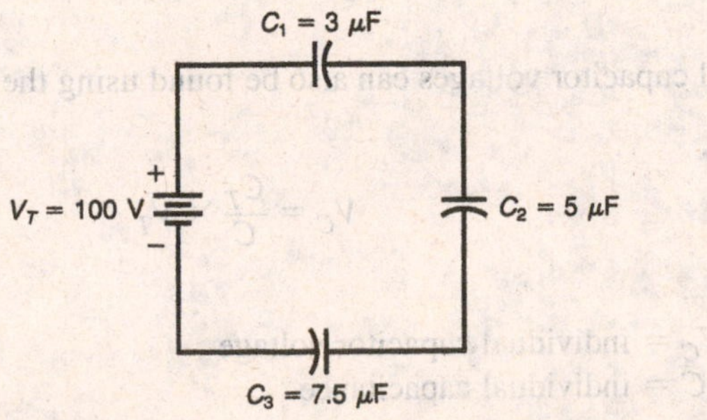

Fig. 16-3 Circuit used to calculate voltage division for series-connected capacitors.

Answer

First, we find C_T.

$$C_T = \frac{1}{1/C_1 + 1/C_2 + 1/C_3}$$
$$= \frac{1}{1/3\ \mu F + 1/5\ \mu F + 1/7.5\ \mu F}$$
$$C_T = 1.5\ \mu F$$

With C_T known, the total charge stored by the capacitors can be found.

$$Q = C \times V$$
$$= 1.5\ \mu F \times 100\ V$$
$$Q = 150\ \mu C$$

This charge is stored by each series-connected capacitor.

To solve for the voltage across each series-connected capacitor, we proceed as follows:

$$V = \frac{Q}{C}$$

$$V_{C_1} = \frac{150\ \mu C}{3\ \mu F}$$
$$V_{C_1} = 50\ V$$
$$V_{C_2} = \frac{150\ \mu C}{5\mu F}$$
$$V_{C_2} = 30\ V$$
$$V_{C_3} = \frac{150\ \mu C}{7.5\ \mu F}$$
$$V_{C_3} = 20\ V$$

Notice that

$$V_T = V_{C_1} + V_{C_2} + V_{C_3}$$
$$= 50\ V + 30\ V + 20\ V$$
$$V_T = 100\ V$$

Individual capacitor voltages can also be found using the equation

$$V_C = \frac{C_T}{C} \times V_T$$

where V_C = individual capacitor voltage
C = individual capacitance

This applies only to series-connected capacitors.

Sec. 16-3 For Figs. 16-4 through 16-11, solve for the unknowns listed.

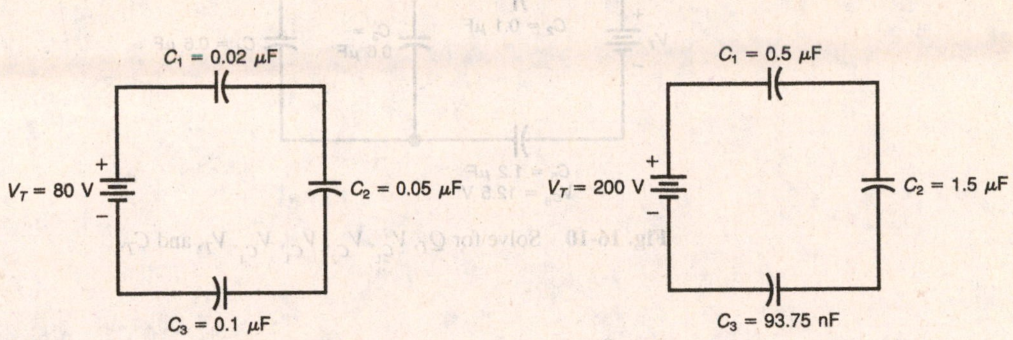

Fig. 16-4 Solve for Q, C_T, V_{C_1}, V_{C_2}, and V_{C_3}.

Fig. 16-5 Solve for Q, C_T, V_{C_1}, V_{C_2}, and V_{C_3}.

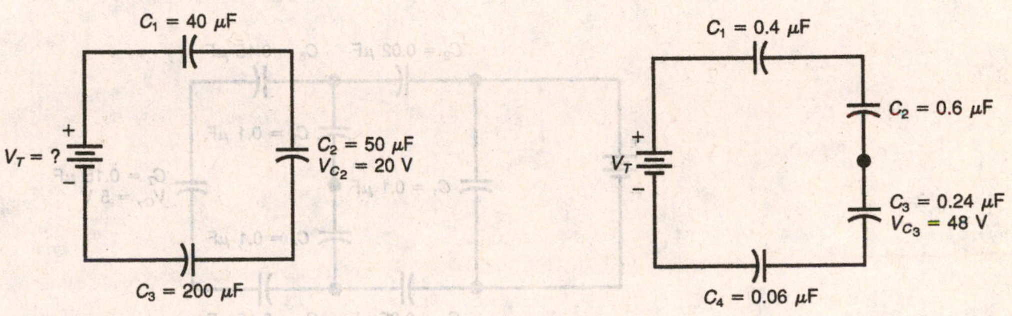

Fig. 16-6 Solve for Q, C_T, V_{C_1}, V_{C_3}, and V_T.

Fig. 16-7 Solve for Q, C_T, V_{C_1}, V_{C_2}, V_{C_4}, and V_T.

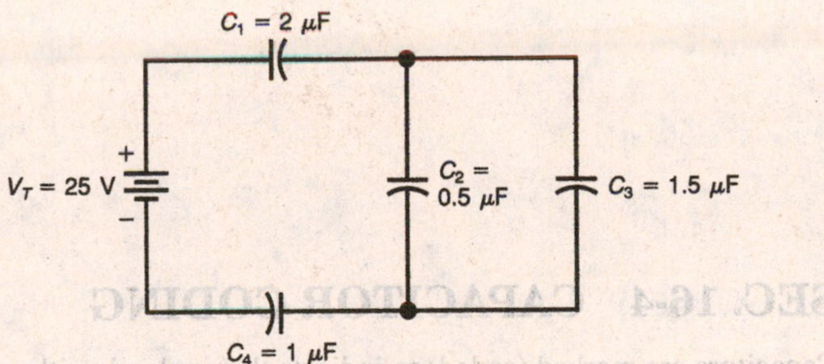

Fig. 16-8 Solve for Q_T, C_T, V_{C_1}, V_{C_2}, V_{C_3}, and V_{C_4}.

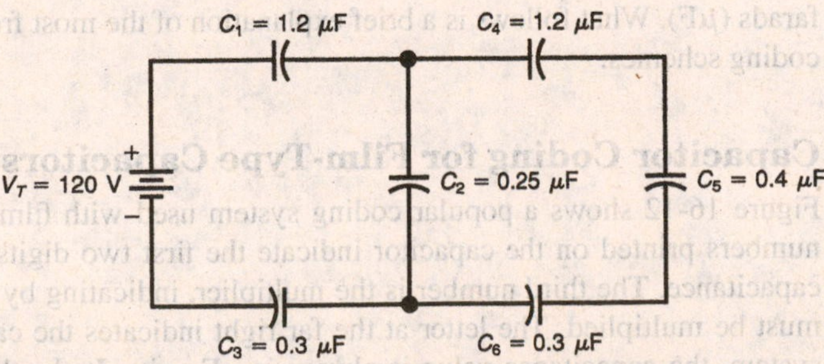

Fig. 16-9 Solve for Q_T, C_T, V_{C_1}, V_{C_2}, V_{C_3}, V_{C_4}, V_{C_5}, and V_{C_6}.

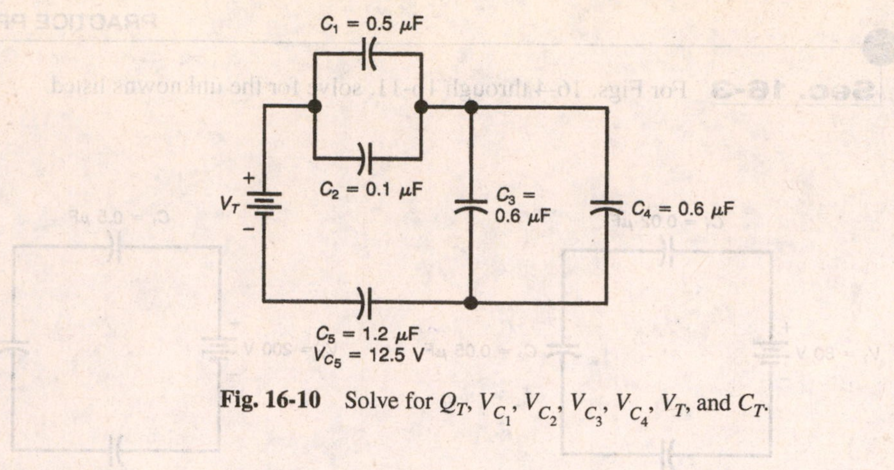

Fig. 16-10 Solve for Q_T, V_{C_1}, V_{C_2}, V_{C_3}, V_{C_4}, V_T, and C_T.

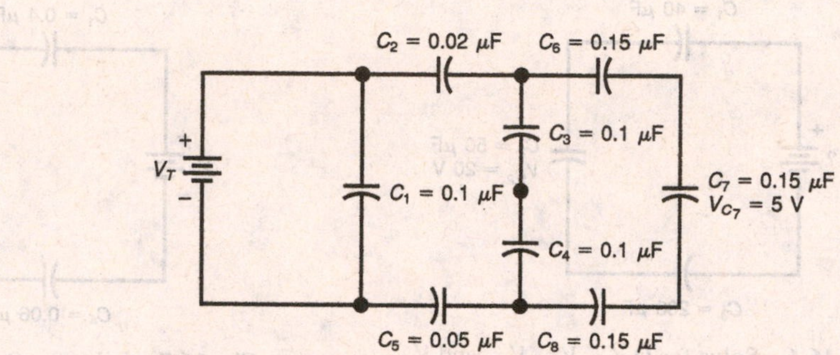

Fig. 16-11 Solve for Q_T, V_{C_1}, V_{C_2}, V_{C_3}, V_{C_4}, V_{C_5}, V_{C_6}, V_{C_8}, V_T, and C_T.

SEC. 16-4 CAPACITOR CODING

Capacitors are marked (coded) to indicate their value in either μF or pF units of capacitance. As a general rule, if a capacitor, other than an electrolytic, is marked using a whole number such as 56, 390, 224, etc. the capacitance, *C*, is in picofarads (pF). Conversely, if a capacitor is marked with a decimal fraction such as 0.01, 0.047, or 0.22, the capacitance, *C*, is in microfarads (μF). What follows is a brief explanation of the most frequently encountered capacitor coding schemes.

Capacitor Coding for Film-Type Capacitors

Figure 16-12 shows a popular coding system used with film-type capacitors. The first two numbers printed on the capacitor indicate the first two digits in the numerical value of the capacitance. The third number is the multiplier, indicating by what factor the first two digits must be multiplied. The letter at the far right indicates the capacitor's tolerance. Using this system, the capacitance value is always in pF units. It should be noted that the capacitor's breakdown voltage rating often appears directly below the coded value of capacitance.

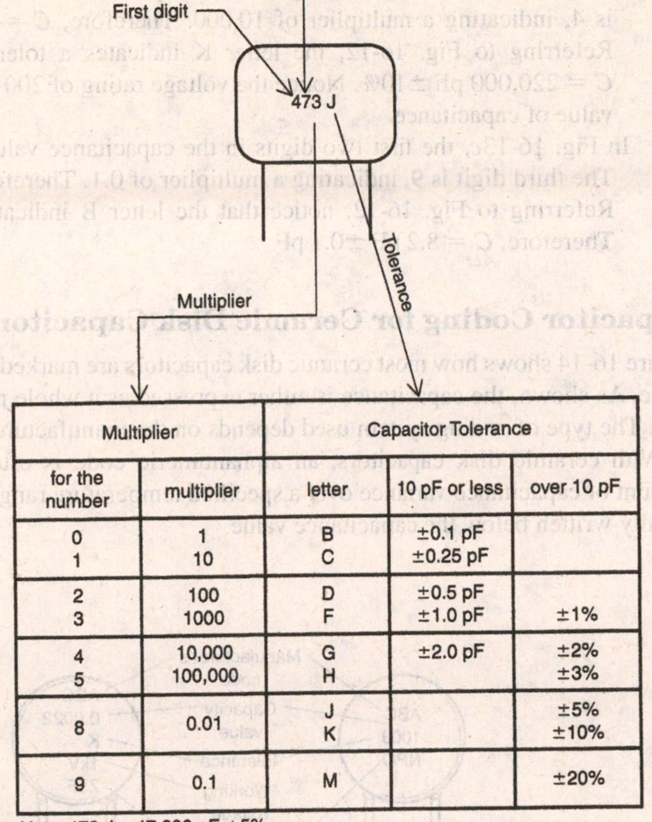

Multiplier		Capacitor Tolerance		
for the number	multiplier	letter	10 pF or less	over 10 pF
0 1	1 10	B C	±0.1 pF ±0.25 pF	
2 3	100 1000	D F	±0.5 pF ±1.0 pF	 ±1%
4 5	10,000 100,000	G H	±2.0 pF	±2% ±3%
8	0.01	J K		±5% ±10%
9	0.1	M		±20%

Note: 473 J = 47,000 pF ±5%

Fig. 16-12 Capacitor coding system for film capacitors.

Solved Problem

Indicate the capacitance and tolerance values for the capacitors in Fig. 16-13.

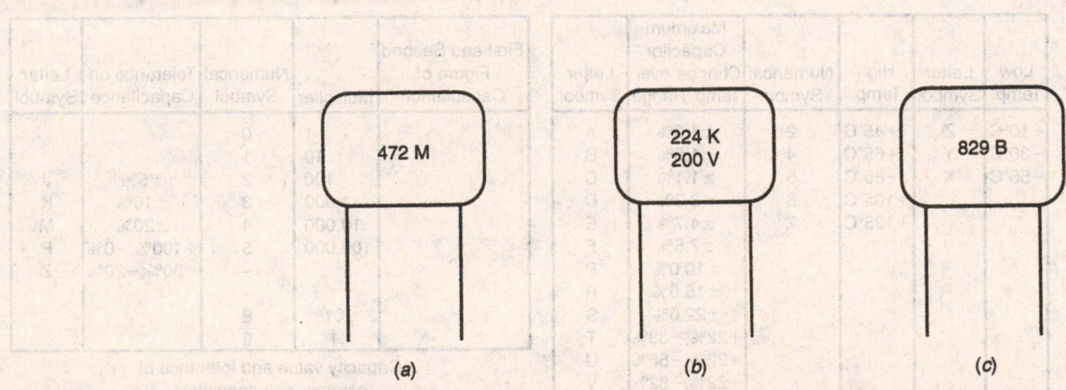

Fig. 16-13 Examples of film capacitor coding scheme.

Answer

In Fig. 16-13*a*, the first two digits in the capacitance value are 4 and 7, respectively. The third digit is 2 indicating a multiplier of 100. Therefore, $C = 47 \times 100 = 4{,}700$ pF. Referring to Fig. 16-12, the letter M represents a tolerance of ±20%. Therefore, $C = 4{,}700$ pF ±20%.

In Fig. 16-13*b*, the first two digits in the capacitance value are 2 and 2. The third digit is 4, indicating a multiplier of 10,000. Therefore, $C = 22 \times 10{,}000 = 220{,}000$ pF. Referring to Fig. 16-12, the letter K indicates a tolerance of ±10%. Therefore, $C = 220{,}000$ pF ±10%. Notice the voltage rating of 200 V is shown below the coded value of capacitance.

In Fig. 16-13*c*, the first two digits in the capacitance value are 8 and 2, respectively. The third digit is 9, indicating a multiplier of 0.1. Therefore, $C = 82 \times 0.1 = 8.2$ pF. Referring to Fig. 16-12, notice that the letter B indicates a tolerance of ±0.1 pF. Therefore, $C = 8.2$ pF ±0.1 pF.

Capacitor Coding for Ceramic Disk Capacitors

Figure 16-14 shows how most ceramic disk capacitors are marked to indicate their capacitance value. As shown, the capacitance is either expressed as a whole number or as a decimal fraction. The type of coding system used depends on the manufacturer.

With ceramic disk capacitors, an alphanumeric code is often included to indicate the amount of capacitance variance over a specified temperature range. The alphanumeric code is usually written below the capacitance value.

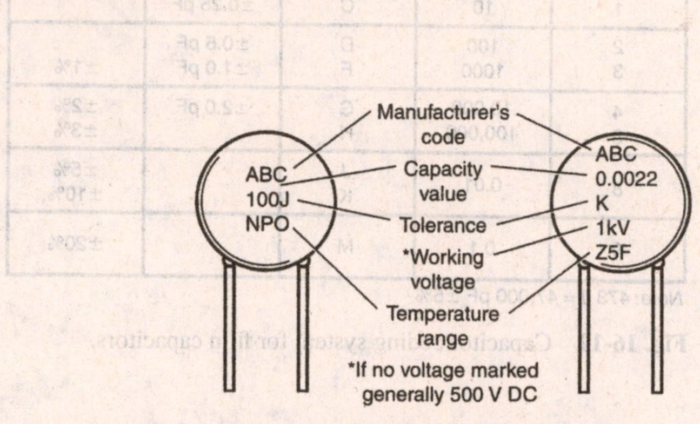

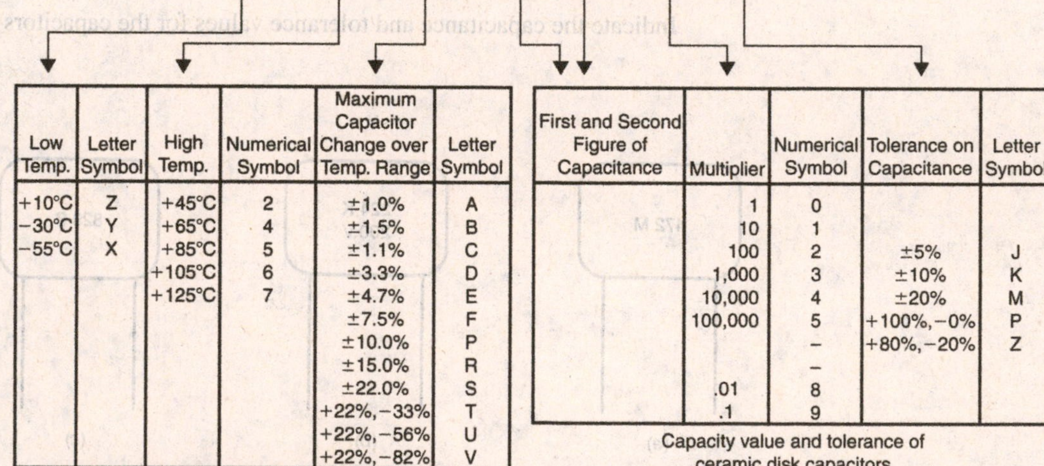

Low Temp.	Letter Symbol	High Temp.	Numerical Symbol	Maximum Capacitor Change over Temp. Range	Letter Symbol
+10°C	Z	+45°C	2	±1.0%	A
−30°C	Y	+65°C	4	±1.5%	B
−55°C	X	+85°C	5	±1.1%	C
		+105°C	6	±3.3%	D
		+125°C	7	±4.7%	E
				±7.5%	F
				±10.0%	P
				±15.0%	R
				±22.0%	S
				+22%, −33%	T
				+22%, −56%	U
				+22%, −82%	V

Temperature range indentification of ceramic disk capacitors

First and Second Figure of Capacitance	Multiplier	Numerical Symbol	Tolerance on Capacitance	Letter Symbol
	1	0		
	10	1		
	100	2	±5%	J
	1,000	3	±10%	K
	10,000	4	±20%	M
	100,000	5	+100%, −0%	P
		–	+80%, −20%	Z
		–		
	.01	8		
	.1	9		

Capacity value and tolerance of ceramic disk capacitors

Fig. 16-14 Capacitor coding scheme for ceramic disk capacitors.

Solved Problem

Indicate the capacitance and tolerance values for the ceramic disk capacitors in Fig. 16-15.

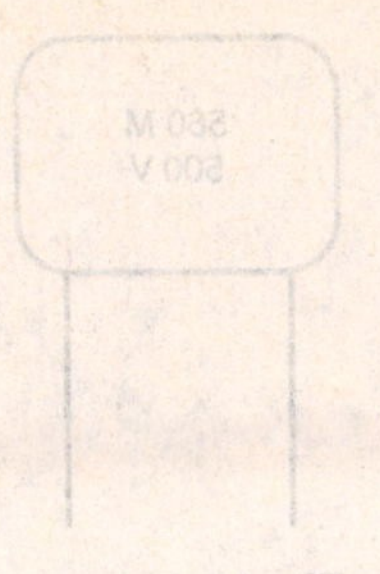

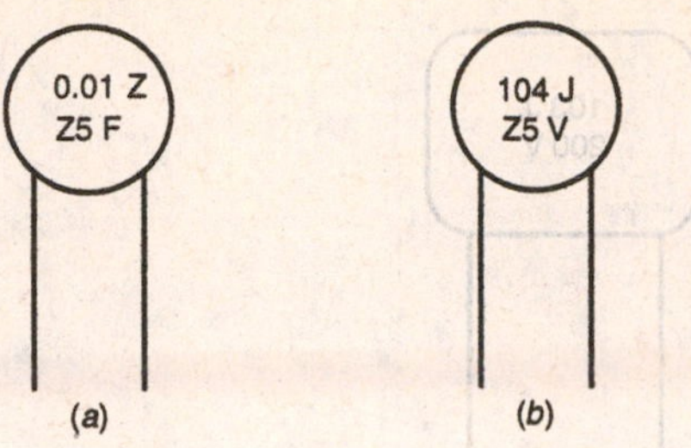

Fig. 16-15 Examples of ceramic disk capacitor coding scheme.

Answer

In Fig. 16-15*a*, the capacitance is expressed as a decimal fraction. This means that the capacitance is marked in μF. Therefore, $C = 0.01\ \mu$F. The letter Z (shown after 0.01) indicates a tolerance of +80%, −20%. The alphanumeric code, Z5F, indicates that C may vary ±7.5% over the temperature range of +10°C to +85°C. Referring to Fig. 16-14, the letter Z indicates a low temperature of +10°C, whereas the digit 5 indicates a high temperature of +85°C. The letter F indicates that the maximum capacitance change will be ±7.5% over the specified temperature range of +10°C to +85°C.

In Fig. 16-15*b*, the capacitance is expressed as a whole number. Therefore, C is marked in pF units. The first two numbers, 1 and 0, indicate the first two digits in the capacitance value. The third digit, 4, indicates a multiplier of 10,000. Therefore, $C = 10 \times 10{,}000 = 100{,}000$ pF. The letter J indicates a tolerance of ±5%. Therefore, $C = 100{,}000$ pF ±5%. The alphanumeric code, Z5V, indicates that C can vary +22%, −82%, over a temperature range of +10°C to +85°C.

PRACTICE PROBLEMS

Sec. 16-4

Indicate the capacitance and tolerance values for the capacitors in Figs. 16-16 through 16-40. For the ceramic disk capacitors, include the information corresponding to the alphanumeric code.

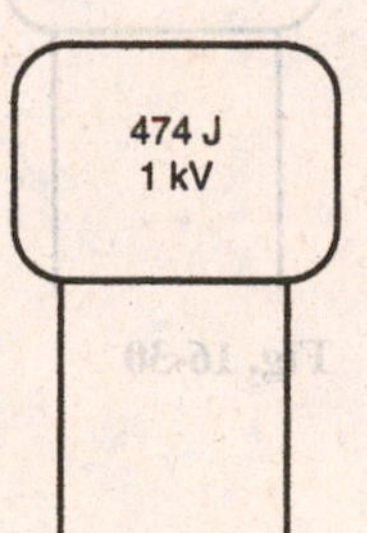

Fig. 16-16

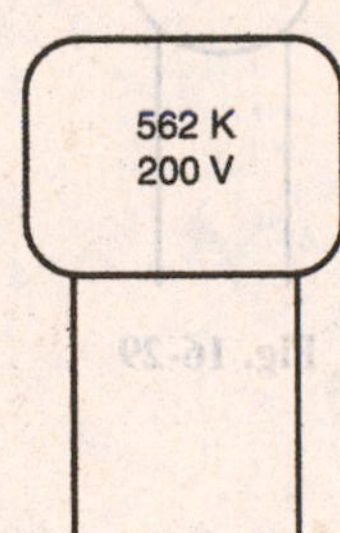

Fig. 16-17

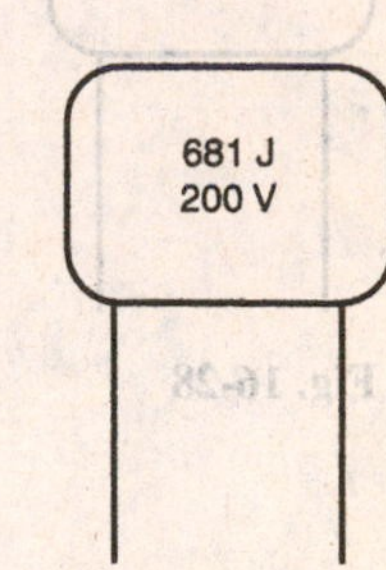

Fig. 16-18

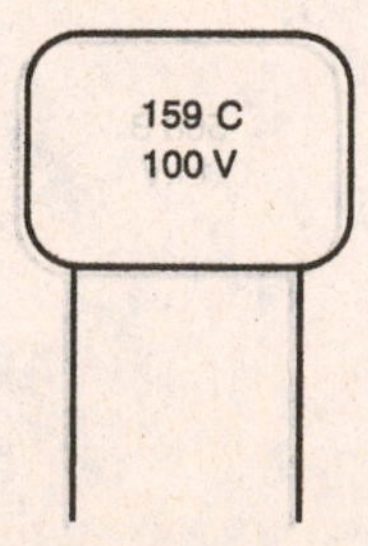

Fig. 16-19

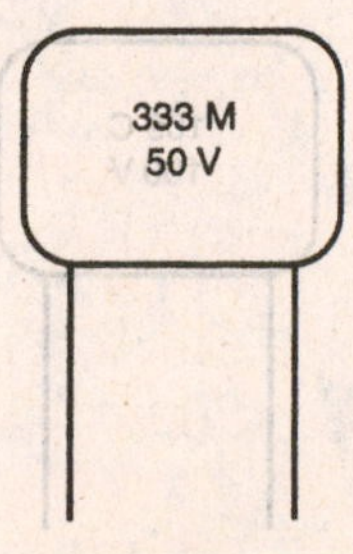

Fig. 16-20

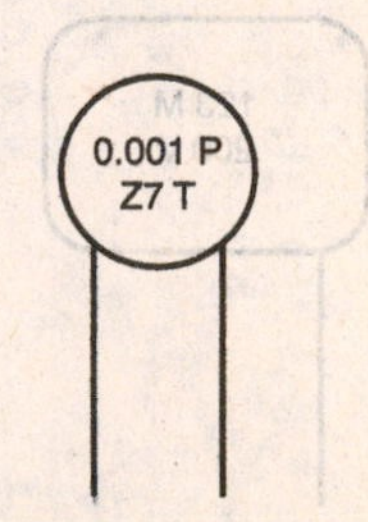

Fig. 16-21

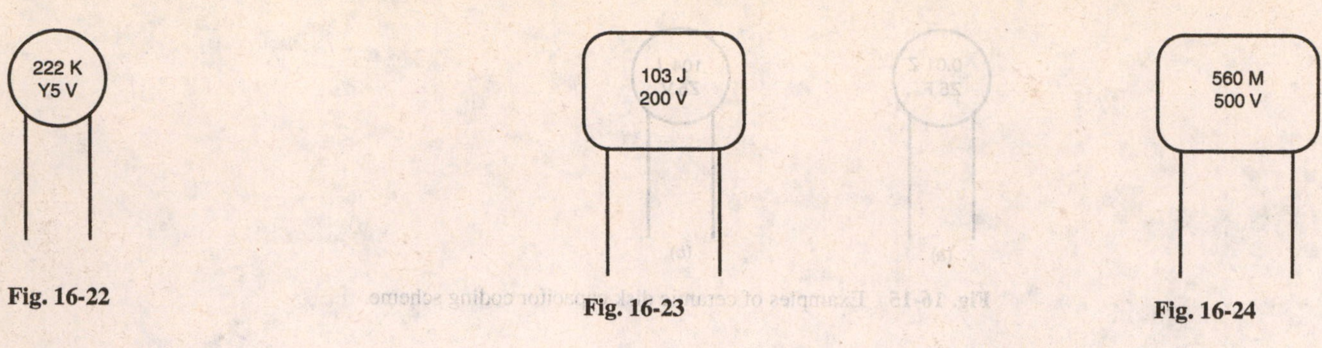

Fig. 16-22

Fig. 16-23

Fig. 16-24

Fig. 16-25

Fig. 16-26

Fig. 16-27

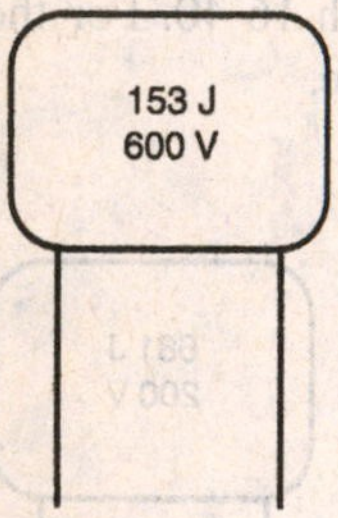

153 J
600 V

Fig. 16-28

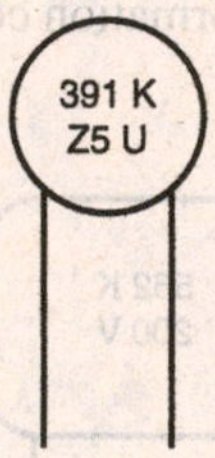

391 K
Z5 U

Fig. 16-29

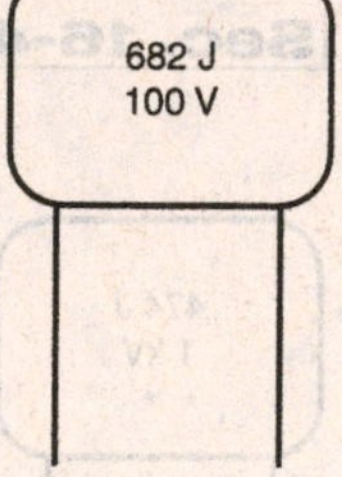

682 J
100 V

Fig. 16-30

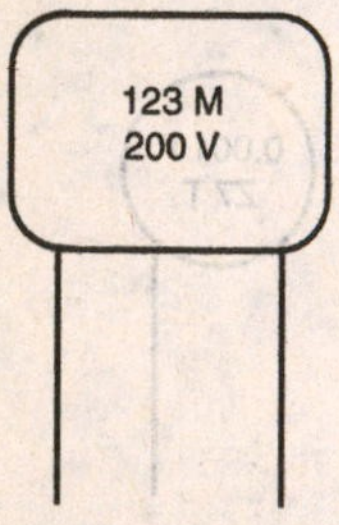

123 M
200 V

Fig. 16-31

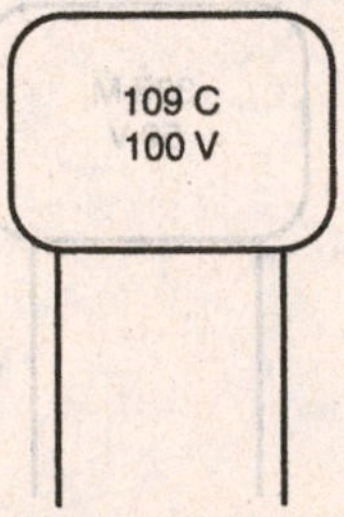

109 C
100 V

Fig. 16-32

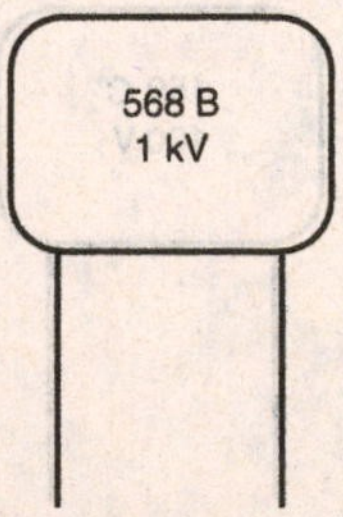

568 B
1 kV

Fig. 16-33

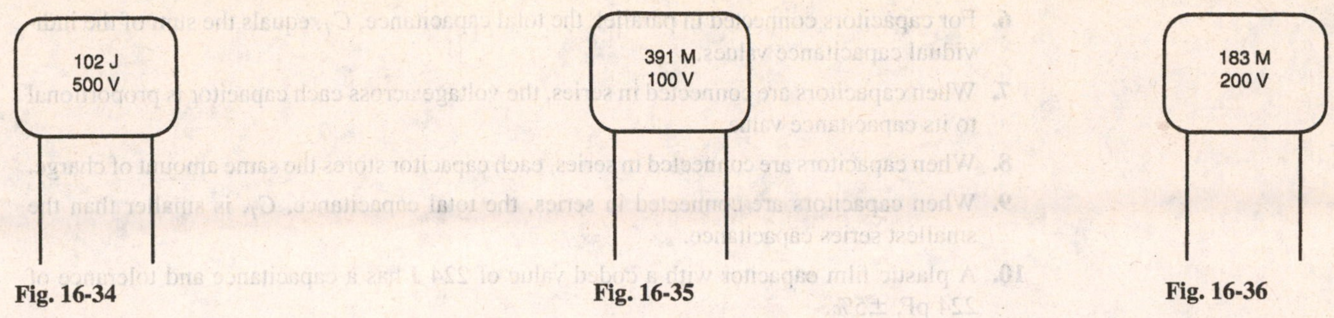

Fig. 16-34 Fig. 16-35 Fig. 16-36

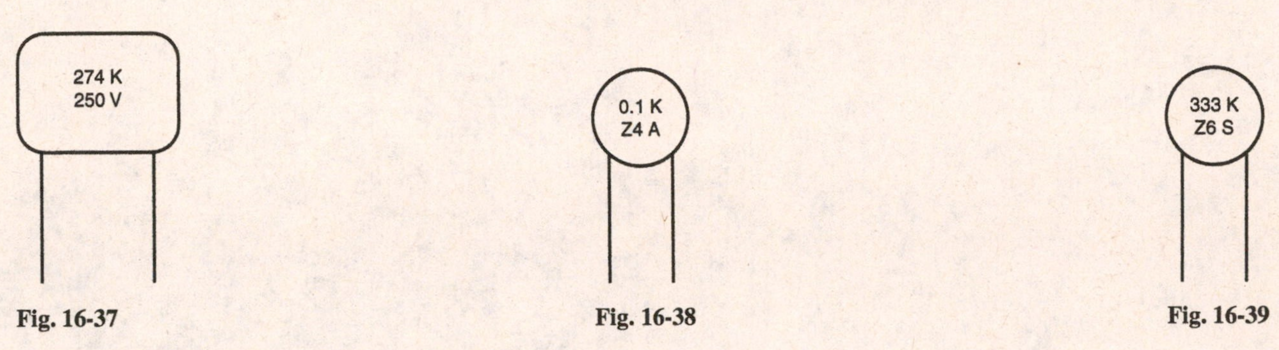

Fig. 16-37 Fig. 16-38 Fig. 16-39

Fig. 16-40

END OF CHAPTER TEST

Chapter 16: Capacitance

Answer True or False.

1. Capacitance is the ability of a dielectric to hold or store an electric charge.
2. A capacitor consists of two metal plates separated by an insulator.
3. The unit of capacitance is the farad.
4. The physical factors affecting the capacitance, C, of a capacitor are the area, A, of the plates, the distance, d, between the plates, and the dielectric constant, K_ϵ.
5. More charge stored for a given amount of voltage corresponds to a larger value of capacitance.

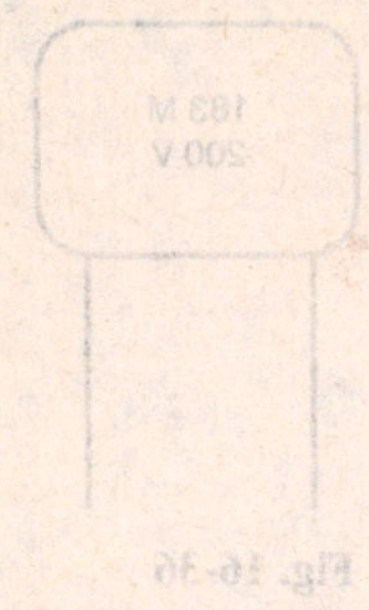

6. For capacitors connected in parallel, the total capacitance, C_T, equals the sum of the individual capacitance values.
7. When capacitors are connected in series, the voltage across each capacitor is proportional to its capacitance value.
8. When capacitors are connected in series, each capacitor stores the same amount of charge.
9. When capacitors are connected in series, the total capacitance, C_T, is smaller than the smallest series capacitance.
10. A plastic film capacitor with a coded value of 224 J has a capacitance and tolerance of 224 pF, ±5%.

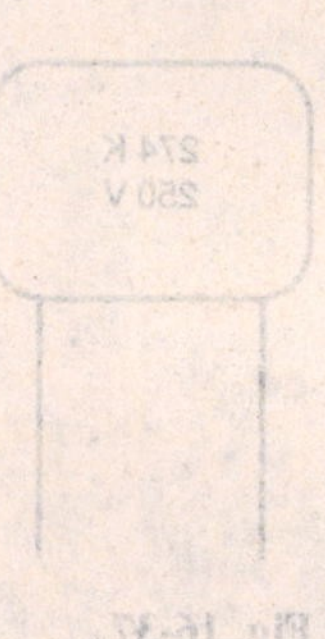

CHAPTER

17

Capacitive Reactance

Capacitive reactance, designated X_C, is a measure of a capacitor's opposition to the flow of sine-wave alternating current. Like resistance, R, X_C is measured in ohms. In this chapter, you will calculate X_C for different values of frequency and capacitance. You will also learn how X_C values combine in series and in parallel.

SEC. 17-1 CAPACITIVE REACTANCE X_C

When an alternating voltage is applied across the plates of a capacitor, the capacitor will alternately charge and discharge. This means there will be charge and discharge current flowing to and from the plates of the capacitor. However, there is no current flowing through the actual dielectric of the capacitor. How much current flows for a given amount of applied voltage is determined by the capacitive reactance, X_C. To determine the capacitive reactance, X_C, for a known capacitance at a given frequency, we use the formula

$$X_C = \frac{1}{2\pi fC}$$

where X_C = capacitive reactance in ohms
f = frequency of alternating voltage in hertz
C = capacitance in farads

The other useful forms of this equation are

$$C = \frac{1}{2\pi f X_C}$$

and

$$f = \frac{1}{2\pi C X_C}$$

It should also be pointed out that X_C is a V/I ratio. Thus, X_C can also be stated as

$$X_C = \frac{V_C}{I}$$

where X_C = capacitive reactance in ohms
V_C = capacitor voltage in volts
I = current in amperes

Solved Problem

Solve for the unknowns shown in Fig. 17-1.

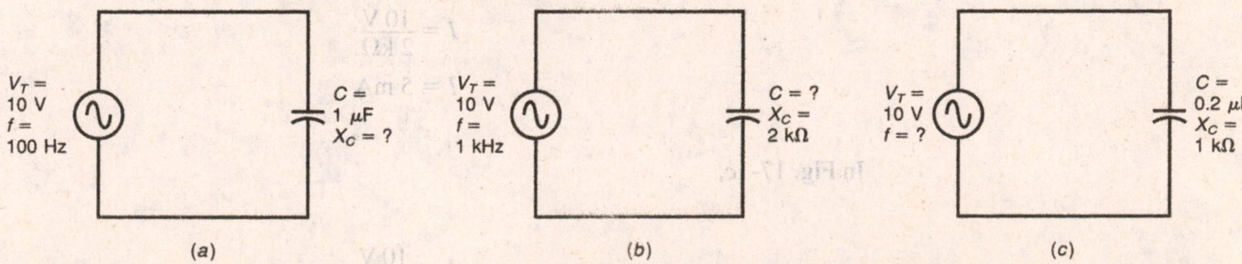

Fig. 17-1 Circuits used to apply the formula $X_C = 1/2\pi fC$. (*a*) Finding X_C when f and C are known. (*b*) Finding C when X_C and f are known. (*c*) Finding f when X_C and C are known.

Answer

In Fig. 17-1*a*, both *C* and *f* are known. We find X_C as follows:

$$X_C = \frac{1}{2\pi fC}$$

$$= \frac{1}{2 \times 3.14 \times 100\text{ Hz} \times 1\ \mu\text{F}}$$

$$X_C = 1{,}591.5\ \Omega$$

In Fig. 17-1*b*, both X_C and *f* are known. We find *C* as follows:

$$C = \frac{1}{2\pi X_C f}$$

$$= \frac{1}{2 \times 3.14 \times 2\text{ k}\Omega \times 1\text{ kHz}}$$

$$C = 79.57\text{ nF}$$

In Fig. 17-1*c*, both *C* and X_C are known. We find *f* as follows:

$$f = \frac{1}{2\pi CX_C}$$

$$= \frac{1}{2 \times 3.14 \times 1\text{ k}\Omega \times 0.2\ \mu\text{F}}$$

$$f = 795.7\text{ Hz}$$

For each circuit in Fig. 17-1,

$$I = \frac{V}{X_C}$$

In Fig. 17-1*a*,

$$I = \frac{10\text{ V}}{1591.5\ \Omega}$$

$$I = 6.28\text{ mA}$$

In Fig. 17-1*b*,

$$I = \frac{10\text{ V}}{2\text{ k}\Omega}$$

$$I = 5\text{ mA}$$

In Fig. 17-1*c*,

$$I = \frac{10\text{ V}}{1\text{ k}\Omega}$$

$$I = 10\text{ mA}$$

Sec. 17-1 For Figs. 17-2 through 17-17, solve for the unknowns listed.

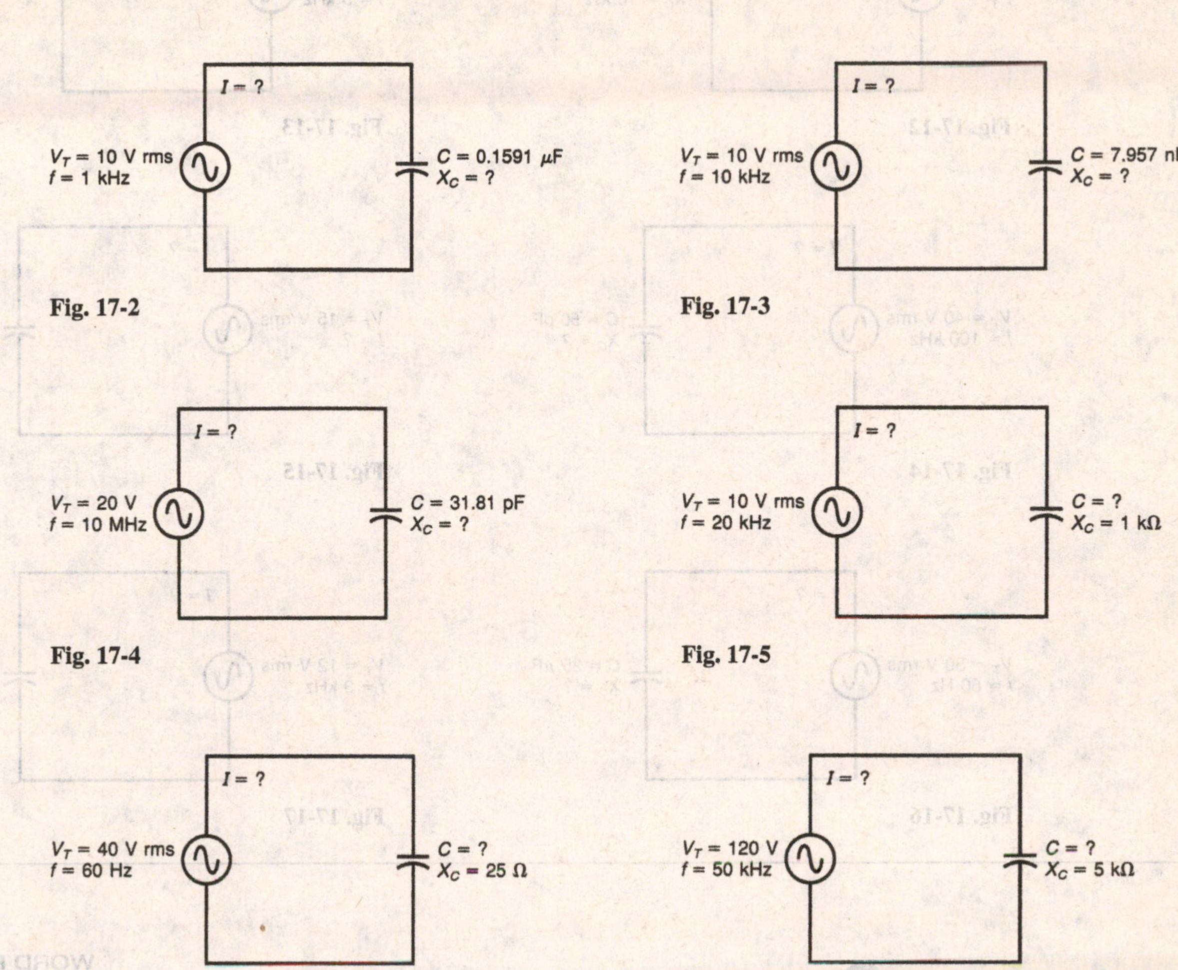

Fig. 17-2

Fig. 17-3

Fig. 17-4

Fig. 17-5

Fig. 17-6

Fig. 17-7

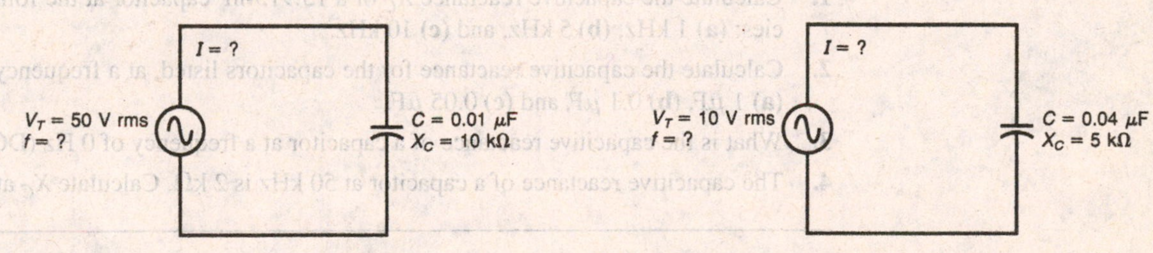

Fig. 17-8

Fig. 17-9

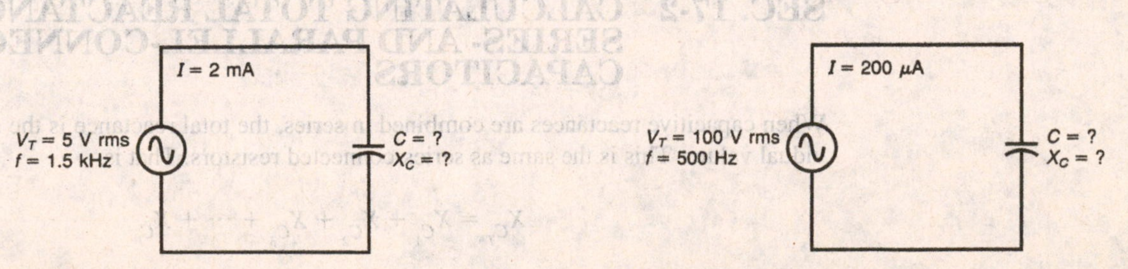

Fig. 17-10

Fig. 17-11

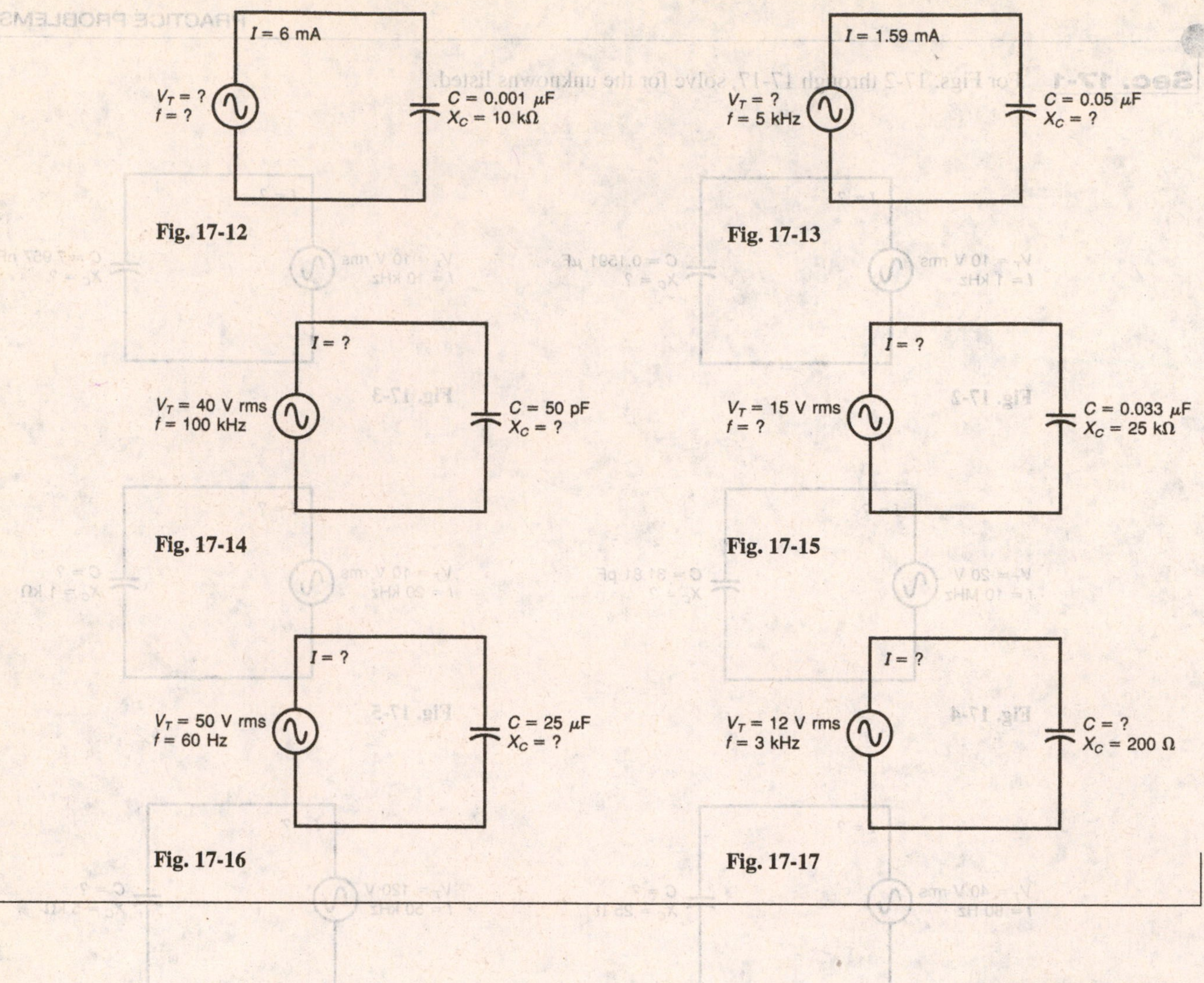

WORD PROBLEMS

Sec. 17-1 Solve the following.

1. Calculate the capacitive reactance X_C of a 15.915-nF capacitor at the following frequencies: **(a)** 1 kHz, **(b)** 5 kHz, and **(c)** 10 kHz.
2. Calculate the capacitive reactance for the capacitors listed, at a frequency of 31.81 kHz: **(a)** 1 μF, **(b)** 0.1 μF, and **(c)** 0.05 μF.
3. What is the capacitive reactance of a capacitor at a frequency of 0 Hz (DC)?
4. The capacitive reactance of a capacitor at 50 kHz is 2 kΩ. Calculate X_C at 10 kHz.

SEC. 17-2 CALCULATING TOTAL REACTANCE FOR SERIES- AND PARALLEL-CONNECTED CAPACITORS

When capacitive reactances are combined in series, the total reactance is the sum of the individual values. This is the same as series-connected resistors. That is,

$$X_{C_T} = X_{C_1} + X_{C_2} + X_{C_3} + \cdots + X_{C_n}$$

where X_{C_T} = total capacitive reactance in ohms

When capacitive reactances are connected in parallel, the total reactance is found using the reciprocal formula

$$X_{C_T} = \frac{1}{1/X_{C_1} + 1/X_{C_2} + 1/X_{C_3} + \cdots + 1/X_{C_n}}$$

This is the same as parallel-connected resistors. The combined parallel reactance will be less than the lowest branch reactance.

Solved Problems

a. For Fig. 17-18*a*, find C_T, X_{C_T}, I, V_{C_1}, and V_{C_2}.

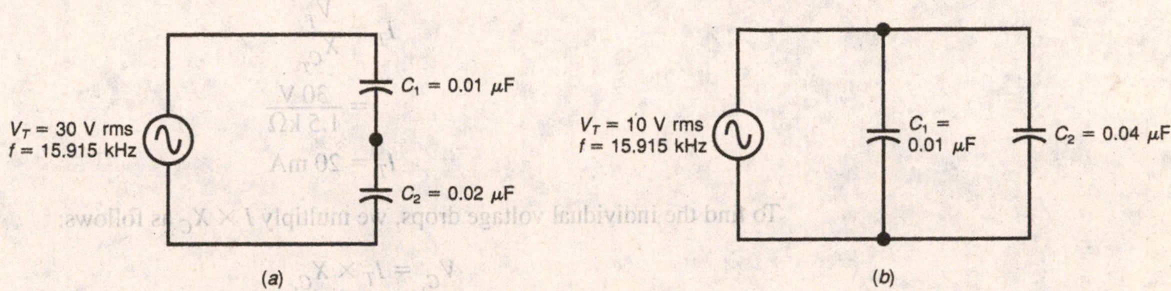

Fig. 17-18 Series and parallel connections of capacitors. (*a*) Capacitors C_1 and C_2 in series. (*b*) Capacitors C_1 and C_2 in parallel.

Answer

First, we find X_{C_1} and X_{C_2}.

$$X_{C_1} = \frac{1}{2\pi f C_1}$$

$$= \frac{1}{2 \times 3.14 \times 15.915 \text{ kHz} \times 0.01 \ \mu\text{F}}$$

$$X_{C_1} = 1 \text{ k}\Omega$$

$$X_{C_2} = \frac{1}{2\pi f C_2}$$

$$= \frac{1}{2 \times 3.14 \times 15.915 \text{ kHz} \times 0.02 \ \mu\text{F}}$$

$$X_{C_2} = 500 \ \Omega$$

Next, we find X_{C_T} which can be found one of two ways. First, since

$$C_T = \frac{1}{1/C_1 + 1/C_2}$$

we have

$$C_T = \frac{1}{1/0.01 \ \mu\text{F} + 1/0.02 \ \mu\text{F}}$$

$$C_T = 6.66 \text{ nF}$$

Then,

$$X_{C_T} = \frac{1}{2\pi f C_T}$$

$$= \frac{1}{2 \times 3.14 \times 15.915 \text{ kHz} \times 6.66 \text{ nF}}$$

$$X_{C_T} = 1.5 \text{ k}\Omega$$

Second,

$$X_{C_T} = X_{C_1} + X_{C_2}$$
$$= 1 \text{ k}\Omega + 500 \text{ }\Omega$$
$$X_{C_T} = 1.5 \text{ k}\Omega$$

To find I, we use Ohm's law. Since the circuit contains only capacitive reactance, $I = V/X_C$. For Fig. 17-18a, we have

$$I_T = \frac{V_T}{X_{C_T}}$$
$$= \frac{30 \text{ V}}{1.5 \text{ k}\Omega}$$
$$I_T = 20 \text{ mA}$$

To find the individual voltage drops, we multiply $I \times X_C$ as follows:

$$V_{C_1} = I_T \times X_{C_1}$$
$$= 20 \text{ mA} \times 1 \text{ k}\Omega$$
$$V_{C_1} = 20 \text{ V}$$
$$V_{C_2} = I_T \times X_{C_2}$$
$$= 20 \text{ mA} \times 500 \text{ }\Omega$$
$$V_{C_2} = 10 \text{ V}$$

Note,

$$V_T = V_{C_1} + V_{C_2}$$
$$= 20 \text{ V} + 10 \text{ V}$$
$$V_T = 30 \text{ V}$$

Also note,

$$V_T = I_T \times X_{C_T}$$
$$= 20 \text{ mA} \times 1.5 \text{ k}\Omega$$
$$V_T = 30 \text{ V}$$

b. For Fig. 17-18b, find C_T, X_{C_T}, I_{C_1}, I_{C_2}, and I_T.

Answer

First, we find X_{C_1} and X_{C_2}.

$$X_{C_1} = \frac{1}{2\pi f C_1}$$

$$= \frac{1}{2 \times 3.14 \times 15.915 \text{ kHz} \times 0.01 \text{ }\mu\text{F}}$$

$$X_{C_1} = 1 \text{ k}\Omega$$

$$X_{C_2} = \frac{1}{2\pi f C_2}$$

$$= \frac{1}{2 \times 3.14 \times 15.915 \text{ kHz} \times 0.04 \text{ }\mu\text{F}}$$

$$X_{C_2} = 250 \text{ }\Omega$$

Next, we find I_{C_1} and I_{C_2}.

$$I_{C_1} = \frac{V_T}{X_{C_1}}$$
$$= \frac{10\text{ V}}{1\text{ k}\Omega}$$
$$I_{C_1} = 10\text{ mA}$$
$$I_{C_2} = \frac{V_T}{X_{C_2}}$$
$$= \frac{10\text{ V}}{250\ \Omega}$$
$$I_{C_2} = 40\text{ mA}$$

The total current, I_T, is found by adding I_{C_1} and I_{C_2}.

$$I_T = I_{C_1} + I_{C_2}$$
$$= 10\text{ mA} + 40\text{ mA}$$
$$I_T = 50\text{ mA}$$

Then, X_{C_T} can be found using the formulas

$$X_{C_T} = \frac{1}{1/X_{C_1} + 1/X_{C_2}}$$
$$= \frac{1}{1/1\text{ k}\Omega + 1/250\ \Omega}$$
$$X_{C_T} = 200\ \Omega$$

X_{C_T} can also be found by using the formula

$$X_{C_T} = \frac{1}{2\pi f C_T}$$

First, C_T must be found. For Fig. 17-18*b*, we have

$$C_T = C_1 + C_2$$
$$= 0.01\ \mu\text{F} + 0.04\ \mu\text{F}$$
$$C_T = 0.05\ \mu\text{F}$$

Then,

$$X_{C_T} = \frac{1}{2\pi f C_T}$$
$$= \frac{1}{2 \times 3.14 \times 15.915\text{ kHz} \times 0.05\ \mu\text{F}}$$
$$X_{C_T} = 200\ \Omega$$

It should be pointed out that I_T can be found using Ohm's law.

$$I_T = \frac{V_T}{X_{C_T}}$$

In Fig. 17-18*b*,

$$I_T = \frac{10\text{ V}}{200\ \Omega}$$
$$I_T = 50\text{ mA}$$

Sec. 17-2 For Figs. 17-19 through 17-30, solve for the unknowns listed.

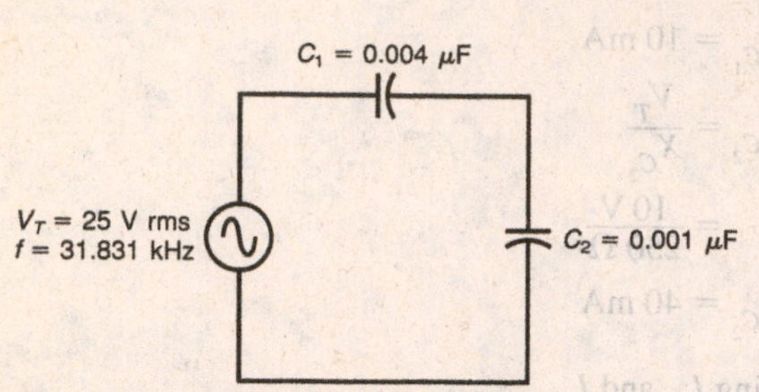

Fig. 17-19 Solve for C_T, X_{C_1}, X_{C_2}, X_{C_T}, I_T, V_{C_1}, and V_{C_2}.

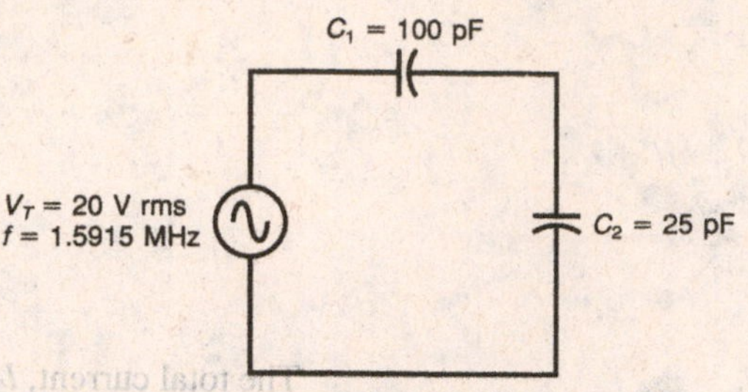

Fig. 17-20 Solve for C_T, X_{C_1}, X_{C_2}, X_{C_T}, I_T, V_{C_1}, and V_{C_2}.

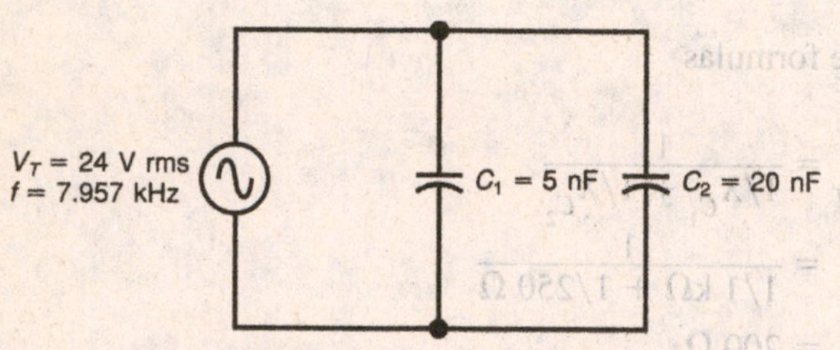

Fig. 17-21 Solve for C_T, X_{C_1}, X_{C_2}, X_{C_T}, I_{C_1}, I_{C_2}, and I_T.

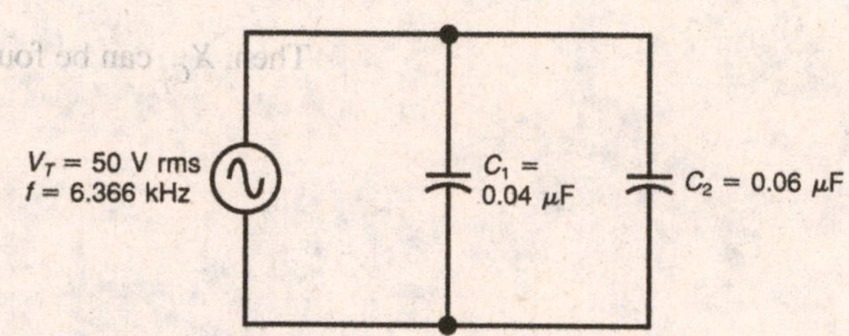

Fig. 17-22 Solve for C_T, X_{C_1}, X_{C_2}, X_{C_T}, I_{C_1}, I_{C_2}, and I_T.

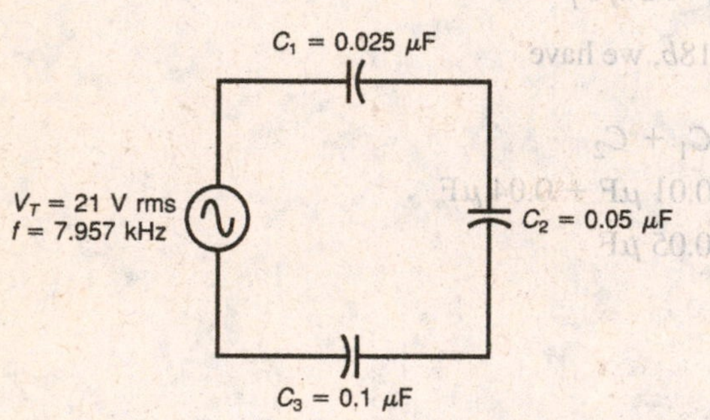

Fig. 17-23 Solve for C_T, X_{C_1}, X_{C_2}, X_{C_3}, X_{C_T}, I_T, V_{C_1}, V_{C_2}, and V_{C_3}.

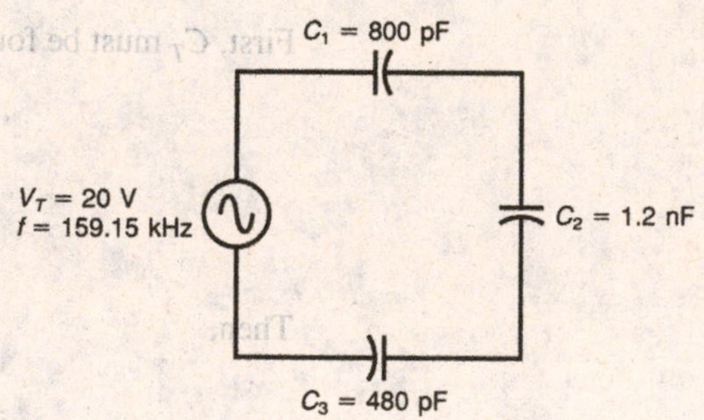

Fig. 17-24 Solve for C_T, X_{C_1}, X_{C_2}, X_{C_3}, X_{C_T}, I_T, V_{C_1}, V_{C_2}, and V_{C_3}.

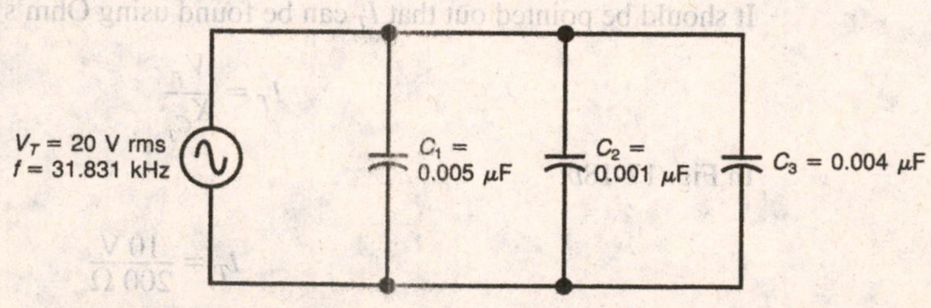

Fig. 17-25 Solve for C_T, X_{C_1}, X_{C_2}, X_{C_3}, X_{C_T}, I_{C_1}, I_{C_2}, I_{C_3}, and I_T.

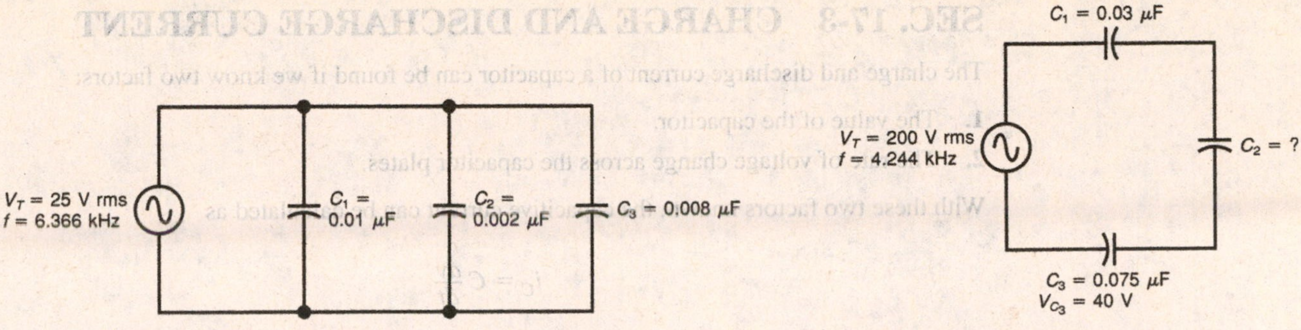

Fig. 17-26 Solve for C_T, X_{C_1}, X_{C_2}, X_{C_3}, X_{C_T}, I_{C_1}, I_{C_2}, I_{C_3}, and I_T.

Fig. 17-27 Solve for C_2, C_T, X_{C_1}, X_{C_2}, X_{C_3}, X_{C_T}, I_T, V_{C_1}, and V_{C_2}.

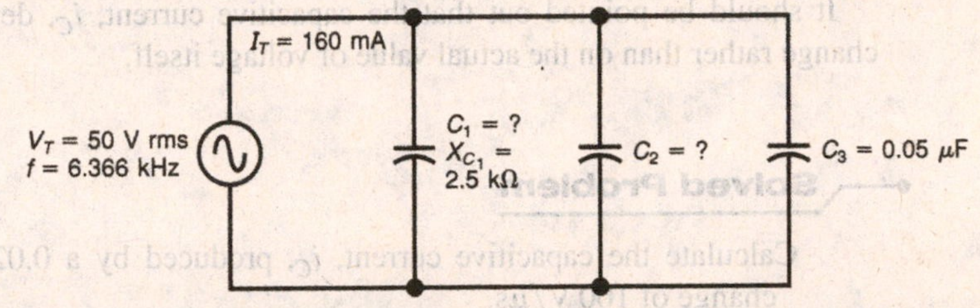

Fig. 17-28 Solve for C_T, C_1, C_2, X_{C_2}, X_{C_3}, X_{C_T}, I_{C_1}, I_{C_2}, and I_{C_3}.

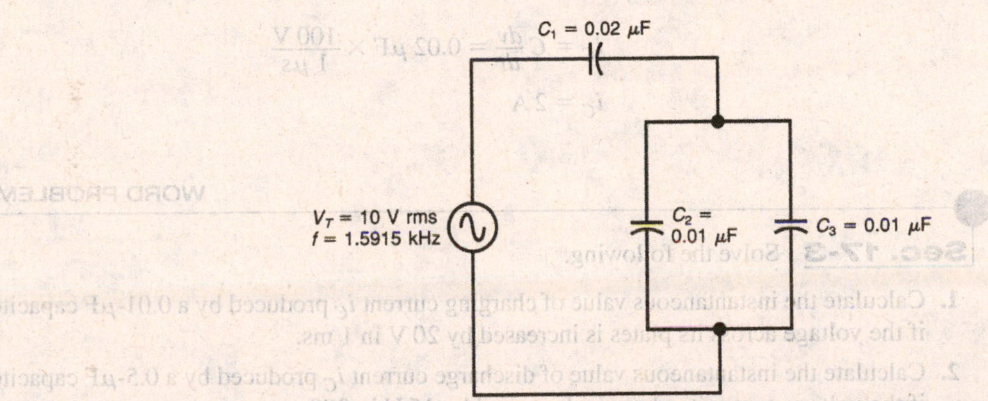

Fig. 17-29 Solve for C_T, X_{C_1}, X_{C_2}, X_{C_3}, X_{C_T}, V_{C_1}, V_{C_2}, V_{C_3}, I_2, I_3, and I_T.

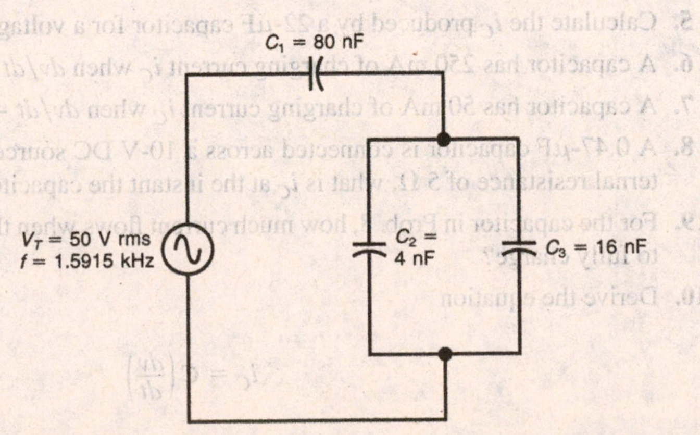

Fig. 17-30 Solve for C_T, X_{C_1}, X_{C_2}, X_{C_3}, X_{C_T}, V_{C_1}, V_{C_2}, V_{C_3}, I_2, I_3, and I_T.

SEC. 17-3 CHARGE AND DISCHARGE CURRENT

The charge and discharge current of a capacitor can be found if we know two factors:

1. The value of the capacitor.
2. The rate of voltage change across the capacitor plates.

With these two factors known, the capacitive current can be calculated as

$$i_C = C\frac{dv}{dt}$$

where i_C = charge or discharge current in amperes
C = capacitance value in farads
$\frac{dv}{dt}$ = rate of voltage change across capacitor plates

It should be pointed out that the capacitive current, i_C, depends on the rate of voltage change rather than on the actual value of voltage itself.

Solved Problem

Calculate the capacitive current, i_C, produced by a 0.02-μF capacitor for a voltage change of 100 V/μs.

Answer

$$i_C = C\frac{dv}{dt} = 0.02\ \mu\text{F} \times \frac{100\ \text{V}}{1\ \mu\text{s}}$$

$$i_C = 2\ \text{A}$$

WORD PROBLEMS

Sec. 17-3 Solve the following.

1. Calculate the instantaneous value of charging current i_C produced by a 0.01-μF capacitor if the voltage across its plates is increased by 20 V in 1 ms.
2. Calculate the instantaneous value of discharge current i_C produced by a 0.5-μF capacitor if the voltage across its plates is decreased by 15 V in 200 μs.
3. Calculate the i_C produced by a 400-pF capacitor for a voltage change of 40 V in 20 μs.
4. Calculate the i_C produced by a 1,000-μF capacitor for a voltage change of 25 V in 1 ms.
5. Calculate the i_C produced by a 22-μF capacitor for a voltage change of 10 V in 4 ms.
6. A capacitor has 250 mA of charging current i_C when dv/dt = 10 V/1 μs. Calculate C.
7. A capacitor has 50 mA of charging current i_C when dv/dt = 4 V/1 ms. Calculate C.
8. A 0.47-μF capacitor is connected across a 10-V DC source. If the DC source has an internal resistance of 5 Ω, what is i_C at the instant the capacitor is connected to the source?
9. For the capacitor in Prob. 8, how much current flows when the capacitor has been allowed to fully charge?
10. Derive the equation

$$i_C = C\left(\frac{dv}{dt}\right)$$

from $Q = CV$.

Chapter 17: Capacitive Reactance

Answer True or False.

1. Capacitive reactance, X_C, is a measure of a capacitor's opposition to the flow of sine-wave alternating current.
2. X_C is measured in farads.
3. For a given capacitance, X_C decreases as the frequency, f, increases.
4. For a capacitor, the charge and discharge current flows to and from the plates but not through the dielectric.
5. For series-connected capacitors, the total capacitive reactance, X_{C_T}, equals the sum of the individual capacitive reactance values.
6. For parallel-connected capacitors, the total capacitive reactance, X_{C_T}, is always smaller than the smallest branch reactance.
7. For a capacitor, the charge and discharge current, i_C, is dependent on the rate of voltage change across the capacitor plates rather than on the absolute value of voltage.
8. Capacitive reactance, X_C, is directly proportional to the capacitance, C.
9. If a capacitor has an X_C value of 1 kΩ at 10 kHz, then at 5 kHz, X_C equals 500 Ω.
10. Two parallel-connected capacitors, each with an X_C value of 200 Ω, have an X_{C_T} of 100 Ω.

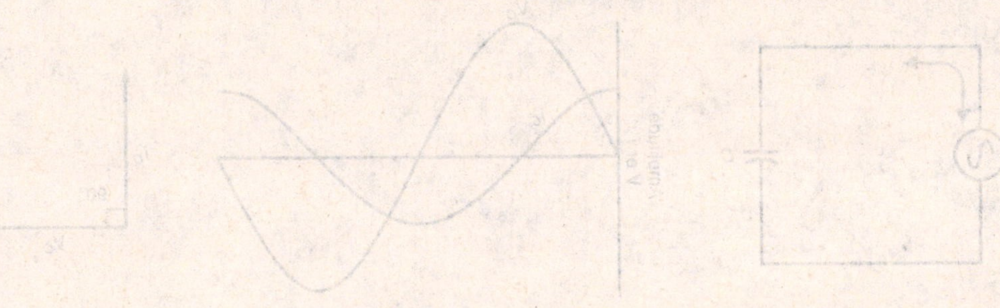

CHAPTER

18

Capacitive Circuits

When a capacitor is used in a sine-wave AC circuit, the charge and discharge current flowing to and from the plates of the capacitor leads the voltage across the capacitor plates by a phase angle of 90°. This 90° phase relationship between voltage and current exists for any value of capacitance or any frequency of sine-wave alternating voltage. The reason for this 90° phase relationship can be seen by examining the formula for charge and discharge current: $i_C = C(dv/dt)$. It should be obvious that the instantaneous value of charge or discharge current is directly proportional to the rate of voltage change dv/dt. If the rate of voltage change is zero, there is no charge or discharge current. The capacitive circuit in Fig. 18-1*a* has the current and voltage waveforms, as shown in Fig. 18-1*b*. The phasor diagram for this voltage-current relationship is shown in Fig. 18-1*c*.

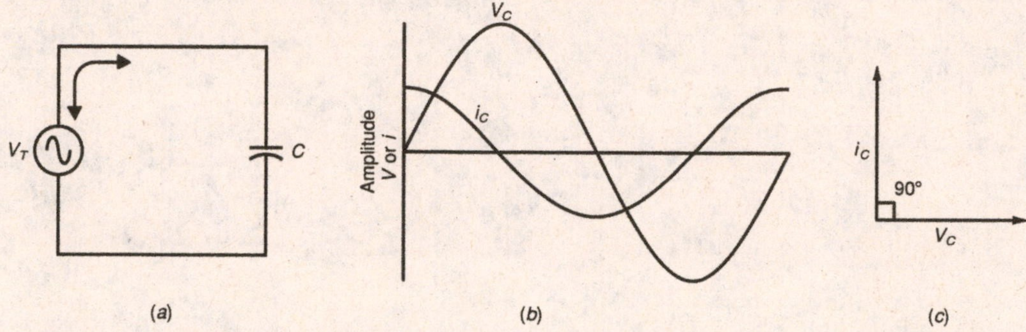

Fig. 18-1 (*a*) Circuit with capacitance *C*. (*b*) Sine wave of V_C lags i_C by 90°. (*c*) Phasor diagram for voltage and current.

SEC. 18-1 ANALYZING SERIES *RC* CIRCUITS

The 90° phase relationship between voltage and current is important to know when we analyze circuits containing both capacitive reactance and resistance. To analyze series *RC* circuits, we use the equations listed in Table 18-1.

TABLE 18-1 FORMULAS USED FOR SERIES *RC* CIRCUITS

$Z_T = \sqrt{R^2 + X_C{}^2}$	$V_T = \sqrt{V_R{}^2 + V_C{}^2}$
$I_T = \dfrac{V_T}{Z_T}$	$\tan \theta_Z = -\dfrac{X_C}{R}$
$V_C = I_T \times X_C$	*or*
$V_R = I_T \times R$	$\tan \theta_V = -\dfrac{V_C}{V_R}$

It is important to realize that V_R and V_C cannot be added algebraically, since they are always 90° out of phase with each other. This is also true for X_C and R. Instead, these quantities must be added phasorally to obtain V_T or Z_T. The phase angle between the applied voltage V_T and total current I_T is specified by θ_Z or θ_V.

Solved Problem

For Fig. 18-2, find Z_T, I_T, V_C, V_R, and θ_V. Draw the phasor voltage triangle.

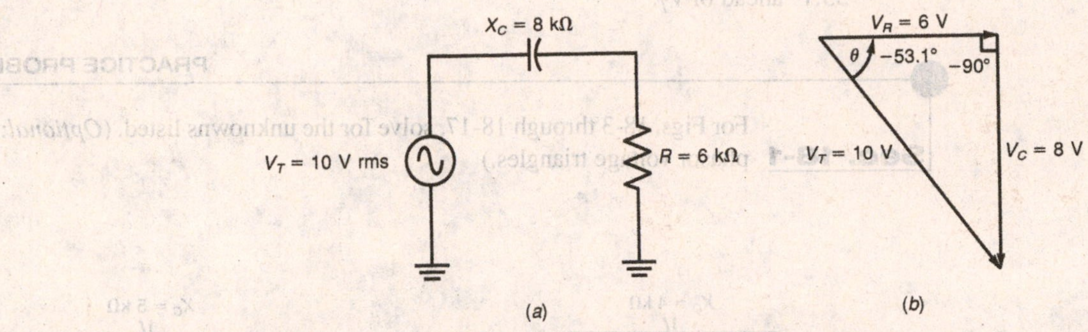

Fig. 18-2 (*a*) Series *RC* circuit. (*b*) Phasor voltage triangle.

Answer

First, we must find Z_T. From Table 18-1, we have

$$Z_T = \sqrt{R^2 + X_C^2}$$
$$= \sqrt{6^2 + 8^2}\ \text{k}\Omega$$
$$Z_T = 10\ \text{k}\Omega$$

This is the total opposition to the flow of alternating current. With Z_T known, we can find I_T as follows:

$$I_T = \frac{V_T}{Z_T}$$
$$= \frac{10\ \text{V}}{10\ \text{k}\Omega}$$
$$I_T = 1\ \text{mA}$$

Knowing I_T allows us to find V_C and V_R.

$$V_C = I_T \times X_C$$
$$= 1\ \text{mA} \times 8\ \text{k}\Omega$$
$$V_C = 8\ \text{V}$$

$$V_R = I_T \times R$$
$$= 1\ \text{mA} \times 6\ \text{k}\Omega$$
$$V_R = 6\ \text{V}$$

To find the phase angle θ for the voltage triangle, we have

$$\tan \theta_V = -\frac{V_C}{V_R}$$
$$= -\frac{8\ \text{V}}{6\ \text{V}}$$
$$\tan \theta_V = -1.33$$

To find the angle corresponding to this tangent value, we take the arctan of −1.33.

$$\theta = \arctan -1.33$$
$$\theta = -53.1°$$

This phase angle tells us that the total voltage, V_T, lags the total current, I_T, by 53.1°. The phasor voltage triangle is shown in Fig. 18-2*b*. Notice that V_R is used as the reference phasor. This is because V_R is in phase with the total current, I_T. Since I_T is the same in all components, the phase relationship between V_R and V_T specifies also the phase angle between total voltage, V_T, and total current, I_T. Also, notice that V_T lags V_R by 53.1°. This means that the total current, I_T, reaches its maximum value 53.1° ahead of V_T.

PRACTICE PROBLEMS

Sec. 18-1 For Figs. 18-3 through 18-17, solve for the unknowns listed. (*Optional*: Draw phasor voltage triangles.)

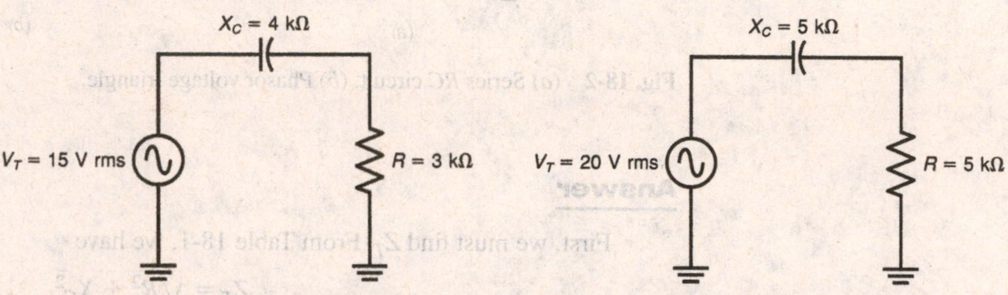

Fig. 18-3 Solve for Z_T, I_T, V_R, V_C, and θ.

Fig. 18-4 Solve for Z_T, I_T, V_R, V_C, and θ.

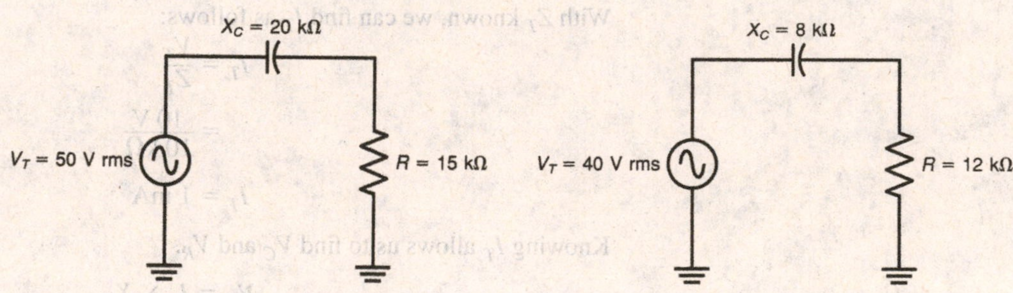

Fig. 18-5 Solve for Z_T, I_T, V_R, V_C, and θ.

Fig. 18-6 Solve for Z_T, I_T, V_R, V_C, and θ.

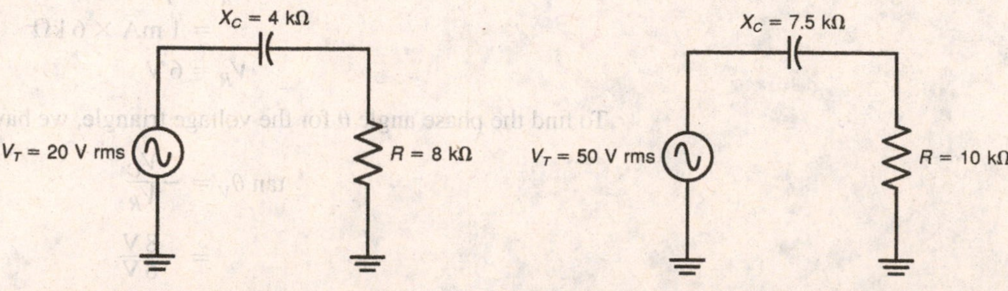

Fig. 18-7 Solve for Z_T, I_T, V_R, V_C, and θ.

Fig. 18-8 Solve for Z_T, I_T, V_R, V_C, and θ.

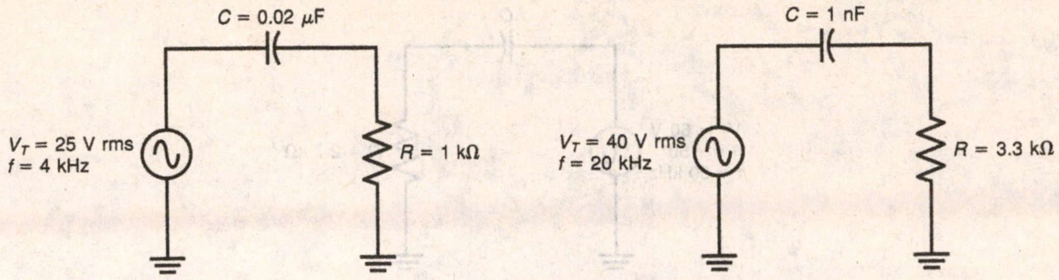

Fig. 18-9 Solve for X_C, Z_T, I_T, V_R, V_C, and θ.

Fig. 18-10 Solve for X_C, Z_T, I_T, V_R, V_C, and θ.

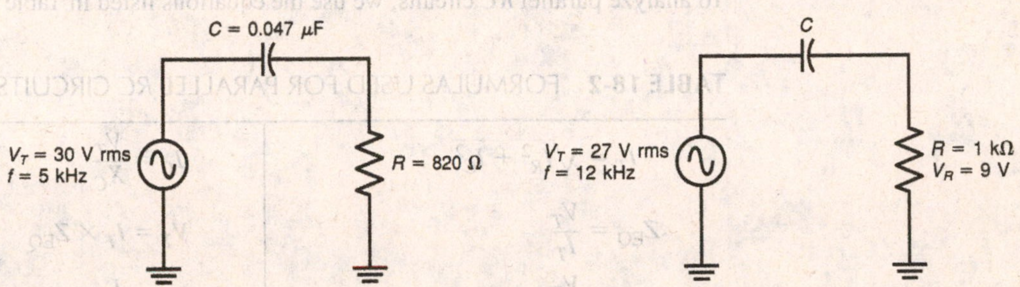

Fig. 18-11 Solve for X_C, Z_T, I_T, V_R, V_C, and θ.

Fig. 18-12 Solve for X_C, Z_T, I_T, V_C, C, and θ.

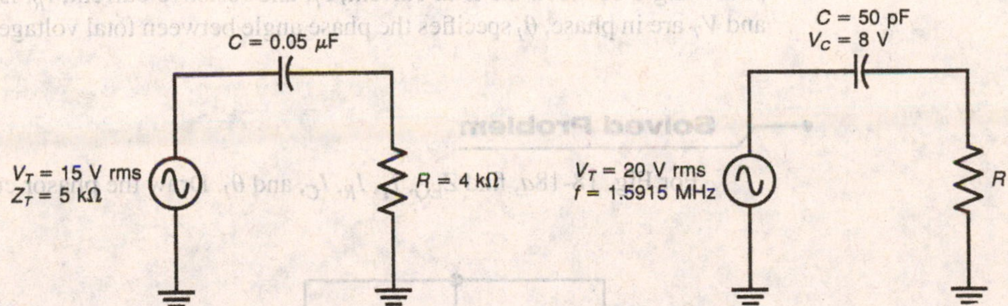

Fig. 18-13 Solve for X_C, I_T, V_R, V_C, f, and θ.

Fig. 18-14 Solve for X_C, Z_T, I_T, V_R, R, and θ.

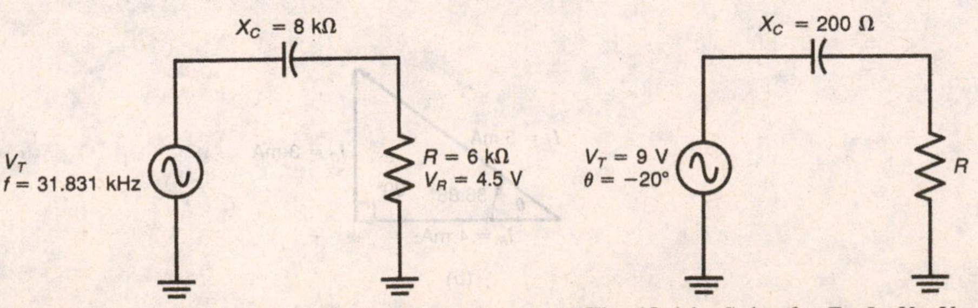

Fig. 18-15 Solve for Z_T, I_T, V_C, V_T, C, and θ.

Fig. 18-16 Solve for Z_T, I_T, V_R, V_C, and R.

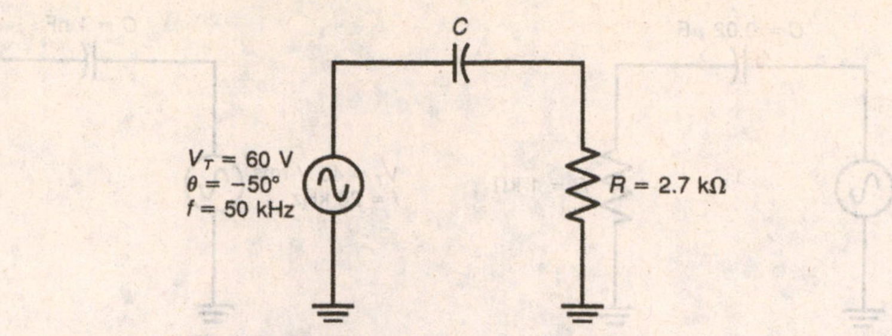

Fig. 18-17 Solve for Z_T, I_T, V_R, V_C, X_C, and C.

SEC. 18-2 ANALYZING PARALLEL *RC* CIRCUITS

To analyze parallel *RC* circuits, we use the equations listed in Table 18-2.

TABLE 18-2 FORMULAS USED FOR PARALLEL *RC* CIRCUITS

$I_T = \sqrt{I_R^2 + I_C^2}$	$I_C = \dfrac{V_T}{X_C}$
$Z_{EQ} = \dfrac{V_T}{I_T}$	$V_T = I_T \times Z_{EQ}$
$I_R = \dfrac{V_T}{R}$	$\tan \theta_I = \dfrac{I_C}{I_R}$

It is important to realize that I_R and I_C cannot be added algebraically because they are always 90° out of phase with each other. Instead, these quantities must be added phasorially. The phase angle between the total current, I_T, and resistive current, I_R, is specified by θ_I. Since I_R and V_T are in phase, θ_I specifies the phase angle between total voltage, V_T, and total current, I_T.

Solved Problem

For Fig. 18-18*a*, find Z_{EQ}, I_T, I_R, I_C, and θ_I. Draw the phasor current triangle.

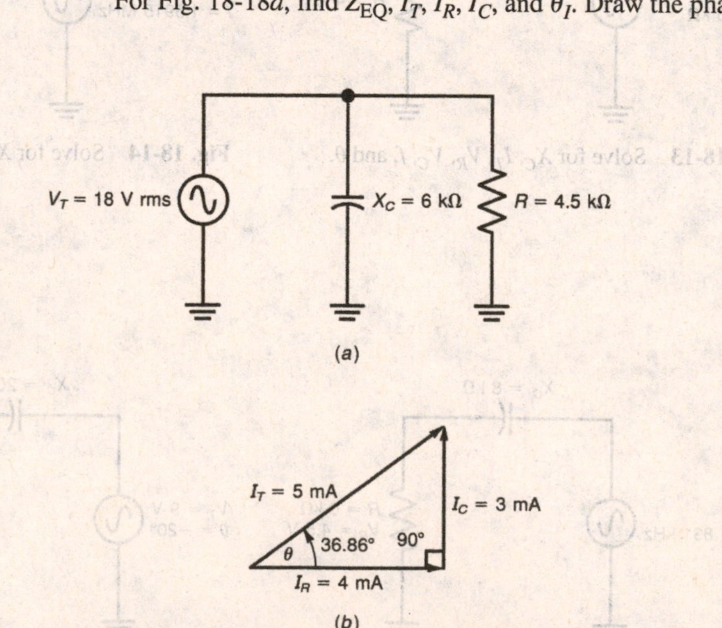

Fig. 18-18 (*a*) Parallel *RC* circuit. (*b*) Phasor current triangle.

Answer

First, we must find I_R and I_C. From Table 18-2, we have

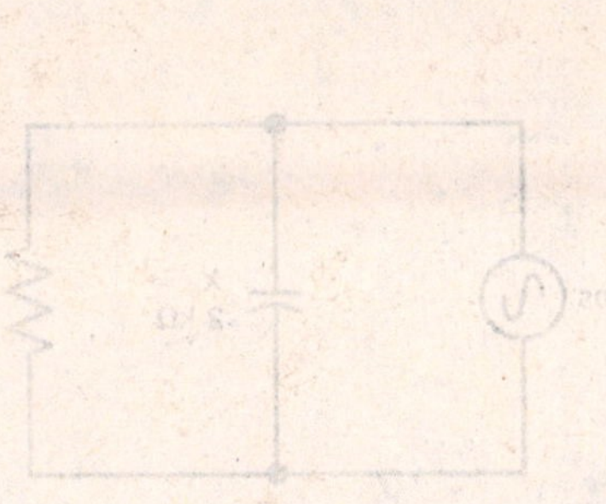

$$I_R = \frac{V_T}{R}$$
$$= \frac{18\text{ V}}{4.5\text{ k}\Omega}$$
$$I_R = 4\text{ mA}$$
$$I_C = \frac{V_T}{X_C}$$
$$= \frac{18\text{ V}}{6\text{ k}\Omega}$$
$$I_C = 3\text{ mA}$$

Next, we can find I_T.

$$I_T = \sqrt{I_R^{\,2} + I_C^{\,2}}$$
$$= \sqrt{4^2 + 3^2}\text{ mA}$$
$$I_T = 5\text{ mA}$$

This is the total current flowing to and from the terminals of the voltage source, V_T. With I_T known, Z_{EQ} can be found.

$$Z_{EQ} = \frac{V_T}{I_T}$$
$$= \frac{18\text{ V}}{5\text{ mA}}$$
$$Z_{EQ} = 3.6\text{ k}\Omega$$

It is important to note that Z_{EQ}, in ohms, is smaller than either branch opposition measured in ohms.

To find the phase angle θ for the current triangle, we have

$$\tan\theta_I = \frac{I_C}{I_R}$$
$$= \frac{3\text{ mA}}{4\text{ mA}}$$
$$\tan\theta_I = 0.75$$

To find the angle to which this tangent value corresponds, we proceed as follows:

$$\theta = \arctan 0.75$$
$$\theta = 36.86°$$

This phase angle tells us that the total current, I_T, leads the resistive current, I_R, by 36.86°. This is shown in Fig. 18-18*b*.

Notice that I_R is used as the reference phasor. This is because V_T and I_R are in phase with each other. Since V_T is the same across all components, the phase relationship between I_R and I_T specifies also the phase angle between total current I_T and total voltage V_T. Notice too that I_T leads I_R by 36.86°. This means that total current reaches its maximum value 36.86° ahead of V_T.

Sec. 18-2 For Figs. 18-19 through 18-33, solve for the unknowns listed. (*Optional*: Draw phasor current triangles.)

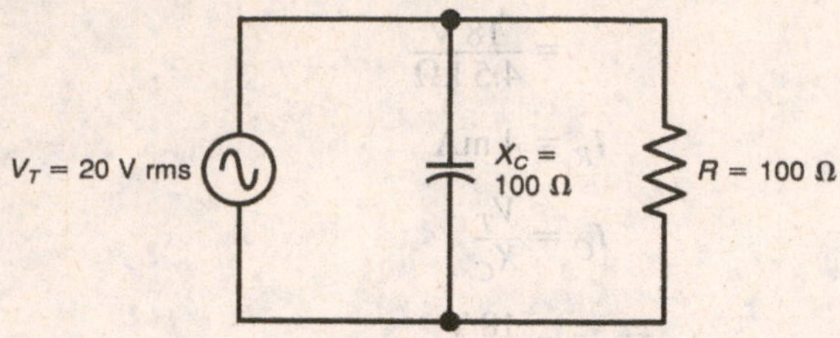

Fig. 18-19 Solve for I_R, I_C, I_T, Z_{EQ}, and θ.

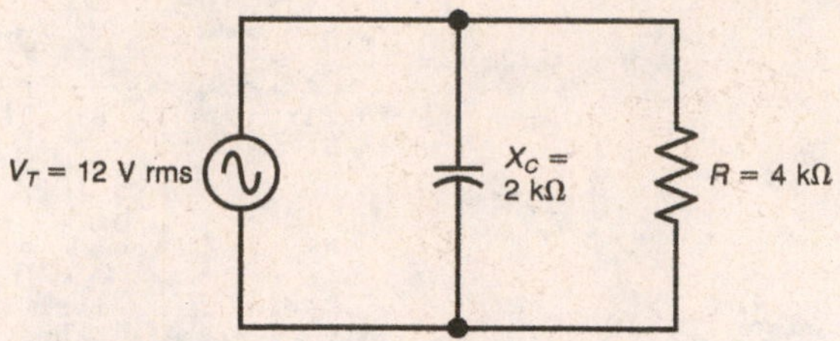

Fig. 18-20 Solve for I_R, I_C, I_T, Z_{EQ}, and θ.

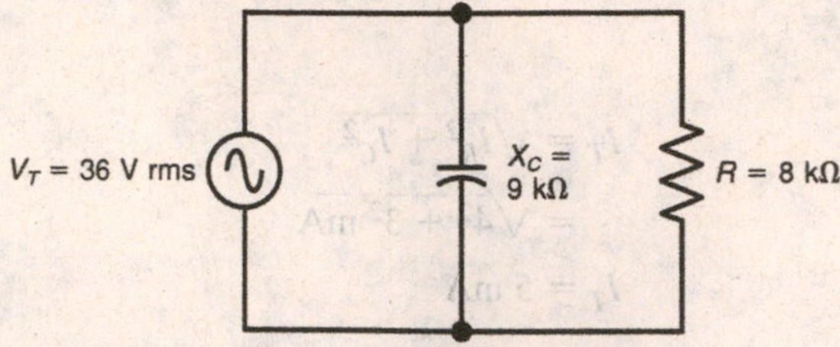

Fig. 18-21 Solve for I_R, I_C, I_T, Z_{EQ}, and θ.

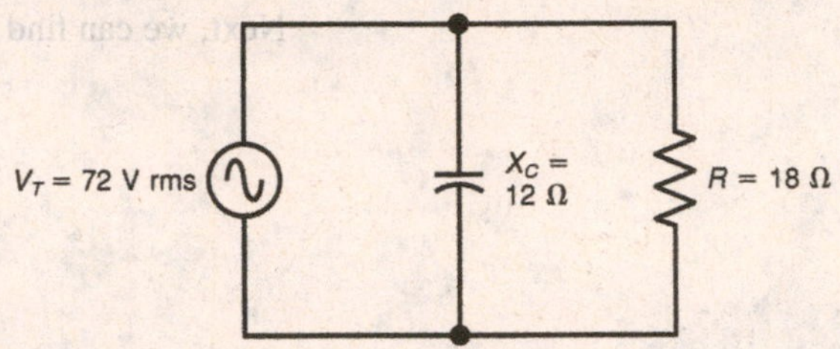

Fig. 18-22 Solve for I_R, I_C, I_T, Z_{EQ}, and θ.

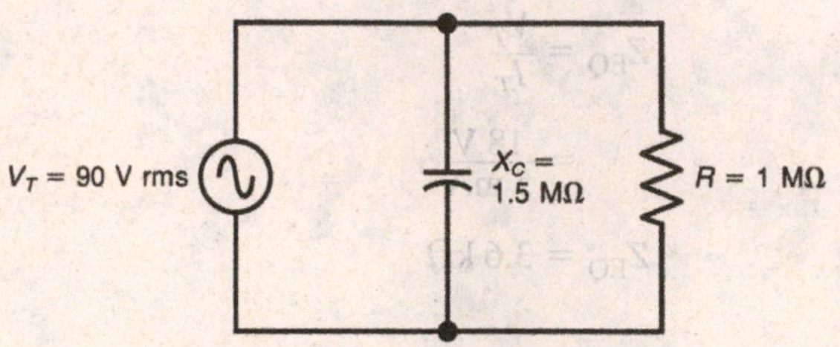

Fig. 18-23 Solve for I_R, I_C, I_T, Z_{EQ}, and θ.

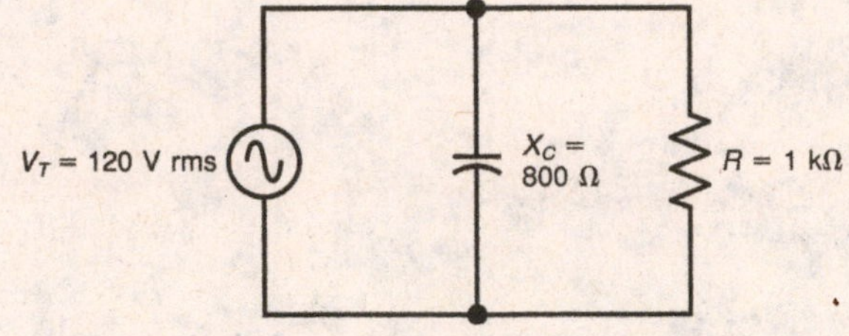

Fig. 18-24 Solve for I_R, I_C, I_T, Z_{EQ}, and θ.

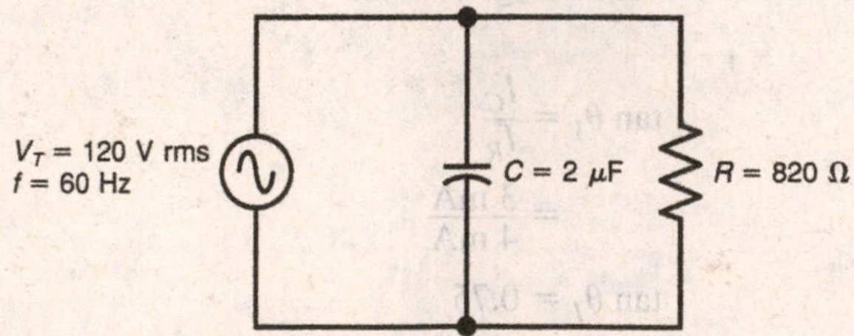

Fig. 18-25 Solve for X_C, I_R, I_C, I_T, Z_{EQ}, and θ.

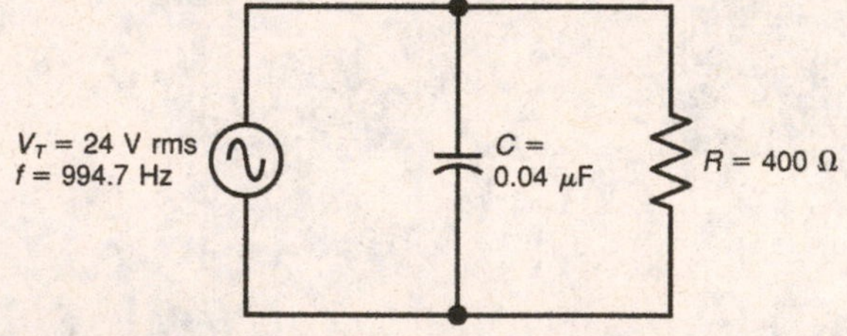

Fig. 18-26 Solve for X_C, I_R, I_C, I_T, Z_{EQ}, and θ.

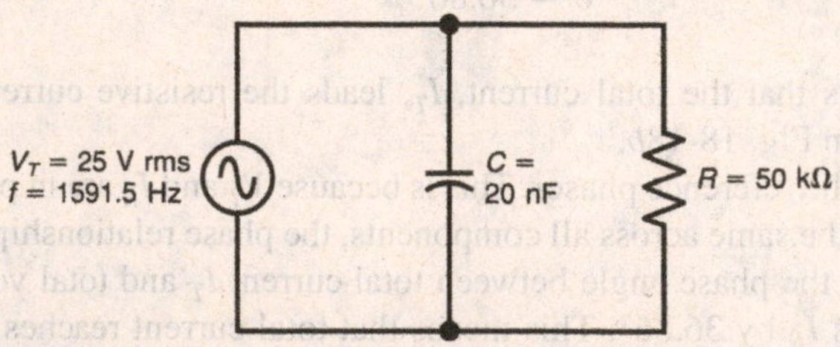

Fig. 18-27 Solve for X_C, I_R, I_C, I_T, Z_{EQ}, and θ.

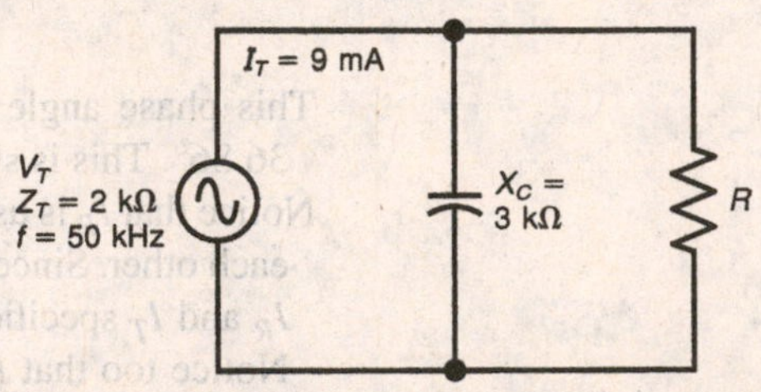

Fig. 18-28 Solve for V_T, I_C, I_R, C, R, and θ.

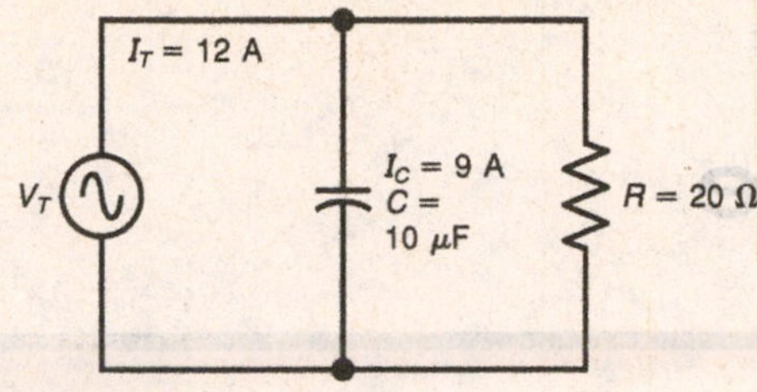

Fig. 18-29 Solve for V_T, I_R, Z_{EQ}, f, X_C, and θ.

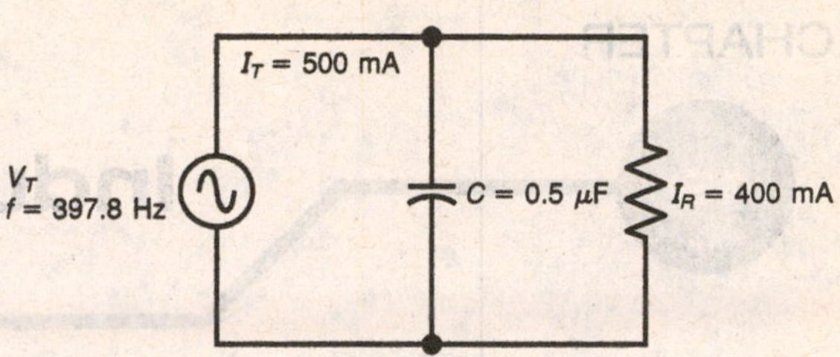

Fig. 18-30 Solve for X_C, I_C, V_T, Z_{EQ}, R, and θ.

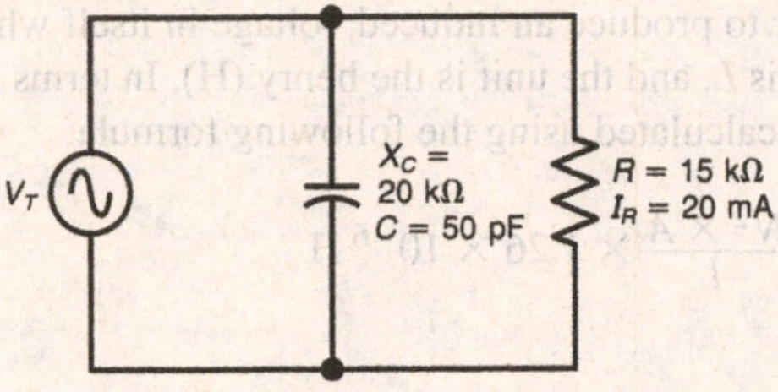

Fig. 18-31 Solve for V_T, I_C, I_T, Z_{EQ}, f, and θ.

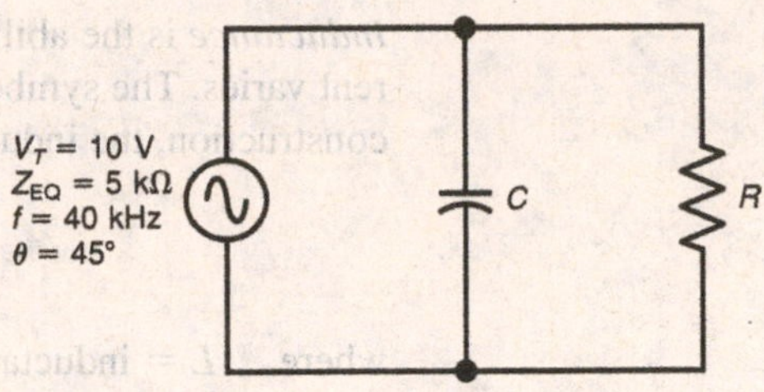

Fig. 18-32 Solve for I_C, I_R, I_T, X_C, R, and C.

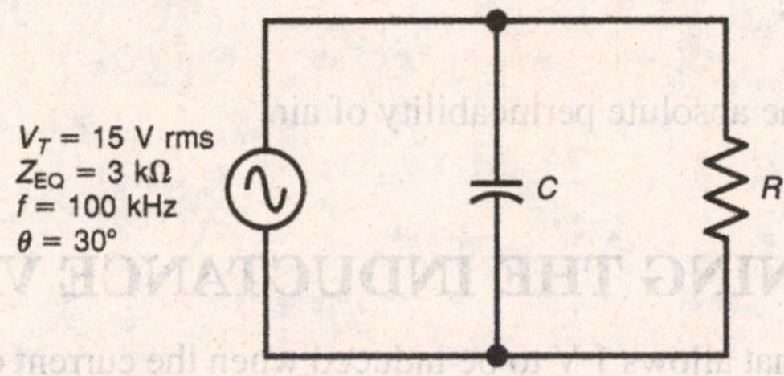

Fig. 18-33 Solve for I_T, I_C, I_R, X_C, R, and C.

END OF CHAPTER TEST

Chapter 18: Capacitive Circuits

Answer True or False.

1. For a capacitor in a sine-wave AC circuit, the charge and discharge current, i_C, leads the capacitor voltage, V_C, by a phase angle of 90°.
2. In a series RC circuit, V_C leads V_R by 90°.
3. In a series RC circuit, the resistor voltage, V_R, and the series current, I, are in phase with each other.
4. In a parallel RC circuit, the capacitor current, I_C, leads the resistor current, I_R, by a phase angle of 90°.
5. In a parallel RC circuit, the applied voltage, V_A, is in phase with the resistor current, I_R.
6. Impedance, Z, is the total opposition to the flow of sine-wave alternating current.
7. In a series RC circuit, the total voltage, V_T, leads the series current, I, by some phase angle θ.
8. In a parallel RC circuit, the total current, I_T, leads the applied voltage, V_A, by some phase angle, θ.
9. If 50 Ω of resistance is in series with a capacitive reactance, X_C of 50 Ω, the total impedance, Z_T, of the circuit is 70.7 Ω.
10. If 50 Ω of resistance is in parallel with a capacitive reactance, X_C, of 50 Ω, the equivalent impedance, Z_{EQ}, of the circuit is 25 Ω.

19 Inductance

Inductance is the ability of a conductor to produce an induced voltage in itself when the current varies. The symbol for inductance is L, and the unit is the henry (H). In terms of physical construction, the inductance, L, can be calculated using the following formula:

$$L = \mu_r \times \frac{N^2 \times A}{l} \times 1.26 \times 10^{-6}\ \text{H}$$

where L = inductance in henries
μ_r = relative permeability of core material
N = number of turns in coil
A = area of each turn in square meters (m^2)
l = length of coil in meters

The constant factor 1.26×10^{-6} is the absolute permeability of air.

SEC. 19-1 DETERMINING THE INDUCTANCE VALUE

An inductance of 1 H is the amount that allows 1 V to be induced when the current changes at the rate of 1 A/s. This tells us that the inductance, L, can be found when the induced voltage, V_L, and rate of current change, di/dt, are known. Expressed as an equation, we have

$$L = \frac{V_L}{di/dt}$$

where V_L = induced voltage in volts
di/dt = current change in amperes per second

To determine the induced voltage, V_L, rearrange the equation.

$$V_L = L \times \frac{di}{dt}$$

Solved Problems

a. Calculate the inductance, L, of a coil that induces 100 V when its current changes by 5 mA in 10 μs.

Answer

$$L = \frac{V_L}{di/dt}$$

$$= \frac{100\ \text{V}}{5\ \text{mA}/10\ \mu\text{s}}$$

$$L = 200\ \text{mH}$$

b. Calculate the inductance L for a coil that has the following characteristics: air core, $N = 500$ turns, $A = 2.5\text{ cm}^2$, $l = 2$ cm.

Answer

$$L = \mu_r \times \frac{N^2 \times A}{l} \times 1.26 \times 10^{-6}\text{ H}$$
$$= 1 \times \frac{500^2 \times 2.5 \times 10^{-4}}{2 \times 10^{-2}} \times 1.26 \times 10^{-6}\text{ H}$$
$$L = 3.94\text{ mH}$$

Note that the relative permeability μ_r of air is 1.

PRACTICE PROBLEMS

Sec. 19-1 Solve the following.

1. The current in a coil changes by 10 mA in 25 μs. The induced voltage produced equals 25 V. Calculate L.
2. The current through a coil changes by 2.5 mA in 5 μs. The induced voltage produced equals 50 V. Calculate L.
3. How much is the self-induced voltage V_L across a 50-mH inductor when the current changes by 20 mA in 50 μs?
4. The current through a 30-mH inductor changes from 50 mA to 75 mA in 5 μs. How much is V_L?
5. The current through an 8-mH inductor changes at the rate of 400 A/s. Calculate V_L.
6. A coil develops an induced voltage of 20 kV when the current changes from 500 mA to 100 mA in 5 μs. Calculate L.
7. Calculate the inductance L for a coil with the following physical characteristics: air core, $N = 150$ turns, $A = 5\text{ cm}^2$, and $l = 5$ cm.
8. Calculate the inductance, L, for a coil with the following physical characteristics: iron core with a μ_r of 2,000, $N = 50$ turns, $A = 1\text{ cm}^2$, $l = 2$ cm.
9. Calculate the inductance, L, for a coil with the following physical characteristics: ferrite core with a μ_r of 600, $N = 20$ turns, $A = 0.5\text{ cm}^2$, $l = 1.5$ cm.
10. An iron core having a relative permeability μ_r of 300 is inserted into the core of the coil in Prob. 7. Calculate L.
11. Calculate the inductance for the coil in Prob. 8 if the iron core is removed.
12. A coil having 200 turns has 5 mH of inductance. How much is L if the number of turns is quadrupled for the same coil length?
13. A coil having 10,000 turns has 2 H of inductance. How much is L if the number of turns is halved for the same coil length?
14. How much is the inductance of a coil that induces 60 mV when its current changes at the rate of 50 mA in 4 ms?
15. How much is the inductance of a coil that induces 180 V when its current changes at the rate of 9 mA in 1.5 μs?

SEC. 19-2 MUTUAL INDUCTANCE L_M

When the current in an inductor changes, the varying flux can cut across any other inductor nearby. Figure 19-1 shows two inductors, L_1 and L_2, placed physically close together. The magnetic flux produced by the changing current in L_1 links L_2. A voltage is induced across L_2, and, in turn, this voltage produces a current flow in R_L. The changing current through R_L

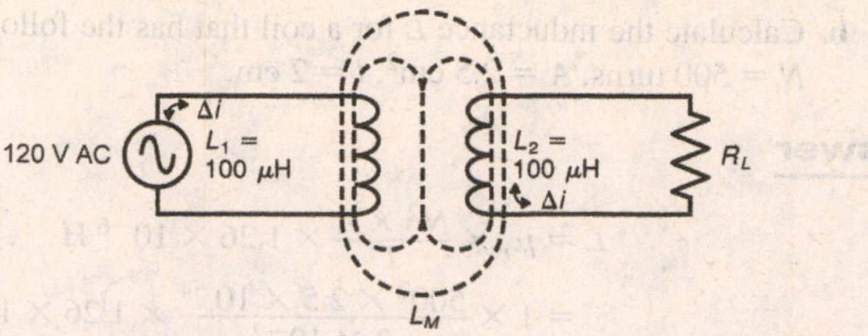

Fig. 19-1 Magnetic flux linking both L_1 and L_2, producing a mutual inductance.

and L_2 has its own magnetic field that induces a voltage across L_1. The two coils, L_1 and L_2, have a mutual inductance L_M because current in one coil can induce a voltage in the other. The fraction of total flux linking coils L_1 and L_2 is the coefficient of coupling, *k*. Expressed as an equation, we have

$$k = \frac{\text{flux linkage between } L_1 \text{ and } L_2}{\text{flux produced by } L_1}$$

There are no units for *k*.

Moving the coils closer together increases the coefficient of coupling, and moving the coils farther apart decreases the coefficient of coupling. When the coils are wound on a common iron core, *k* increases to 1 or unity.

The mutual inductance for two coils can be found when the coefficient of coupling and inductance values are known. Expressed as a formula,

$$L_M = k\sqrt{L_1 \times L_2}$$

where L_M = mutual inductance in henries
k = coefficient of coupling

A transformer is an important application of mutual inductance. Larger transformers used for power supplies and impedance matching use an iron core to obtain a coefficient of coupling, *k*, equal to 1 or unity. A basic transformer is illustrated in Fig. 19-2. The 120-V AC power line voltage is connected to the primary, and the load resistor, R_{L_1}, is connected across the secondary. The primary voltage produces alternating current in the primary, thus creating the magnetic flux.

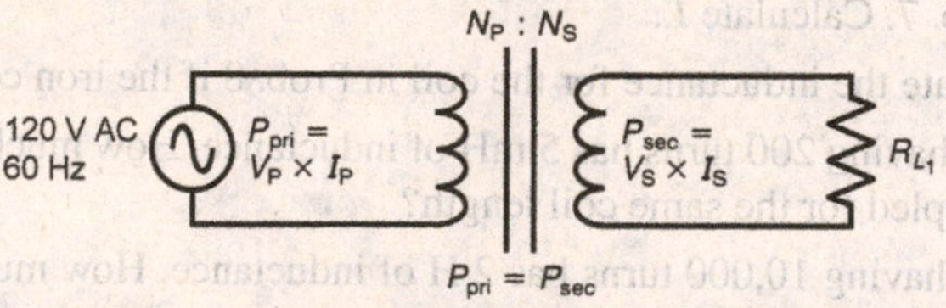

Fig. 19-2 Iron-core transformer used to step up or step down primary voltage.

Since the primary and secondary windings are wound on a common iron core, each turn in the secondary has the same amount of induced voltage as each turn in the primary. This fact allows us to develop the following relationship:

$$\frac{N_P}{N_S} = \frac{V_P}{V_S}$$

where N_P = number of turns in primary
N_S = number of turns in secondary
V_P = voltage across primary
V_S = voltage across secondary

The ratio N_P/N_S is called the *turns ratio*.

With no losses assumed for the transformer in Fig. 19-2, the power in the primary must equal the total power in the secondary. This is shown as

$$V_P \times I_P = V_S \times I_S$$

or

$$P_{pri} = P_{sec}$$

This relationship is true for a transformer whose coefficient of coupling k is 1. Rearranging this equation gives us

$$\frac{V_P}{V_S} = \frac{I_S}{I_P}$$

Notice that the current ratio is the inverse of the voltage ratio.

Solved Problems

a. Find L_M in Fig. 19-1 if $k = 0.5$.

Answer

Using the formula for L_M we proceed as follows:

$$L_M = k\sqrt{L_1 \times L_2}$$

$$= 0.5\sqrt{100\ \mu\text{H} \times 100\ \mu\text{H}}$$

$$L_M = 50\ \mu\text{H}$$

b. For Fig. 19-3, solve for $V_{sec\ 1}$, $V_{sec\ 2}$, $I_{R_{L1}}$, $I_{R_{L2}}$, I_P, P_{pri}, $P_{sec\ 1}$, and $P_{sec\ 2}$.

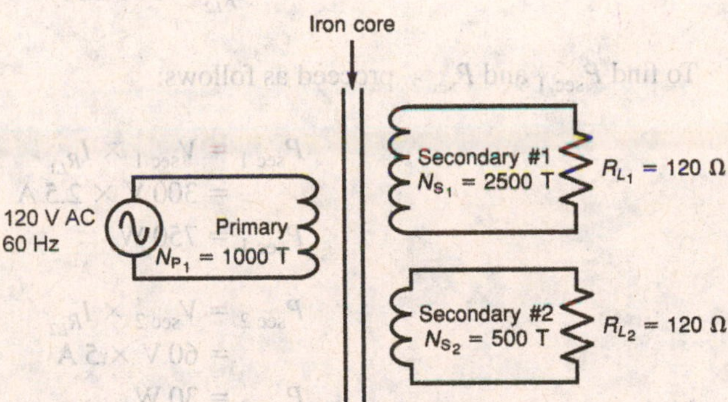

Fig. 19-3 Transformer with two secondary windings.

Answer

To determine $V_{sec\ 1}$, we find the turns ratio.

$$\frac{V_P}{V_{sec\ 1}} = \frac{N_P}{N_{S_1}}$$

Substituting values from Fig. 19-3, we have

$$\frac{V_P}{V_{sec\ 1}} = \frac{1{,}000\text{ T}}{2{,}500\text{ T}}$$

$$\frac{V_P}{V_{sec\ 1}} = \frac{1}{2.5}$$

Solving for $V_{sec\ 1}$, we have

$$V_{sec\ 1} = V_P \times 2.5$$
$$= 120\text{ V} \times 2.5$$
$$V_{sec\ 1} = 300\text{ V}$$

The general equation for a secondary voltage is

$$V_{sec} = \frac{N_S}{N_P} \times V_P$$

To determine $V_{sec\ 2}$, we proceed as follows:

$$V_{sec\ 2} = \frac{N_{S_2}}{N_P} \times V_P$$
$$= \frac{500\text{ T}}{1{,}000\text{ T}} \times 120\text{ V}$$
$$V_{sec\ 2} = 60\text{ V}$$

To find $I_{R_{L1}}$ and $I_{R_{L2}}$, use Ohm's law.

$$I_{R_{L1}} = \frac{V_{sec\ 1}}{R_{L_1}}$$
$$= \frac{300\text{ V}}{120\ \Omega}$$
$$I_{R_{L1}} = 2.5\text{ A}$$

$$I_{R_{L2}} = \frac{V_{sec\ 2}}{R_{L_2}}$$
$$= \frac{60\text{ V}}{120\ \Omega}$$
$$I_{R_{L2}} = 500\text{ mA}$$

To find $P_{sec\ 1}$ and $P_{sec\ 2}$, proceed as follows:

$$P_{sec\ 1} = V_{sec\ 1} \times I_{R_{L1}}$$
$$= 300\text{ V} \times 2.5\text{ A}$$
$$P_{sec\ 1} = 750\text{ W}$$

$$P_{sec\ 2} = V_{sec\ 2} \times I_{R_{L2}}$$
$$= 60\text{ V} \times .5\text{ A}$$
$$P_{sec\ 2} = 30\text{ W}$$

To find P_{pri}, we add $P_{sec\ 1}$ and $P_{sec\ 2}$. Remember that $P_{pri} = P_{sec}$. For Fig. 19-2, this can be shown as

$$P_{pri} = P_{sec\ 1} + P_{sec\ 2}$$
$$= 750\text{ W} + 30\text{ W}$$
$$P_{pri} = 780\text{ W}$$

Finally, to find I_P, we remember that $P_{pri} = V_P \times I_P$. Solving for I_P, we have

$$I_P = \frac{P_{pri}}{V_P}$$
$$= \frac{780\text{ W}}{120\text{ V}}$$
$$I_P = 6.5\text{ A}$$

Sec. 19-2 For the transformer circuits shown in Figs. 19-4 through 19-13, solve for the unknowns listed.

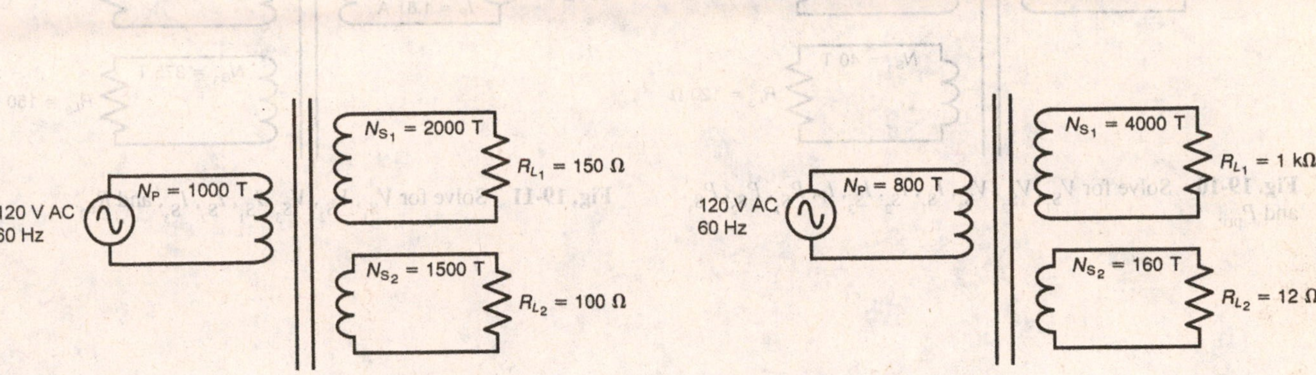

Fig. 19-4 Solve for V_{S_1}, V_{S_2}, I_{S_1}, I_{S_2}, P_{S_1}, P_{S_2}, P_{pri}, and I_P.

Fig. 19-5 Solve for V_{S_1}, V_{S_2}, I_{S_1}, I_{S_2}, P_{S_1}, P_{S_2}, P_{pri}, and I_P.

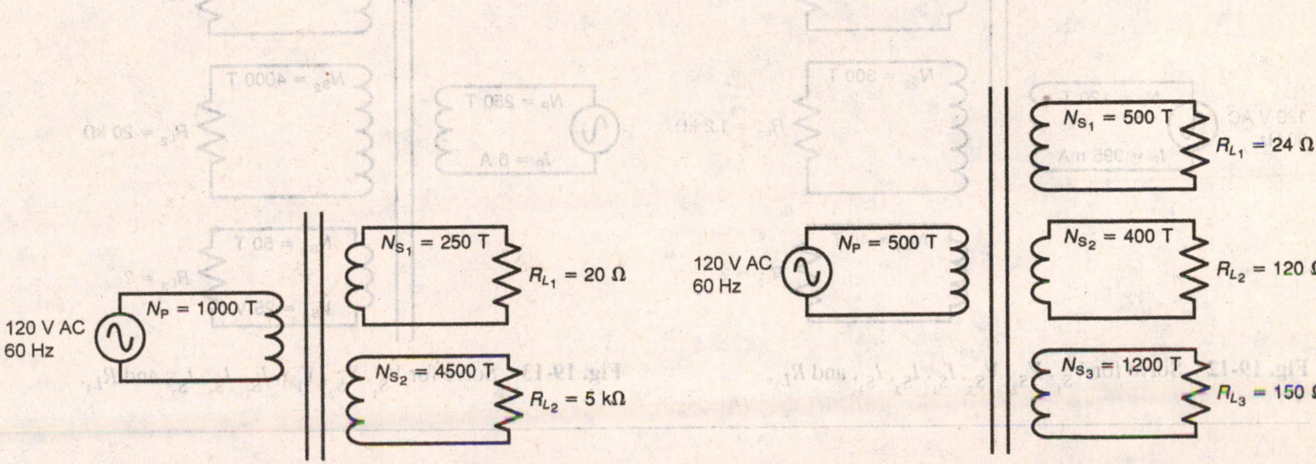

Fig. 19-6 Solve for V_{S_1}, V_{S_2}, I_{S_1}, I_{S_2}, P_{S_1}, P_{S_2}, P_{pri}, and I_P.

Fig. 19-7 Solve for V_{S_1}, V_{S_2}, V_{S_3}, I_{S_1}, I_{S_2}, I_{S_3}, I_P, P_{S_1}, P_{S_2}, P_{S_3}, and P_{pri}.

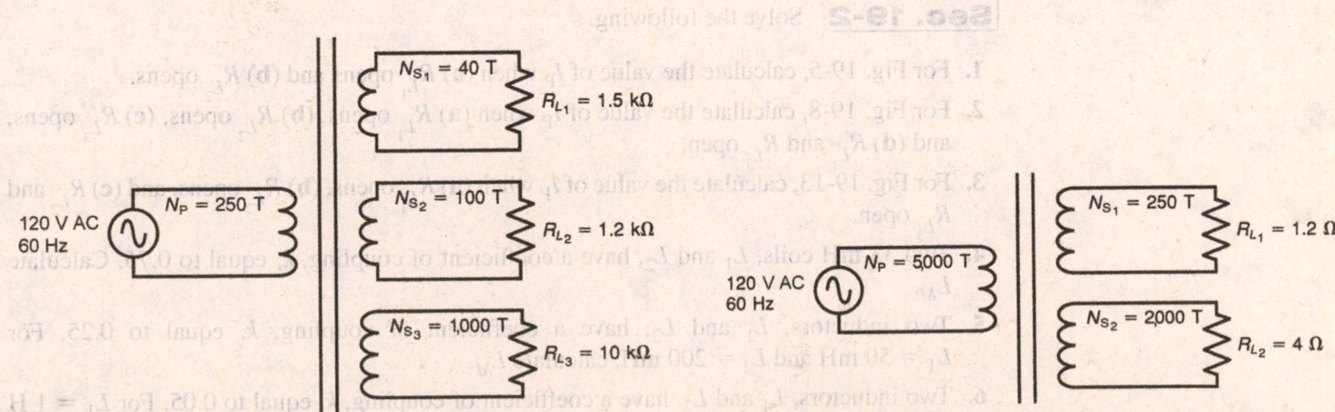

Fig. 19-8 Solve for V_{S_1}, V_{S_2}, V_{S_3}, I_{S_1}, I_{S_2}, I_{S_3}, I_P, P_{S_1}, P_{S_2}, P_{S_3}, and P_{pri}.

Fig. 19-9 Solve for V_{S_1}, V_{S_2}, I_{S_1}, I_{S_2}, I_P, P_{S_1}, P_{S_2}, and P_{pri}.

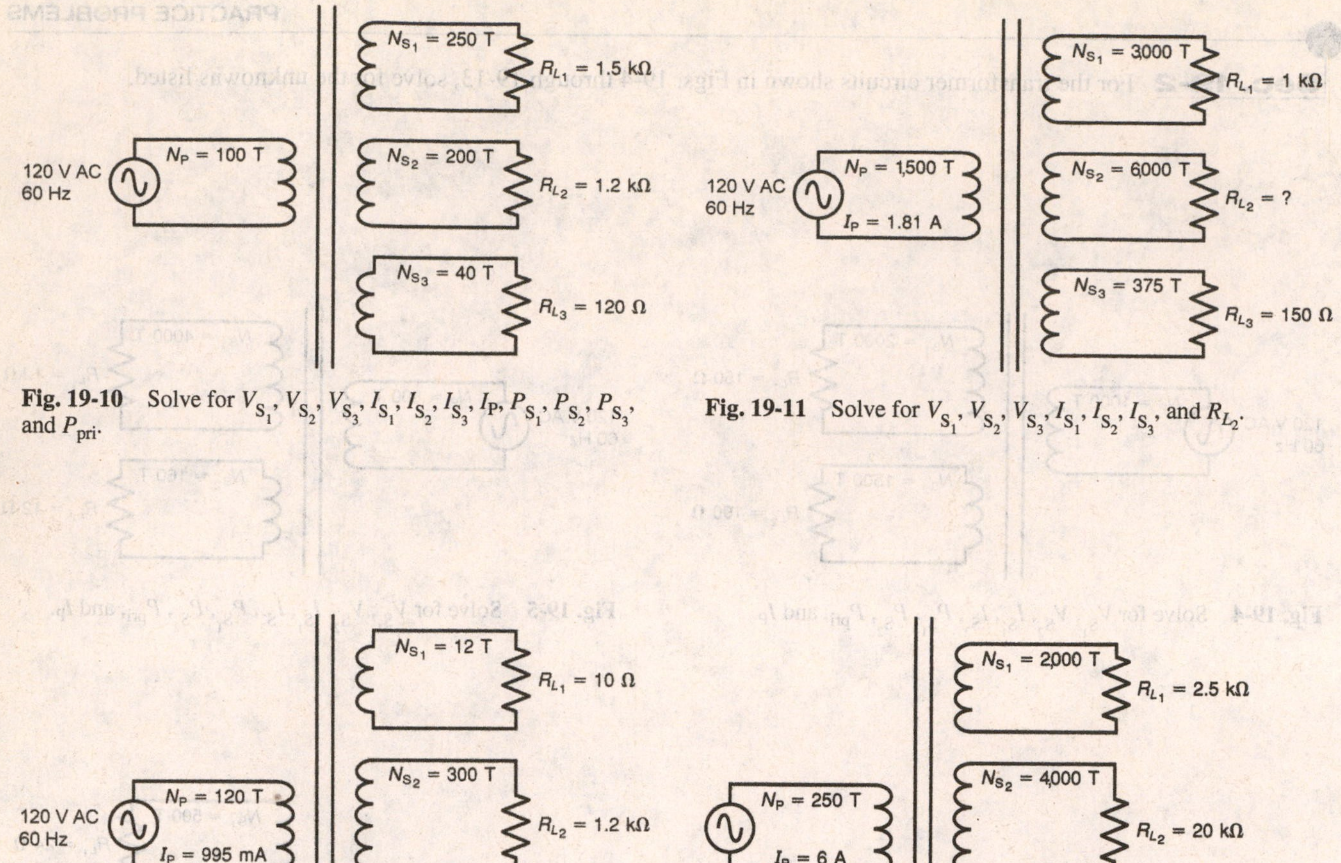

Fig. 19-10 Solve for V_{S_1}, V_{S_2}, V_{S_3}, I_{S_1}, I_{S_2}, I_{S_3}, I_P, P_{S_1}, P_{S_2}, P_{S_3}, and P_{pri}.

Fig. 19-11 Solve for V_{S_1}, V_{S_2}, V_{S_3}, I_{S_1}, I_{S_2}, I_{S_3}, and R_{L_2}.

Fig. 19-12 Solve for V_{S_1}, V_{S_2}, V_{S_3}, I_{S_1}, I_{S_2}, I_{S_3}, and R_{L_3}.

Fig. 19-13 Solve for V_{S_1}, V_{S_2}, V_{pri}, I_{S_1}, I_{S_2}, I_{S_3}, and R_{L_3}.

WORD PROBLEMS

Sec. 19-2 Solve the following.

1. For Fig. 19-5, calculate the value of I_P when (**a**) R_{L_1} opens and (**b**) R_{L_2} opens.
2. For Fig. 19-8, calculate the value of I_P when (**a**) R_{L_1} opens, (**b**) R_{L_2} opens, (**c**) R_{L_3} opens, and (**d**) R_{L_1} and R_{L_3} open.
3. For Fig. 19-13, calculate the value of I_P when (**a**) R_{L_1} opens, (**b**) R_{L_2} opens, and (**c**) R_{L_2} and R_{L_3} open.
4. Two 33-mH coils, L_1 and L_2, have a coefficient of coupling, k, equal to 0.75. Calculate L_M.
5. Two inductors, L_1 and L_2, have a coefficient of coupling, k, equal to 0.25. For L_1 = 50 mH and L_2 = 200 mH, calculate L_M.
6. Two inductors, L_1 and L_2, have a coefficient of coupling, k, equal to 0.05. For L_1 = 1 H and L_2 = 4 H, calculate L_M.
7. Two 50-mH coils have a mutual inductance, L_M, of 37.5 mH. Calculate k.

8. Two inductors, L_1 and L_2, have a mutual inductance of 10 mH. For L_1 = 400 mH and L_2 = 80 mH, calculate k.
9. What is meant by the terms *tight coupling* and *loose coupling*?
10. Why is the iron core of a power transformer made of laminated sheets insulated from each other?
11. In soft iron, which frequency produces more hysteresis losses: 60 Hz or 1 MHz?
12. An inductor uses a slug that can move in or out to vary L. Will L be maximum when the slug is in or out of the coil?

SEC. 19-3 TRANSFORMERS USED FOR IMPEDANCE MATCHING

Transformers are often used to transform a secondary load resistance to a desired value as seen by the primary. Before we examine how a transformer does this, let us review why it is desirable. In Chap. 12, we discussed the matched load condition. For this condition, we have a maximum transfer of power from the generator to the load. If the load, R_L, and the internal resistance, r_i, are not matched, R_L will receive less power than it will for the matched load condition. This condition of power transfer applies to AC circuits also. Fig. 19-14 shows three different loads connected to the same 10-V AC source.

The load, R_L, receives maximum power when $R_L = r_i$. This is shown in Fig. 19-14*b*. For the circuits shown in Fig. 19-14*a* and *c*, R_L receives less power. We can see that if it is desirable to obtain maximum transfer of power, r_i must equal R_L!

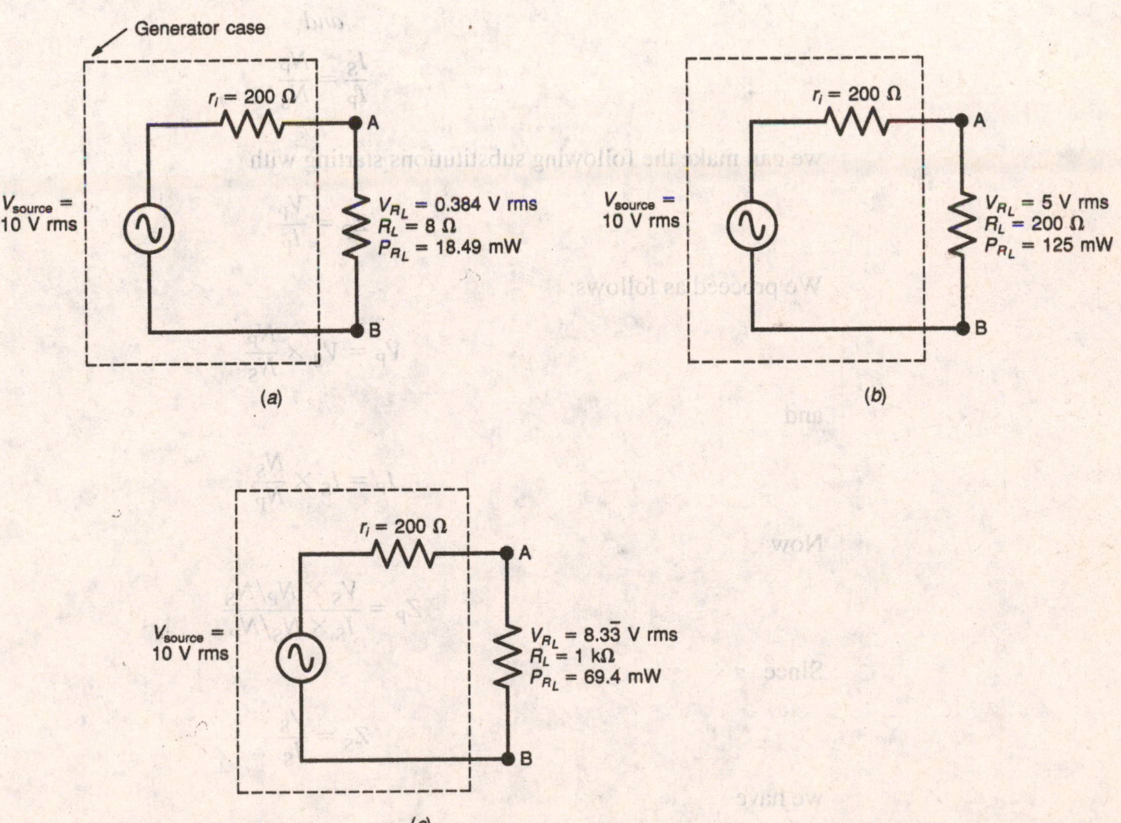

Fig. 19-14 Circuits used to illustrate transfer of power.

A transformer can make a load resistance, R_L, appear to the source as a value larger or smaller than its actual resistance value. For example, it may be desirable to transfer 125 mW of power to an 8-Ω load using the voltage source of Fig. 19-14*a*. We know that R_L must somehow be transformed to a value of 200 Ω in order to transfer maximum power from generator to load. We know from Fig. 19-14*b* that when $R_L = r_i$, maximum power transfer does occur.

A transformer can be connected, as shown in Fig. 19-15, to transform the 8-Ω resistive load connected to the secondary to 200 Ω in the primary. Here we will use the term *impedance,* with the symbol Z to represent opposition to current flow in both the primary and secondary. Impedance is measured in ohms, just like resistance.

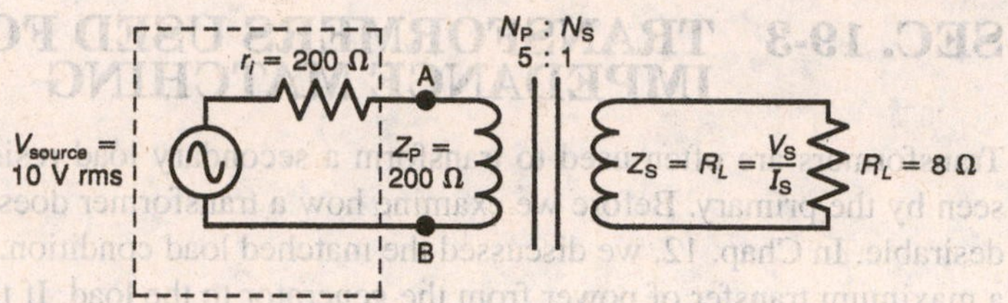

Fig. 19-15 Transformer used to match the 8-Ω load, R_L, to the generator's internal resistance, r_i, of 200 Ω.

The impedance of a primary can be stated as a V/I ratio: $Z_P = V_P/I_P$. Likewise, the impedance of a secondary can also be stated as a V/I ratio: $Z_S = V_S/I_S$. Since

$$\frac{N_P}{N_S} = \frac{V_P}{V_S}$$

and

$$\frac{I_S}{I_P} = \frac{N_P}{N_S}$$

we can make the following substitutions starting with

$$Z_P = \frac{V_P}{I_P}$$

We proceed as follows:

$$V_P = V_S \times \frac{N_P}{N_S}$$

and

$$I_P = I_S \times \frac{N_S}{N_P}$$

Now

$$Z_P = \frac{V_S \times N_P/N_S}{I_S \times N_S/N_P}$$

Since

$$Z_S = \frac{V_S}{I_S}$$

we have

$$Z_P = Z_S \times \frac{{N_P}^2}{{N_S}^2}$$

It is often common to see this equation as

$$\frac{Z_P}{Z_S} = \left(\frac{N_P}{N_S}\right)^2$$

or

$$\frac{N_P}{N_S} = \sqrt{\frac{Z_P}{Z_S}}$$

This tells us that the impedance ratio equals the turns ratio squared, or the square root of the impedance ratio equals the turns ratio. Therefore, it can be seen that a transformer is able to make the load connected to its secondary appear as a new value determined by the transformer's turn ratio.

To calculate Z_P in Fig. 19-15, we substitute the given values into the formulas for Z_P.

$$Z_P = Z_S \times \left(\frac{N_P}{N_S}\right)^2$$

$$= 8\ \Omega \times \left(\frac{5}{1}\right)^2$$

$$Z_P = 200\ \Omega$$

Since $Z_P = r_i$, maximum power is transferred from the source to the primary. Since $P_{pri} = P_{sec}$, we see that maximum power is also supplied to the load, R_L. It should be pointed out that Z_S in Fig. 19-15 has the same value as R_L. It is also true that both Z_P and Z_S are nearly purely resistive values. For Fig. 19-15,

$$P_{pri} = V_P \times I_P$$

$$= 5\ V_{rms} \times 25\ \text{mA}$$

$$P_{pri} = 125\ \text{mW}$$

Since $P_{sec} = P_{pri}$, we have

$$P_{sec} = 125\ \text{mW}$$

Solved Problem

Refer to Fig. 19-16. Find Z_P.

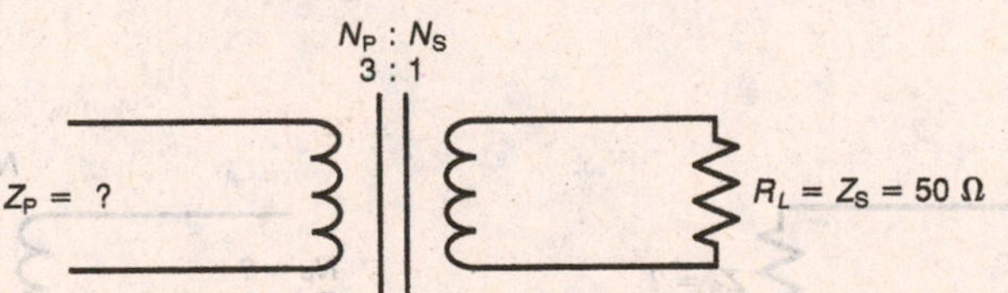

Fig. 19-16 Circuit used to calculate primary impedance, Z_P.

Answer

$$Z_P = Z_S \times \left(\frac{N_P}{N_S}\right)^2$$

$$= 50\ \Omega \times \left(\frac{3}{1}\right)^2$$

$$Z_P = 450\ \Omega$$

Sec. 19-3 For Figs. 19-17 through 19-31, solve for the unknown listed.

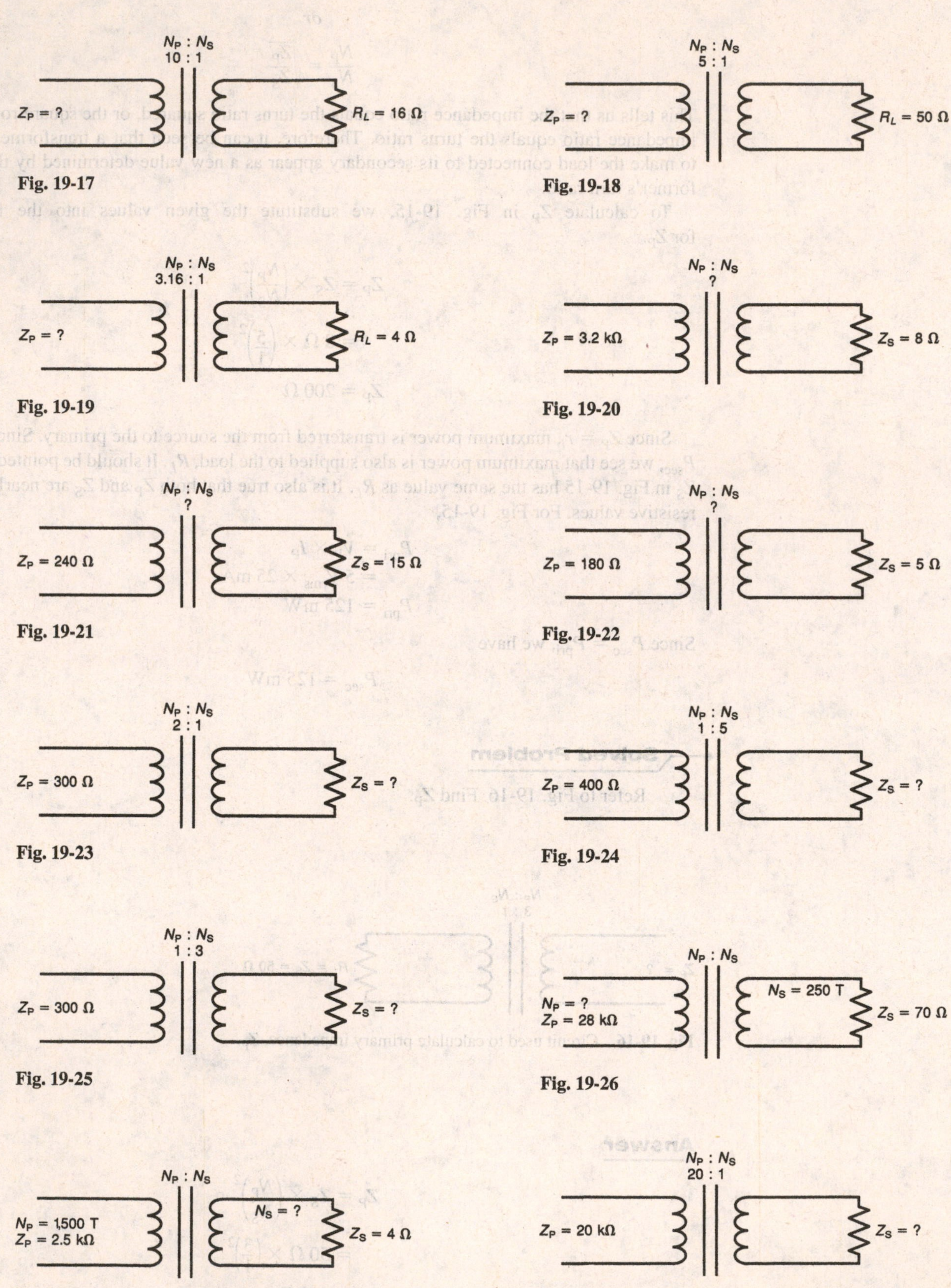

Np : Ns

Np = 1,500 T
Zp = 2.5 kΩ

Ns = ?

Zs = 4 Ω

Fig. 19-27

Np : Ns
20 : 1

Zp = 20 kΩ

Zs = ?

Fig. 19-28

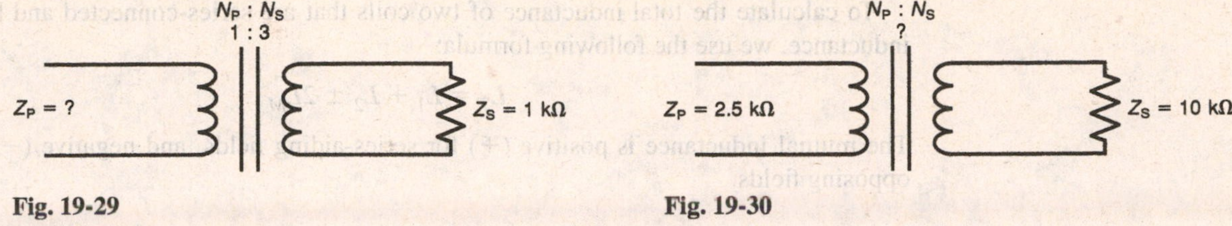

Fig. 19-29 **Fig. 19-30**

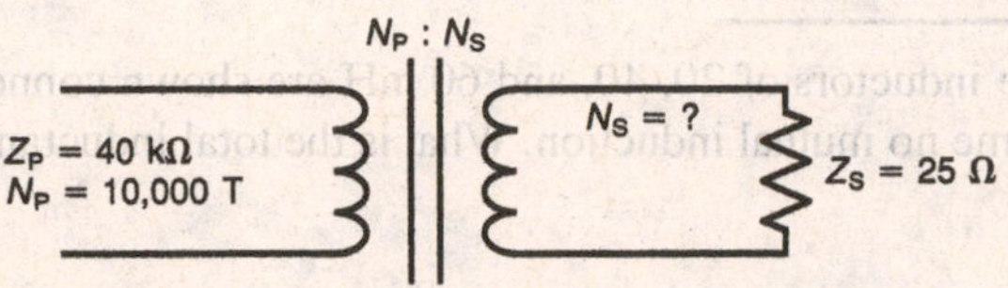

Fig. 19-31

For the next three problems, refer to Fig. 19-32.

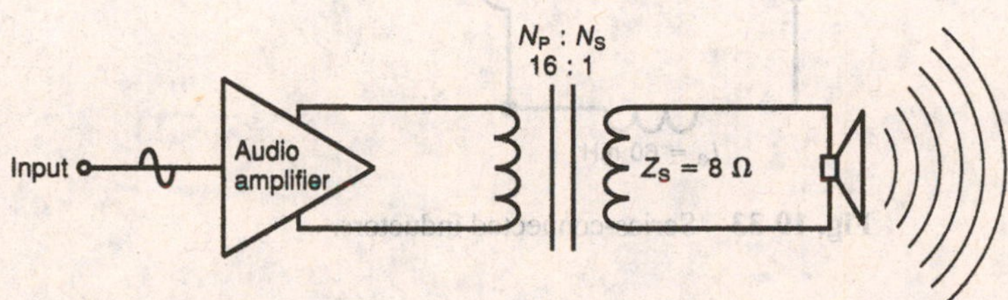

Fig. 19-32 Circuit used to illustrate maximum transfer of power from amplifier output to the speaker load.

1. What should be the amplifier's output impedance for maximum transfer of power?
2. With a signal of 32 V p-p across the primary, what would be the primary current?
3. How much power is delivered to the speaker load with 128 V p-p across the primary?

SEC. 19-4 CALCULATING TOTAL INDUCTANCE FOR SERIES- AND PARALLEL-CONNECTED INDUCTORS

Series-connected inductors having no mutual inductance, L_M, have a total inductance equal to their sum. That is,

$$L_T = L_1 + L_2 + L_3 + \cdots + L_n$$

where L_T = total inductance in henries

When coils are connected in parallel, the total inductance is found using the reciprocal formula. (Again, assume no mutual inductance.)

$$L_T = \frac{1}{1/L_1 + 1/L_2 + 1/L_3 + \cdots + 1/L_n}$$

where L_T = total inductance in henries

When series-connected coils have mutual inductance, we need to determine whether the magnetic fields are series aiding or series opposing. The coupling depends on the coil connections and the directions of the windings.

To calculate the total inductance of two coils that are series-connected and have mutual inductance, we use the following formula:

$$L_T = L_1 + L_2 \pm 2L_M$$

The mutual inductance is positive (+) for series-aiding fields, and negative (−) for series-opposing fields.

Solved Problems

a. Three inductors of 20, 40, and 60 mH are shown connected in series in Fig. 19-33. Assume no mutual induction. What is the total inductance?

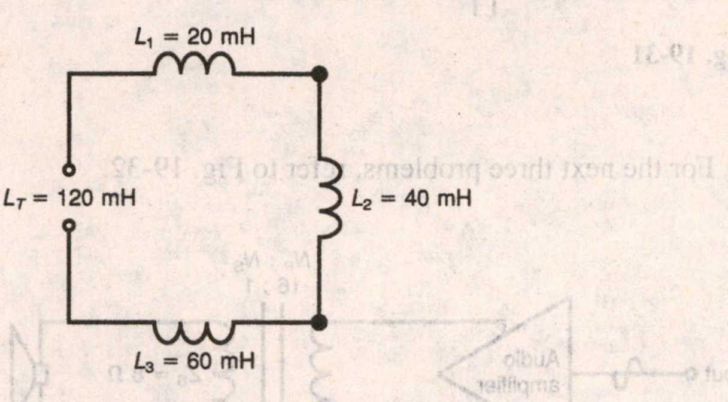

Fig. 19-33 Series-connected inductors.

Answer

We solve for total inductance as follows:

$$L_T = L_1 + L_2 + L_3 + \cdots + L_n$$
$$= 20 \text{ mH} + 40 \text{ mH} + 60 \text{ mH}$$
$$L_T = 120 \text{ mH}$$

b. Three inductors of 20, 30, and 60 mH are shown connected in parallel in Fig. 19-34. Assume no mutual induction. Calculate the total inductance.

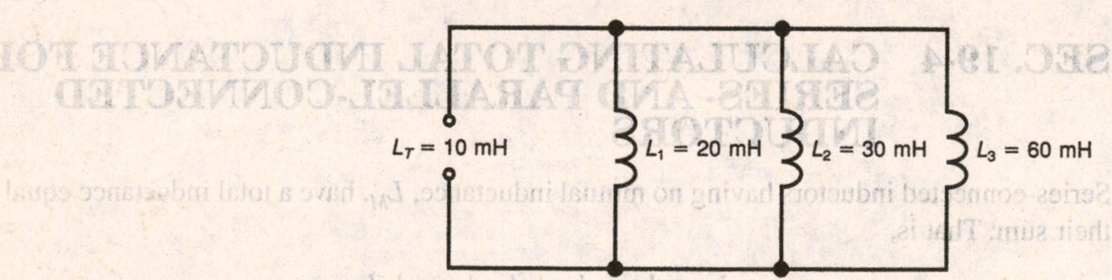

Fig. 19-34 Parallel-connected inductors.

Answer

We solve for the total inductance as follows:

$$L_T = \frac{1}{1/L_1 + 1/L_2 + 1/L_3 + \cdots + 1/L_n}$$
$$= \frac{1}{1/20 \text{ mH} + 1/30 \text{ mH} + 1/60 \text{ mH}}$$
$$L_T = 10 \text{ mH}$$

c. Two coils of 10 H each are connected in series aiding and have a mutual inductance of 0.75 H. Calculate L_T.

Answer

We proceed as follows:

$$L_T = L_1 + L_2 \pm 2L_M$$

Since the fields are series-aiding, we have

$$L_T = 10\text{ H} + 10\text{ H} + (2 \times 0.75\text{ H})$$
$$L_T = 21.5\text{ H}$$

PRACTICE PROBLEMS

Sec. 19-4 Solve the following

1. A 500-μH coil and a 1-mH coil are in series without L_M. Calculate L_T.
2. A 2-mH and a 6-mH coil are in parallel without L_M. Calculate L_T.
3. Three 33-mH inductors are in parallel without L_M. Calculate L_T.
4. Three inductors of 15, 8, and 10 H are in series without L_M. Calculate L_T.
5. Three inductors (L_1, L_2, and L_3) are in series without L_M. If $L_2 = 3L_1$ and $L_3 = 2L_2$, what is L_1, L_2, and L_3 if $L_T = 450\ \mu$H?
6. Two inductors in series without L_M have a total inductance L_T of 150 μH. If $L_1/L_2 = 0.1$, what is L_1 and L_2?
7. Two inductors in series without L_M have a total inductance of 26 μH. If $L_2/L_1 = 0.3$, what is L_1 and L_2?
8. How much inductance must be connected in parallel with a 40-μH inductor to obtain a total inductance of 24 μH?
9. Two coils of 60-μH each are connected in series with $L_M = 10\ \mu$H series opposing. Calculate L_T. Also calculate the coefficient of coupling, k.
10. For Prob. 9, calculate L_T if L_M is series aiding.

SEC. 19-5 INDUCTOR CODING

Many axial lead inductors use a color-coding scheme, similar to resistors, to indicate their inductance value in microhenries (μH). Figure 19-35 shows an inductor coding scheme that uses colored stripes or bands. The first color stripe is a wide silver band to indicate the component is an inductor and not a resistor. The second and fourth color stripes indicate the first two digits in the numerical value of the inductance. The third color stripe indicates the multiplier, whereas the fifth color stripe indicates the inductor's tolerance.

Figure 19-36 shows another coding scheme used with inductors having radial leads. In this system, the top two color stripes indicate the first two digits in the numerical value of the inductance, whereas the third color stripe indicates the multiplier. With this system, there is no tolerance band.

It should be noted that the two inductor coding schemes, as shown in Figs. 19-35 and 19-36, are by no means the only two coding schemes you will encounter. When a strange or unusual code is found, it is best to consult the manufacturer's service literature to determine the inductors value.

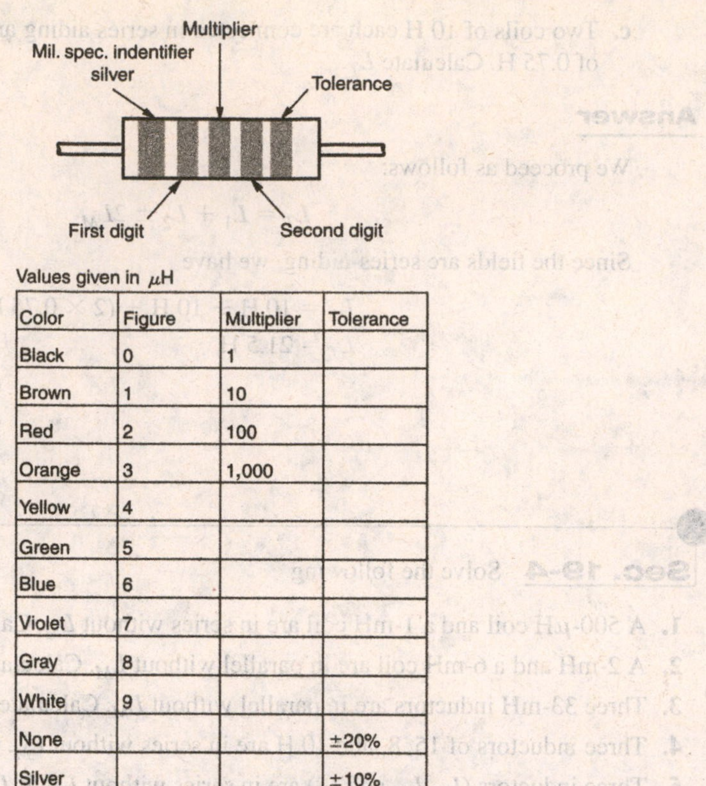

Values given in μH

Color	Figure	Multiplier	Tolerance
Black	0	1	
Brown	1	10	
Red	2	100	
Orange	3	1,000	
Yellow	4		
Green	5		
Blue	6		
Violet	7		
Gray	8		
White	9		
None			±20%
Silver			±10%
Gold			±5%

Fig. 19-35 Coding scheme for axial lead inductors.

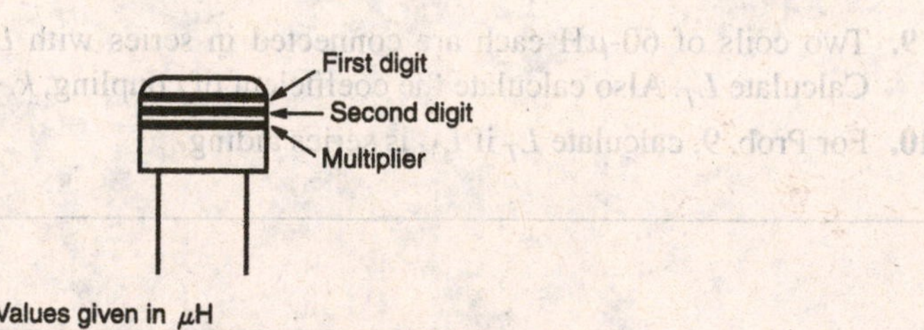

Values given in μH

Color	First digit First stripe	Second digit Second stripe	Multiplier Third stripe
Black (or Blank)	0	0	1
Brown	1	1	10
Red	2	2	100
Orange	3	3	1,000
Yellow	4	4	10,000
Green	5	5	100,000
Blue	6	6	
Violet	7	7	
Gray	8	8	
White	9	9	
Gold			× 0.1
Silver			× 0.01

Fig. 19-36 Coding scheme for radial lead inductors.

Solved Problem

In Fig. 19-37, determine the inductance and tolerance of each inductor.

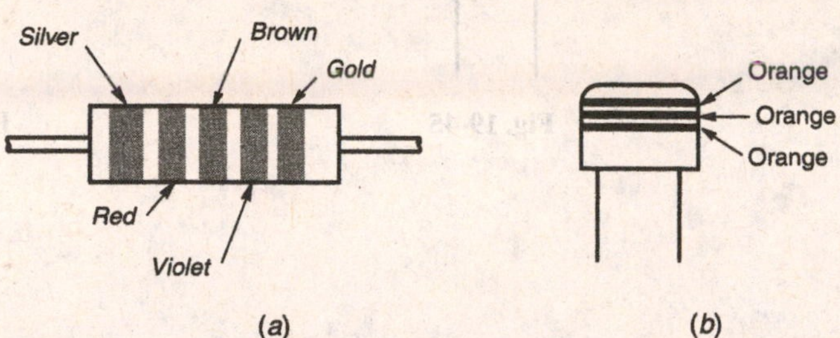

Fig. 19-37 Examples of color-coded inductors.

Answer

In Fig. 19-37*a*, the second color stripe is red for 2 and the fourth color stripe is violet for 7. Therefore, the first two digits in the inductance value are 2 and 7, respectively. The third color stripe is brown, indicating a multiplier of 10. Therefore, $L = 27 \times 10 = 270\ \mu\text{H}$. The fifth color stripe is gold indicating a tolerance of $\pm 5\%$. Therefore, $L = 270\ \mu\text{H} \pm 5\%$.

In Fig. 19-37*b*, the first two color stripes are both orange representing the digits 3 and 3. The third color stripe is also orange indicating a multiplier of 1,000. Therefore, $L = 33 \times 1{,}000 = 33{,}000\ \mu\text{H}$. There is no tolerance band with this system.

PRACTICE PROBLEMS

Sec. 19-5

Determine the inductance value for each of the inductors in Figs. 19-38 through 19-57. (Indicate the tolerance values for the axial lead inductors.)

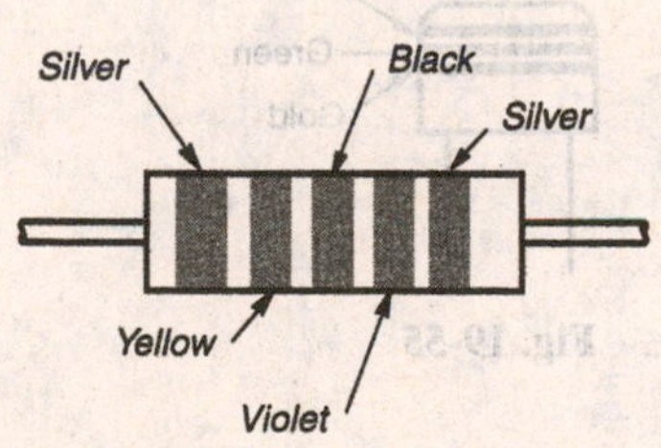

Fig. 19-38

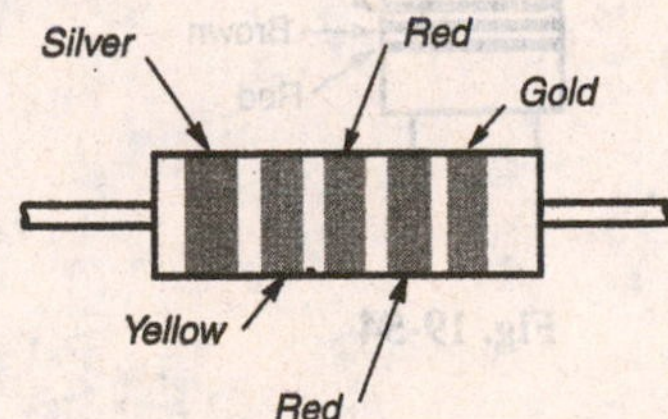

Fig. 19-39

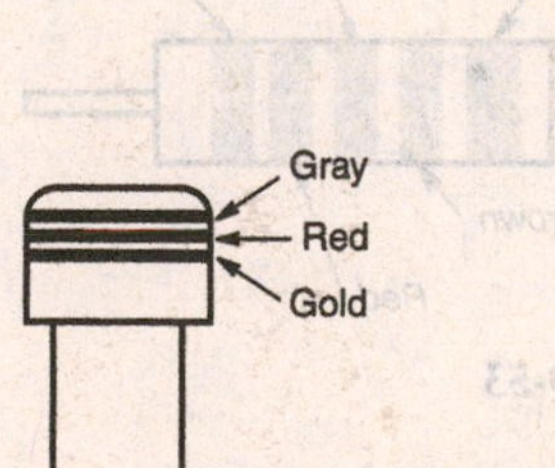

Fig. 19-40

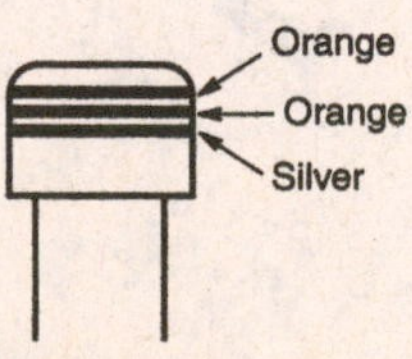

Fig. 19-41

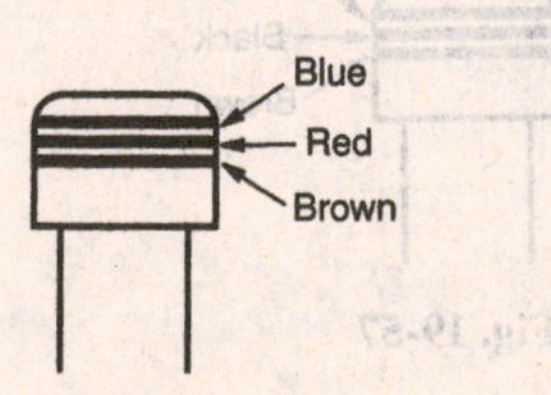

Fig. 19-42

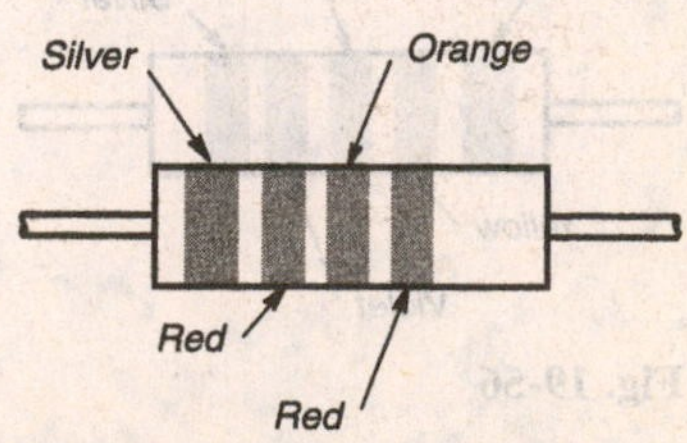

Fig. 19-43

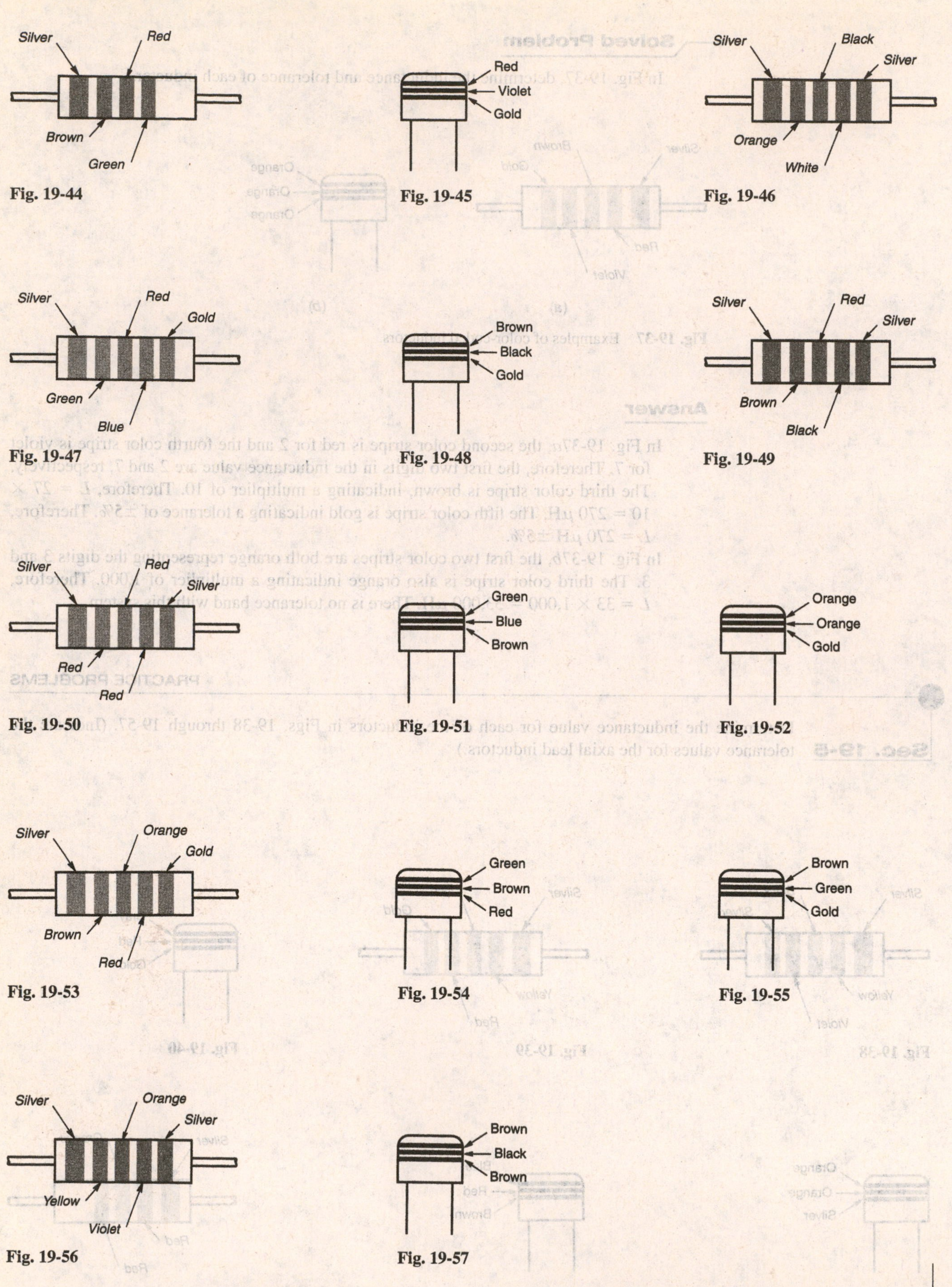

Fig. 19-44

Fig. 19-45

Fig. 19-46

Fig. 19-47

Fig. 19-48

Fig. 19-49

Fig. 19-50

Fig. 19-51

Fig. 19-52

Fig. 19-53

Fig. 19-54

Fig. 19-55

Fig. 19-56

Fig. 19-57

Chapter 19: Inductance

Answer True or False.

1. Inductance is the ability of a conductor to produce an induced voltage in itself when the current varies.
2. The unit of inductance is the henry (H).
3. The inductance, L, of an inductor is dependent on the number of coil turns, the length of the coil, the area enclosed by each coil turn, and the permeability of the core.
4. For an inductor, doubling the number of coil turns in the same length doubles the inductance, L.
5. The amount of induced voltage in a coil is not dependent on the rate of current change through the coil.
6. Two coils have mutual inductance, L_M, if the current in one coil induces a voltage in the other.
7. A transformer is an important application of mutual inductance, L_M.
8. A transformer can step up or step down an AC voltage.
9. An iron core transformer has a coefficient of coupling, k, approximately equal to 1 or unity.
10. In an iron core transformer with unity coupling, $\frac{N_P}{N_S} = \frac{I_P}{I_S}$.
11. In an ideal transformer with zero losses, $P_P = P_S$.
12. For an iron core transformer, $\sqrt{\frac{Z_P}{Z_S}} = \frac{N_P}{N_S}$.
13. Assuming no mutual inductance, the total inductance, L_T, of a 100 and 150 mH inductor in series is 250 mH.
14. Assuming no mutual inductance, the total inductance, L_T, of a 100 and 150 mH inductor in parallel is 60 mH.
15. For a color-coded axial-lead inductor, the first colored stripe is a wide silver band to indicate it is an inductor and not a resistor.

CHAPTER

Inductive Reactance

Inductive reactance, designated X_L, is a measure of a coil's opposition to the flow of sine-wave alternating current. Like resistance, R, X_L is measured in ohms. In this chapter, you will calculate X_L for different values of frequency and inductance. You will also learn how X_L values combine in series and in parallel.

SEC. 20-1 INDUCTIVE REACTANCE X_L OPPOSES CURRENT

When alternating current flows in an inductance L, the amount of current is much less than the resistance of the coil windings alone would allow. The reason is that an inductance will oppose a change in current. The varying current through the inductance will produce an induced voltage. Remember that $V_L = L \times di/dt$. The polarity of the induced voltage will always oppose an increase or decrease in current. This additional opposition to the flow of current is called *inductive reactance*. The symbol for reactance is X. The unit of reactance is the ohm. The symbol for inductive reactance is X_L. To determine the inductive reactance, X_L, for a known inductance at a given frequency, we use the formula

$$X_L = 2\pi fL$$

where X_L = inductive reactance in ohms
f = frequency of alternating current in hertz
L = inductance in henries

The other useful forms of this equation are

$$L = \frac{X_L}{2\pi f}$$

and

$$f = \frac{X_L}{2\pi L}$$

It should also be pointed out that reactance is a V/I ratio. Therefore, the inductive reactance, X_L, can also be stated as

$$X_L = \frac{V_L}{I}$$

where X_L = inductive reactance in ohms
V_L = induced voltage in volts
I = current in amperes

Solved Problem

Solve for the unknowns shown in Fig. 20-1.

Answer

In Fig. 20-1*a*, both L and f are known. We find X_L as follows:

$$\begin{aligned} X_L &= 2\pi fL \\ &= 2 \times 3.14 \times 1\text{ kHz} \times 30\text{ mH} \\ X_L &= 188.5\ \Omega \end{aligned}$$

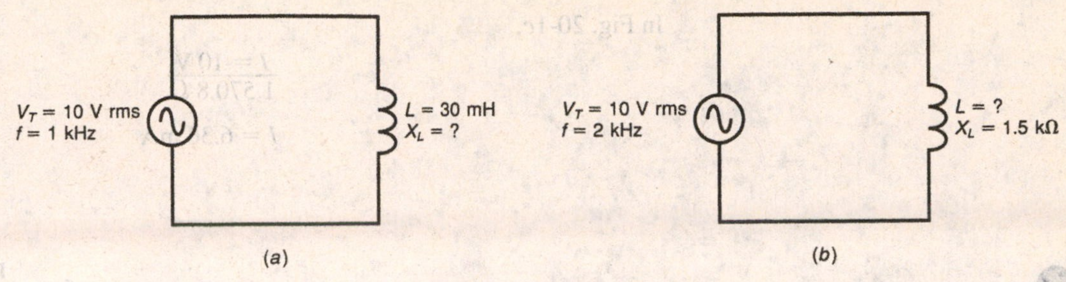

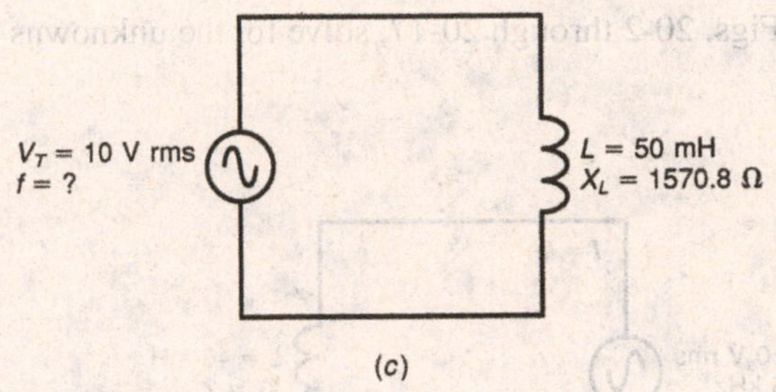

Fig. 20-1 Circuits used to apply the formula $X_L = 2\pi fL$. (*a*) Finding X_L when *f* and *L* are known. (*b*) Finding *L* when X_L and *f* are known. (*c*) Finding *f* when X_L and *L* are known.

In Fig. 20-1*b*, both X_L and *f* are known. We find *L* as follows:

$$L = \frac{X_L}{2\pi f}$$

$$= \frac{1.5\text{k}\Omega}{2 \times 3.14 \times 2\text{ kHz}}$$

$$L = 119.4\text{ mH}$$

In Fig. 20-1*c*, both *L* and X_L are known. We find *f* as follows:

$$f = \frac{X_L}{2\pi L}$$

$$= \frac{1570.8\ \Omega}{2 \times 3.14 \times 50\text{ mH}}$$

$$f = 5\text{ kHz}$$

For each circuit in Fig. 20-1,

$$I = \frac{V}{X_L}$$

In Fig. 20-1*a*,

$$I = \frac{10\text{ V}}{188.5\ \Omega}$$

$$I = 53\text{ mA}$$

In Fig. 20-1*b*,

$$I = \frac{10\text{ V}}{1.5\text{ k}\Omega}$$

$$I = 6.66\text{ mA}$$

In Fig. 20-1*c*,

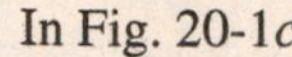

$$I = \frac{10\text{ V}}{1{,}570.8\ \Omega}$$

$$I = 6.36\text{ mA}$$

PRACTICE PROBLEMS

Sec. 20-1 For Figs. 20-2 through 20-17, solve for the unknowns listed.

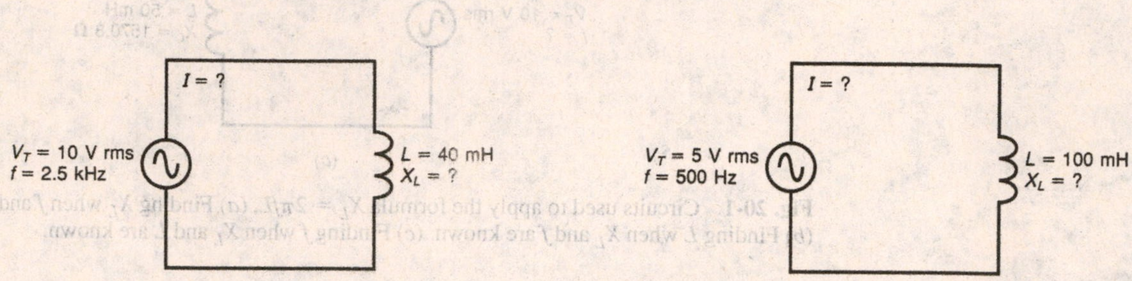

Fig. 20-2

Fig. 20-3

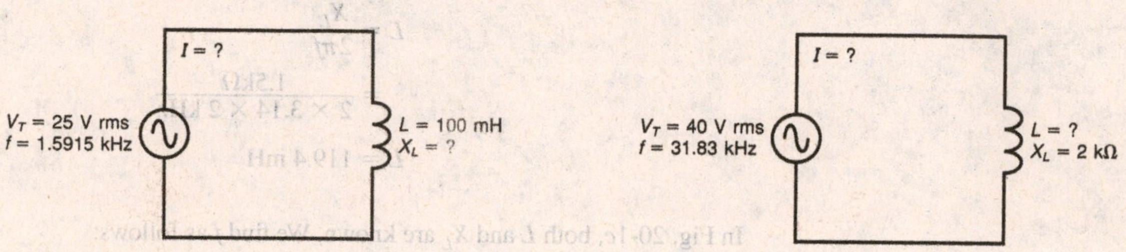

Fig. 20-4

Fig. 20-5

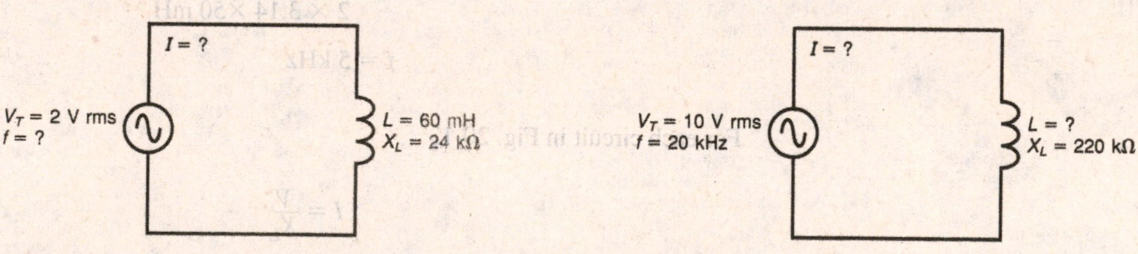

Fig. 20-6

Fig. 20-7

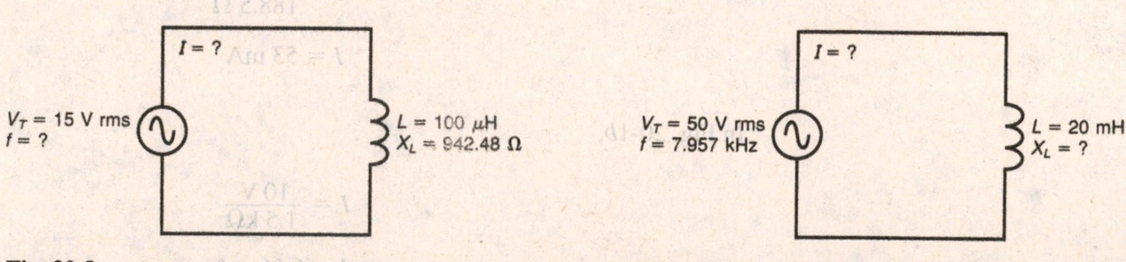

Fig. 20-8

Fig. 20-9

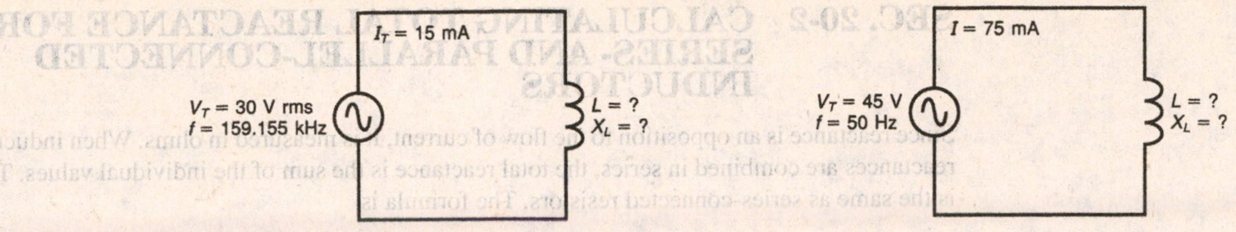

Fig. 20-10 **Fig. 20-11**

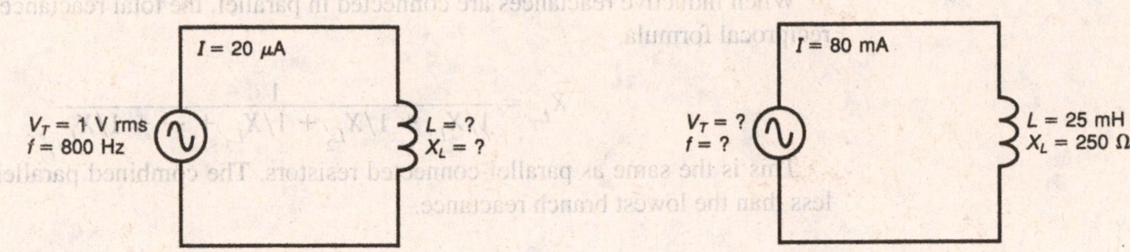

Fig. 20-12 **Fig. 20-13**

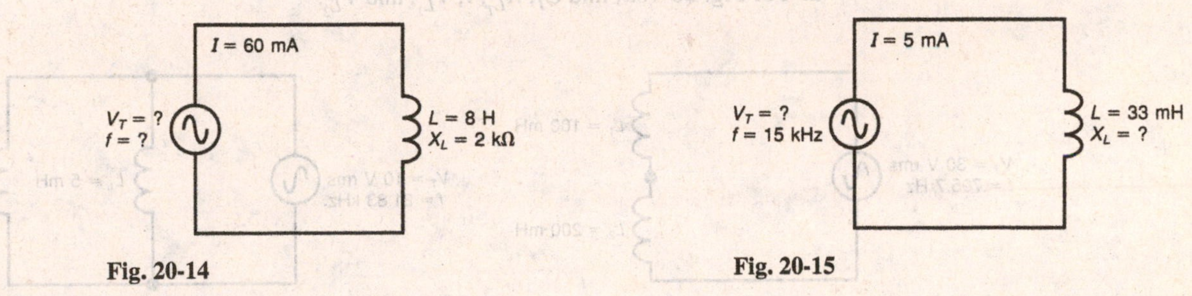

Fig. 20-14 **Fig. 20-15**

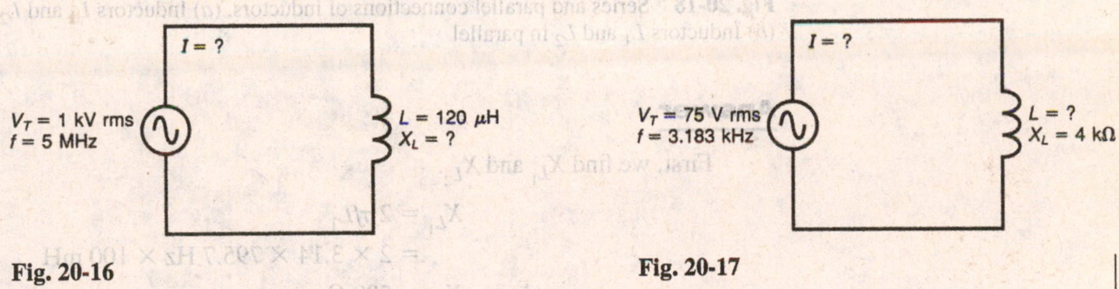

Fig. 20-16 **Fig. 20-17**

WORD PROBLEMS

Sec. 20-1 Solve the following

1. Calculate the inductive reactance of a 100-mH coil at the following frequencies: **(a)** 1,591.59 Hz, **(b)** 3,183 Hz, and **(c)** 6,366 Hz.
2. Calculate the inductive reactance for the coils listed, at a frequency of 1,591.59 Hz: **(a)** 10 mH, **(b)** 20 mH, and **(c)** 30 mH.
3. What is the inductive reactance of a coil at a frequency of 0 Hz (DC)?
4. The inductive reactance of a coil at 2 MHz is 10 kΩ. Calculate X_L at 200 kHz.

SEC. 20-2 CALCULATING TOTAL REACTANCE FOR SERIES- AND PARALLEL-CONNECTED INDUCTORS

Since reactance is an opposition to the flow of current, it is measured in ohms. When inductive reactances are combined in series, the total reactance is the sum of the individual values. This is the same as series-connected resistors. The formula is

$$X_{L_T} = X_{L_1} + X_{L_2} + X_{L_3} + \cdots + X_{L_n}$$

where X_{L_T} = total inductive reactance in ohms

When inductive reactances are connected in parallel, the total reactance is found using the reciprocal formula.

$$X_{L_T} = \frac{1}{1/X_{L_1} + 1/X_{L_2} + 1/X_{L_3} + \cdots + 1/X_{L_n}}$$

This is the same as parallel-connected resistors. The combined parallel reactance will be less than the lowest branch reactance.

Solved Problem

a. For Fig. 20-18*a*, find L_T, X_{L_T}, I, V_{L_1}, and V_{L_2}.

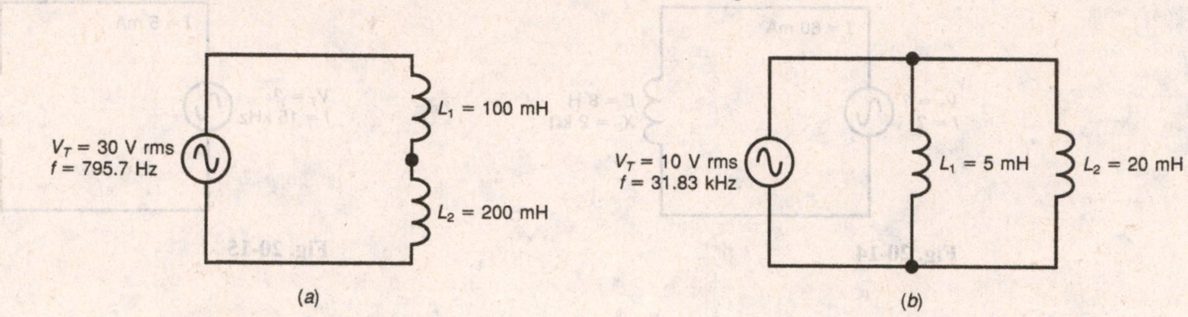

Fig. 20-18 Series and parallel connections of inductors. (*a*) Inductors L_1 and L_2 in series. (*b*) Inductors L_1 and L_2 in parallel.

Answer

First, we find X_{L_1} and X_{L_2}.

$$\begin{aligned}
X_{L_1} &= 2\pi f L_1 \\
&= 2 \times 3.14 \times 795.7 \text{ Hz} \times 100 \text{ mH} \\
X_{L_1} &= 500\ \Omega \\
X_{L_2} &= 2\pi f L_2 \\
&= 2 \times 3.14 \times 795.7 \text{ Hz} \times 200 \text{ mH} \\
X_{L_2} &= 1\ \text{k}\Omega
\end{aligned}$$

Next, we find X_{L_T} which can be calculated one of two ways. First, since $L_T = L_1 + L_2$, we have

$$\begin{aligned}
L_T &= 100 \text{ mH} + 200 \text{ mH} \\
L_T &= 300 \text{ mH}
\end{aligned}$$

Then,

$$\begin{aligned}
X_{L_T} &= 2\pi f L_T \\
&= 2 \times 3.14 \times 795.7 \text{ Hz} \times 300 \text{ mH} \\
X_{L_T} &= 1.5\ \text{k}\Omega
\end{aligned}$$

Also,

$$X_{L_T} = X_{L_1} + X_{L_2}$$
$$= 500\ \Omega + 1\ \text{k}\Omega$$
$$X_{L_T} = 1.5\ \text{k}\Omega$$

To find I, we use Ohm's law. Since the circuit contains only inductive reactance, $I = V/X_L$. For Fig. 20-18a, we have

$$I_T = \frac{V_T}{X_{L_T}}$$
$$= \frac{30\ \text{V}}{1.5\ \text{k}\Omega}$$
$$I_T = 20\ \text{mA}$$

To find the individual voltage drops, we multiply $I \times X_L$ as follows:

$$V_{L_1} = I_T \times X_{L_1}$$
$$= 20\ \text{mA} \times 500\ \Omega$$
$$V_{L_1} = 10\ \text{V}$$
$$V_{L_2} = I_T \times X_{L_2}$$
$$= 20\ \text{mA} \times 1\ \text{k}\Omega$$
$$V_{L_2} = 20\ \text{V}$$

Note,

$$V_T = V_{L_1} + V_{L_2}$$
$$= 10\ \text{V} + 20\ \text{V}$$
$$V_T = 30\ \text{V}$$

Also note,

$$V_T = I_T \times X_{L_T}$$
$$= 20\ \text{mA} \times 1.5\ \text{k}\Omega$$
$$V_T = 30\ \text{V}$$

b. For Fig. 20-18b, find L_T, X_{L_T}, I_{L_1}, I_{L_2}, and I_T.

Answer

First, we find X_{L_1} and X_{L_2}.

$$X_{L_1} = 2\pi f L_1$$
$$= 2 \times 3.14 \times 31.830\ \text{kHz} \times 5\ \text{mH}$$
$$X_{L_1} = 1\ \text{k}\Omega$$
$$X_{L_2} = 2\pi f L_2$$
$$= 2 \times 3.14 \times 31.830\ \text{kHz} \times 20\ \text{mH}$$
$$X_{L_2} = 4\ \text{k}\Omega$$

Next, we find I_{L_1} and I_{L_2}.

$$I_{L_1} = \frac{V_T}{X_{L_1}}$$
$$= \frac{10\ \text{V}}{1\ \text{k}\Omega}$$
$$I_{L_1} = 10\ \text{mA}$$
$$I_{L_2} = \frac{V_T}{X_{L_2}}$$
$$= \frac{10\ \text{V}}{4\ \text{k}\Omega}$$
$$I_{L_2} = 2.5\ \text{mA}$$

Total current is found by adding I_{L_1} and I_{L_2}.

$$I_T = I_{L_1} + I_{L_2}$$
$$= 10 \text{ mA} + 2.5 \text{ mA}$$
$$I_T = 12.5 \text{ mA}$$

Next, we find X_{L_T} which can be calculated one of two ways. First,

$$X_{L_T} = \frac{1}{1/X_{L_1} + 1/X_{L_2}}$$
$$= \frac{1}{1/1 \text{ k}\Omega + 1/4 \text{ k}\Omega}$$
$$X_{L_T} = 800 \ \Omega$$

Also, X_{L_T} can be found by using the formula $X_{L_T} = 2\pi f L_T$. First, L_T must be found. For Fig. 20-18b, we have

$$L_T = \frac{1}{1/L_1 + 1/L_2}$$
$$= \frac{1}{1/5 \text{ mH} + 1/20 \text{ mH}}$$
$$L_T = 4 \text{ mH}$$

We can now proceed as follows:

$$X_{L_T} = 2\pi f L_T$$
$$= 2 \times 3.14 \times 31.830 \text{ kHz} \times 4 \text{ mH}$$
$$X_{L_T} = 800 \ \Omega$$

It should be pointed out that I_T can be found using the following formula:

$$I_T = \frac{V_T}{X_{L_T}}$$

In Fig. 20-18*b*,

$$I_T = \frac{10 \text{ V}}{800 \ \Omega}$$
$$I_T = 12.5 \text{ mA}$$

PRACTICE PROBLEMS

Sec. 20-2 For Figs. 20-19 through 20-30, solve for the unknowns listed. (Assume no mutual induction.)

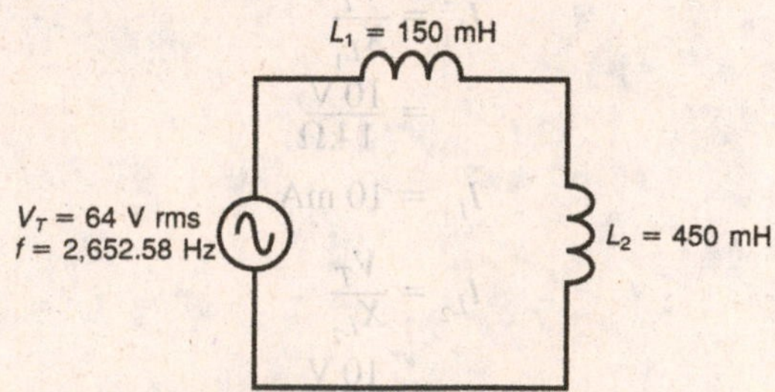

Fig. 20-19 Solve for L_T, X_{L_1}, X_{L_2}, X_{L_T}, I_T, V_{L_1}, and V_{L_2}.

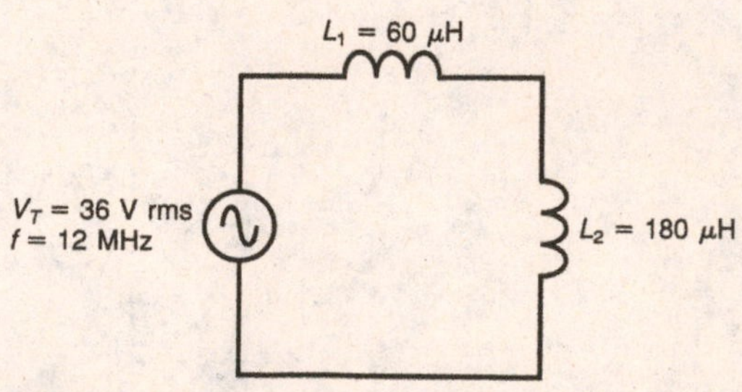

Fig. 20-20 Solve for L_T, X_{L_1}, X_{L_2}, X_{L_T}, I_T, V_{L_1}, and V_{L_2}.

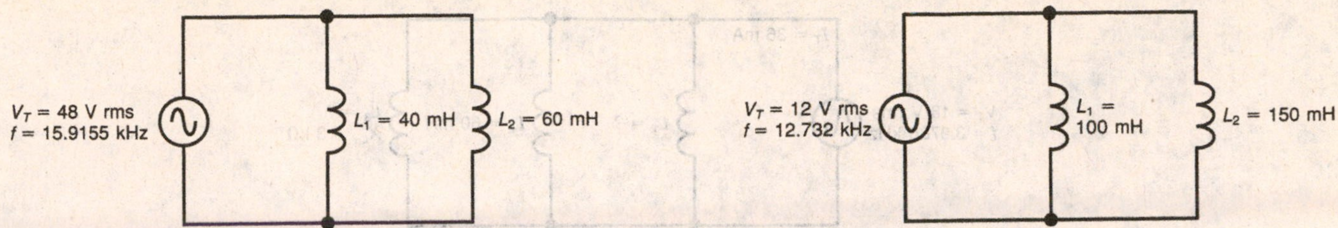

Fig. 20-21 Solve for L_T, X_{L_1}, X_{L_2}, X_{L_T}, I_1, I_2, and I_T.

Fig. 20-22 Solve for L_T, X_{L_1}, X_{L_2}, X_{L_T}, I_1, I_2, and I_T.

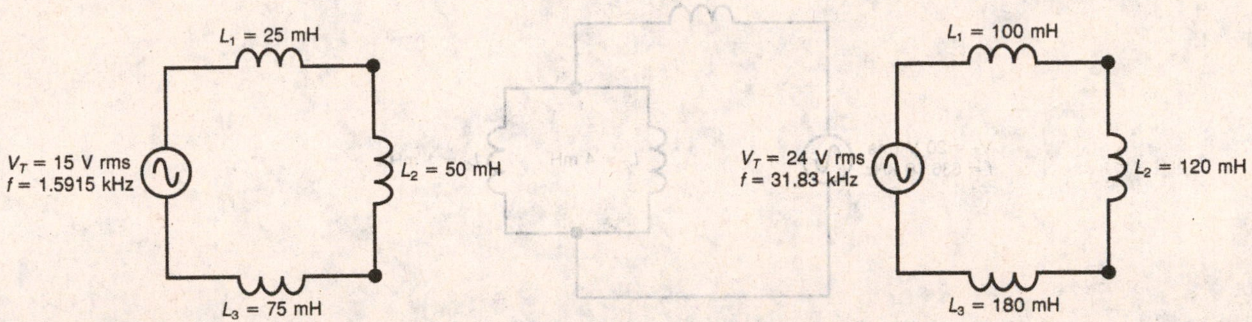

Fig. 20-23 Solve for L_T, X_{L_1}, X_{L_2}, X_{L_3}, X_{L_T}, I_T, V_{L_1}, V_{L_2}, and V_{L_3}.

Fig. 20-24 Solve for L_T, X_{L_1}, X_{L_2}, X_{L_3}, X_{L_T}, I_T, V_{L_1}, V_{L_2}, and V_{L_3}.

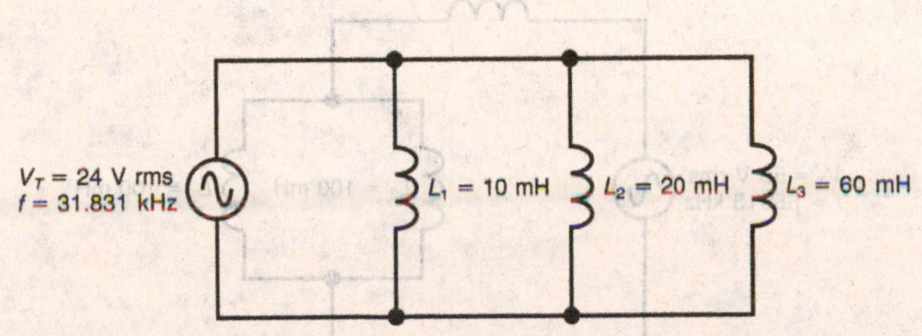

Fig. 20-25 Solve for L_T, X_{L_1}, X_{L_2}, X_{L_3}, X_{L_T}, I_1, I_2, I_3, and I_T.

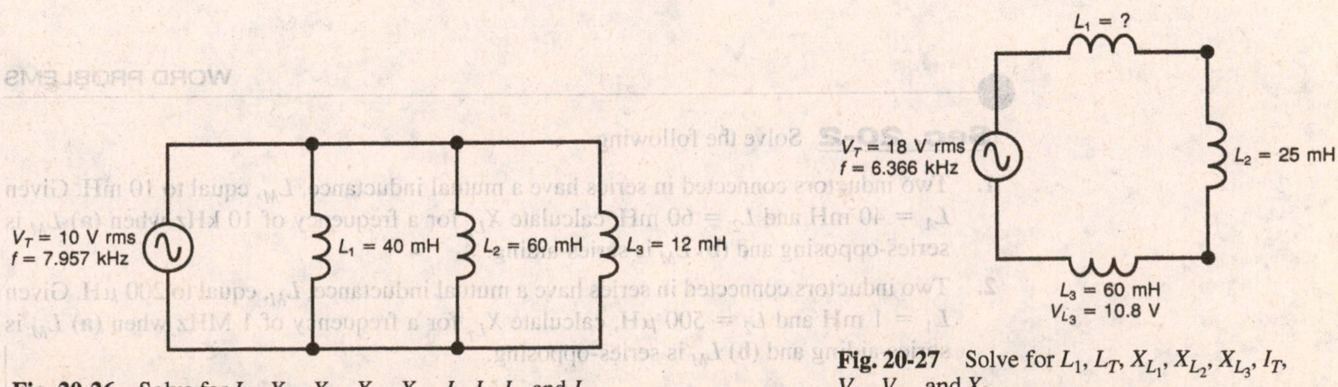

Fig. 20-26 Solve for L_T, X_{L_1}, X_{L_2}, X_{L_3}, X_{L_T}, I_1, I_2, I_3, and I_T.

Fig. 20-27 Solve for L_1, L_T, X_{L_1}, X_{L_2}, X_{L_3}, I_T, V_{L_1}, V_{L_2}, and X_{L_T}.

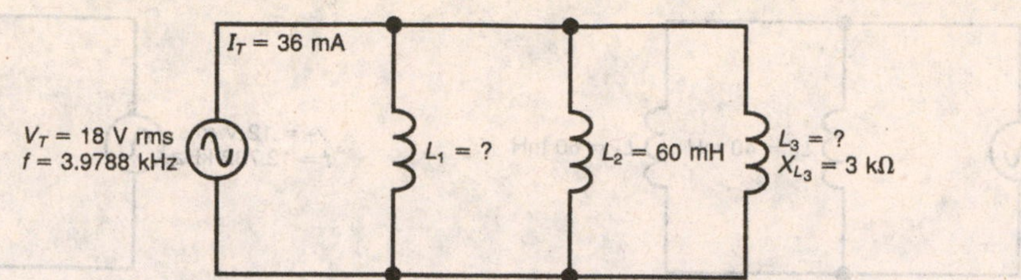

Fig. 20-28 Solve for L_T, L_1, L_3, X_{L_1}, X_{L_2}, I_1, I_2, and I_3.

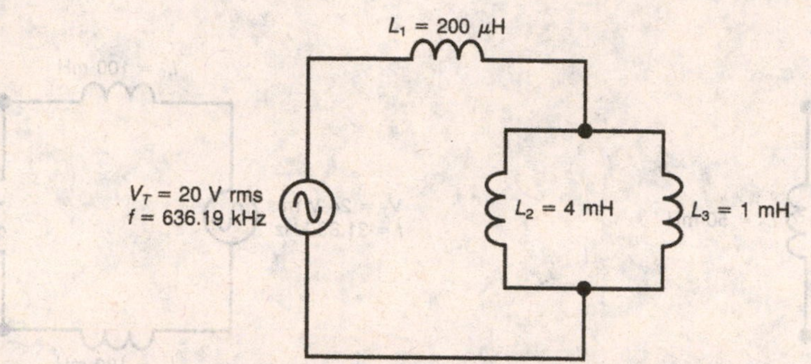

Fig. 20-29 Solve for L_T, X_{L_1}, X_{L_2}, X_{L_3}, X_{L_T}, V_{L_1}, V_{L_2}, V_{L_3}, I_2, I_3, and I_T.

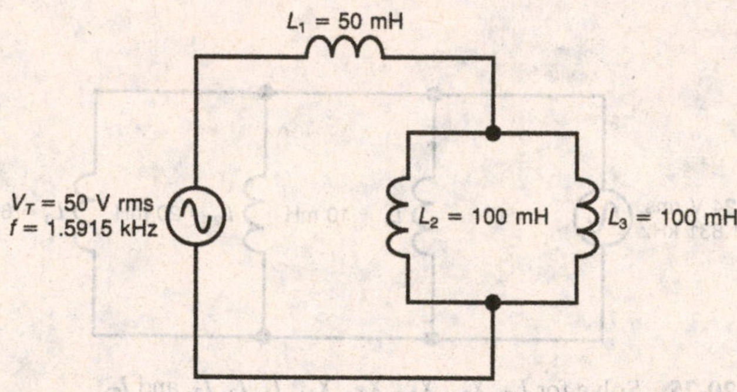

Fig. 20-30 Solve for L_T, X_{L_1}, X_{L_2}, X_{L_3}, X_{L_T}, V_{L_1}, V_{L_2}, V_{L_3}, I_2, I_3, and I_T.

WORD PROBLEMS

Sec. 20-2 Solve the following

1. Two inductors connected in series have a mutual inductance, L_M, equal to 10 mH. Given $L_1 = 40$ mH and $L_2 = 60$ mH, calculate X_{L_T} for a frequency of 10 kHz when **(a)** L_M is series-opposing and **(b)** L_M is series-aiding.
2. Two inductors connected in series have a mutual inductance, L_M, equal to 200 μH. Given $L_1 = 1$ mH and $L_2 = 500$ μH, calculate X_{L_T} for a frequency of 1 MHz when **(a)** L_M is series-aiding and **(b)** L_M is series-opposing.

Chapter 20: Inductive Reactance

Answer True or False.

1. Inductive reactance, X_L, is a measure of an inductor's opposition to the flow of sine-wave alternating current.
2. X_L is measured in ohms.
3. For an inductor, the inductive reactance, X_L, decreases as the frequency, f, increases.
4. The polarity of an induced voltage will always oppose an increase or decrease in current.
5. For series-connected inductors, the total inductive reactance, X_{L_T}, equals the sum of the individual inductive reactance values.
6. For parallel-connected inductors, the total inductive reactance, X_{L_T}, is calculated the same as it is with parallel-connected resistors.
7. The inductive reactance, X_L, of an inductor at 0 Hz (DC) is 0 Ω.
8. At a given frequency, increasing the inductance, L, increases the inductive reactance, X_L.
9. If an inductor has an X_L value of 1 kΩ at 10 kHz, then at 5 kHz, X_L equals 500 Ω.
10. Two parallel-connected inductors, each with an X_L value of 200 Ω, have an X_{L_T} of 400 Ω.

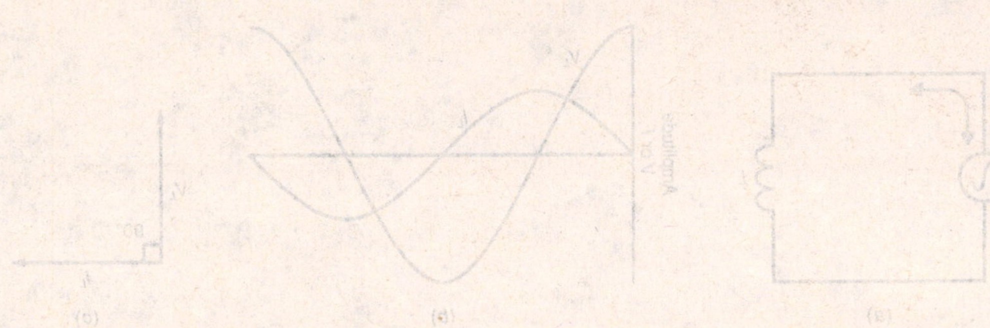

CHAPTER 21

Inductive Circuits

When an inductor is used in a sine-wave AC circuit, the voltage across the inductor leads the current through the inductor by a phase angle of 90°. This 90° phase relationship between voltage and current exists for any value of inductance or any frequency of sine-wave alternating current. The reason for this 90° phase relationship can be seen by examining the formula for induced voltage: $V_L = L(di/dt)$. It should be obvious that the induced voltage is directly proportional to the rate of current change, di/dt. If the rate of current change is zero, there is no induced voltage. The inductive circuit in Fig. 21-1*a* has the current and voltage waveforms, as shown in Fig. 21-1*b*. The phasor diagram for this voltage-current relationship is shown in Fig. 21-1*c*.

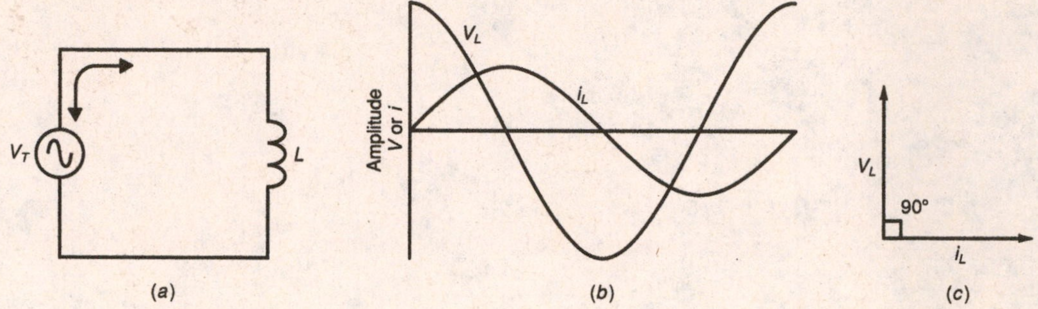

Fig. 21-1 (*a*) Circuit with inductance *L*. (*b*) Sine wave of i_L lags V_L by 90°. (*c*) Phasor diagram for voltage and current.

SEC. 21-1 ANALYZING SERIES *RL* CIRCUITS

The 90° phase relationship between voltage and current is important to know when we analyze circuits containing both inductive reactance and resistance. To analyze series *RL* circuits, we use the equations listed in Table 21-1.

TABLE 21-1 FORMULAS USED FOR SERIES *RL* CIRCUITS

$Z_T = \sqrt{R^2 + X_L{}^2}$ $I_T = \frac{V_T}{Z_T}$ $V_L = I_T \times X_L$ $V_R = I_T \times R$ $V_T = \sqrt{V_R{}^2 + V_L{}^2}$	$\tan \theta_Z = \frac{X_L}{R}$ *or* $\tan \theta_V = \frac{V_L}{V_R}$

It is important to realize that V_R and V_L cannot be added algebraically because they are always 90° out of phase with each other. This is also true for X_L and R. Instead, these quantities must be added using phasors to obtain V_T or Z_T. The phase angle between the applied voltage V_T and total current I_T is specified by θ_Z or θ_V.

Solved Problem

For Fig. 21-2, find Z_T, I_T, V_L, V_R, and θ_V. Draw the phasor voltage triangle.

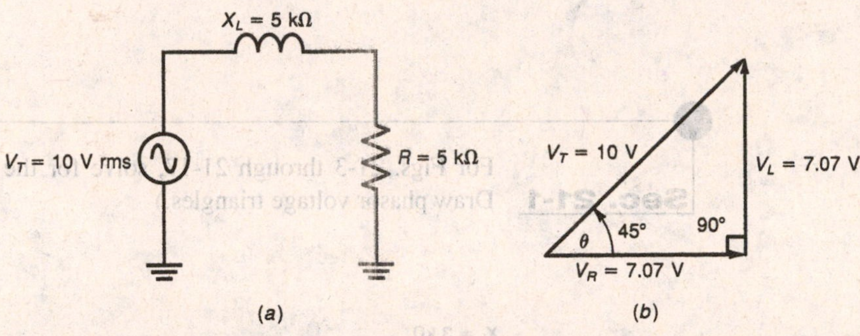

Fig. 21-2 (*a*) Series *RL* circuit. (*b*) Phasor voltage triangle.

Answer

First, we must find Z_T. From Table 21-1, we have

$$Z_T = \sqrt{R^2 + X_L{}^2}$$
$$= \sqrt{5^2 + 5^2}\ \text{k}\Omega$$
$$Z_T = 7.07\ \text{k}\Omega$$

This is the total opposition to the flow of alternating current. With Z_T known, we can find I_T as follows:

$$I_T = \frac{V_T}{Z_T}$$
$$= \frac{10\ \text{V}}{7.07\ \text{k}\Omega}$$
$$I_T = 1.414\ \text{mA}$$

Knowing I_T allows us to find V_L and V_R.

$$V_L = I_T \times X_L$$
$$= 1.414\ \text{mA} \times 5\ \text{k}\Omega$$
$$V_L = 7.07\ \text{V}$$

$$V_R = I_T \times R$$
$$= 1.414\ \text{mA} \times 5\ \text{k}\Omega$$
$$V_R = 7.07\ \text{V}$$

To find the phase angle θ for the voltage triangle, we have

$$\tan \theta_V = \frac{V_L}{V_R}$$
$$= \frac{7.07\ \text{V}}{7.07\ \text{V}}$$
$$\tan \theta_V = 1$$

To find the angle corresponding to this tangent value, we take the arctan of 1.

$$\theta = \arctan 1$$
$$\theta = 45°$$

This phase angle tells us that the total voltage, V_T, leads the total current, I_T, by 45°. The phasor voltage triangle is shown in Fig. 21-2*b*. Notice that V_R is used for the reference phasor. This is because V_R is in phase with the total current, I_T. Since I_T is the same in all components, the phase relationship between V_R and V_T specifies also the phase angle between total voltage, V_T, and total current, I_T. Notice that V_T leads V_R by 45°. This means that the total voltage, V_T, reaches its maximum value 45° ahead of I_T.

PRACTICE PROBLEMS

Sec. 21-1

For Figs. 21-3 through 21-17, solve for the unknowns listed. (*Optional:* Draw phasor voltage triangles.)

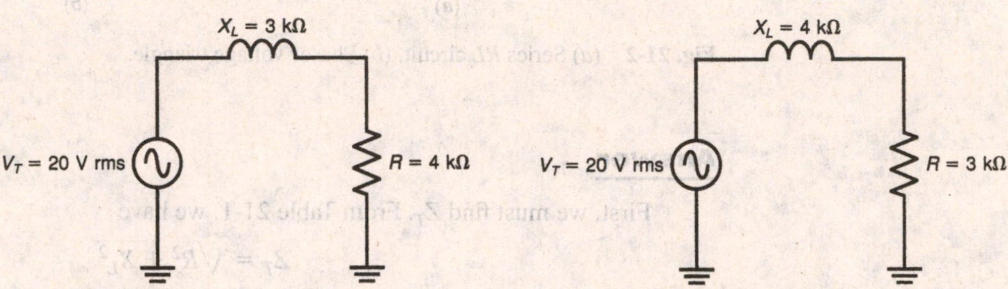

Fig. 21-3 Solve for Z_T, I_T, V_R, V_L, and θ.

Fig. 21-4 Solve for Z_T, I_T, V_R, V_L, and θ.

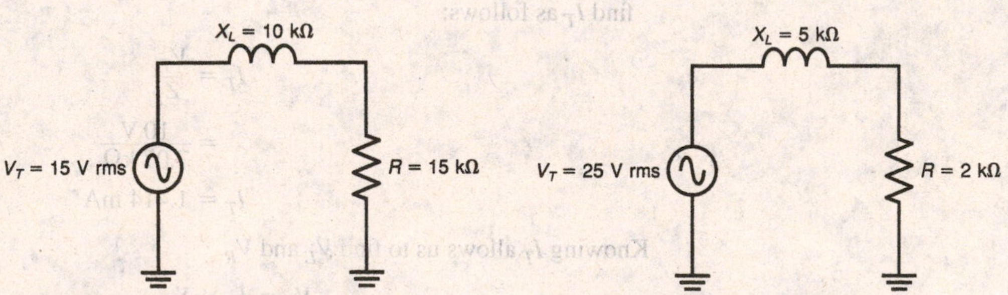

Fig. 21-5 Solve for Z_T, I_T, V_R, V_L, and θ.

Fig. 21-6 Solve for Z_T, I_T, V_R, V_L, and θ.

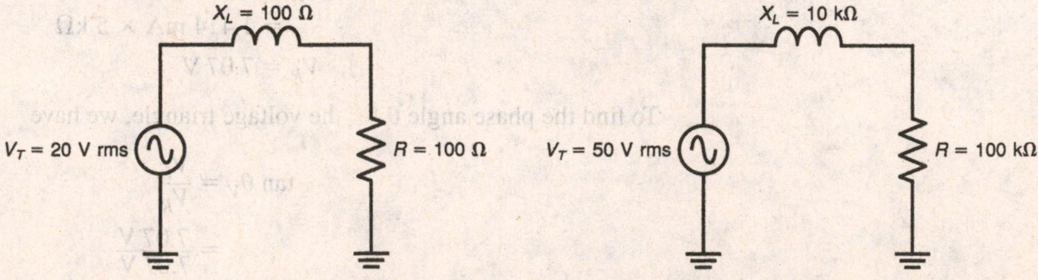

Fig. 21-7 Solve for Z_T, I_T, V_R, V_L, and θ.

Fig. 21-8 Solve for Z_T, I_T, V_R, V_L, and θ.

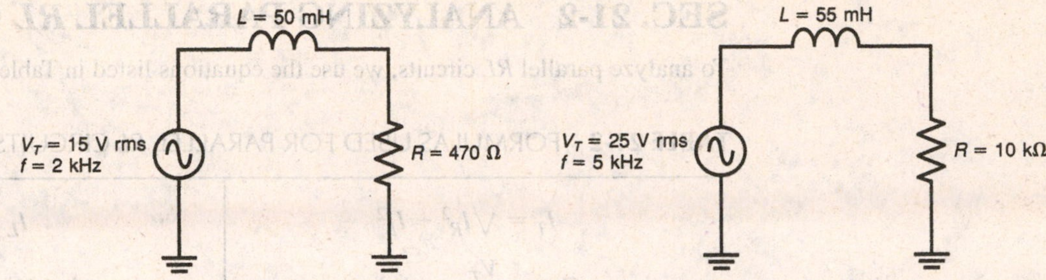

Fig. 21-9 Solve for X_L, Z_T, I_T, V_R, V_L, and θ.

Fig. 21-10 Solve for X_L, Z_T, I_T, V_R, V_L, and θ.

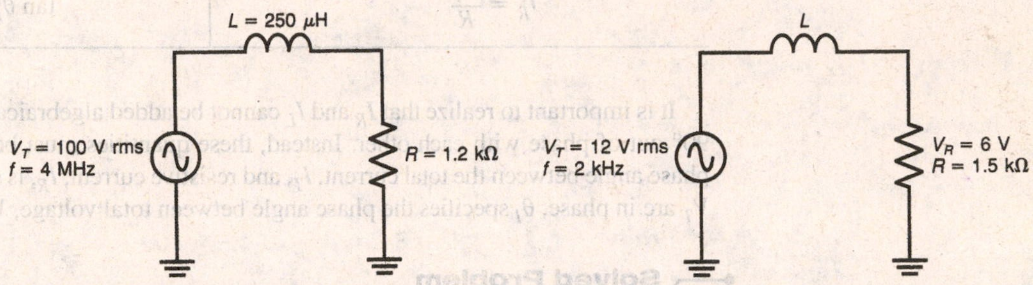

Fig. 21-11 Solve for X_L, Z_T, I_T, V_R, V_L, and θ.

Fig. 21-12 Solve for Z_T, I_T, X_L, V_L, L, and θ.

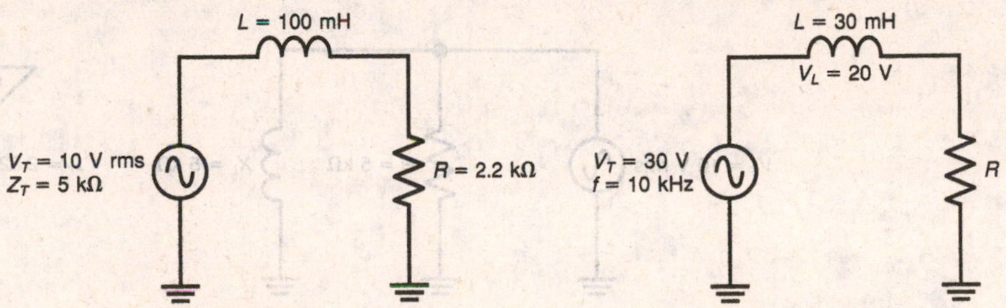

Fig. 21-13 Solve for I_T, X_L, V_R, V_L, f, and θ.

Fig. 21-14 Solve for X_L, I_T, Z_T, R, V_R, and θ.

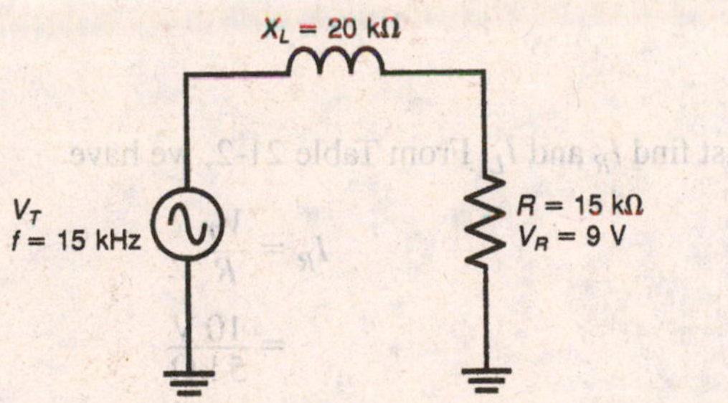

Fig. 21-15 Solve for Z_T, I_T, V_L, V_T, L, and θ.

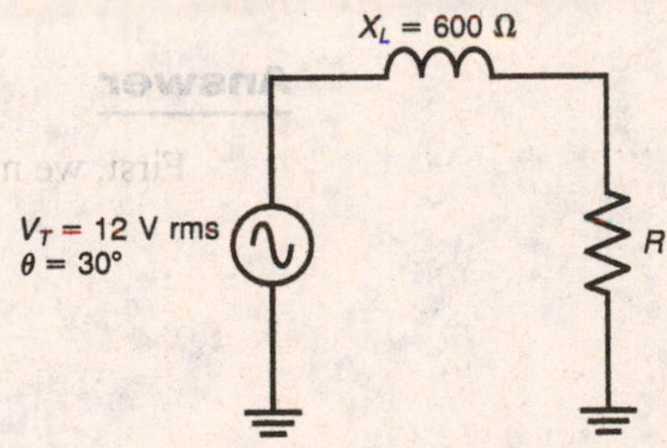

Fig. 21-16 Solve for Z_T, I_T, V_R, V_L, and R.

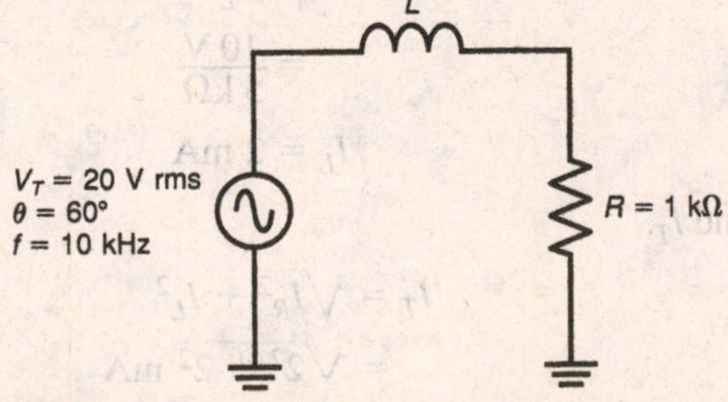

Fig. 21-17 Solve for Z_T, I_T, V_R, V_L, X_L, and L.

SEC. 21-2 ANALYZING PARALLEL *RL* CIRCUITS

To analyze parallel *RL* circuits, we use the equations listed in Table 21-2.

TABLE 21-2 FORMULAS USED FOR PARALLEL *RL* CIRCUITS

$I_T = \sqrt{I_R^2 + I_L^2}$	$I_L = \dfrac{V_T}{X_L}$
$Z_{EQ} = \dfrac{V_T}{I_T}$	$V_T = I_T \times Z_{EQ}$
$I_R = \dfrac{V_T}{R}$	$\tan \theta_I = -\dfrac{I_L}{I_R}$

It is important to realize that I_R and I_L cannot be added algebraically because they are always 90° out of phase with each other. Instead, these quantities must be added using phasors. The phase angle between the total current, I_T, and resistive current, I_R, is specified by θ_I. Since I_R and V_T are in phase, θ_I specifies the phase angle between total voltage, V_T, and total current, I_T.

Solved Problem

For Fig. 21-18, find Z_{EQ}, I_T, I_R, I_L, and θ_I. Draw the phasor current triangle.

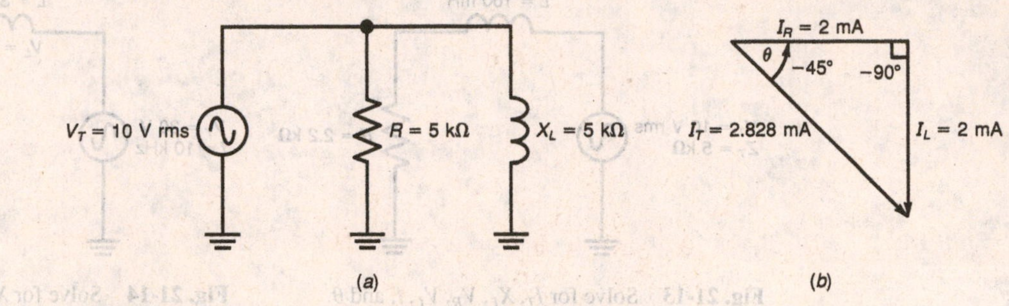

Fig. 21-18 (*a*) Parallel *RL* circuit. (*b*) Phasor current triangle.

Answer

First, we must find I_R and I_L. From Table 21-2, we have

$$I_R = \frac{V_T}{R} = \frac{10\ \text{V}}{5\ \text{k}\Omega}$$

$$I_R = 2\ \text{mA}$$

$$I_L = \frac{V_T}{X_L} = \frac{10\ \text{V}}{5\ \text{k}\Omega}$$

$$I_L = 2\ \text{mA}$$

Next, we can find I_T.

$$I_T = \sqrt{I_R^2 + I_L^2} = \sqrt{2^2 + 2^2}\ \text{mA}$$

$$I_T = 2.828\ \text{mA}$$

This is the total current flowing to and from the terminals of the voltage source, V_T.

With I_T known, Z_{EQ} can be found.

$$Z_{EQ} = \frac{V_T}{I_T}$$
$$= \frac{10\text{ V}}{2.828\text{ mA}}$$
$$Z_{EQ} = 3.535\text{ k}\Omega$$

It is important to note that Z_{EQ} in ohms is smaller than either branch opposition measured in ohms. This is the same principle realized for parallel-connected resistors; that is, the total resistance, R_T, must always be smaller than the smallest branch resistance. Here the total impedance, Z_{EQ}, is always smaller than the smallest branch opposition in ohms.

To find the phase angle θ for the current triangle, we have

$$\tan\theta_I = -\frac{I_L}{I_R}$$
$$= -\frac{2\text{ mA}}{2\text{ mA}}$$
$$\tan\theta_I = -1$$

To find the angle to which this tangent value corresponds, we take the arctan as follows:

$$\theta_I = \arctan - 1$$
$$\theta_I = -45°$$

This phase angle tells us that the total current I_T lags the resistive current I_R by 45°. The phasor current triangle is shown in Fig. 21-18*b*. Notice that I_R is used for the reference phasor. This is because V_T and I_R are in phase with each other. Since V_T is the same across all components, the phase relationship between I_R and I_T also specifies the phase angle between total current, I_T, and total voltage, V_T. Also, notice that V_T leads I_T by 45°. This means that total current reaches its maximum value 45° behind V_T.

PRACTICE PROBLEMS

Sec. 21-2 For Figs. 21-19 through 21-33, solve for the unknowns listed. (*Optional:* Draw phasor current triangles.)

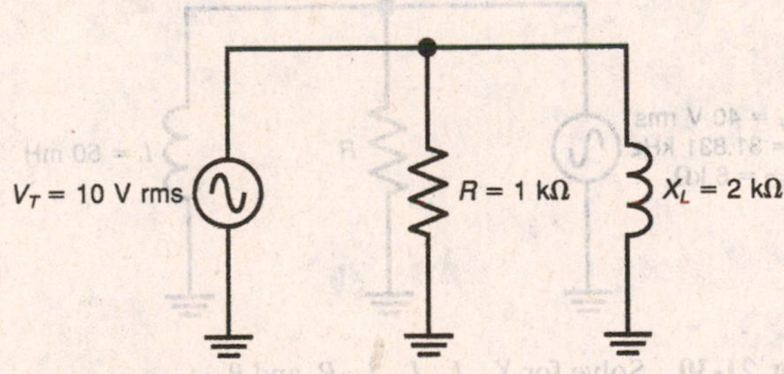

Fig. 21-19 Solve for I_R, I_L, I_T, Z_{EQ}, and θ_I.

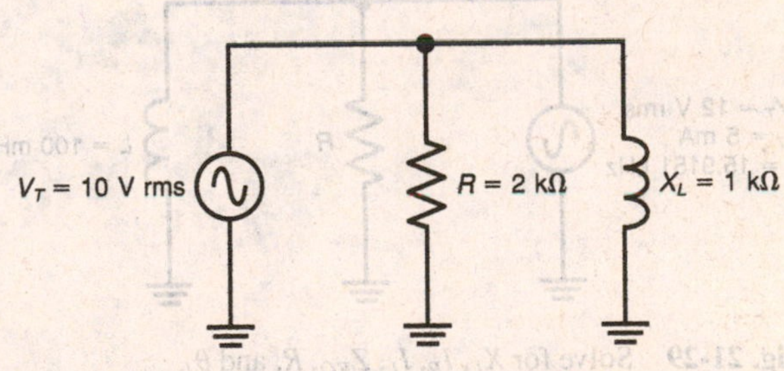

Fig. 21-20 Solve for I_R, I_L, I_T, Z_{EQ}, and θ_I.

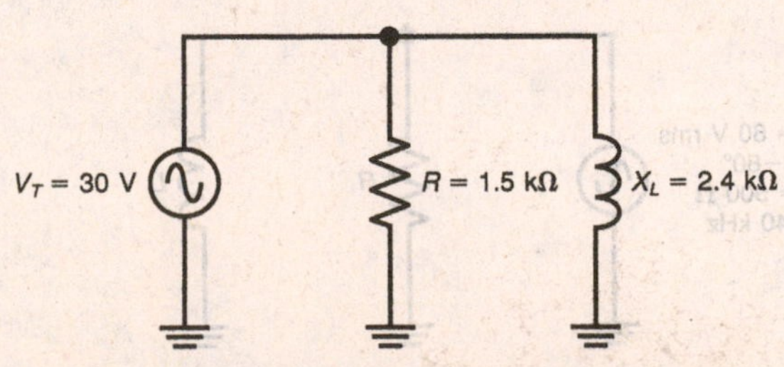

Fig. 21-21 Solve for I_R, I_L, I_T, Z_{EQ}, and θ_I.

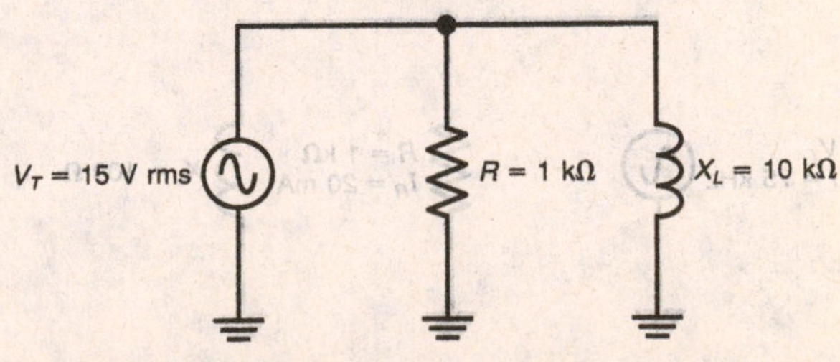

Fig. 21-22 Solve for I_R, I_L, I_T, Z_{EQ}, and θ_I.

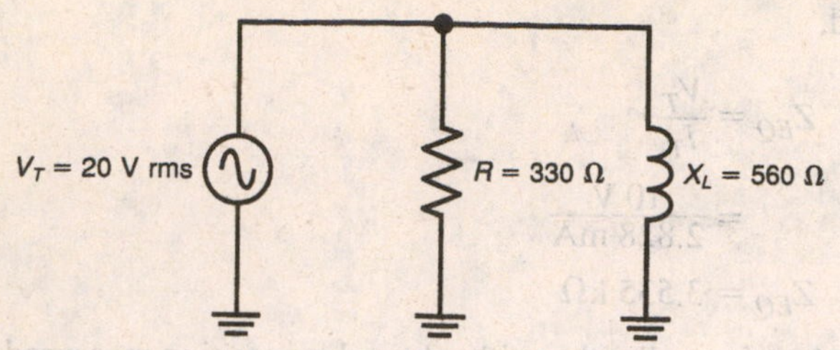

Fig. 21-23 Solve for I_R, I_L, I_T, Z_{EQ}, and θ_I.

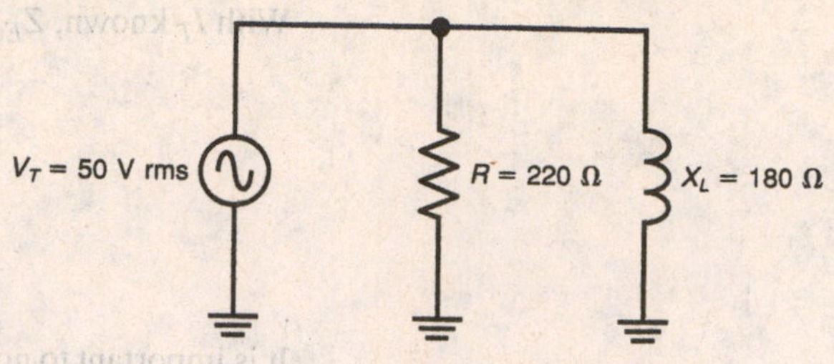

Fig. 21-24 Solve for I_R, I_L, I_T, Z_{EQ}, and θ_I.

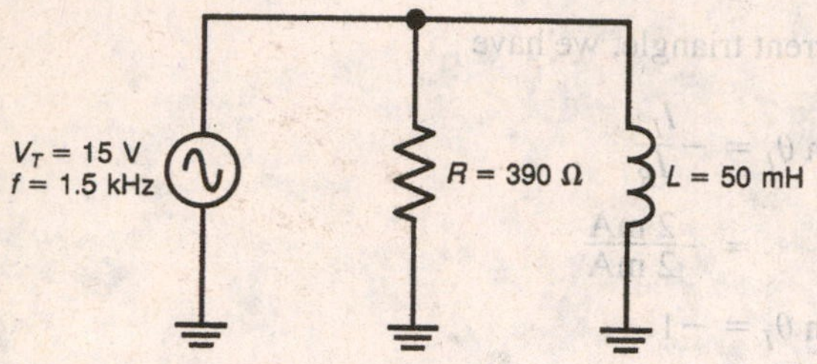

Fig. 21-25 Solve for I_R, X_L, I_L, I_T, Z_{EQ}, and θ_I.

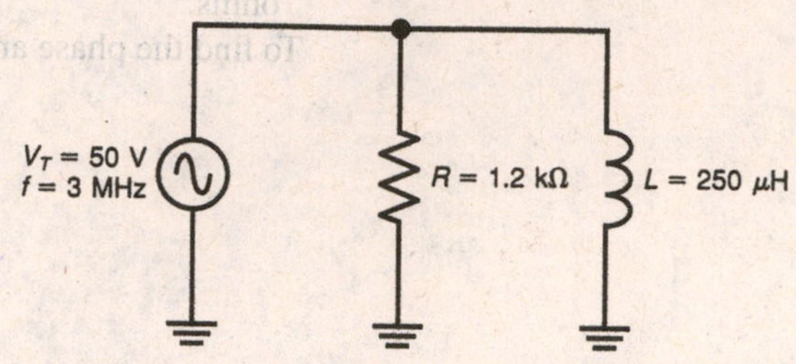

Fig. 21-26 Solve for I_R, X_L, I_L, I_T, Z_{EQ}, and θ_I.

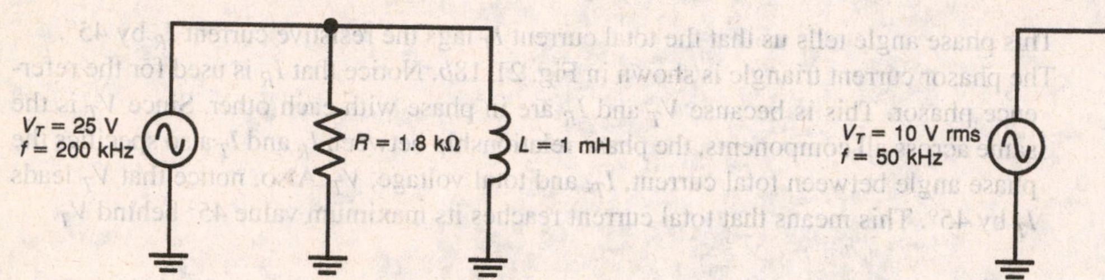

Fig. 21-27 Solve for I_R, X_L, I_L, I_T, Z_{EQ}, and θ_I.

Fig. 21-28 Solve for I_R, X_L, I_L, I_T, Z_{EQ}, and θ_I.

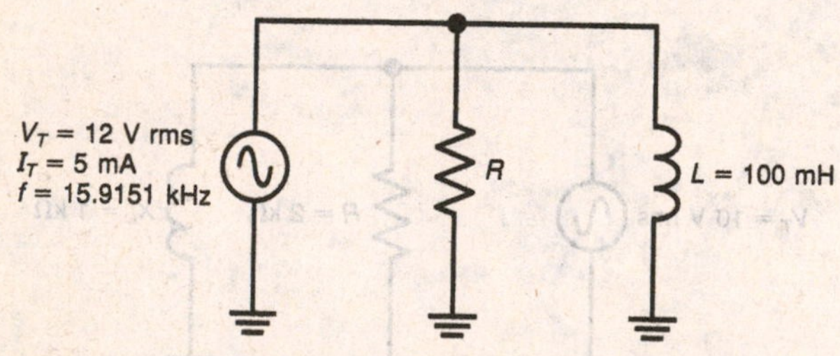

Fig. 21-29 Solve for X_L, I_R, I_L, Z_{EQ}, R, and θ_I.

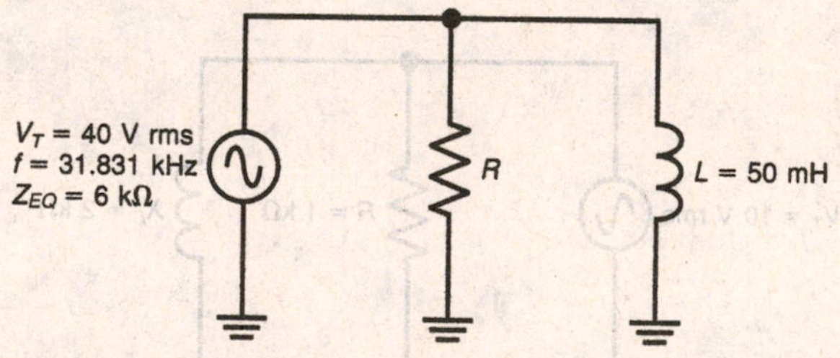

Fig. 21-30 Solve for X_L, I_T, I_R, I_L, R, and θ_I.

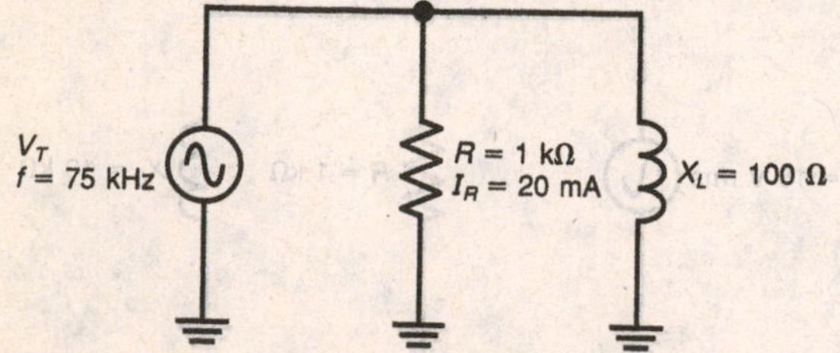

Fig. 21-31 Solve for V_T, I_T, Z_{EQ}, I_L, L, and θ_I.

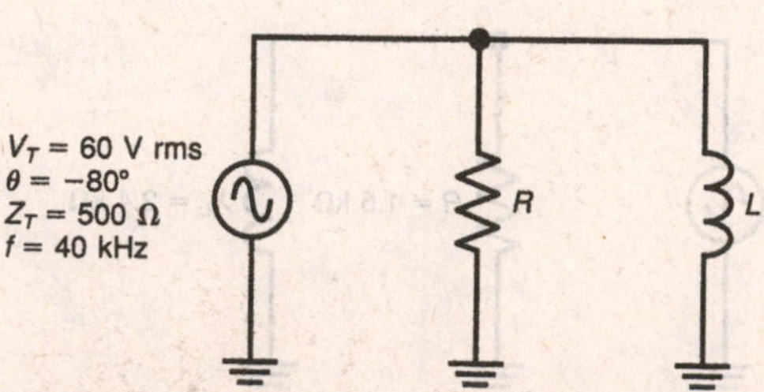

Fig. 21-32 Solve for I_T, I_R, I_L, X_L, and L.

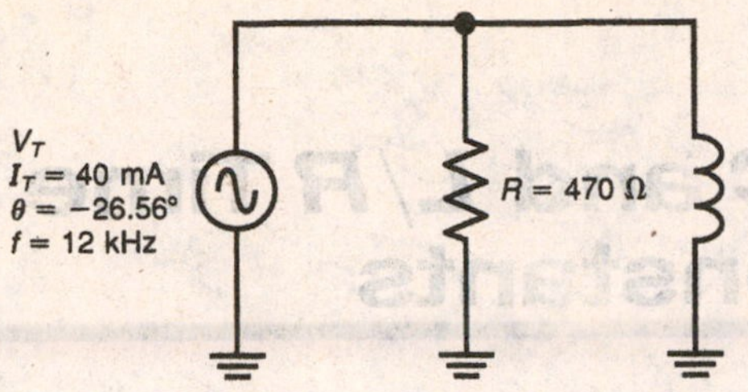

Fig. 21-33 Solve for I_R, I_L, V_T, Z_{EQ}, X_L, and L.

END OF CHAPTER TEST

Chapter 21: Inductive Circuits

Answer True or False.

1. For an inductor in a sine-wave AC circuit, the inductor voltage, V_L, lags the current, I, by a phase angle of −90°.
2. In a series *RL* circuit, V_L leads V_R by a phase angle of 90°.
3. In a series *RL* circuit, the resistor voltage, V_R, and the series current, I, are in phase with each other.
4. In a parallel *RL* circuit, the inductor current, I_L, leads the resistor current, I_R, by a phase angle of 90°.
5. In a parallel *RL* circuit, the applied voltage, V_A, is in phase with the resistor current, I_R.
6. In a series *RL* circuit, the total voltage, V_T, leads the series current, I, by some phase angle, θ.
7. In a parallel *RL* circuit, the total current, I_T, lags the applied voltage, V_A, by some phase angle, θ.
8. If 50 Ω of resistance is in series with an inductive reactance, X_L of 50 Ω, the total impedance, Z_T, of the circuit is 100 Ω.
9. If 50 Ω of resistance is in parallel with an inductive reactance, X_L, of 50 Ω, the equivalent impedance, Z_{EQ}, of the circuit is 35.35 Ω.
10. In a parallel *RL* circuit, the total current, I_T, is the phasor sum of I_L and I_R.

CHAPTER

22

RC and L/R Time Constants

Inductors and capacitors are also used in circuits containing nonsinusoidal-type waveforms. However, we cannot use reactance as a means of measuring the amount of opposition to current flow because reactance applies only to sine waves. Instead, we measure the transient response of the circuit in time constants.

SEC. 22-1 *L/R* TIME CONSTANT

A circuit that contains both resistance and inductance has an L/R time constant. Let us examine the transient response for the circuit in Fig. 22-1. Assume L to be an ideal inductor having an internal resistance of 0 Ω.

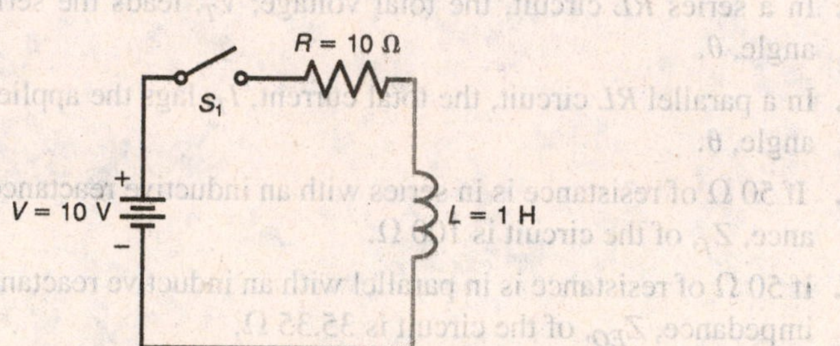

Fig. 22-1 Transient response of *RL* circuit. When S_1 is closed, I rises gradually from 0 to its steady-state value of 1 A.

When S_1 is closed, the current, I, builds up gradually from zero. Eventually, I will reach its steady state value of 1 A, equal to the battery voltage of 10 V divided by the circuit resistance of 10 Ω. The current buildup is gradual because the inductor, L, opposes the increase in current. Remember that an inductor produces an induced voltage when the current changes. The polarity of V_L produces an opposing current during buildup.

The time constant of a circuit is the time it takes for a 63.2% change to occur. For the circuit shown in Fig. 22-1, the transient response is measured in terms of the L/R ratio, which is the time constant for an inductive circuit. To calculate the time constant, we have

$$T = \frac{L}{R}$$

where T = time constant in seconds
L = inductance in henries
R = resistance in ohms

From Fig. 22-1, we have

$$T = \frac{L}{R}$$

$$= \frac{1\ \text{H}}{10\ \Omega}$$

$$T = 0.1\ \text{s}$$

This 0.1 s is a measure of how long it takes for a 63.2% change in current to occur. After five L/R constants have elapsed, the current is practically equal to its steady-state value of 1 A. For Fig. 22-1, five time constants correspond to a time of 0.5 s.

The universal time constant chart in Fig. 22-2 shows how I builds up over a period of five time constants. This is represented by curve A. Notice in Fig. 22-2 that I increases to 63.2% of its maximum value after one time constant. For each one-time-constant interval, a 63.2% change in current will occur.

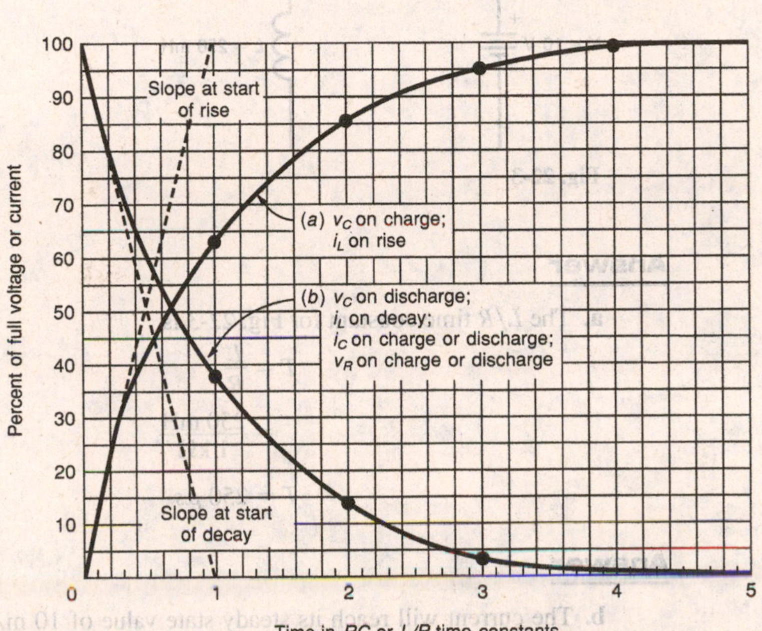

Fig. 22-2 Universal time constant chart for *RC* or *RL* circuits.

The formula used to determine the current value at any time during buildup is

$$i = I_{\text{max}}(1 - \epsilon^{-t/(L/R)})$$

where i = instantaneous value of current flow
I_{max} = steady state value of current
e = base of natural logarithms
t = elapsed time from start of buildup

The formula used to determine the current value at any time during current decay is

$$i = I_{\text{max}}\epsilon^{-t/(L/R)}$$

where t = elapsed time from start of decay

Curve B in Fig. 22-2 shows how I decays over a period of five time constants.

When the switch in Fig. 22-1 opens, the time constant for current decay becomes very short because L/R becomes smaller with the very high resistance of the open switch. The result is that a high value of induced voltage, V_L, will exist across the inductor. This value of induced voltage can be much greater than the applied voltage. Since the high resistance of the switch in Fig. 22-1 is unknown, we are unable to determine the transient response for the decay of current.

Solved Problems

In Fig. 22-3 calculate the following:

a. The L/R time constant during current buildup.

b. The length of time required for the current to reach its steady state value after S_1 has been closed.

c. The instantaneous value of current 173.25 μs after S_1 has been closed.

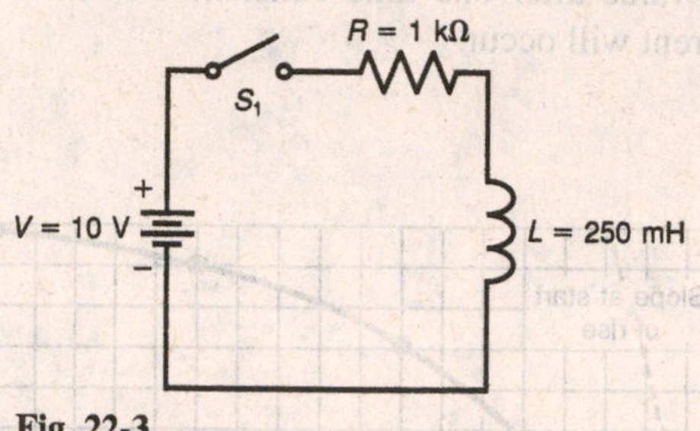

Fig. 22-3

Answer

a. The L/R time constant for Fig. 22-3 is

$$T = \frac{L}{R}$$
$$= \frac{250 \text{ mH}}{1 \text{ k}\Omega}$$
$$T = 250 \ \mu s$$

Answer

b. The current will reach its steady state value of 10 mA five time constants after the switch, S_1, has been closed. This corresponds to a time of

$$5 \times (L/R) = 5 \times 250 \ \mu s$$
$$5 \times (L/R) = 1.25 \text{ ms}$$

Answer

c. The instantaneous value of current is found using the equation for current buildup.

$$i = I_{max}(1 - \epsilon^{-t/(L/R)})$$

To determine i 173.25 μs after S_1 has been closed, we insert the values from Fig. 22-3 as follows. Since $-\dfrac{t}{(L/R)} = -\dfrac{173.25 \ \mu s}{250 \ \mu s} = -0.693$, then

$$i = 10 \text{ mA}(1 - \epsilon^{-0.693})$$
$$= 10 \text{ mA}(1 - 0.5)$$
$$= 10 \text{ mA} \times 0.5$$
$$i = 5 \text{ mA}$$

It should be pointed out at this time that the value of resistor voltage can be determined when i is known: $V_R = i \times R$. It is also true that V_T must equal $V_L + V_R$ for any instant in time. Therefore, if V_R is known, V_L can be found as follows $V_L = V_T - V_R$.

Sec. 22-1 Problems 1–7 refer to Fig. 22-4.

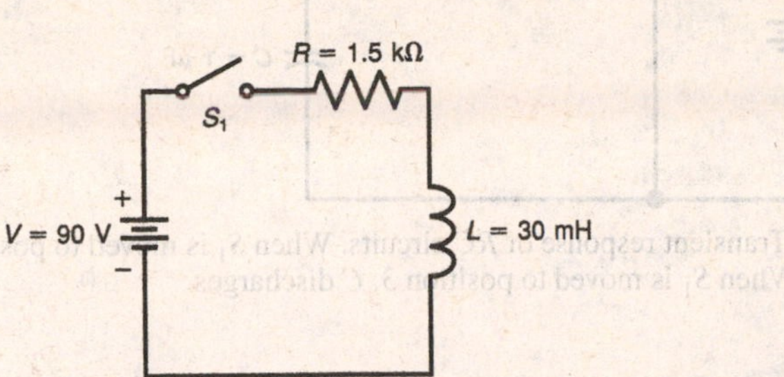

Fig. 22-4

1. Calculate the L/R time constant for the buildup of current.
2. **(a)** How long will it take the current I to reach its steady-state value after S_1 has been closed? **(b)** What is the steady-state value of current?
3. What is the value of current exactly two time constants after the closing of the switch?
4. Calculate i for the following time intervals after S_1 has been closed: **(a)** 0 s, **(b)** 14 μs, **(c)** 30 μs, **(d)** 50 μs, and **(e)** 100 μs.
5. Calculate the value of resistor voltage V_R for the following time intervals after S_1 has been closed: **(a)** 20 μs, **(b)** 25 μs, **(c)** 35 μs, **(d)** 40 μs, and **(e)** 75 μs.
6. Calculate the value of inductor voltage V_L for the time intervals listed in Prob. 5.
7. A 1-MΩ resistor is placed across the closed switch contacts of S_1. **(a)** Calculate V_L the instant S_1 is opened. **(b)** Calculate the required di/dt rate to produce this peak inductor voltage.
8. How much time must elapse before an *RL* circuit with a 100-mH inductor and a 10-kΩ resistance reaches its maximum current?
9. An *RL* circuit has a time constant of 40 μs and a resistance value of 1 kΩ. Calculate *L*.
10. An *RL* circuit has a time constant of 5 ms and an inductance of 180 mH. Calculate *R*.
11. An *RL* circuit has a time constant of 50 μs and a resistance of 1.2 kΩ. Calculate *L*.
12. How can the time constant for an *RL* circuit be **(a)** increased and **(b)** decreased?

SEC. 22-2 *RC* TIME CONSTANT

For capacitive circuits, the transient response is measured in terms of the product $R \times C$. To calculate the time constant, we use

$$T = R \times C$$

where T = time constant in seconds
R = resistance in ohms
C = capacitance in farads

Let us examine the transient response for the circuit shown in Fig. 22-5.

When S_1 is moved to position 1, the voltage across the capacitor builds up gradually from zero. This buildup is gradual because the capacitor opposes a change in voltage. The

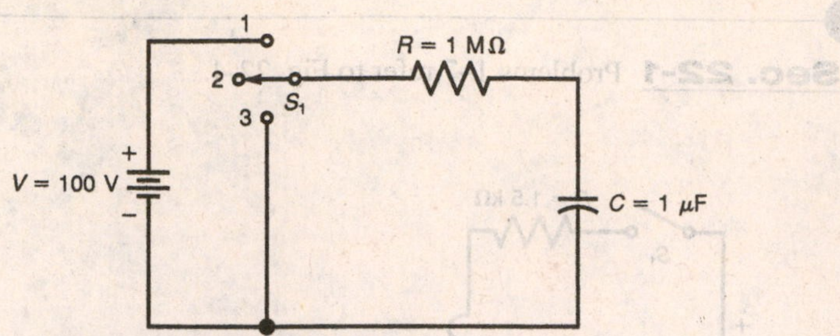

Fig. 22-5 Transient response of *RC* circuits. When S_1 is moved to position 1, *C* charges. When S_1 is moved to position 3, *C* discharges.

voltage across the capacitance cannot increase until the charging current has stored enough charge in *C*.

For the circuit shown in Fig. 22-5, the time constant is calculated as

$$T = R \times C$$
$$= 1\ M\Omega \times 1\ \mu F$$
$$T = 1\ s$$

This time constant of 1 s is a measure of how long it takes for a 63.2% change to occur. After five *RC* time constants have elapsed, the voltage across *C* is practically equal to its steady-state value of 100 V. For Fig. 22-5, five time constants correspond to a time of 5 s.

During the time capacitor *C* is charging, the voltage across resistor *R* is decaying. Note that on charge $V_R = V_T - V_C$. The universal time constant chart in Fig. 22-2 shows how the capacitor voltage builds up over a period of five time constants. This is represented by curve A. Curve B shows how the resistor voltage decays over a period of five time constants.

The formula used to determine the voltage across capacitor *C* at any time during charge is

$$V_C = V_{max}(1 - \epsilon^{-(t/RC)})$$

where V_C = instantaneous value of capacitor voltage
V_{max} = value of voltage applied to circuit
ϵ = base for natural logarithms
t = elapsed time from beginning of charge

The formula used to determine the voltage across resistor *R* at any time during charge or discharge is

$$V_R = V_{max} \times \epsilon^{-(t/RC)}$$

where V_R = resistor voltage in volts
t = elapsed time from beginning of discharge

If, in Fig. 22-5, capacitor *C* is allowed to fully charge, and then switch S_1 is moved to position 3, the capacitor will discharge. The voltage across both *R* and *C* will be identical for this switch position because *R* is in parallel with *C*. After five *RC* time constants, the capacitor *C* is almost completely discharged and both V_R and V_C equal 0 V. Use the formula $V_R = V_{max}\epsilon^{-(t/RC)}$ for determining the value of resistor or capacitor voltage at any time, *t*, during discharge.

Solved Problems

For Fig. 22-6, calculate

a. The RC time constant during charge or discharge.

b. The length of time required for V_C to reach 100 V.

c. The instantaneous value of V_C and V_R 30 ms after S_1 is put in position 1.

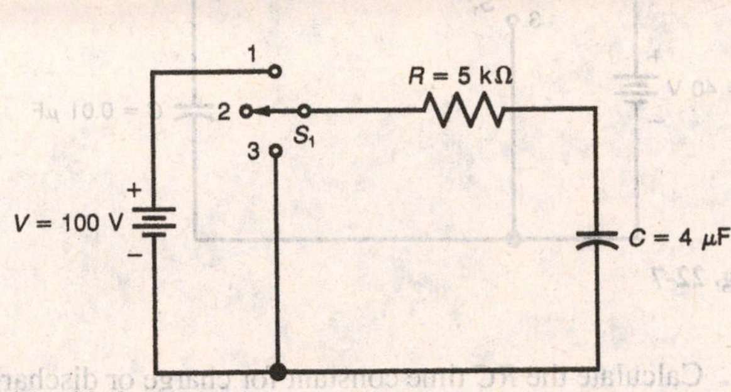

Fig. 22-6

Answer

a. The RC time constant for Fig. 22-6 is

$$T = R \times C$$
$$= 5\ \text{k}\Omega \times 4\ \mu\text{F}$$
$$T = 20\ \text{ms}$$

Answer

b. The voltage V_C will reach 100 V (approximately) five time constants after S_1 is put in position 1. This corresponds to a time of

$$5 \times RC = 5 \times 20\ \text{ms}$$
$$5 \times RC = 100\ \text{ms}$$

Answer

c. The instantaneous value of capacitor voltage, V_C, is found using the equation for capacitor charge.

$$V_C = V_{\text{max}}(1 - \epsilon^{-(t/RC)})$$

To determine V_C 30 ms after S_1 is put in position 1, we insert the values from Fig. 22-6.

$$V_C = 100\ \text{V}(1 - \epsilon^{-1.5})$$
$$= 100\ \text{V}(1 - 0.223)$$
$$= 100\ \text{V} \times 0.776$$
$$V_C = 77.7\ \text{V}$$

Since $V_T = V_C + V_R$, V_R can be found by subtracting V_C from V_T.

$$V_R = V_T - V_C$$
$$= 100\ \text{V} - 77.7\ \text{V}$$
$$V_R = 22.3\ \text{V}$$

It is important to note that V_R can also be found using the following formula:

$$V_R = V_{\text{max}} \times \epsilon^{-(t/RC)}$$
$$= 100\ \text{V} \times \epsilon^{-1.5}$$
$$= 100\ \text{V} \times 0.223$$
$$V_R = 22.3\ \text{V}$$

Sec. 22-2 Problems 1–8 refer to Fig. 22-7.

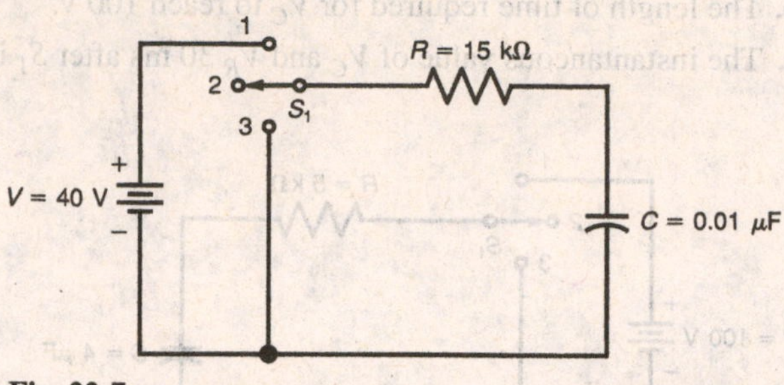

Fig. 22-7

1. Calculate the *RC* time constant for charge or discharge.
2. **(a)** How long will it take the capacitor voltage to reach 40 V after S_1 has been placed in position 1? **(b)** What is the resistor voltage at the first instant S_1 is moved from position 2 to position 1?
3. What is the instantaneous value of charge current exactly two time constants after S_1 has been put in position 1?
4. Calculate V_C at the following time intervals after S_1 has been moved to position 1: **(a)** 0 s, **(b)** 50 μs, **(c)** 100 μs, **(d)** 300 μs, and **(e)** 600 μs.
5. Calculate V_R at the following time intervals after S_1 has been moved to position 1: **(a)** 0 s, **(b)** 103.95 μs, **(c)** 200 μs, **(d)** 400 μs, and **(e)** 750 μs.
6. Calculate the instantaneous values of charge current for the time intervals listed in Prob. 4.
7. Capacitance *C* is allowed to fully charge in position 1. Calculate the required *dv*/*dt* rate to produce the initial discharge current when S_1 is moved to position 3.
8. Capacitance *C* is allowed to fully charge in position 1. Calculate V_C for the following time intervals after S_1 has been placed in position 3: **(a)** 50 μs, **(b)** 105 μs, **(c)** 175 μs, **(d)** 250 μs, **(e)** 750 μs.
9. The time constant of an *RC* circuit is 50 μs. If $R = 1$ kΩ, calculate *C*.
10. The time constant of an *RC* circuit is 1 ms. If $C = 0.05$ μF, calculate *R*.
11. The time constant of an *RC* circuit is 40 ms. If $R = 200$ kΩ, calculate *C*.
12. The time constant of an *RC* circuit is 600 μs. If $C = 0.015$ μF, calculate *R*.

SEC. 22-3 ADVANCED TIME CONSTANT ANALYSIS

If a capacitor begins charging from an initial voltage other than zero, which is often the case, another formula is used to calculate the value of the capacitor voltage, V_C, at any time, *t*. This formula is

$$V_C = (V_F - V_i)(1 - \epsilon^{-(t/RC)}) + V_i$$

Where V_i represents the initial voltage the capacitor begins charging from and V_F represents the final voltage the capacitor eventually charges to. The quantity, $(V_F - V_i)$, represents the net, or remaining charging voltage.

In Fig. 22-8, assume the capacitor, *C*, is fully charged to 25 V with the switch S_1, in position 2. Now, if S_1 is moved up to position 3, the capacitor begins charging toward 100 V. The *RC* time constant is

$$\begin{aligned} T &= RC \\ &= 100\ \text{k}\Omega \times 0.1\ \mu\text{F} \\ &= 10\ \text{ms} \end{aligned}$$

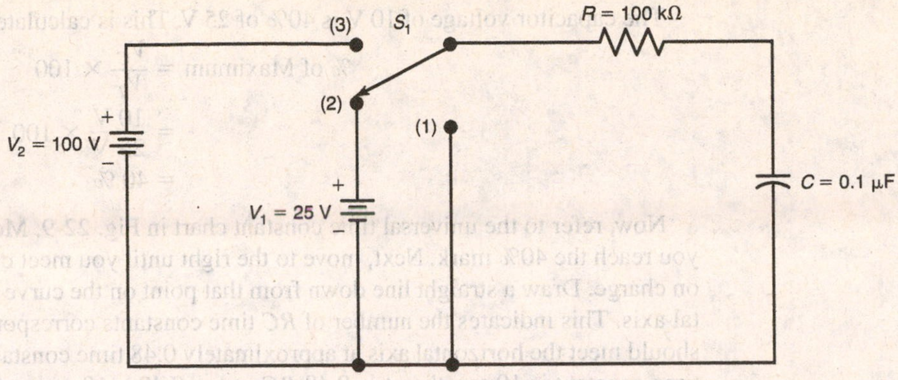

Fig. 22-8 *RC* circuit used for analyzing the condition where *C* charges from an initial voltage other than zero.

To determine the capacitor voltage, V_C, 10 ms after S_1 is moved up to position 3, proceed as follows. Since t = 10 ms and RC = 10 ms, $-\frac{t}{RC} = -1$. Inserting values into the formula for V_C gives us

$$\begin{aligned} V_C &= (V_F - V_i)(1 - \epsilon^{-(t/RC)}) + V_i \\ &= (100\text{ V} - 25\text{ V})(1 - \epsilon^{-1}) + 25\text{ V} \\ &= 75\text{ V}\,(1 - \epsilon^{-1}) + 25\text{ V} \\ &= 75\text{ V}\,(0.632) + 25\text{ V} \\ &= 47.4\text{ V} + 25\text{ V} \\ &= 72.4\text{ V} \end{aligned}$$

It is important to note that the universal time constant chart could also be used to solve for the capacitor voltage, V_C, in this case. Close examination of curve *A* on the universal time constant chart in Fig. 22-2 reveals that one time constant corresponds to a 63% change (approximately). Therefore, to find V_C we multiply 0.63 by 75 V, which is the net or remaining charging voltage, and then add this answer to the initial capacitor voltage of 25 V. The calculations are

$$\begin{aligned} V_C &= (0.63 \times 75\text{ V}) + 25\text{ V} \\ &= 47.25\text{ V} + 25\text{ V} \\ &= 72.25\text{ V} \end{aligned}$$

Notice this answer is reasonably close to the value obtained with the formula used earlier. In most cases, the value obtained from the universal time constant chart is accurate enough.

Solving for the Elapsed Time, *t*

It is often of interest to know how long it will take for the voltage or current in an *RC* or *RL* circuit to reach a certain value. When this is the case, either the universal time constant chart can be used to get an approximate answer or the equation for V_C, V_R, or i can be rearranged to get an exact answer for the elapsed time, *t*.

In Fig. 22-8, assume the capacitor, *C*, is fully discharged with the switch, S_1 in position 1. How long will it take for the capacitor voltage, V_C, to reach 10 V, after S_1 is moved to position 2? The answer could be found one of two ways. We could use the universal time constant chart to get an approximate answer, or we could solve for t in the equation $V_C = (1 - \epsilon^{-(t/RC)})$ to get an exact answer. Let's solve it both ways, starting with the universal time constant chart.

The capacitor voltage of 10 V is 40% of 25 V. This is calculated as

$$\begin{aligned}\% \text{ of Maximum} &= \frac{V_C}{V_T} \times 100 \\ &= \frac{10\text{ V}}{25\text{ V}} \times 100 \\ &= 40\ \%\end{aligned}$$

Now, refer to the universal time constant chart in Fig. 22-9. Move up the vertical axis until you reach the 40% mark. Next, move to the right until you meet curve *A*, which represents V_C on charge. Draw a straight line down from that point on the curve until you reach the horizontal axis. This indicates the number of *RC* time constants corresponding to a 40% change. You should meet the horizontal axis at approximately 0.48 time constants, as shown. Since one *RC* time constant is 10 ms, then $t = 0.48\ RC$ or $t = 0.48 \times 10\text{ ms} = 4.8\text{ ms}$.

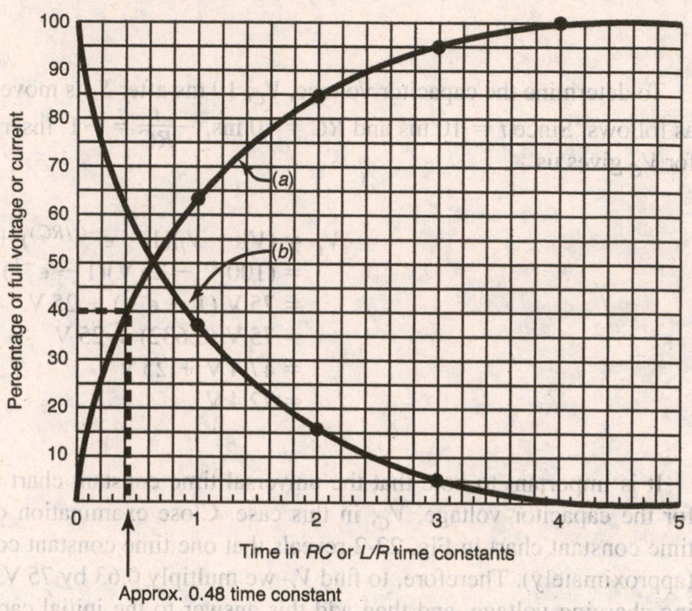

Fig. 22-9 Universal time constant chart being used to calculate the time required for the capacitor voltage to reach 40% of V_{max}.

A more accurate way to determine how long it takes the capacitor voltage to reach 10 V would be to solve for *t* in the equation for V_C on charge. This formula is

$$t = \ln\left(\frac{1}{1 - \frac{V_C}{V_{max}}}\right) RC$$

Inserting values give us

$$\begin{aligned}t &= \ln\left(\frac{1}{1 - \frac{10\text{ V}}{25\text{ V}}}\right) 100\text{ k}\Omega \times 0.1\ \mu\text{F} \\ &= \ln(1.67)10\text{ ms} \\ &= 0.511 \times 10\text{ ms} \\ &= 5.11\text{ ms}\end{aligned}$$

Notice this answer is slightly different (more accurate) from the value obtained by using the universal time constant chart. This will usually be the case. In most cases, however, the answer obtained from the universal time constant chart is accurate enough.

Solved Problems

a. In Fig. 22-8, assume the capacitor, C, is fully charged to 25 V with the switch, S_1, in position 2. What is the capacitor voltage, V_C, 15 ms after S_1 is placed into position 3?

Answer

Recall that the RC time constant equals 10 ms. Therefore, $-\frac{t}{RC} = -\frac{15 \text{ ms}}{10 \text{ ms}} = -1.5$.

Because the capacitor begins charging from an initial voltage other than zero, we use the formula for V_C, which includes the initial voltage, V_i.

$$
\begin{aligned}
V_C &= (V_F - V_i)(1 - \epsilon^{-(t/RC)}) + V_i \\
&= (100 \text{ V} - 25 \text{ V})(1 - \epsilon^{-1.5}) + 25 \text{ V} \\
&= 75 \text{ V}(0.777) + 25 \text{ V} \\
&= 58.28 \text{ V} + 25 \text{ V} \\
&= 83.3 \text{ V}
\end{aligned}
$$

It is important to note that the universal time constant chart could also be used to get a reasonably accurate answer for V_C.

b. In Fig. 22-8, assume the capacitor, C, is fully discharged with the switch, S_1, in position 1. How long will it take the capacitor voltage, V_C, to reach 20 V after S_1 is placed into position 2?

Answer

Use the formula as given earlier for t.

$$t = \ln\left(\frac{1}{1 - \frac{V_C}{V_{max}}}\right) RC$$

$$
\begin{aligned}
t &= \ln\left(\frac{1}{1 - \frac{20 \text{ V}}{25 \text{ V}}}\right) 100 \text{ k}\Omega \times 0.1 \, \mu\text{F} \\
&= \ln(5) 10 \text{ ms} \\
&= 1.61 \times 10 \text{ ms} \\
&= 16.1 \text{ ms}
\end{aligned}
$$

It is important to note that the universal time constant chart could also be used to get a reasonably accurate value for the elapsed time, t.

c. In Fig. 22-8, assume the capacitor, C, is fully charged to 25 V with the switch, S_1, in position 2. If the switch, S_1, is moved to position 1, how long will it take the capacitor voltage, V_C, to decrease to 12.5 V?

Answer

Recall that the equation for V_R on charge or discharge is $V_R = V_{max}\epsilon^{-(t/RC)}$. This is also the same formula used to find V_C on discharge (on discharge R and C are in parallel). Solving for the elapsed time, t, in this equation, gives us

$$t = \ln\left(\frac{V_{max}}{V_R}\right) RC$$

Inserting values we have

$$
\begin{aligned}
t &= \ln\left(\frac{25 \text{ V}}{12.5 \text{ V}}\right) 100 \text{ k}\Omega \times 0.1 \, \mu\text{F} \\
&= \ln(2) \times 10 \text{ ms} \\
&= 0.693 \times 10 \text{ ms} \\
t &= 6.93 \text{ ms}
\end{aligned}
$$

If curve B on the universal time constant chart is examined, we see that a 50% change in voltage corresponds to approximately 0.7 time constants, or 0.7×10 ms $= 7$ ms in this case. This is reasonably close to the exact value of 6.93 ms calculated with the formula.

PRACTICE PROBLEMS

Sec. 22-3 Solve the following.

1. In Fig. 22-10, assume the capacitor, C, is fully discharged with the switch, S_1, in position 1. If S_1 is moved to position 2, how long will it take for the capacitor voltage, V_C, to reach **(a)** 5 V, **(b)** 10 V, **(c)** 25 V, **(d)** 35 V, and **(e)** 45 V?
2. In Fig. 22-10, assume the capacitor, C, is fully charged to 50 V with the switch, S_1, in position 2. If S_1 is now moved to position 1, how long will it take the capacitor voltage, V_C, to decrease to **(a)** 40 V, **(b)** 30 V, **(c)** 20 V, **(d)** 15 V, and **(e)** 10 V?
3. In Fig. 22-10, assume, the capacitor, C, is fully charged with the switch, S_1, in position 2. If S_1 is now moved to position 3, how much is the capacitor voltage, V_C, after **(a)** 200 μs, **(b)** 350 μs, **(c)** 500 μs, **(d)** 1 ms, and **(e)** 2.5 ms?
4. In Fig. 22-10, assume the capacitor, C, is fully charged with the switch, S_1, in position 2. If S_1 is now moved to position 3, how much is the charging current, i_C, after **(a)** 100 μs, **(b)** 400 μs, **(c)** 1 ms, **(d)** 1.5 ms, and **(e)** 2.5 ms?

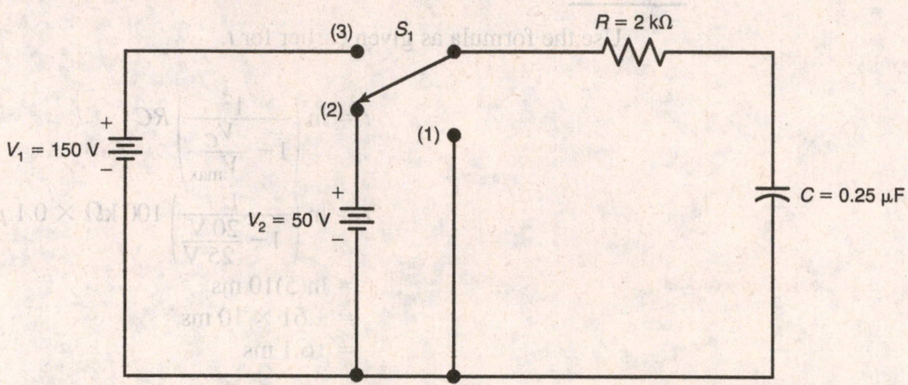

Fig. 22-10 *RC* circuit used for Probs. 1–4.

5. In Fig. 22-11, assume the capacitor, C, is fully discharged with the switch, S_1, in position 1. If S_1 is moved to position 2, how long will it take the capacitor voltage, V_C, to reach **(a)** 3 V, **(b)** 10 V, **(c)** 15 V, **(d)** 18 V, and **(e)** 25 V?
6. In Fig. 22-11, assume the capacitor, C, is fully charged to 30 V with the switch, S_1, in position 2. If S_1 is now moved to position 1, how long will it take the capacitor voltage, V_C, to decrease to **(a)** 27 V, **(b)** 24 V, **(c)** 20 V, **(d)** 10 V, and **(e)** 5 V?

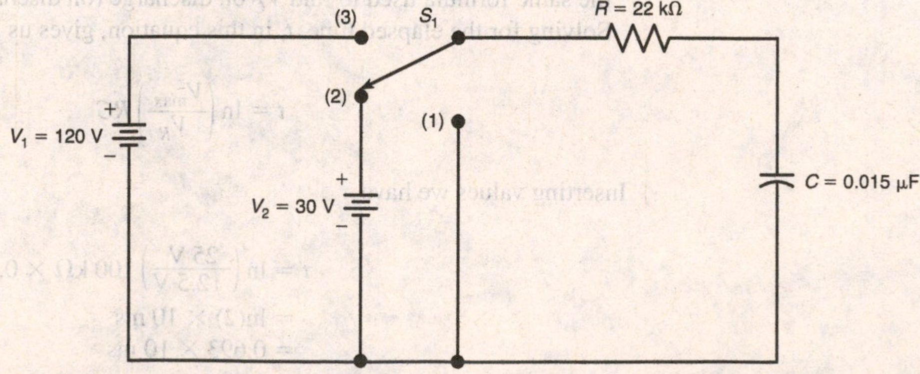

Fig. 22-11 *RC* circuit used for Probs. 5–7.

7. In Fig. 22-11, assume the capacitor, *C*, is fully charged to 30 V in position 2. If S_1 is now moved to position 3, how much is the capacitor voltage, V_C, after **(a)** 231 μs, **(b)** 330 μs, **(c)** 450 μs, **(d)** 600 μs, and **(e)** 1 ms?

8. In Fig. 22-12, assume the capacitor, *C*, is allowed to fully charge with the switch, S_1 in position 1. If S_1 is now moved to position 2, how much is the capacitor voltage, V_C after **(a)** 10 ms, **(b)** 20 ms, **(c)** 30 ms, **(d)** 40 ms, **(e)** 80 ms, and **(f)** 200 ms? (Note the polarities of V_1 and V_2.)

9. In Prob. 8, how much is the resistor voltage, V_R, at the first instant S_1 is moved to position 2?

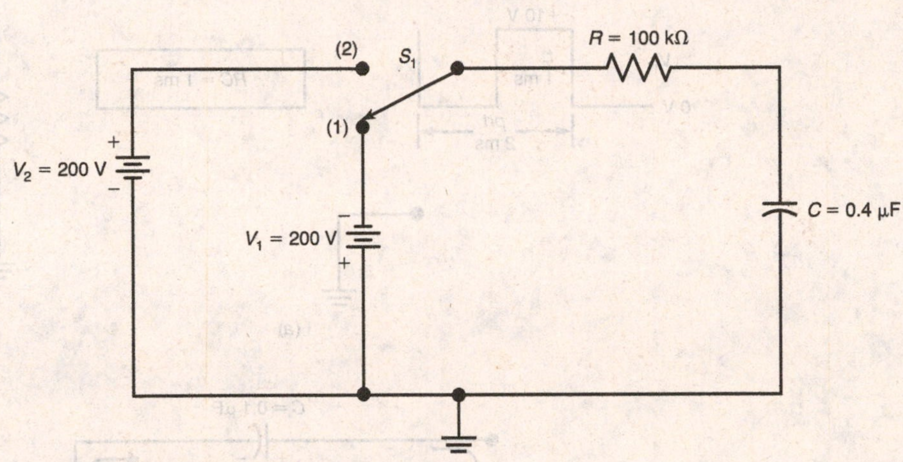

Fig. 22-12 *RC* circuit used for Probs. 8 and 9.

10. In Fig. 22-13, how long after S_1 is closed does it take for the current, *I*, to build up to **(a)** 5 mA, **(b)** 10 mA, **(c)** 25 mA, **(d)** 32.5 mA, and **(e)** 45 mA? (Try using the universal time constant chart to obtain your answers.)

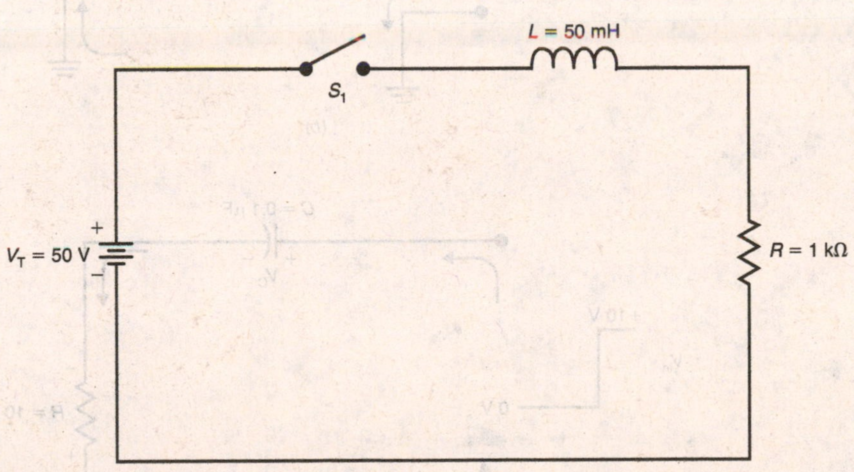

Fig. 22-13 *RL* circuit used for Prob. 10.

SEC. 22-4 NONSINUSOIDAL WAVEFORMS IN *RC* CIRCUITS

In many cases, the input to an *RC* circuit is a nonsinusoidal waveform, such as a square wave. When this is the case, the resistor and capacitor voltage waveforms have a waveshape that is dependant on the relationship of the pulse time, *tp*, and the pulse repetition time, *prt*, to the circuit time constant. Fig. 22-14*a* shows an *RC* circuit driven by an input square wave. The time constant of the circuit is 1 ms, which is calculated as follows:

$$T = RC$$
$$= 10\ \text{k}\Omega \times 0.1\ \mu\text{F}$$
$$= 1\ \text{ms}$$

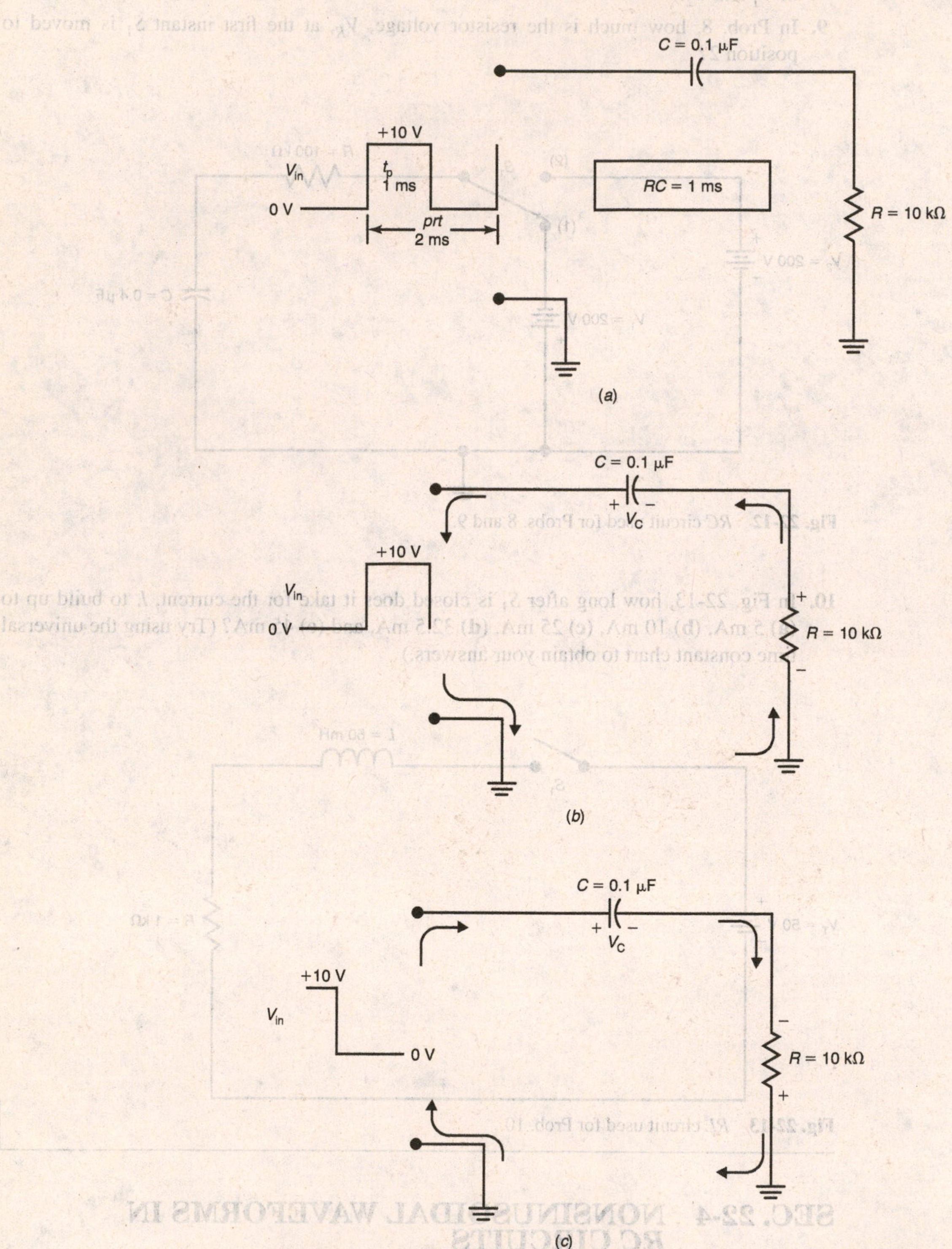

Fig. 22-14 *RC* circuit driven by an input square wave. (*a*) *RC* circuit driven by a 0- to +10-V square wave; *tp* = 1 ms, *prt* = 2 ms. (*b*) When V_{in} is positive, "*C*" charges. The polarity of V_R is positive with respect to ground because electron flow is up through *R*. (*c*) When V_{in} is 0 V, "*C*" discharges. The polarity of V_R is negative with respect to ground because electron flow is down through *R*.

The input waveform is a 0- to +10-V square wave (50% duty cycle) whose pulse time, *tp* is 1 ms and pulse repetition time, *prt*, is 2 ms. The average or DC value of the input waveform is +5 V, which is calculated as follows:

$$V\,\text{DC} = \frac{tp}{prt} \times V_{\text{pk}}$$

$$= \frac{1\ \text{ms}}{2\ \text{ms}} \times 10\ \text{V}$$

$$= +5\ \text{V}$$

The circuit waveforms will be analyzed beginning at the first instant the input voltage goes from 0 V to +10 V. Refer to the waveforms in Fig. 22-15. Before time, t_0, the input voltage is 0 V and the resulting resistor and capacitor voltages are also 0 V. At time t_0, however, V_{in} switches to +10 V and the capacitor begins charging. The pulse of +10 V lasts for 1 ms, until t_1, which is exactly one time constant. The voltage across C at time t_1 is calculated as

$$V_C = V_{\max}(1 - \epsilon^{-(t/RC)})$$

Since

$$\frac{-t}{RC} = -1$$

we have

$$V_C = 10\ \text{V}(1 - \epsilon^{-1})$$
$$= 10\ \text{V}(1 - 0.368)$$
$$= 10\ \text{V} \times 0.632$$
$$V_C = 6.32\ \text{V}$$

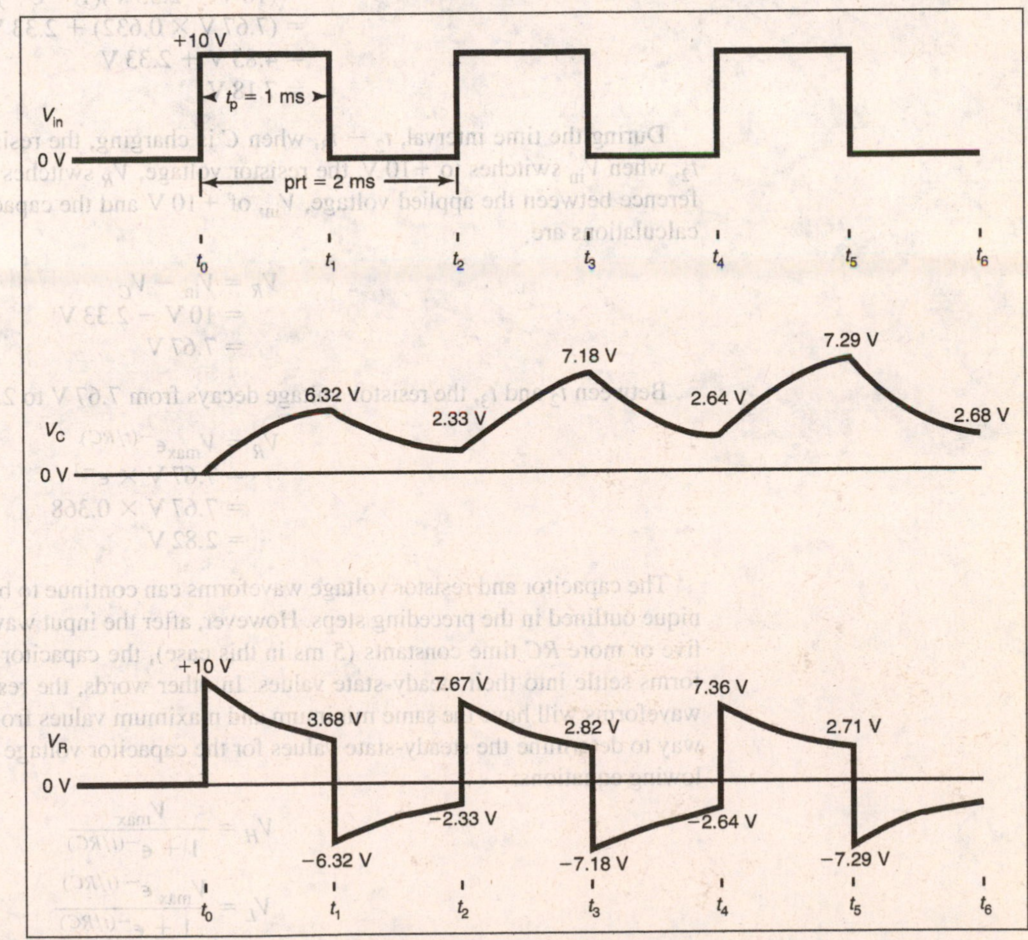

Fig. 22-15 Waveforms V_{in}, V_C, and V_R for the circuit in Fig. 22-14.

During the same time interval (t_0 to t_1), the resistor voltage, V_R, decays from +10 V. The calculations for V_R at t_1 are as follows:

$$V_R = V_{max}\epsilon^{-(t/RC)}$$
$$= 10\text{ V} \times \epsilon^{-1}$$
$$= 10\text{ V} \times 0.368$$
$$V_R = 3.68\text{ V}$$

Fig. 22-14*b* shows the direction of electron flow during the time that *C* is charging. Note that the polarity of V_R is positive with respect to ground.

At time t_1, V_{in} switches back to 0 V. This causes the capacitor to discharge because V_{in} serves as a short (ideally). The capacitor is allowed to discharge for 1 ms until time t_2. The value of V_C at t_2 equals

$$V_C = V_{max}\epsilon^{-(t/RC)}$$
$$= 6.32\text{ V} \times \epsilon^{-1}$$
$$= 6.32\text{ V} \times 0.368$$
$$= 2.33\text{ V}$$

Examine the resistor voltage waveform in Fig. 22-15 during the time interval, $t_1 - t_2$, when *C* is discharging. Notice that the resistor voltage, V_R, is the same as the capacitor voltage, V_C, during this time, except that the resistor voltage is negative. The reason that V_R is negative is due to the fact that when *C* is discharging, electron flow is downward through *R* as shown in Fig. 22-14*c*.

At time t_2, V_{in} switches back to +10 V and the capacitor begins charging again. Unlike at t_0, however, the capacitor begins charging from 2.33 V instead of 0 V. To calculate the voltage across *C* at t_3, 1 ms after charging, we proceed as follows:

$$V_C = (V_F - V_i)(1 - \epsilon^{-(t/RC)}) + V_i$$
$$= (10\text{ V} - 2.33\text{ V})(1 - \epsilon^{-1}) + 2.33\text{ V}$$
$$= (7.67\text{ V} \times 0.632) + 2.33\text{ V}$$
$$= 4.85\text{ V} + 2.33\text{ V}$$
$$= 7.18\text{ V}$$

During the time interval, $t_2 - t_3$, when *C* is charging, the resistor voltage is decaying. At t_2, when V_{in} switches to +10 V, the resistor voltage, V_R switches to 7.67 V, which is the difference between the applied voltage, V_{in}, of +10 V and the capacitor voltage of 2.33 V. The calculations are

$$V_R = V_{in} - V_C$$
$$= 10\text{ V} - 2.33\text{ V}$$
$$= 7.67\text{ V}$$

Between t_2 and t_3, the resistor voltage decays from 7.67 V to 2.82 V. This is calculated as

$$V_R = V_{max}\epsilon^{-(t/RC)}$$
$$= 7.67\text{ V} \times \epsilon^{-1}$$
$$= 7.67\text{ V} \times 0.368$$
$$= 2.82\text{ V}$$

The capacitor and resistor voltage waveforms can continue to be calculated using the technique outlined in the preceding steps. However, after the input waveform has been present for five or more *RC* time constants (5 ms in this case), the capacitor and resistor voltage waveforms settle into their steady-state values. In other words, the resistor and capacitor voltage waveforms will have the same minimum and maximum values from this point on. The easiest way to determine the steady-state values for the capacitor voltage waveform is to use the following equations:

$$V_H = \frac{V_{max}}{1 + \epsilon^{-(t/RC)}}$$

$$V_L = \frac{V_{max}\,\epsilon^{-(t/RC)}}{1 + \epsilon^{-(t/RC)}}$$

$$V_{p\text{-}p} = V_H - V_L$$

The value, V_H, represents the maximum steady-state capacitor voltage, whereas V_L represents the minimum steady-state capacitor voltage. $V_{P\text{-}P}$ represents the steady-state peak-to-peak capacitor voltage. In the equations for V_H and V_L, t corresponds with tp.

In Fig. 22-14, the steady-state values for V_H, V_L, and $V_{P\text{-}P}$ are calculated as follows:

$$\begin{aligned} V_H &= \frac{V_{max}}{1+\epsilon^{-(t/RC)}} \\ &= \frac{10\text{ V}}{1+\epsilon^{-1}} \\ &= \frac{10\text{ V}}{1.368} \\ &= 7.31\text{ V} \end{aligned}$$

$$\begin{aligned} V_L &= \frac{V_{max}\epsilon^{-(t/RC)}}{1+\epsilon^{-(t/RC)}} \\ &= \frac{10\text{ V}\times 0.368}{1.368} \\ &= 2.69\text{ V} \end{aligned}$$

$$\begin{aligned} V_{P\text{-}P} &= V_H - V_L \\ &= 7.31\text{ V} - 2.69\text{ V} \\ &= 4.62\text{ V} \end{aligned}$$

The steady-state waveforms for V_C and V_R are shown in Fig. 22-16. Notice that when C is charging, $V_R = V_{in} - V_C$. Also, when C is discharging, $V_R = -V_C$. One more point: The average or DC value of the capacitor waveform is +5 V, the same as the DC value of the input voltage. In any RC circuit, the capacitor will always charge to the average value of the input waveform. To prove that the average value of V_C is +5 V, just divide $V_{P\text{-}P}$ by 2 and add this value to V_L. The calculations are

$$\begin{aligned} V_{C(DC)} &= V_L + \frac{V_{P\text{-}P}}{2} \\ &= 2.69\text{ V} + \frac{4.62\text{ V}}{2} \\ &= 2.69\text{ V} + 2.31\text{ V} \\ &= 5\text{ V} \end{aligned}$$

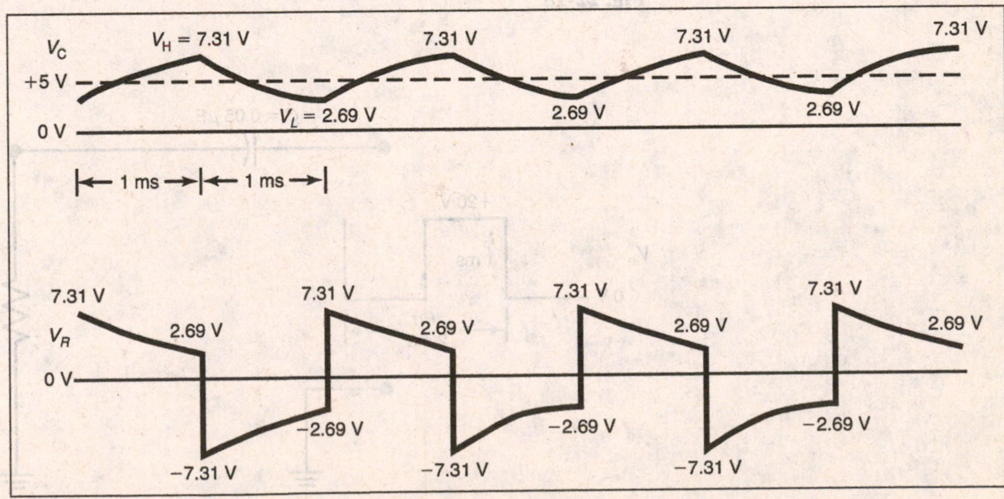

Fig. 22-16 Steady-state values for V_C and V_R in Fig. 22-14.

Sec. 22-4

In Figs. 22-17 through 22-26, draw the steady-state voltage waveforms for V_C and V_R. Be sure to include the values V_H, V_L, $V_{P\text{-}P}$, and $V_{C(DC)}$ on the V_C waveform. On the V_R waveform, be sure to include the voltage values at the beginning and end of both the charge and discharge times. Be sure that both the V_C and V_R waveforms are drawn with the proper time relationship with respect to the input voltage, V_{in}.

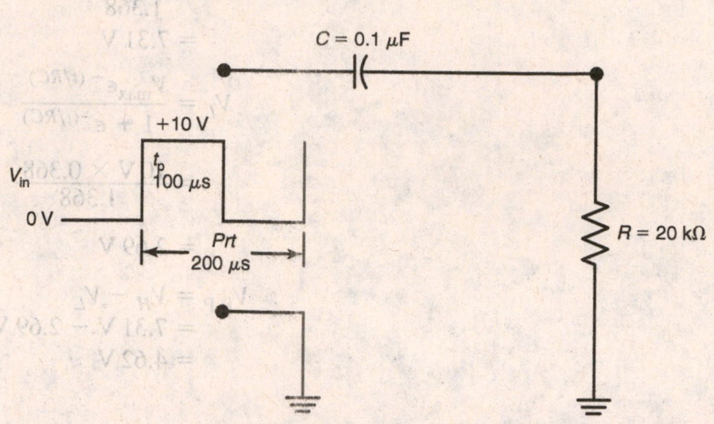

C = 0.1 μF
+10 V
Vin
tp
100 μs
0 V
Prt
200 μs
R = 20 kΩ

Fig. 22-17

C = 0.15 μF
+5 V
Vin
tp
50 μs
0 V
Prt
100 μs
R = 1.2 kΩ

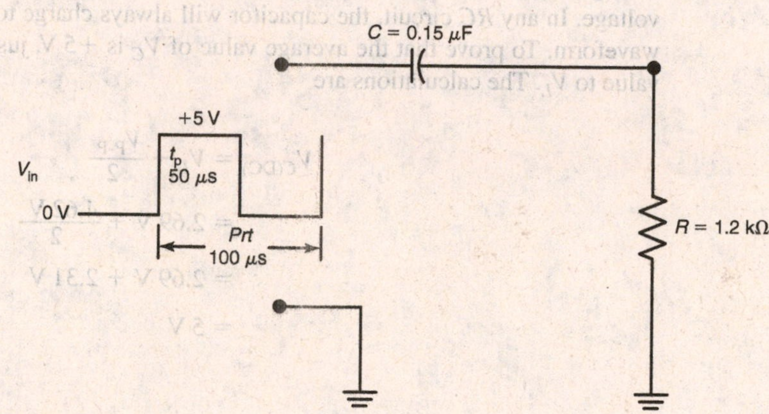

Fig. 22-18

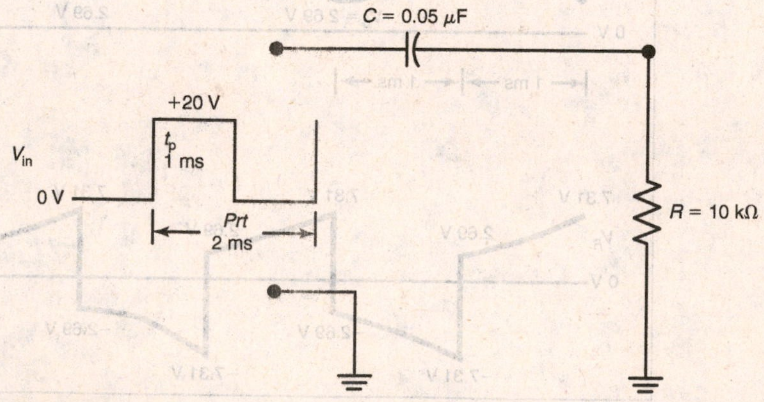

C = 0.05 μF
+20 V
Vin
tp
1 ms
0 V
Prt
2 ms
R = 10 kΩ

Fig. 22-19

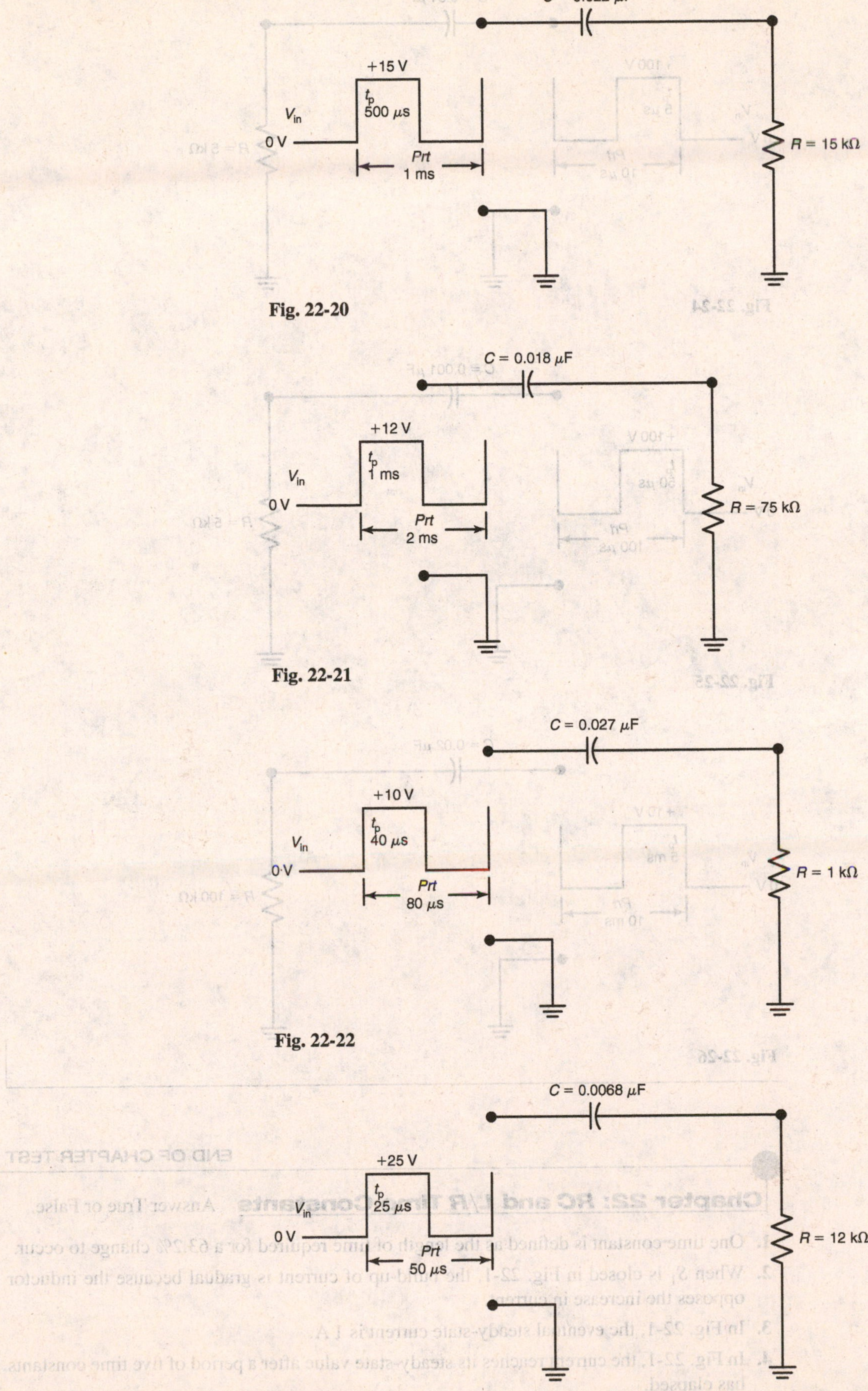

Fig. 22-20

Fig. 22-21

Fig. 22-22

Fig. 22-23

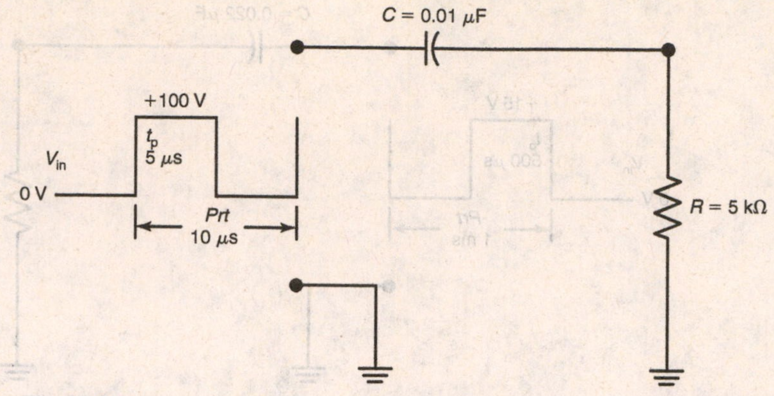

Fig. 22-24

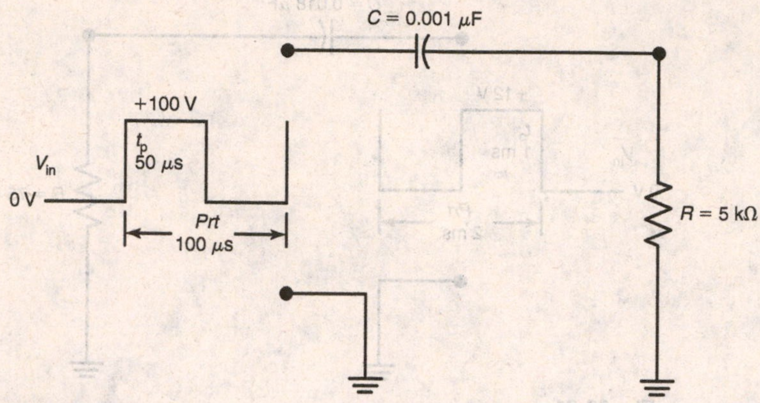

Fig. 22-25

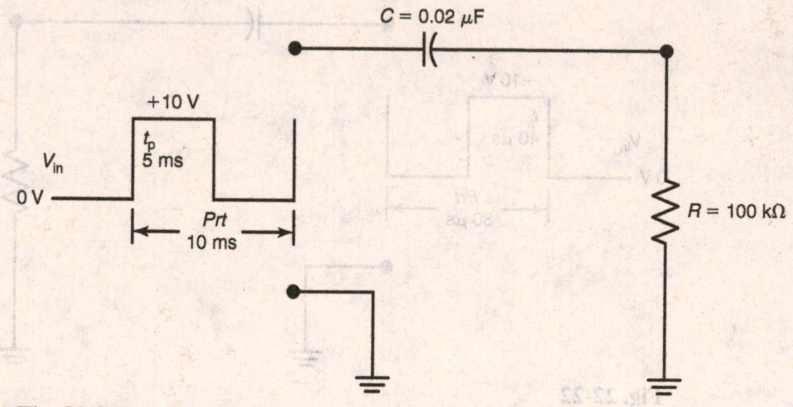

Fig. 22-26

END OF CHAPTER TEST

Chapter 22: *RC* and *L/R* Time Constants

Answer True or False.

1. One time constant is defined as the length of time required for a 63.2% change to occur.
2. When S_1 is closed in Fig. 22-1, the build-up of current is gradual because the inductor opposes the increase in current.
3. In Fig. 22-1, the eventual steady-state current is 1 A.
4. In Fig. 22-1, the current reaches its steady-state value after a period of five time constants has elapsed.

5. In Fig. 22-1, opening the switch, S_1, results in a high value of induced voltage across the inductor because the high resistance of the open switch results in an extremely short time constant.
6. In Fig. 22-1, the time constant could be increased by either increasing the resistance or decreasing the inductance.
7. In Fig. 22-5, the capacitor voltage is 0 V at the first instant S_1 is moved to position 1.
8. In Fig. 22-5, the resistor voltage equals 100 V if S_1 has been in position 1 for five or more time constants.
9. In Fig. 22-5, the resistor voltage is 100 V at the first instant S_1 is placed in position 1.
10. After S_1 is moved to position 2 in Fig. 22-5, it takes approximately 0.7 time constants for the capacitor to discharge to 50 V.
11. With S_1 in position 1 in Fig. 22-5, the eventual steady-state current is 100 μA.
12. To increase the time constant in Fig. 22-5, increase either the resistance or capacitance value.
13. With S_1 in position 2 in Fig. 22-5, the resistor and capacitor voltages are the same.
14. The universal time constant chart cannot be used to solve for the capacitor and resistor voltages in an *RC* circuit if the capacitor begins charging from an initial voltage other than zero.
15. When a nonsinusoidal waveform is applied to an *RC* circuit, the relationship between the pulse time, t_p, of the applied square wave and the *RC* time constant affects the waveshape of the resistor and capacitor voltage waveforms.

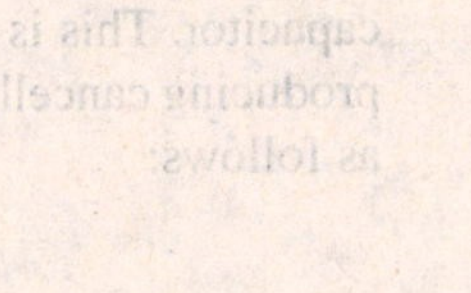

CHAPTER

AC Circuits

When inductive reactance, X_L, and capacitive reactance, X_C, are combined in the same circuit, the opposite phase angles enable one to cancel the effect of the other. For X_L and X_C in series, the net reactance is the difference between the two reactance values. In parallel circuits, where I_L and I_C have opposite phase angles, the net line current is the difference between the two branch currents.

SEC. 23-1 COMBINING X_L AND X_C

Solved Problem

For Fig. 23-1*a*, solve for X_T, I_T, V_L, V_C, and θ.

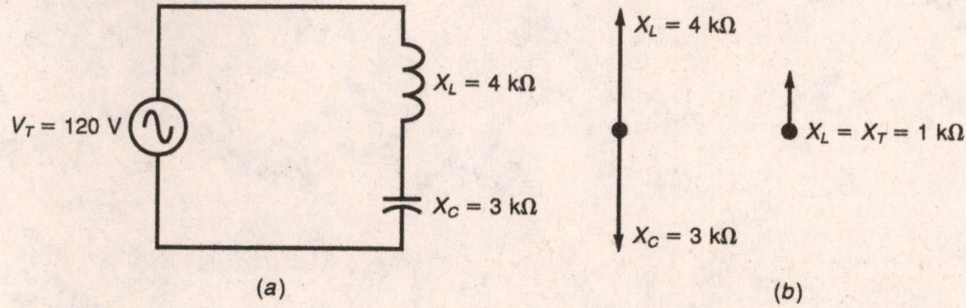

Fig. 23-1 When X_L and X_C are in series, their ohms of reactance cancel. (*a*) Series *LC* circuit. (*b*) Phasors for X_L and X_C with net resultant.

Answer

The phasor diagram of reactances for Fig. 23-1*a* is shown in Fig. 23-1*b*. Since X_L and X_C are 180° out of phase with each other, the net reactance is $X_L - X_C$. For Fig. 23-1*a*, we have

$$X_T = X_L - X_C$$
$$= 4\text{ k}\Omega - 3\text{ k}\Omega$$
$$X_T = 1\text{ k}\Omega$$

Since X_L is larger than X_C, the net reactance is inductive. Notice that the total opposition to the flow of current is, in fact, smaller than the reactance of either the inductor or the capacitor. This is the result of the reactances being 180° out of phase and, therefore, producing cancellation of some portion of reactance. To find I_T, V_L, and V_C, we proceed as follows:

$$I_T = \frac{V_T}{X_T}$$
$$= \frac{120\text{ V}}{1\text{ k}\Omega}$$
$$I_T = 120\text{ mA}$$

$$V_L = I_T \times X_L$$
$$= 120\text{ mA} \times 4\text{ k}\Omega$$
$$V_L = 480\text{ V}$$

$$V_C = I_T \times X_C$$
$$= 120\text{ mA} \times 3\text{ k}\Omega$$
$$V_C = 360\text{ V}$$

Notice that the individual voltage drops are larger than the applied voltage. The phasor sum of the two voltages, however, still equals 120 V, which is the value of V_T. The voltages V_L and V_C are 180° out of phase with each other, which means that they are always, at any instant in time, going to have series-opposing voltage drops. For Fig. 23-1*a*,

$$V_T = V_L - V_C$$
$$= 480\text{ V} - 360\text{ V}$$
$$V_T = 120\text{ V}$$

The circuit of Fig. 23-1*a* can be replaced with its equivalent circuit in Fig. 23-2*a*.

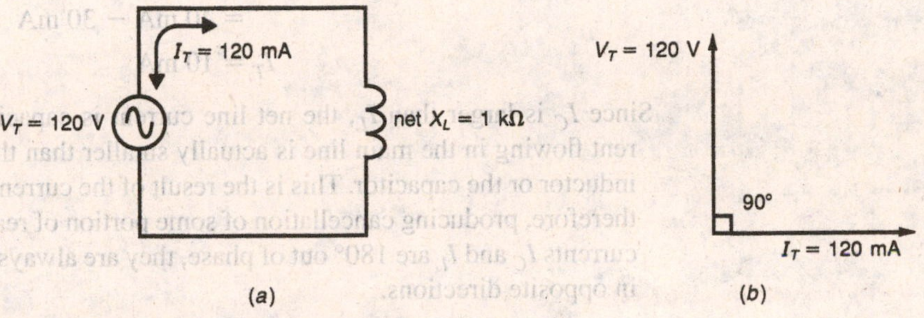

Fig. 23-2 (*a*) Equivalent circuit for Fig. 23-1*a* with net reactance $X_L = 1\text{ k}\Omega$. (*b*) Phasor diagram between V_T and I_T.

Since the circuit in Fig. 23-1*a* can be represented as a purely inductive circuit in Fig. 23-2*a*, the phase angle θ between applied voltage, V_T, and total current, I_T, is 90°, just as it would be for a single inductor connected across the terminals of the generator. The phase relationship between V_T and I_T is shown in Fig. 23-2*b*. Since Fig. 23-1*a* is a series-type circuit, we use the total current, I_T, as the reference phasor.

Solved Problem

For Fig. 23-3*a*, solve for I_L, I_C, I_T, X_T, and θ.

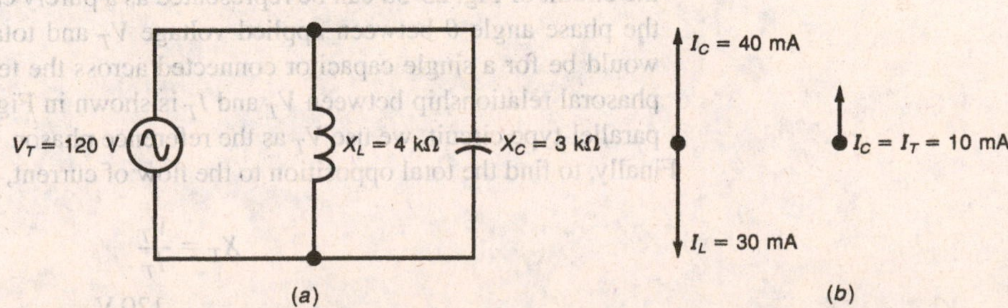

Fig. 23-3 When X_L and X_C are in parallel, their branch currents cancel. (*a*) Parallel *LC* circuit. (*b*) Phasors for I_L and I_C with net resultant.

Answer

The individual branch currents for Fig. 23-3 are found as follows:

$$I_L = \frac{V_T}{X_L}$$
$$= \frac{120\text{ V}}{4\text{ k}\Omega}$$
$$I_L = 30\text{ mA}$$
$$I_C = \frac{V_T}{X_C}$$
$$= \frac{120\text{ V}}{3\text{ k}\Omega}$$
$$I_C = 40\text{ mA}$$

The phasor diagram of branch currents for Fig. 23-3*a* is shown in Fig. 23-3*b*. Since I_C and I_L are 180° out of phase with each other, the net line current is $I_C - I_L$. For Fig. 23-3*a*, we have

$$I_T = I_C - I_L$$
$$= 40\text{ mA} - 30\text{ mA}$$
$$I_T = 10\text{ mA}$$

Since I_C is larger than I_L, the net line current is capacitive. Notice that the total current flowing in the main line is actually smaller than the branch currents of either the inductor or the capacitor. This is the result of the currents being 180° out of phase and, therefore, producing cancellation of some portion of reactive current. Since the branch currents I_C and I_L are 180° out of phase, they are always, at any instant in time, flowing in opposite directions.

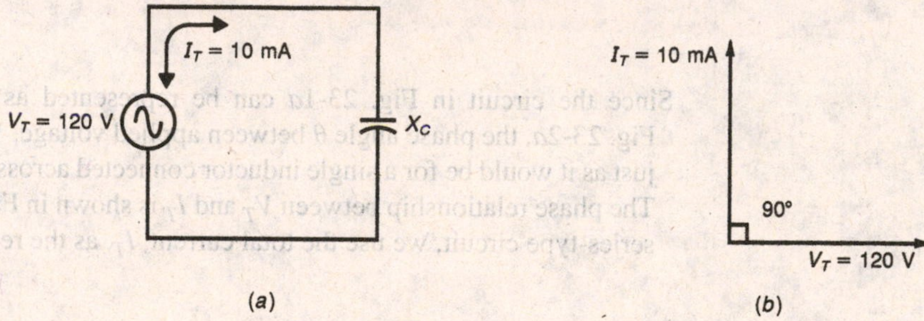

Fig. 23-4 (*a*) Equivalent circuit for Fig. 23-3*a*. (*b*) Phasor diagram between V_T and I_T.

The circuit of Fig. 23-3*a* can be replaced with its equivalent circuit in Fig. 23-4*a*. Since the circuit of Fig. 23-3*a* can be represented as a purely capacitive circuit in Fig. 23-4*a*, the phase angle θ between applied voltage V_T and total current I_T is 90°, just as it would be for a single capacitor connected across the terminals of the generator. The phasoral relationship between V_T and I_T is shown in Fig. 23-4*b*. Since Fig. 23-3*a* is a parallel-type circuit, we use V_T as the reference phasor.

Finally, to find the total opposition to the flow of current, we have

$$X_T = \frac{V_T}{I_T}$$
$$= \frac{120\text{ V}}{10\text{ mA}}$$
$$X_T = 12\text{ k}\Omega$$

Since I_T leads V_T, the total reactance is capacitive. Notice that X_T is larger than the value of either branch reactance. This is the result of the canceling of currents.

PRACTICE PROBLEMS

Sec. 23-1 For Figs. 23-5 through 23-14, solve for the unknowns listed.

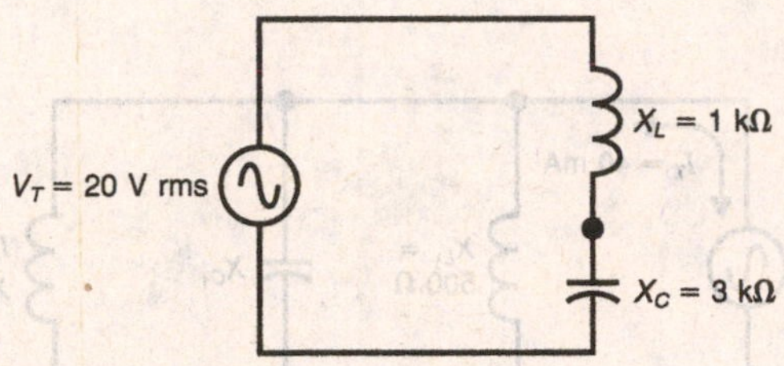

Fig. 23-5 Solve for X_T, I_T, V_L, V_C, and θ.

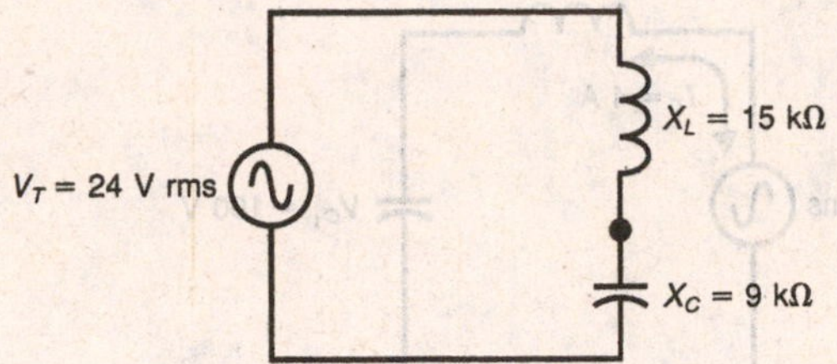

Fig. 23-6 Solve for X_T, I_T, V_L, V_C, and θ.

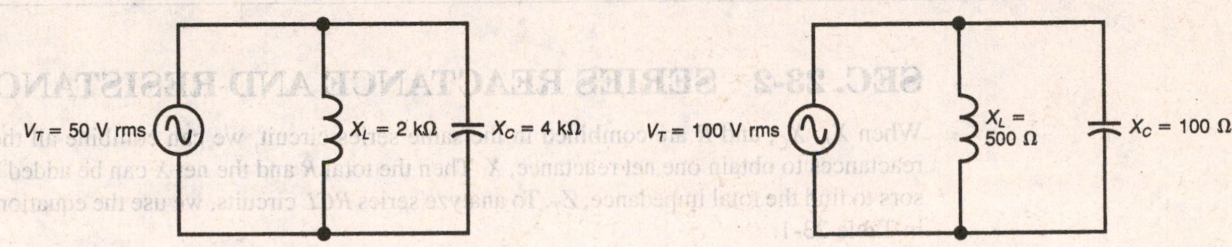

Fig. 23-7 Solve for I_L, I_C, I_T, X_T, and θ.

Fig. 23-8 Solve for I_L, I_C, I_T, X_T, and θ.

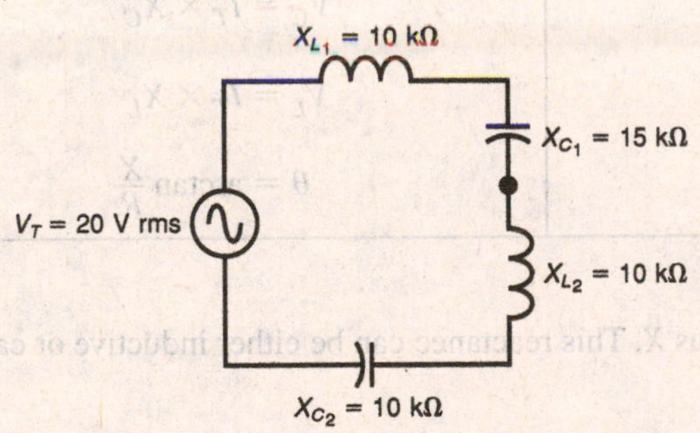

Fig. 23-9 Solve for X_T, I_T, V_{C_1}, V_{C_2}, V_{L_1}, V_{L_2}, and θ.

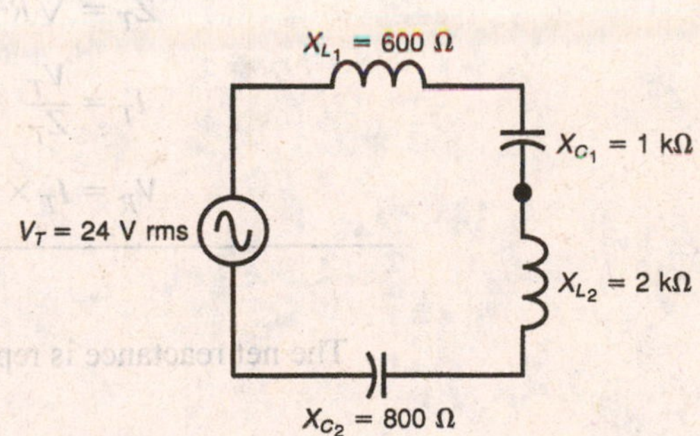

Fig. 23-10 Solve for X_T, I_T, V_{C_1}, V_{C_2}, V_{L_1}, V_{L_2}, and θ.

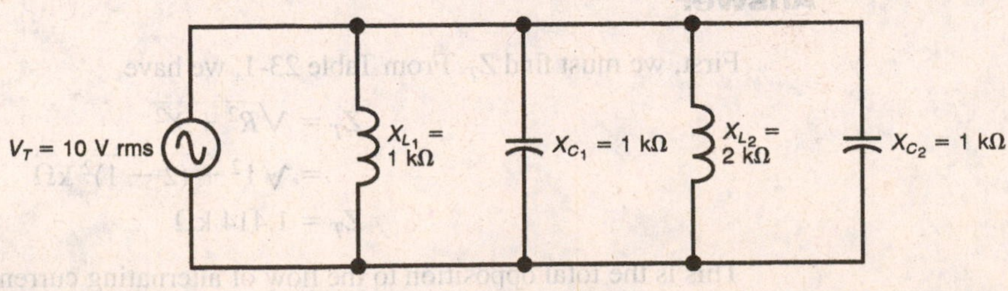

Fig. 23-11 Solve for I_{C_1}, I_{C_2}, I_{L_1}, I_{L_2}, I_T, X_T, and θ.

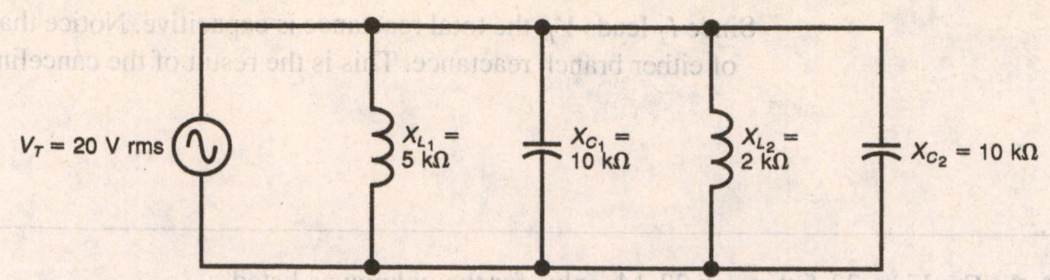

Fig. 23-12 Solve for I_{C_1}, I_{C_2}, I_{L_1}, I_{L_2}, I_T, X_T, and θ.

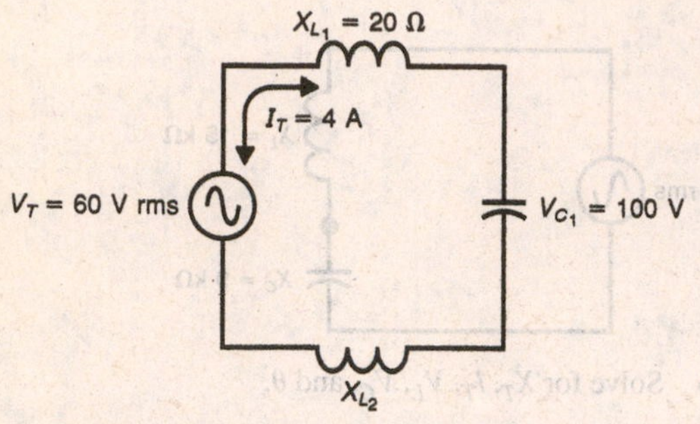

Fig. 23-13 Solve for V_{L_1}, X_{C_1}, X_{L_2}, V_{L_2}, X_T, and θ.

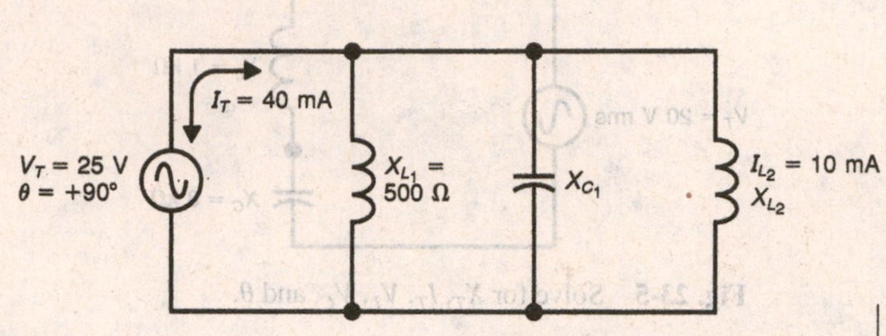

Fig. 23-14 Solve for I_{L_1}, I_{C_1}, X_{C_1}, X_{L_2}, and X_T.

SEC. 23-2 SERIES REACTANCE AND RESISTANCE

When X_L, X_C, and R are combined in the same series circuit, we can combine all the series reactances to obtain one net reactance, X. Then the total R and the net X can be added by phasors to find the total impedance, Z_T. To analyze series RCL circuits, we use the equations listed in Table 23-1.

TABLE 23-1 FORMULAS USED FOR SERIES *RCL* CIRCUITS

$Z_T = \sqrt{R^2 + X^2}$	$V_C = I_T \times X_C$
$I_T = \dfrac{V_T}{Z_T}$	$V_L = I_T \times X_L$
$V_R = I_T \times R$	$\theta = \arctan \dfrac{X}{R}$

The net reactance is represented as X. This reactance can be either inductive or capacitive.

Solved Problem

For Fig. 23-15, find Z_T, I_T, V_L, V_C, V_R, and θ_V. Also draw the phasor voltage triangle.

Answer

First, we must find Z_T. From Table 23-1, we have

$$Z_T = \sqrt{R^2 + X^2}$$
$$= \sqrt{1^2 + (2 - 1)^2}\ \text{k}\Omega$$
$$Z_T = 1.414\ \text{k}\Omega$$

This is the total opposition to the flow of alternating current. The circuit in Fig. 23-15*b* shows the equivalent circuit with one net reactance.

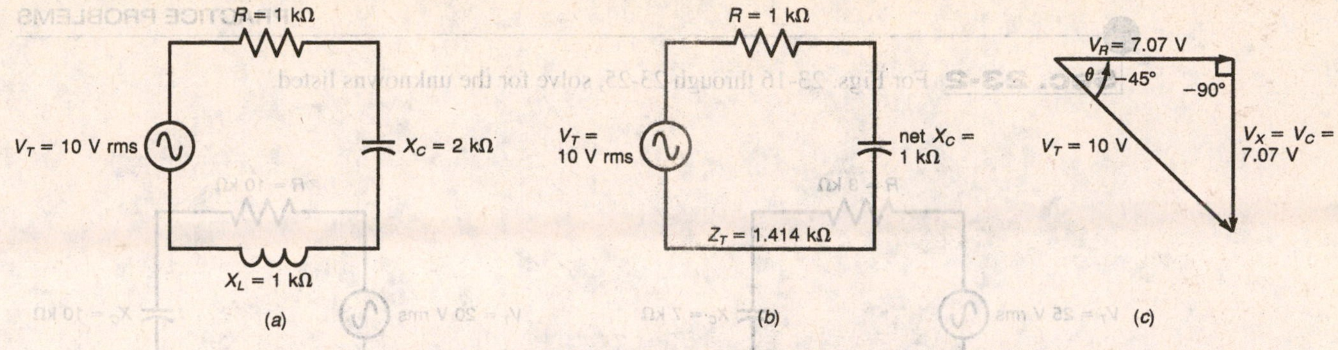

Fig. 23-15 Calculating impedance Z in a series *RCL* circuit. (*a*) Series *RCL* circuit. (*b*) Equivalent circuit with one net reactance. (*c*) Phasor diagram for equivalent circuit.

With Z_T known, I_T can be found. We proceed as follows:

$$I_T = \frac{V_T}{Z_T}$$
$$= \frac{10\text{ V}}{1.414\text{ k}\Omega}$$
$$I_T = 7.07\text{ mA}$$

Knowing I_T allows us to find V_C, V_L, and V_R.

$$V_L = I_T \times X_L$$
$$= 7.07\text{ mA} \times 1\text{ k}\Omega$$
$$V_L = 7.07\text{ V}$$

$$V_C = I_T \times X_C$$
$$= 7.07\text{ mA} \times 2\text{ k}\Omega$$
$$V_C = 14.14\text{ V}$$

$$V_R = I_T \times R$$
$$= 7.07\text{ mA} \times 1\text{ k}\Omega$$
$$V_R = 7.07\text{ V}$$

Since V_C and V_L are 180° out of phase, the net reactive voltage is 14.14 V − 7.07 V = 7.07 V. Since V_C is greater than V_L, the net reactive voltage is capacitive.

To find the phase angle θ_V for the voltage triangle, we have

$$\tan \theta_V = \frac{V_X}{V_R}$$
$$= -\frac{V_C}{V_R}$$
$$= -\frac{7.07\text{ V}}{7.07\text{ V}}$$
$$\tan \theta_V = -1$$

To find the angle corresponding to this tangent value, we take the arctan of −1.

$$\theta = \arctan -1$$
$$\theta = -45°$$

This phase angle tells us that the total voltage, V_T lags the total current, I_T, by 45°. The phasor voltage triangle is shown in Fig. 23-15*c*.

Sec. 23-2 For Figs. 23-16 through 23-25, solve for the unknowns listed.

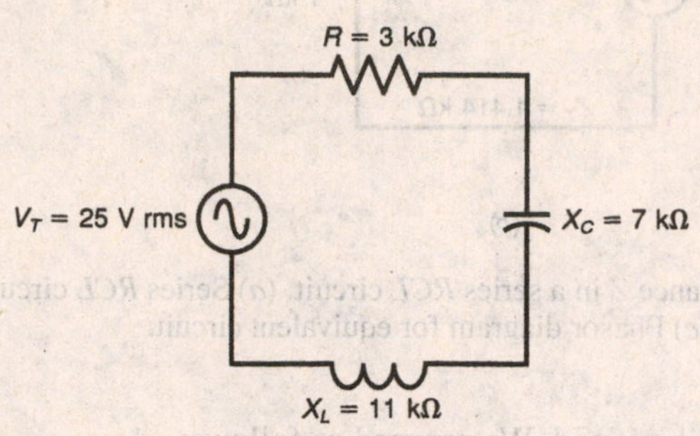

Fig. 23-16 Solve for Z_T, I_T, V_R, V_C, V_L, and θ_V.

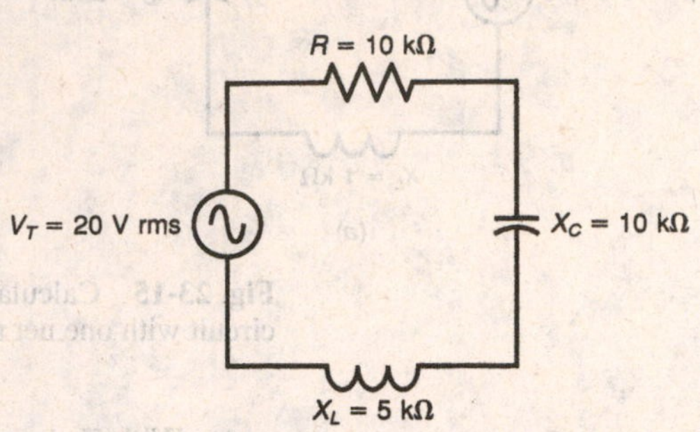

Fig. 23-17 Solve for Z_T, I_T, V_R, V_C, V_L, and θ_V.

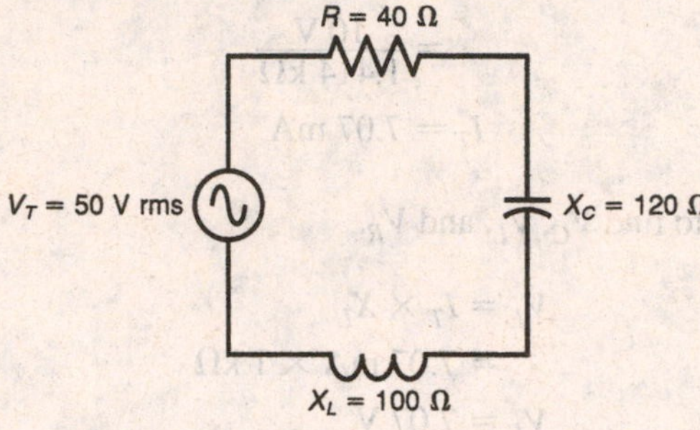

Fig. 23-18 Solve for Z_T, I_T, V_R, V_C, V_L, and θ_V.

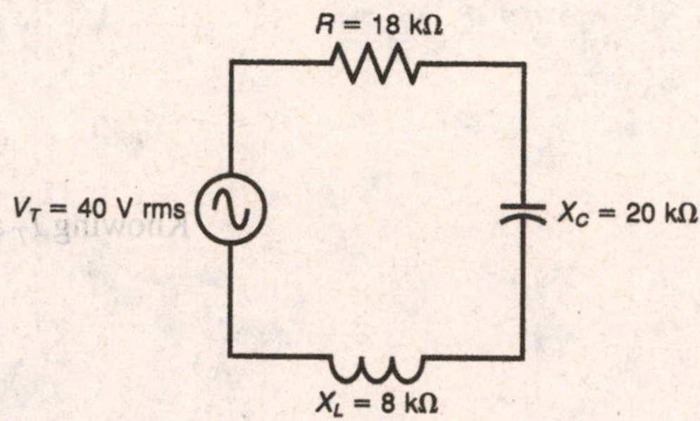

Fig. 23-19 Solve for Z_T, I_T, V_R, V_C, V_L, and θ_V.

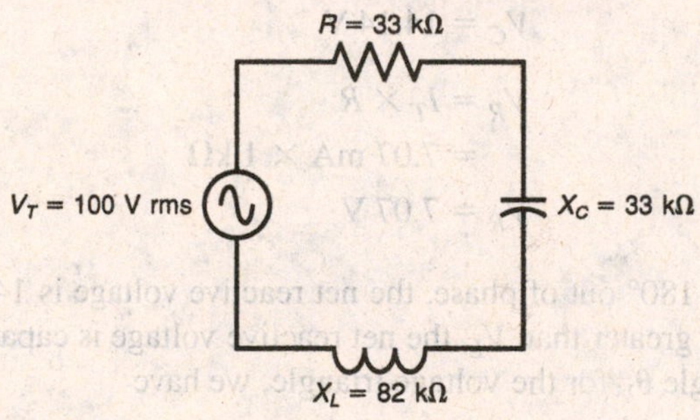

Fig. 23-20 Solve for Z_T, I_T, V_R, V_C, V_L, and θ_V.

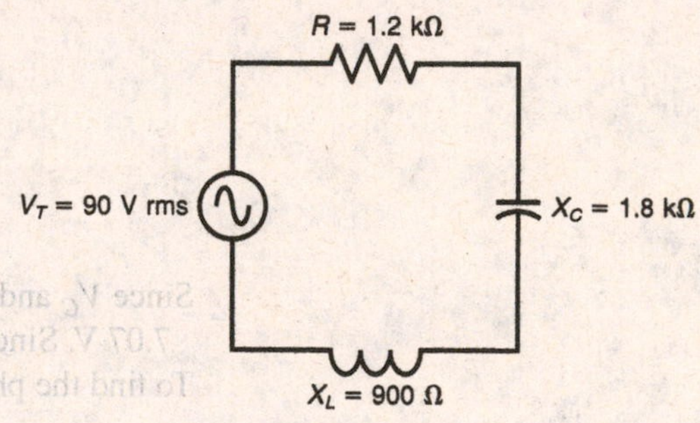

Fig. 23-21 Solve for Z_T, I_T, V_R, V_C, V_L, and θ_V.

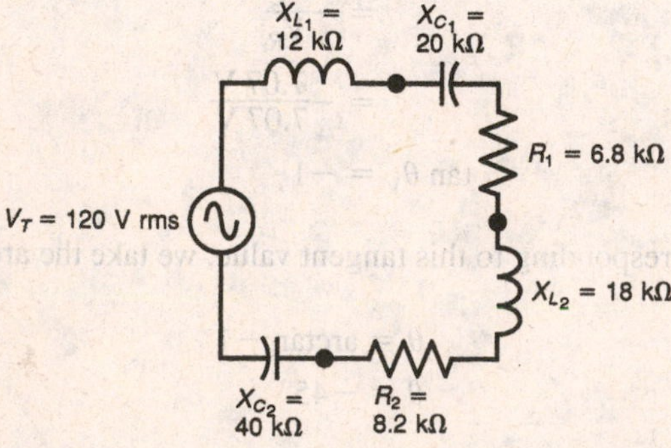

Fig. 23-22 Solve for Z_T, I_T, V_{C_1}, V_{C_2}, V_{L_1}, V_{L_2}, V_{R_1}, V_{R_2}, and θ_V.

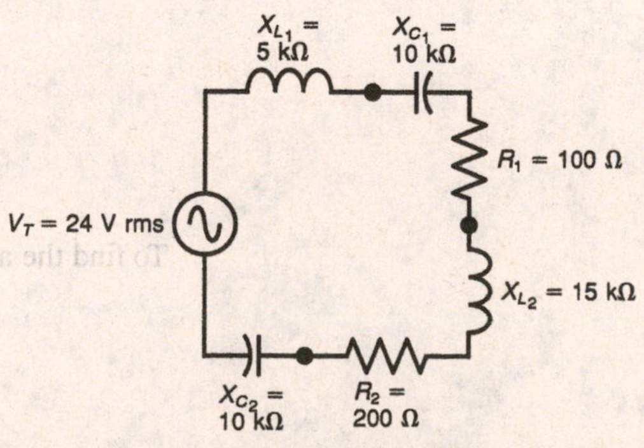

Fig. 23-23 Solve for Z_T, I_T, V_{C_1}, V_{C_2}, V_{L_1}, V_{L_2}, V_{R_1}, V_{R_2}, θ_V.

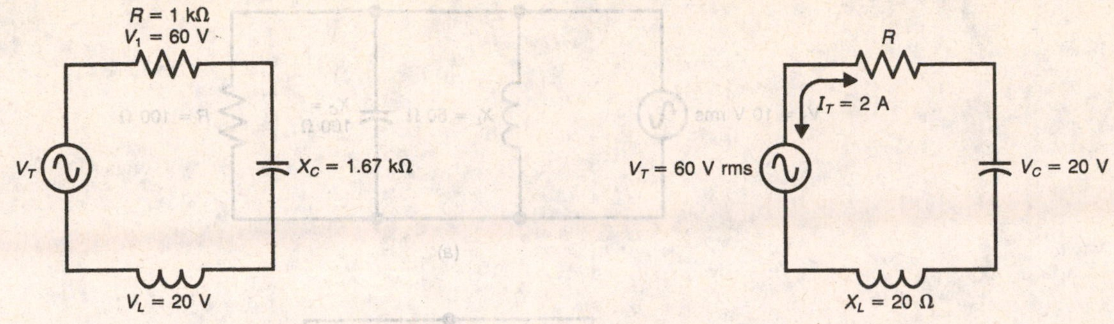

Fig. 23-24 Solve for Z_T, I_T, V_T, V_C, X_L, and θ.

Fig. 23-25 Solve for Z_T, V_R, X_C, V_L, θ, and R.

SEC. 23-3 PARALLEL REACTANCE AND RESISTANCE

When X_L, X_C, and R are combined in the same parallel circuit, we can combine all the reactive branch currents to obtain one net current, I_X. Then the resistive current, I_R, and the net reactive current, I_X, can be added by phasors to obtain I_T. To analyze parallel *RCL* circuits, we use the equations listed in Table 23-2.

TABLE 23-2 FORMULAS USED FOR PARALLEL *RCL* CIRCUITS

$I_T = \sqrt{I_R^2 + I_X^2}$	$I_C = \dfrac{V_T}{X_C}$
$Z_{EQ} = \dfrac{V_T}{I_T}$	$V_T = I_T \times Z_{EQ}$
$I_R = \dfrac{V_T}{R}$	$\theta = \arctan \dfrac{I_X}{I_R}$
$I_L = \dfrac{V_T}{X_L}$	

The net reactive current is represented by I_X. This net reactive current can be inductive or capacitive.

Solved Problem

For Fig. 23-26, on the next page, find I_R, I_C, I_L, I_T, Z_{EQ}, and θ. Also, draw the phasor current triangle.

Answer

First, we must find I_R, I_L, and I_C. From Table 23-2, we have

$$I_R = \frac{V_T}{R}$$
$$= \frac{10\text{ V}}{100\ \Omega}$$
$$I_R = 100\text{ mA}$$
$$I_L = \frac{V_T}{X_L}$$
$$= \frac{10\text{ V}}{50\ \Omega}$$
$$I_L = 200\text{ mA}$$

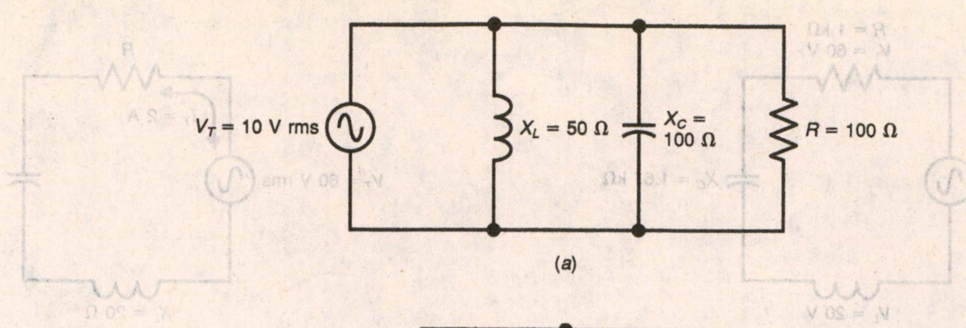

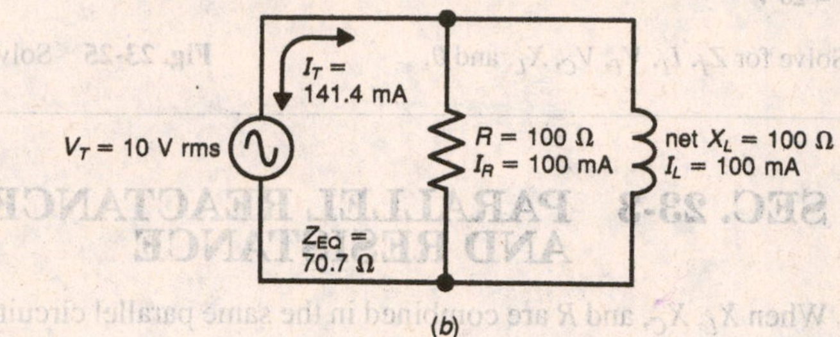

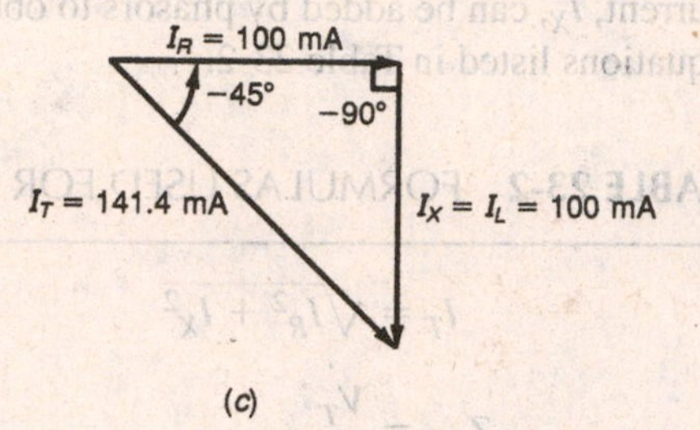

Fig. 23-26 Calculating total current in a parallel *RCL* circuit. (*a*) Parallel *RCL* circuit. (*b*) Equivalent circuit with one net reactance. (*c*) Phasor diagram for equivalent circuit.

$$I_C = \frac{V_T}{X_C}$$

$$= \frac{10 \text{ V}}{100 \ \Omega}$$

$$I_C = 100 \text{ mA}$$

To find I_T, we proceed as follows:

$$I_T = \sqrt{I_R{}^2 + I_X{}^2}$$

$$= \sqrt{100^2 + (200 - 100)^2} \text{ mA}$$

$$I_T = 141.4 \text{ mA}$$

This is the total current flowing to and from the terminals of the voltage source. With I_T known, Z_{EQ} can be found.

$$Z_{EQ} = \frac{V_T}{I_T}$$

$$= \frac{10 \text{ V}}{141.14 \text{ mA}}$$

$$Z_{EQ} = 70.7 \ \Omega$$

To find the phase angle θ_I for the current triangle, we have

$$\tan \theta_I = \frac{I_X}{I_R}$$

Since I_L is greater than I_C, the net reactive current is inductive. We then have

$$\tan\theta_I = -\frac{I_L}{I_R}$$
$$= -\frac{100\text{ mA}}{100\text{ mA}}$$
$$\tan\theta_I = -1$$

To find the angle corresponding to this tangent value, we take the arctan of −1.

$$\theta = \arctan -1$$
$$\theta = -45°$$

This phase angle tells us that the total voltage, V_T, leads total current, I_T, by 45°. The phasor current triangle is shown in Fig. 23-26*c*.

PRACTICE PROBLEMS

Sec. 23-3 For Figs. 23-27 through 23-36, solve for the unknowns listed.

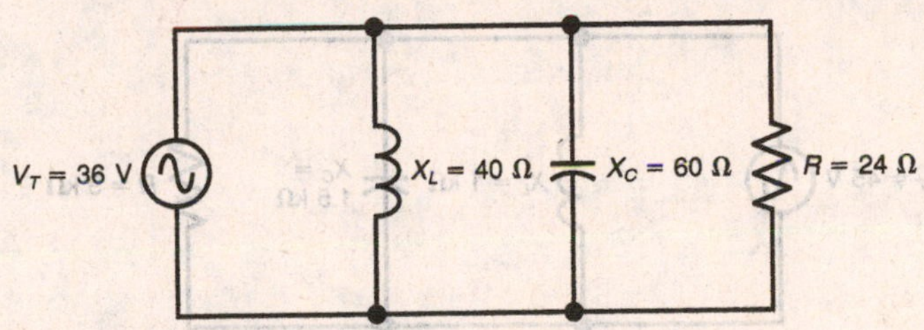

Fig. 23-27 Solve for I_L, I_C, I_R, I_T, Z_{EQ}, and θ.

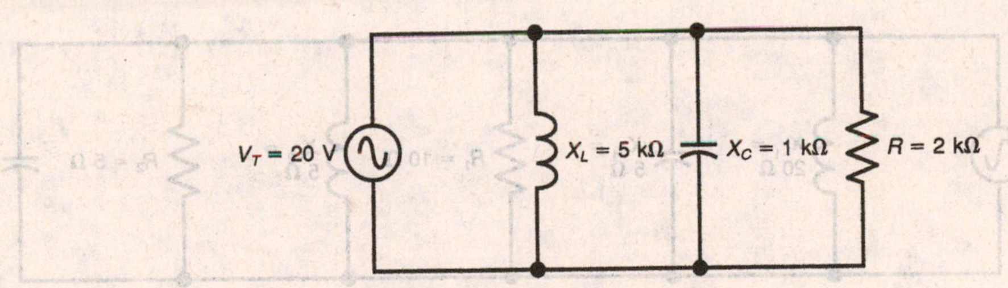

Fig. 23-28 Solve for I_L, I_C, I_R, I_T, Z_{EQ}, and θ.

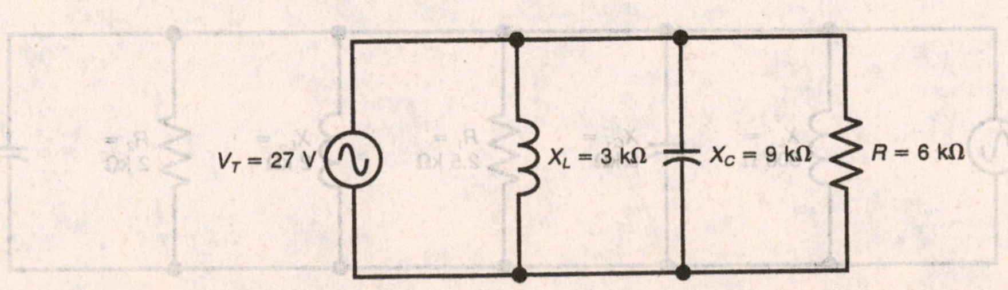

Fig. 23-29 Solve for I_L, I_C, I_R, I_T, Z_{EQ}, and θ.

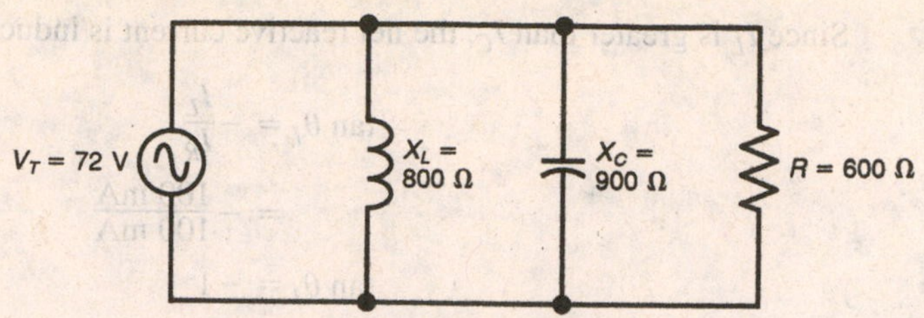

Fig. 23-30 Solve for I_L, I_C, I_R, I_T, Z_{EQ}, and θ.

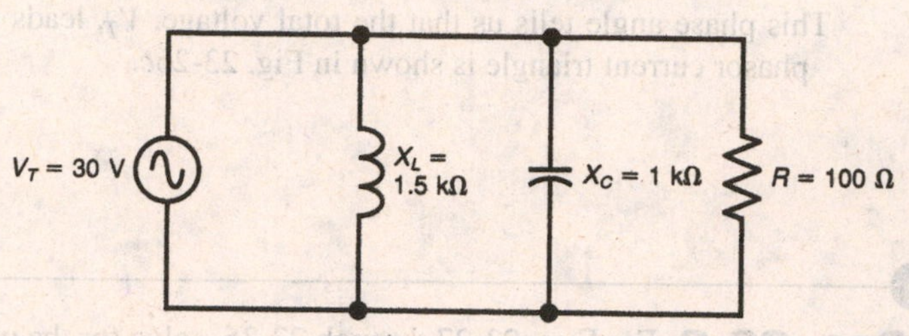

Fig. 23-31 Solve for I_L, I_C, I_R, I_T, Z_{EQ}, and θ.

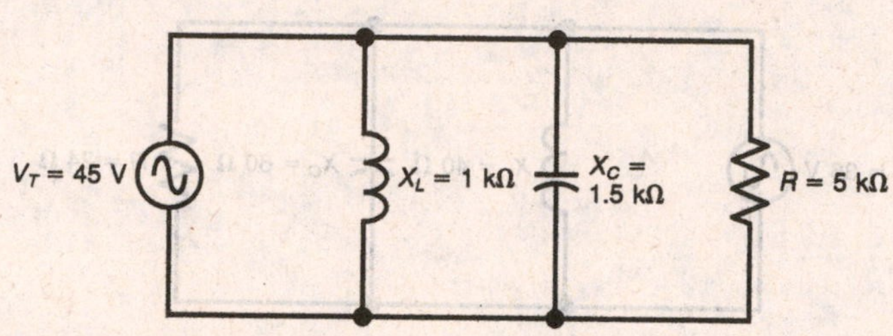

Fig. 23-32 Solve for I_L, I_C, I_R, I_T, Z_{EQ}, and θ.

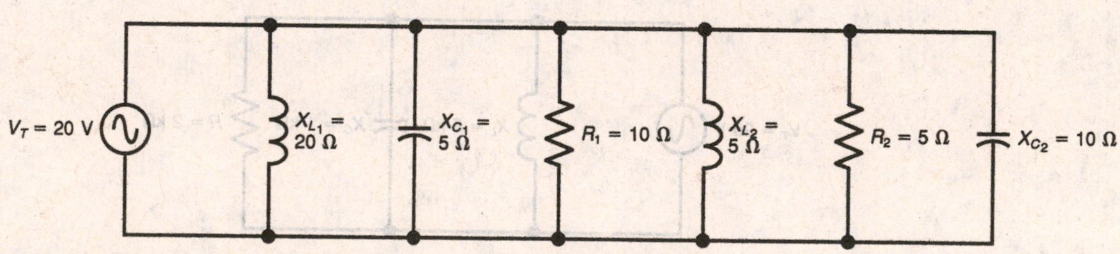

Fig. 23-33 Solve for I_{L_1}, I_{L_2}, I_{C_1}, I_{C_2}, I_{R_1}, I_{R_2}, I_T, Z_{EQ}, and θ.

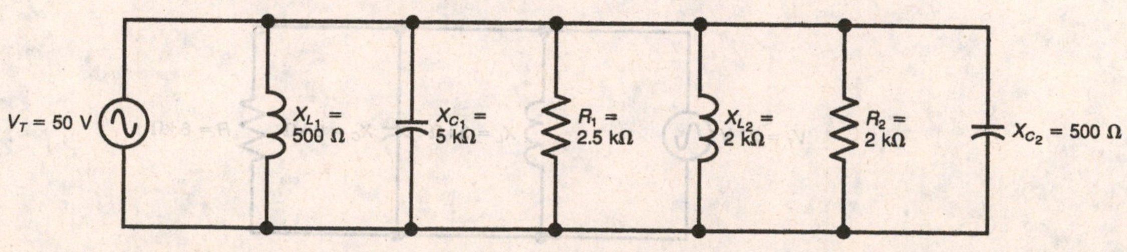

Fig. 23-34 Solve for I_{L_1}, I_{L_2}, I_{C_1}, I_{C_2}, I_{R_1}, I_{R_2}, I_T, Z_{EQ}, and θ.

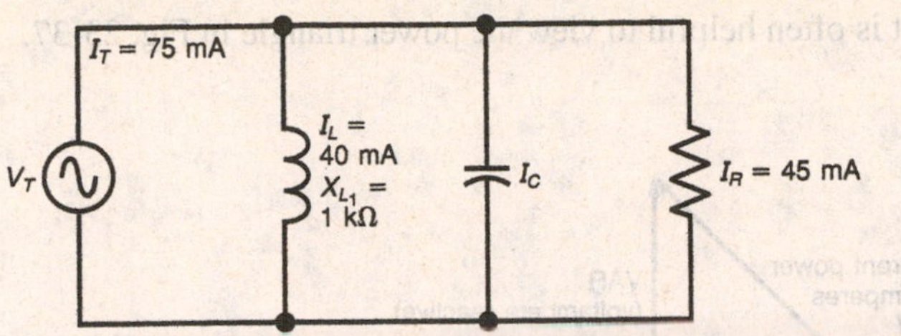

Fig. 23-35 Solve for V_T, Z_{EQ}, I_C, X_C, θ, and R.

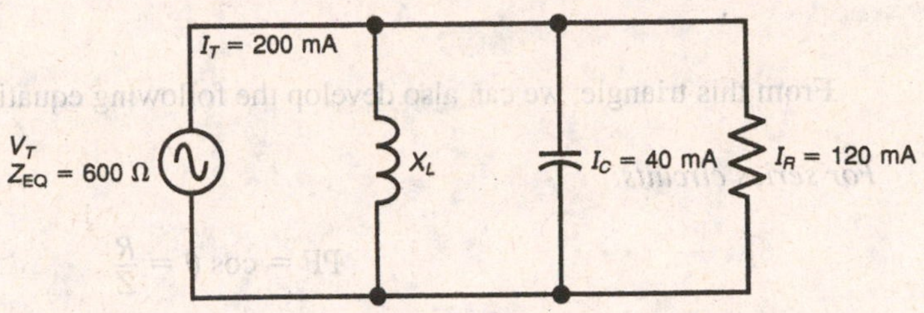

Fig. 23-36 Solve for V_T, X_L, I_L, X_C, R, and θ.

SEC. 23-4 POWER IN AC CIRCUITS

Resistors dissipate power in AC circuits just as they do in DC circuits, since the voltage and current are always in phase for resistive components. Reactive components (inductors and capacitors) will ideally consume no power from the generator. This is because the reactive components return as much power as they take from the generator for each cycle of applied voltage. The power is returned in the form of a discharge current for a capacitor or a collapsing magnetic field for an inductor.

For an *RCL* circuit—whether it be series, parallel, or series-parallel—the real power consumed by the circuit can always be calculated as I^2R.

$$\text{Real power} = I^2R$$

Since the reactive components dissipate no power (ideally), the resistor is the only component that will consume power. Real power is often referred to as true power. The unit of real or true power is the watt.

The product of voltage and current for a reactive component is called *voltampere reactive,* abbreviated VAR. Specifically, VARs are voltamperes at 90°.

The product of total voltage, V_T, and total current, I_T, for an AC circuit is called *apparent power*. Since voltage and current can be out of phase, the apparent power and real power can be different. The unit of apparent power is the voltampere (VA) instead of watts, since the voltampere product is not an indication of the actual power consumed by the AC circuit.

For an AC circuit, the real power is a fractional part of apparent power. That fraction is called *power factor,* abbreviated PF. The power factor is found as shown.

$$\text{PF} = \frac{\text{real power}}{\text{apparent power}}$$

The power factor will be a number between 0 and 1.

It is often helpful to view the power triangle in Fig. 23-37.

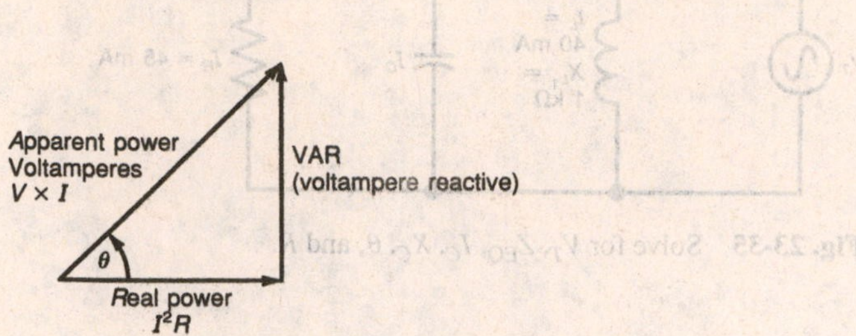

Fig. 23-37 Power triangle.

From this triangle, we can also develop the following equations for the power factor PF.

For series circuits:

$$\text{PF} = \cos\theta = \frac{R}{Z}$$

For parallel circuits:

$$\text{PF} = \cos\theta = \frac{I_R}{I_T}$$

Notice that the formulas for PF really tell us what fraction of the total impedance or total current is resistive. The equations have been obtained by expanding the terms representing the three sides of the power triangle.

Solved Problem

For Fig. 23-38*a* and *b*, find the real power, apparent power, and the PF.

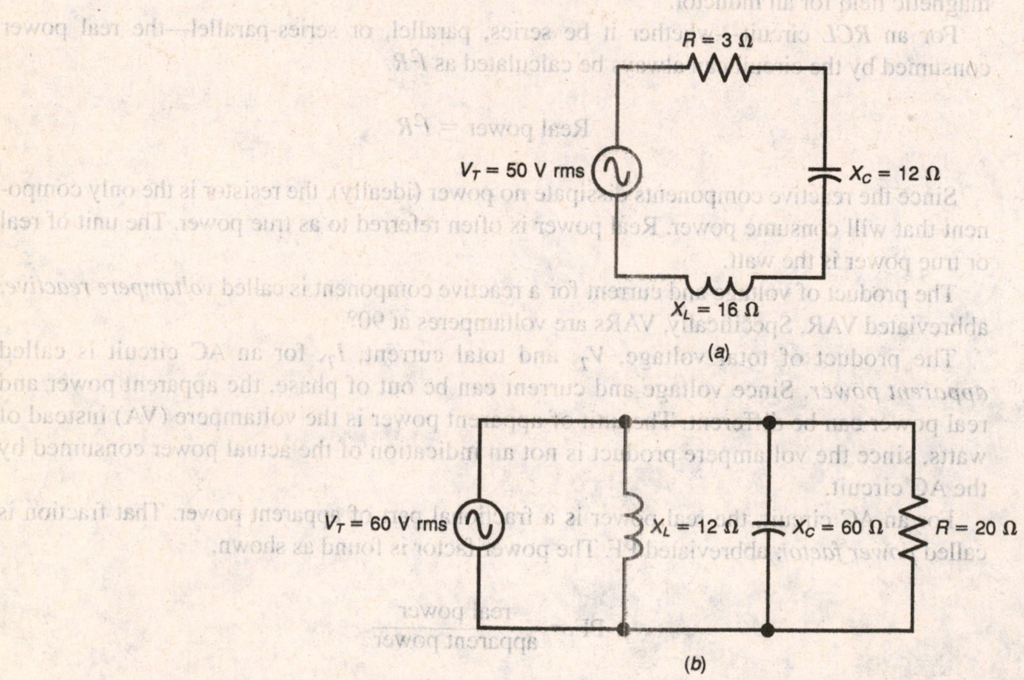

Fig. 23-38

Answer

In Fig. 23-38*a*, we will start by showing Z_T, I_T, V_L, V_C, V_R, and θ. These values are found using Table 23-1.

$$Z_T = 5\ \Omega$$
$$I_T = 10\ \text{A}$$
$$V_L = 160\ \text{V}$$
$$V_C = 120\ \text{V}$$
$$V_R = 30\ \text{V}$$
$$\theta = 53.1^\circ$$

To find real power, we have

$$\begin{aligned}\text{Real power} &= I^2R \\ &= (10\ \text{A})^2 \times 3\ \Omega \\ \text{Real power} &= 300\ \text{W}\end{aligned}$$

Apparent power is found by multiplying $V_T \times I_T$.

$$\begin{aligned}\text{Apparent power} &= V_T \times I_T \\ &= 50\ \text{V} \times 10\ \text{A} \\ \text{Apparent power} &= 500\ \text{VA}\end{aligned}$$

Notice the voltampere unit for apparent power.

The power factor is calculated as

$$\begin{aligned}\text{PF} &= \frac{\text{real power}}{\text{apparent power}} \\ &= \frac{300\ \text{W}}{500\ \text{VA}} \\ \text{PF} &= 0.6\end{aligned}$$

The power factor is also found by

$$\begin{aligned}\text{PF} &= \cos\theta = \frac{R}{Z} \\ &= \cos 53.1^\circ = \frac{3\ \Omega}{5\ \Omega} \\ \text{PF} &= 0.6\end{aligned}$$

In Fig. 23-38*b*, we will start by showing I_L, I_C, I_R, I_T, Z_{EQ}, and θ. These values are found using Table 23-2.

$$I_L = 5\ \text{A}$$
$$I_C = 1\ \text{A}$$
$$I_R = 3\ \text{A}$$
$$I_T = 5\ \text{A}$$
$$Z_{EQ} = 12\ \Omega$$
$$\theta = -53.13^\circ$$

To find real power, we have

$$\begin{aligned}\text{Real power} &= I^2\text{R} \\ &= (3\ \text{A})^2 \times 20\ \Omega \\ \text{Real power} &= 180\ \text{W}\end{aligned}$$

Apparent power is found by multiplying $V_T \times I_T$.

$$\begin{aligned}\text{Apparent power} &= V_T \times I_T \\ &= 60\ \text{V} \times 5\ \text{A} \\ \text{Apparent power} &= 300\ \text{VA}\end{aligned}$$

The power factor is calculated as

$$PF = \frac{\text{real power}}{\text{apparent power}}$$

$$= \frac{180 \text{ W}}{300 \text{ VA}}$$

$$PF = 0.6$$

The power factor is also found by

$$PF = \cos \theta = \frac{I_R}{I_T}$$

$$= \cos(-53.1°) = \frac{3 \text{ A}}{5 \text{ A}}$$

$$PF = 0.6$$

PRACTICE PROBLEMS

Sec. 23-4 Calculate real power, apparent power, and PF for the circuits identified in each problem.

1. Figure 23-5.
2. Figure 23-6.
3. Figure 23-7.
4. Figure 23-8.
5. Figure 23-17.
6. Figure 23-23.
7. Figure 23-25.
8. Figure 23-27.
9. Figure 23-28.
10. Figure 23-32.

END OF CHAPTER TEST

Chapter 23: AC Circuits Answer True or False.

1. In a series *RCL* circuit, X_L and X_C are 90° out of phase.
2. For X_L and X_C in series, the net reactance, X, is the difference between the two reactance values.
3. In a parallel *RCL* circuit, I_L and I_C are 180° out of phase.
4. For X_L and X_C in parallel, the net line current, I_T, is the difference between the two branch currents, I_L and I_C.
5. The unit of real or true power is the watt (W).
6. The unit of apparent power is the voltampere (VA).
7. Ideally, inductors and capacitors dissipate no real power.
8. For a sine-wave AC circuit, the power factor is equal to the tangent of the phase angle.
9. The power factor, *PF*, is a ratio of real to apparent power.
10. The power factor tells us what fraction of apparent power is actually real power dissipated as heat in a resistance.

CHAPTER

Complex Numbers for AC Circuits

Complex numbers form a numerical system that includes the phase angle of a quantity, along with its magnitude. Complex numbers become very useful when analyzing series-parallel circuits that contain both resistance and reactance in one or more branches.

SEC. 24-1 RECTANGULAR AND POLAR FORM OF A COMPLEX NUMBER

The rectangular form of a complex number specifies the rectangular coordinates of the real and imaginary terms. The real number is written first, followed by the $\pm j$ term. The polar form of a complex number specifies the phasor sum of the rectangular coordinates, as well as the phase angle that exists with respect to the horizontal axis for this phasor sum.

It is often necessary to convert from rectangular form to polar form, or vice versa. Here are some basic rules which make the conversions possible.

To convert from rectangular form to polar form, we use the following rules:

1. Find the magnitude by phasor addition of the $\pm j$ term and the real term.
2. Find the angle whose tangent value is the $\pm j$ term divided by the real term.

To convert from polar form to rectangular form, we use the following rules:

1. Multiply the magnitude of the resultant phasor by the cos θ to obtain the real term.
2. Multiply the magnitude of the resultant phasor by the sin θ to obtain the $\pm j$ term.

Solved Problems

a. For the circuit shown in Fig. 24-1*a*, state the impedance, Z_T, in both rectangular and polar form.

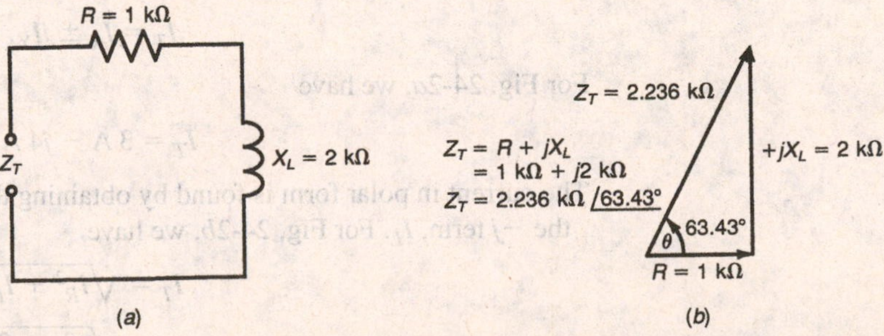

Fig. 24-1 (*a*) Series *RL* circuit. (*b*) Phasor diagram for (*a*) showing the rectangular components of Z_T and their phasor sum.

Answer

The general form used to state the impedance, Z_T, in rectangular form is $Z_T = R \pm jX$. For Fig. 24-1*a*, we have

$$Z_T = 1\ \text{k}\Omega + j2\ \text{k}\Omega$$

The impedance, Z_T, in polar form is found by obtaining the phasor sum of the real term R and the j term, X_L. We have

$$Z_T = \sqrt{R^2 + X_L^2}$$
$$= \sqrt{1^2 + 2^2}\ \text{k}\Omega$$
$$Z_T = 2.236\ \text{k}\Omega$$

The phase angle of the impedance is found by dividing the j term, X_L, by the real term R and then by finding the angle that corresponds to this tangent value. We proceed as follows:

$$\theta = \arctan \frac{X_L}{R}$$
$$= \arctan \frac{2\ \text{k}\Omega}{1\ \text{k}\Omega}$$
$$= \arctan 2$$
$$\theta = 63.43°$$

Figure 24-1*b* shows the rectangular components of Z_T and their phasor sum. In polar form, Z_T is shown as

$$Z_T = 2.236\ \text{k}\Omega \angle 63.43°$$

b. For the circuit shown in Fig. 24-2*a*, state the current, I_T, in both rectangular and polar form.

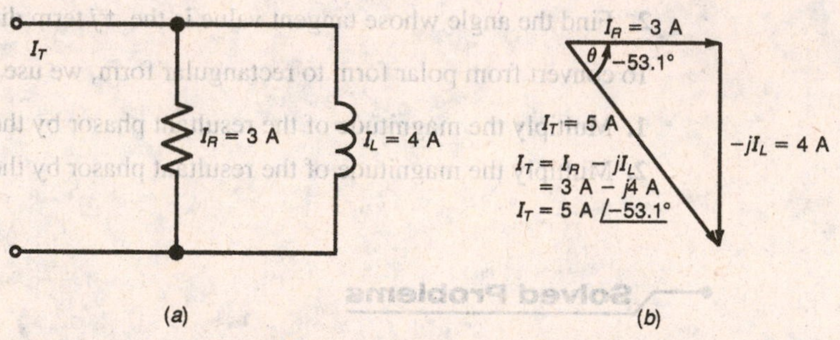

Fig. 24-2 (*a*) Parallel *RL* circuit. (*b*) Phasor diagram for (*a*) showing the rectangular components of I_T and their phasor sum.

Answer

The general form used to state I_T in rectangular form is

$$I_T = I_R \pm jI_X$$

For Fig. 24-2*a*, we have

$$I_T = 3\ \text{A} - j4\ \text{A}$$

The current in polar form is found by obtaining the phasor sum of the real term, I_R, and the $-j$ term, I_L. For Fig. 24-2*b*, we have

$$I_T = \sqrt{I_R^2 + I_L^2}$$
$$= \sqrt{3^2 + 4^2}\ \text{A}$$
$$I_T = 5\ \text{A}$$

The phase angle of the current, I_T, is found by dividing the $-j$ term, I_L, by the real term, I_R, and then by finding the angle that corresponds to this tangent value.

$$\theta = \arctan -\frac{I_L}{I_R}$$
$$= \arctan -\frac{4\text{ A}}{3\text{ A}}$$
$$= \arctan -1.33$$
$$\theta = -53.1°$$

Figure 24-2*b* shows the rectangular components of I_T and their phasor sum. In polar form, I_T is shown as

$$I_T = 5\text{ A} \angle -53.1°$$

c. For Fig. 24-3*a*, state Z_T in rectangular form. Draw the series-equivalent circuit that corresponds to this impedance value.

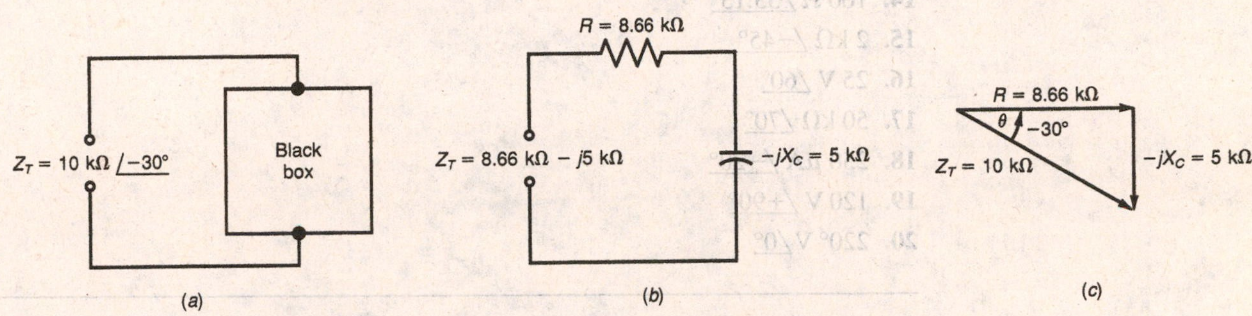

Fig. 24-3 Converting from polar form to rectangular form. (*a*) Black box diagram. (*b*) Series-equivalent *RC* circuit for (*a*). (*c*) Phasor diagram for (*a*) and (*b*).

Answer

To convert $Z_T = 10\text{ k}\Omega \angle -30°$ to rectangular form, we use the rules stated earlier. To find the real term, we multiply the magnitude of Z_T by the cos θ.

$$\text{Real term} = Z_T \times \cos\theta$$
$$= 10\text{ k}\Omega \times \cos -30°$$
$$= 10\text{ k}\Omega \times 0.866$$
$$\text{Real term} = 8.66\text{ k}\Omega$$

To find the *j* term, we multiply the magnitude, Z_T, by the sin θ.

$$j\text{ term} = Z_T \times \sin\theta$$
$$= 10\text{ k}\Omega \times \sin -30°$$
$$= 10\text{ k}\Omega \times -0.5$$
$$j\text{ term} = -5\text{ k}\Omega$$

The minus sign indicates that this *j* term is at an angle of −90°. This is stated as $-j5\text{ k}\Omega$. The $-j$ is used to represent capacitive reactance, X_C. The series-equivalent circuit for the impedance, Z_T, in Fig. 24-3*a* is shown in *b*. The phasor diagram for Fig. 24-3*a* and *b* is shown in *c*.

PRACTICE PROBLEMS

Sec. 24-1 Convert from rectangular form to polar form.

1. 6 A + *j*4 A
2. 400 mA + *j*300 mA

3. 10 V − j12 V
4. 5 A − j0 A
5. 20 V + j5 V
6. 0 Ω + j5 kΩ
7. 15 V − j20 V
8. 10 kΩ − j10 kΩ
9. 8 kΩ + j12 kΩ
10. 25 kΩ − j10 kΩ

Convert from polar form to rectangular form.

11. 10 A $\underline{/20°}$
12. 6 kΩ $\underline{/36.86°}$
13. 15 V $\underline{/-90°}$
14. 100 Ω $\underline{/53.13°}$
15. 2 kΩ $\underline{/-45°}$
16. 25 V $\underline{/60°}$
17. 50 kΩ $\underline{/70°}$
18. 250 μA $\underline{/-20°}$
19. 120 V $\underline{/+90°}$
20. 220° V $\underline{/0°}$

SEC. 24-2 MATHEMATICAL OPERATIONS WITH COMPLEX NUMBERS

Real terms and j terms cannot be combined directly because they are 90° out of phase with each other. The following rules apply:

1. When adding or subtracting complex numbers, add or subtract the real and j terms separately. The answer should be in the form $Z_T = R \pm jX$ for impedance, $V_T = V_R \pm jV_X$ for voltage, and $I_T = I_R \pm jI_X$ for currents.
2. When multiplying or dividing complex numbers, multiply and divide in polar form. For multiplication, multiply the magnitudes but add the phase angles. For division, divide the magnitudes but subtract the angle in the denominator from the angle in the numerator. The answer should be in the form of $Z_T \underline{/\theta}$, $V_T \underline{/\theta}$, or $I_T \underline{/\theta}$.

It should be noted that multiplication and division can be done in rectangular form, but the polar form is much more convenient.

We must use the rules stated for mathematical operations on complex numbers when we combine series- and parallel-connected impedances. First, let us state the general formulas for series- and parallel-connected impedances. For series impedances, we have

$$Z_T = Z_1 + Z_2 + Z_3 + \cdots + Z_n$$

For parallel impedances, we have

$$Z_T = \frac{1}{1/Z_1 + 1/Z_2 + 1/Z_3 + \cdots + 1/Z_n}$$

For just two parallel-connected impedances, we have

$$Z_T = \frac{Z_1 \times Z_2}{Z_1 + Z_2}$$

We apply these rules and formulas in the problems that follow.

Solved Problems

a. For Fig. 24-4*a*, state Z_T in both rectangular and polar form. Also, state the polar form for each of the following: I_T, V_C, and V_L. Draw the phasor diagram for all voltages and currents using V_T as the reference phasor at an angle of 0°.

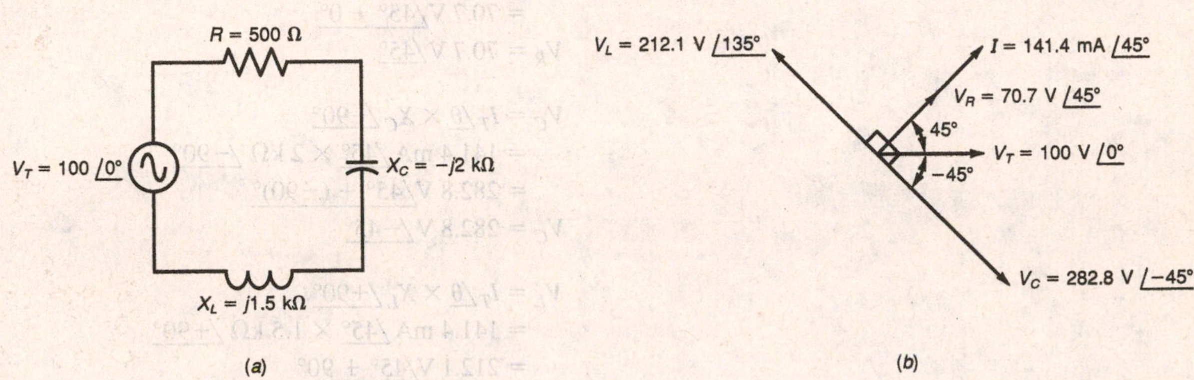

Fig. 24-4 Complex numbers applied to AC circuits. (*a*) Series *RCL* circuit. (*b*) Phasor diagram for all voltage and current values.

Answer

The total impedance, Z_T, is found using the formula for series impedances. In fact, this formula just involves adding or subtracting the real and *j* terms separately. For the rectangular form of Z_T in Fig. 24-4*a*, we have

$$Z_T = 500\ \Omega - j2\ \text{k}\Omega + j1.5\ \text{k}\Omega$$
$$Z_T = 500\ \Omega - j500\ \Omega$$

We can find Z_T in polar form as follows:

$$Z_T = \sqrt{R^2 + X_C{}^2}$$
$$= \sqrt{500^2 + 500^2}\ \Omega$$
$$Z_T = 707\ \Omega$$

$$\theta = \arctan -\frac{X_C}{R}$$
$$= \arctan -\frac{500\ \Omega}{500\ \Omega}$$
$$= \arctan -1$$
$$\theta = -45°$$

In polar form, we state Z_T as

$$Z_T = 707\ \Omega \angle -45°$$

To calculate I_T, we use Ohm's law.

$$I_T\angle\theta = \frac{V_T\angle 0^\circ}{Z_T\angle\theta}$$
$$= \frac{100\ \text{V}\angle 0^\circ}{707\ \Omega\ \angle -45^\circ}$$
$$= 141.4\ \text{mA}\ \angle 0-(-45)^\circ$$
$$I_T = 141.4\ \text{mA}\ \angle 45^\circ$$

Notice that Z_T has the negative phase angle of -45°, but that the sign changes to $+45^\circ$ for I_T because of the division into a quantity with the angle of 0°. The positive angle for I shows that the series circuit is capacitive with leading current.

To find V_R, V_C, and V_L in polar form, we proceed as follows:

$$V_R = I_T\angle\theta \times R\angle 0^\circ$$
$$= 141.4\ \text{mA}\ \angle 45^\circ \times 500\ \Omega\ \angle 0^\circ$$
$$= 70.7\ \text{V}\angle 45^\circ + 0^\circ$$
$$V_R = 70.7\ \text{V}\angle 45^\circ$$

$$V_C = I_T\angle\theta \times X_C\angle -90^\circ$$
$$= 141.4\ \text{mA}\ \angle 45^\circ \times 2\ \text{k}\Omega\ \angle -90^\circ$$
$$= 282.8\ \text{V}\angle 45^\circ + (-90)^\circ$$
$$V_C = 282.8\ \text{V}\angle -45^\circ$$

$$V_L = I_T\angle\theta \times X_L\angle +90^\circ$$
$$= 141.4\ \text{mA}\ \angle 45^\circ \times 1.5\ \text{k}\Omega\ \angle +90^\circ$$
$$= 212.1\ \text{V}\angle 45^\circ + 90^\circ$$
$$V_L = 212.1\ \text{V}\angle 135^\circ$$

The phasors for all voltages and currents are shown in Fig. 24-4*b*. Notice that V_R and V_C, as well as V_R and V_L, are 90° out of phase with each other. Also, V_C and V_L are 180° out of phase with each other.

b. For Fig. 24-5*a*, find Z_T for the two complex impedances, Z_1 and Z_2, in parallel. State Z_T in both rectangular and polar form. Draw the series-equivalent circuit that corresponds to Fig. 24-5*a*.

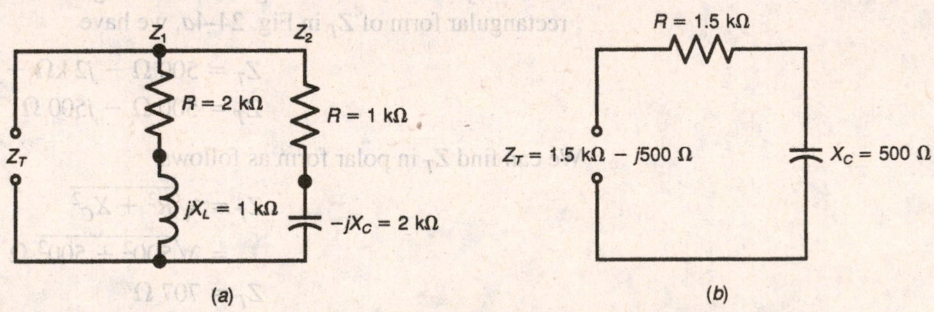

Fig. 24-5 Finding Z_T for two complex impedances, Z_1 and Z_2, in parallel. (*a*) Circuit. (*b*) Series equivalent for (*a*).

Answer

To find Z_T, we use the formula for two parallel-connected impedances.

$$Z_T = \frac{Z_1 \times Z_2}{Z_1 + Z_2}$$

We start by stating Z_1 and Z_2 in both rectangular and polar form. We then add Z_1 and Z_2 and convert to polar form.

$$Z_1 = 2\ \text{k}\Omega + j1\ \text{k}\Omega = 2.236\ \text{k}\Omega \underline{/26.56^\circ}$$

$$Z_2 = 1\ \text{k}\Omega - j2\ \text{k}\Omega = 2.236\ \text{k}\Omega \underline{/-63.43^\circ}$$

$$Z_1 + Z_2 = 3\ \text{k}\Omega - j1\ \text{k}\Omega = 3.162\ \text{k}\Omega \underline{/-18.43^\circ}$$

Notice that when we add Z_1 and Z_2, the real terms and j terms are added separately. We can now multiply and divide in polar form because the polar form of $Z_1 + Z_2$ is known. We have

$$Z_T = \frac{2.236\ \text{k}\Omega \underline{/26.56^\circ} \times 2.236\ \text{k}\Omega \underline{/-63.43^\circ}}{3.162\ \text{k}\Omega \underline{/-18.43^\circ}}$$

$$= \frac{5\ \text{M}\Omega \underline{/-36.87^\circ}}{3.162\ \text{k}\Omega \underline{/-18.43^\circ}}$$

$$Z_T = 1.581\ \text{k}\Omega \underline{/-18.43^\circ}$$

To convert Z_T back into rectangular form, to find the series-equivalent circuit, we use the rules stated earlier. We proceed as follows:

$$\text{Real term} = Z_T \times \cos\theta$$

$$= 1.581\ \text{k}\Omega \times \cos(-18.43^\circ)$$

$$\text{Real term} = 1.5\ \text{k}\Omega$$

$$j\ \text{term} = Z_T \times \sin\theta$$

$$= 1.581\ \text{k}\Omega \times \sin(-18.43^\circ)$$

$$j\ \text{term} = -500\ \Omega$$

or

$$-jX_C = 500\ \Omega$$

The series-equivalent circuit for Fig. 24-5*a* is shown in Fig. 24-5*b*. Notice that the series-parallel circuit of Fig. 24-5*a* can be replaced with the series-equivalent circuit in Fig. 24-5*b*.

PRACTICE PROBLEMS

Sec. 24-2 For Figs. 24-6 through 24-11, state Z_T and I_T in polar form. Also, state the polar form for all voltage drops. Use V_T as the zero reference phasor. For Figs. 24-12 through 24-21, state Z_T in both rectangular and polar form.

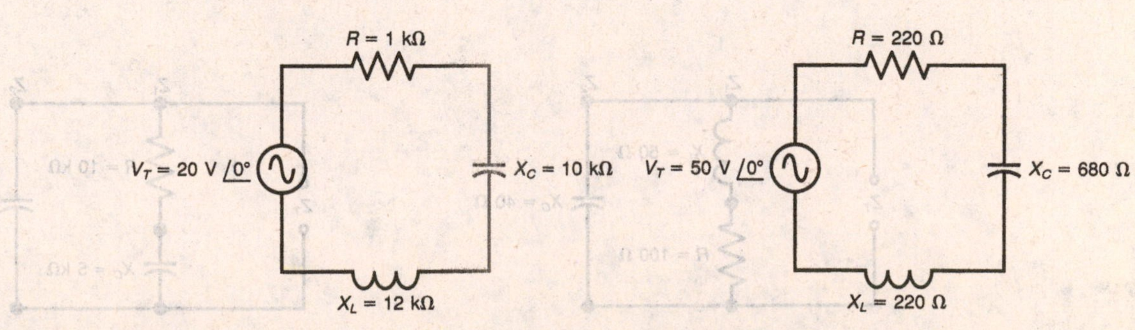

Fig. 24-6

Fig. 24-7

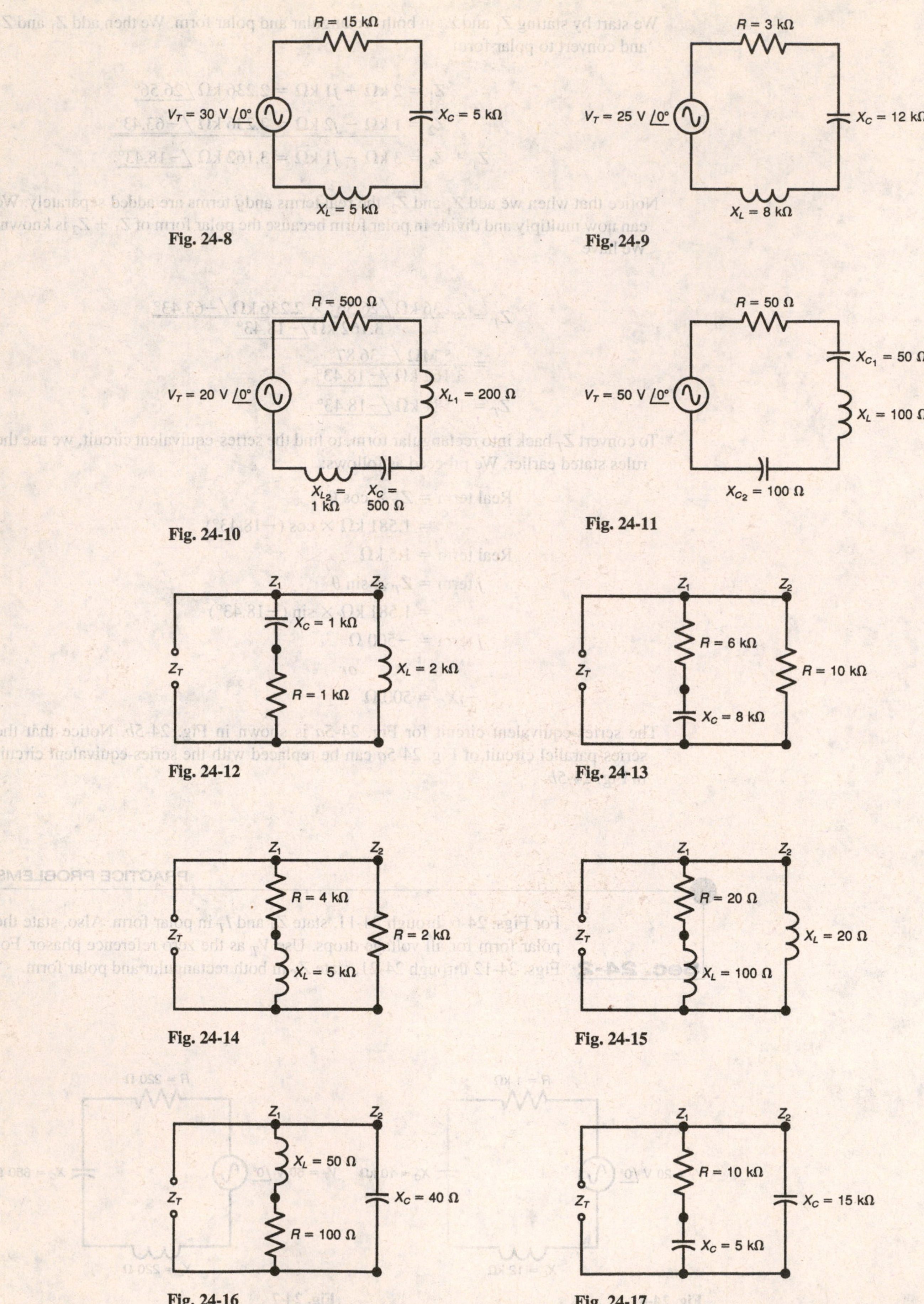

Fig. 24-8

Fig. 24-9

Fig. 24-10

Fig. 24-11

Fig. 24-12

Fig. 24-13

Fig. 24-14

Fig. 24-15

Fig. 24-16

Fig. 24-17

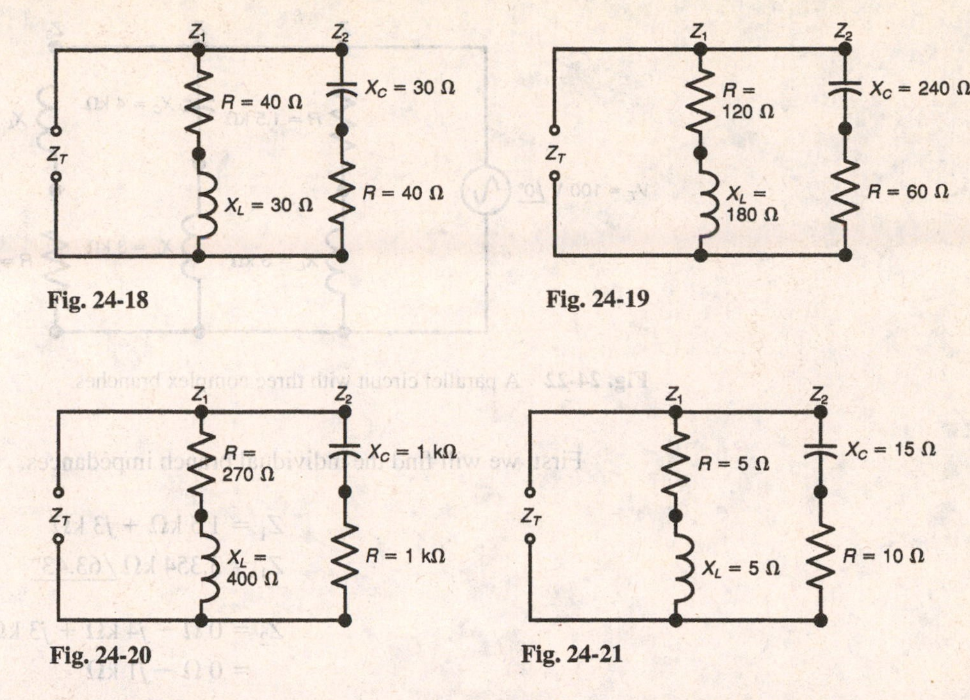

Fig. 24-18

Fig. 24-19

Fig. 24-20

Fig. 24-21

SEC. 24-3 PARALLEL CIRCUITS WITH THREE COMPLEX BRANCHES

In parallel circuits, it is usually easier to add branch currents than it is to combine reciprocal impedances. Reciprocal terms can be defined for complex impedances, just as branch conductance G is often used instead of branch resistance, where $G = 1/R$. The reciprocals, which are often used with complex impedances, are summarized here.

$$\text{Conductance} = G = 1/R \quad (\text{S})$$
$$\text{Susceptance} = B = 1/\pm X \quad (\text{S})$$
$$\text{Admittance} = Y = 1/Z \quad (\text{S})$$

Note that the unit for all reciprocals is the siemens (S). The phase angle for B or Y is the same as for current. An inductive branch has susceptance $-jB$, whereas a capacitive branch has susceptance $+jB$. In polar form, we state the total admittance as

$$Y_T\underline{/\theta}$$

in rectangular form, we have

$$Y_T = G \pm jB$$

Solved Problem

For Fig. 24-22, find Z_T, I_T, I_1, I_2, and I_3. State these quantities in polar form.

Answer

We will use two methods of solution for Z_T. Let us start by using the reciprocal formula.

$$Z_T = \frac{1}{1/Z_1 + 1/Z_2 + 1/Z_3 + \cdots + 1/Z_n}$$

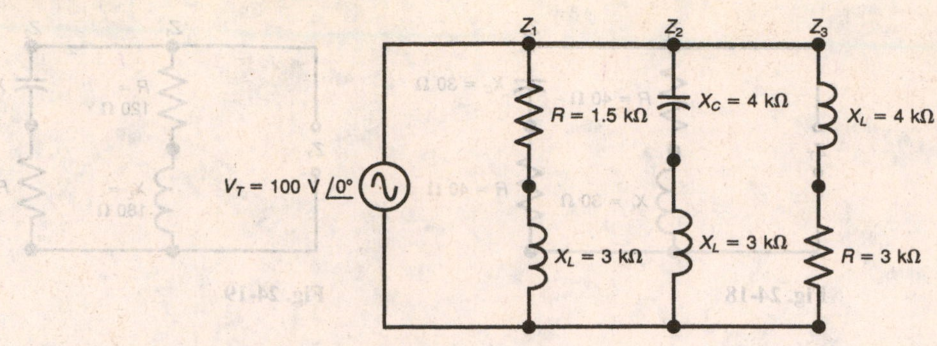

Fig. 24-22 A parallel circuit with three complex branches.

First, we will find the individual branch impedances.

$$Z_1 = 1.5\ \text{k}\Omega + j3\ \text{k}\Omega$$
$$Z_1 = 3.354\ \text{k}\Omega\ \underline{/63.43^\circ}$$

$$Z_2 = 0\ \Omega - j4\ \text{k}\Omega + j3\ \text{k}\Omega$$
$$= 0\ \Omega - j1\ \text{k}\Omega$$
$$Z_2 = 1\ \text{k}\Omega\ \underline{/-90^\circ}$$

$$Z_3 = 3\ \text{k}\Omega + j4\ \text{k}\Omega$$
$$Z_3 = 5\ \text{k}\Omega\ \underline{/53.13^\circ}$$

We now convert each impedance to admittance Y in siemens. We will state Y in both polar and rectangular form.

$$Y_1\ \underline{/\theta} = \frac{1\ \underline{/0^\circ}}{Z_1\ \underline{/\theta}}$$
$$= \frac{1\ \underline{/0^\circ}}{3.354\ \text{k}\Omega\ \underline{/63.43^\circ}}$$
$$Y_1\ \underline{/\theta} = 298.15\ \mu\text{S}\ \underline{/-63.43^\circ}$$

We find Y_1 in rectangular form as follows:

$$G = Y_1 \times \cos\theta$$
$$= 298.15\ \mu\text{S} \times \cos -63.43^\circ$$
$$G = 133.36\ \mu\text{S}$$

$$B = Y_1 \times \sin\theta$$
$$= 298.15\ \mu\text{S} \times \sin -63.43^\circ$$
$$B = -266.6\ \mu\text{S}$$

or

$$-jB = 266.6\ \mu\text{S}$$

In rectangular form,

$$Y_1 = 133.36\ \mu\text{S} - j266.6\ \mu\text{S}$$

$$Y_2\ \underline{/\theta} = \frac{1\ \underline{/0^\circ}}{Z_2\ \underline{/\theta}}$$
$$= \frac{1\ \underline{/0^\circ}}{1\ \text{k}\Omega\ \underline{/-90^\circ}}$$
$$Y_2\ \underline{/\theta} = 1\ \text{mS}\ \underline{/+90^\circ}$$

We find Y_2 in rectangular form as follows:

$$G = Y_2 \times \cos\theta$$
$$= 1\text{ mS} \times \cos 90°$$
$$G = 0\text{ S}$$

$$B = Y_2 \times \sin\theta$$
$$= 1\text{ mS} \times \sin 90°$$
$$B = 1\text{ mS}$$

In rectangular form,

$$Y_2 = 0 + j1\text{ mS}$$

$$Y_3\underline{/\theta} = \frac{1\underline{/0°}}{Z_3\underline{/\theta}}$$
$$= \frac{1\underline{/0°}}{5\text{ k}\Omega\underline{/53.13°}}$$
$$Y_3\underline{/\theta} = 200\ \mu\text{S}\underline{/-53.13°}$$

We find Y_3 in rectangular form as follows:

$$G = Y_3 \times \cos\theta$$
$$= 200\ \mu\text{S} \times \cos -53.13°$$
$$G = 120\ \mu\text{S}$$

$$B = Y_3 \times \sin\theta$$
$$= 200\ \mu\text{S} \times \sin -53.13°$$
$$-jB = 160\ \mu\text{S}$$

In rectangular form,

$$Y_3 = 120\ \mu\text{S} - j160\ \mu\text{S}$$

To find Y_T, we must add the individual admittance values.

$$Y_T = Y_1 + Y_2 + Y_3$$

We must add these admittance values in rectangular form.

$$\begin{aligned}
Y_1 &= 133.36\ \mu\text{S} - j266.6\ \mu\text{S} \\
Y_2 &= 0 \qquad\qquad + j1\text{ mS} \\
Y_3 &= 120\ \mu\text{S} \quad - j160\ \mu\text{S} \\
\hline
Y_T &= 253.36\ \mu\text{S} + j573.3\ \mu\text{S} \text{ (rectangular form)} \\
Y_T &= 626.8\ \mu\text{S}\underline{/66.15°} \text{ (polar form)}
\end{aligned}$$

To find Z_T, we have

$$Z_T\underline{/\theta} = \frac{1\underline{/0°}}{Y_T\underline{/\theta}}$$
$$= \frac{1\underline{/0°}}{626.8\ \mu\text{S}\underline{/66.15°}}$$
$$Z_T\underline{/\theta} = 1595.4\ \Omega\underline{/-66.15°}$$

To find I_T, we have

$$I_T\angle\theta = \frac{V_T\angle 0°}{Z_T\angle\theta}$$
$$= \frac{100\ \text{V}\angle 0°}{1595.4\ \Omega\angle -66.15°}$$
$$I_T\angle\theta = 62.68\ \text{mA}\angle +66.15°$$

To find each branch current, we have

$$I_1\angle\theta = \frac{V_T\angle 0°}{Z_1\angle\theta}$$
$$= \frac{100\ \text{V}\angle 0°}{3.354\ \text{k}\Omega\angle 63.43°}$$
$$I_1\angle\theta = 29.815\ \text{mA}\angle -63.43°$$

In rectangular form,

$$I_1 = 13.3\ \text{mA} - j26.6\ \text{mA}$$
$$I_2\angle\theta = \frac{V_T\angle 0°}{Z_2\angle\theta}$$
$$= \frac{100\ \text{V}\angle 0°}{1\ \text{k}\Omega\angle -90°}$$
$$I_2\angle\theta = 100\ \text{mA}\angle +90°$$

In rectangular form,

$$I_2 = 0\ \text{mA} + j100\ \text{mA}$$
$$I_3\angle\theta = \frac{V_T\angle 0°}{Z_3\angle\theta°}$$
$$= \frac{100\ \text{V}\angle 0°}{5\ \text{k}\Omega\angle 53.13°}$$
$$I_3\angle\theta = 20\ \text{mA}\angle -53.13°$$

In rectangular form,

$$I_3 = 12\ \text{mA} - j16\ \text{mA}$$

Adding I_1, I_2, and I_3, we have

$$\begin{array}{ll} I_1 = 13.3\ \text{mA} & -\,j26.66\ \text{mA} \\ I_2 = 0\ \text{mA} & +\,j100\ \text{mA} \\ I_3 = 12\ \text{mA} & -\,j16\ \text{mA} \\ \hline I_T = 25.33\ \text{mA} & +\,j57.33\ \text{mA} \quad \text{(rectangular form)} \\ I_T = 62.68\ \text{mA}\angle 66.15° & \quad \text{(polar form)} \end{array}$$

We can also find Z_T as shown.

$$Z_T\angle\theta = \frac{V_T\angle 0°}{I_T\angle\theta}$$
$$= \frac{100\ \text{V}\angle 0°}{62.68\ \text{mA}\angle 66.15°}$$
$$Z_T\angle\theta = 1595.4\ \Omega\angle -66.15°$$

Sec. 24-3 For Figs. 24-23 through Fig. 24-25, solve for Z_T, I_T, I_1, I_2, and I_3. For Fig. 24-26, solve for Z_T, I_T, I_A, and I_B in polar form.

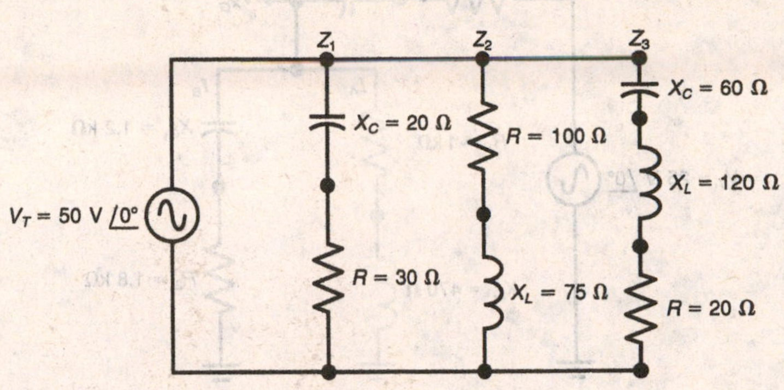

Fig. 24-23

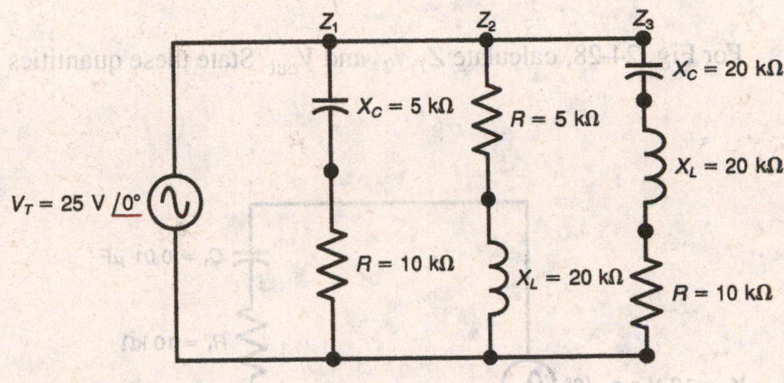

Fig. 24-24

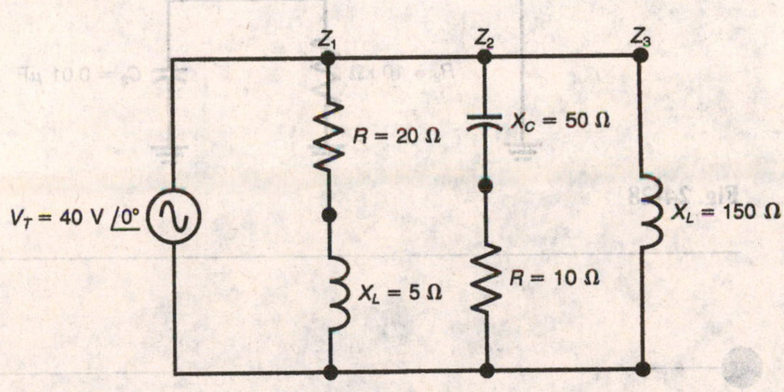

Fig. 24-25

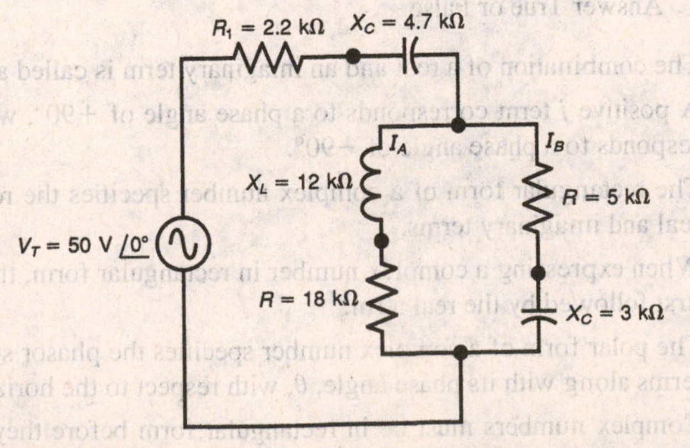

Fig. 24-26

For Fig. 24-27, solve for Z_T, I_T, V_{R_1}, V_{R_2}, V_{R_3}, V_{C_1}, V_{C_2}, V_{L_1}, V_{X_G}, I_A, and I_B. State these quantities in polar form.

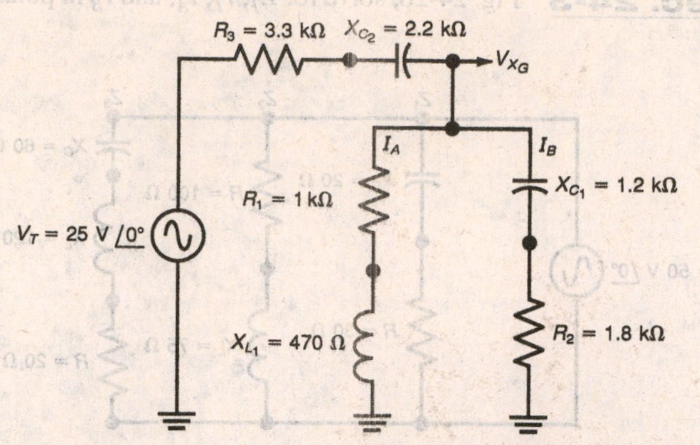

Fig. 24-27

For Fig. 24-28, calculate Z_T, I_T, and V_{out}. State these quantities in polar form.

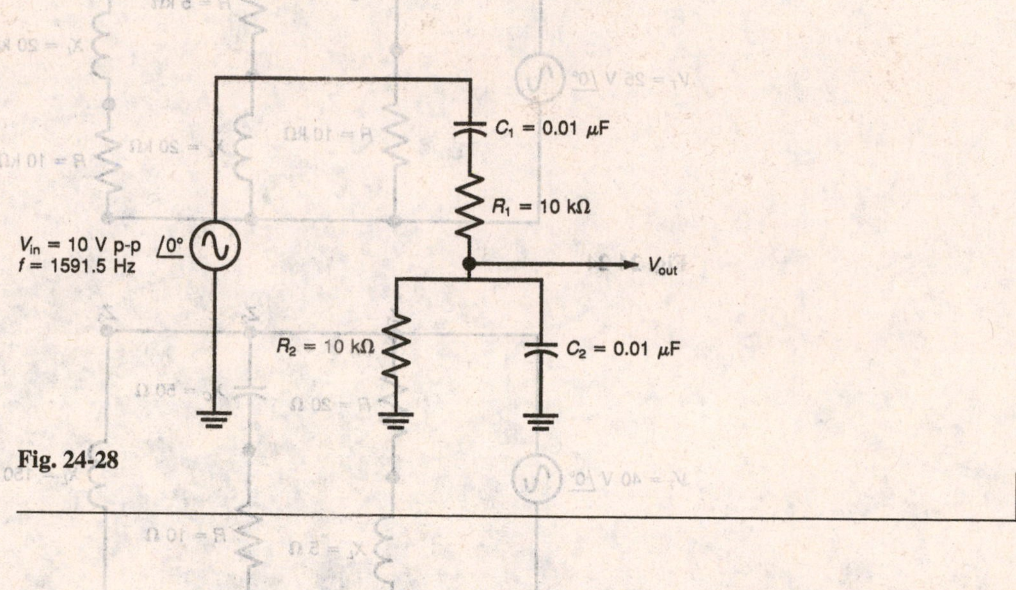

Fig. 24-28

END OF CHAPTER TEST

Chapter 24: Complex Numbers for AC Circuits

Answer True or False.

1. The combination of a real and an imaginary term is called a complex number.
2. A positive j term corresponds to a phase angle of $+90°$, whereas a negative j term corresponds to a phase angle of $-90°$.
3. The rectangular form of a complex number specifies the rectangular coordinates of the real and imaginary terms.
4. When expressing a complex number in rectangular form, the $\pm j$ term is usually written first followed by the real term.
5. The polar form of a complex number specifies the phasor sum of the real and imaginary terms along with its phase angle, θ, with respect to the horizontal axis.
6. Complex numbers must be in rectangular form before they can be added or subtracted because the real and imaginary terms must be added or subtracted separately.

7. When multiplying or dividing complex numbers, it is much more convenient to use the rectangular form rather than the polar form of a complex number.
8. With complex numbers, an inductive reactance of 25 Ω is shown as $j25$ Ω.
9. With complex numbers, an inductive branch current of 25 mA is shown as $-j25$ mA.
10. Susceptance, B, is the reciprocal of reactance, X.
11. Admittance, Y, is the reciprocal of impedance, Z.
12. The unit of susceptance and admittance is the siemens (S).
13. $6 - j8\ \Omega = 10\angle -53.13°\ \Omega$.
14. $25\angle 36.87°$ mA = $20 + j5$ mA.
15. $Y_T = G \pm jB$

CHAPTER 25

Resonance

For series or parallel *LC* circuits, we have a special effect when the inductive reactance, X_L, equals the capacitive reactance, X_C. For a series *LC* circuit, the inductive reactance, X_L, and capacitive reactance, X_C, will be equal at some frequency and cancel each other leaving only the internal resistance of the coil to limit current flow. For a parallel *LC* circuit, the inductive and capacitive branch currents will be equal at some frequency as a result of X_L and X_C being equal. The equal but opposite branch currents, therefore, will cancel each other resulting in a net or total line current that is nearly zero. The formula used to determine the frequency where $X_L = X_C$ is

$$f_r = \frac{1}{2\pi\sqrt{LC}}$$

This frequency is called the *resonant frequency* and is designated f_r.

The resonant frequency is important for series *LC* circuits because we have a sharp increase in current at this frequency.

For parallel *LC* circuits at resonance, the effect is a sharp increase in total impedance.

Frequencies near the resonant frequency for both series and parallel *LC* circuits are also useful. Any resonant circuit has an associated band of frequencies that produce the resonant effect. The width of the resonant band is determined by the circuit *Q*. The bandwidth of a resonant circuit is defined as those frequencies for which the resonant effect is 70.7% or more of its maximum value at resonance. The edge frequencies are defined specifically as those points at which the resonant effect has been reduced to 70.7% of its maximum value.

SEC. 25-1 SERIES RESONANT CIRCUITS

Figure 25-1 shows a series *RCL* circuit. Notice that the resistance in the circuit is, in fact, the coil's own internal resistance, designated r_i. It is not possible to obtain an *LC* circuit with zero resistance due to the coil's r_i. The quality or figure of merit for a coil is a measure

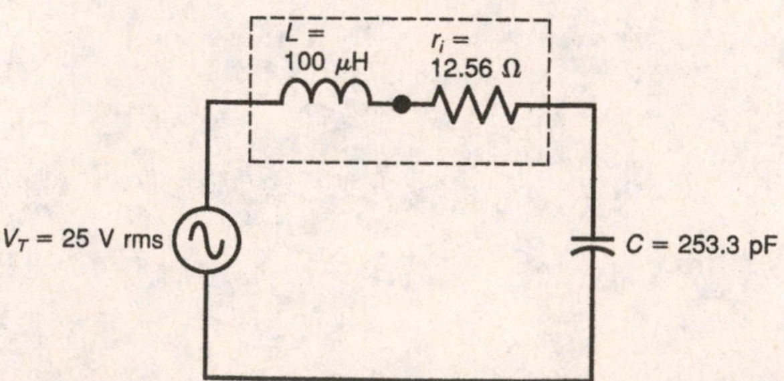

Fig. 25-1 Series *LC* circuit showing internal resistance of coil.

of its reactance to its resistance ratio at a given frequency. We call this the Q of a coil. The formula used to determine the Q of a coil is

$$Q = \frac{X_L}{r_i}$$

It appears from the formula for Q that its value can increase without limit as X_L increases for higher frequencies. However, Q cannot increase without limit because such effects as skin effect, eddy currents, and hysteresis losses produce an increase in the coil's resistance at higher frequencies.

For a series RCL circuit at resonance, we have the following characteristics:

$$f_r = \frac{1}{2\pi\sqrt{LC}}$$

$$Z_T = R \text{ (usually the } r_i \text{ of the coil)}$$

$$I_T = \frac{V_T}{R}$$

$$V_C = V_L = Q \times V_T$$

$$\text{BW} = \frac{f_r}{Q}$$

$$f_1 = f_r - \frac{\text{BW}}{2}$$

$$f_2 = f_r + \frac{\text{BW}}{2}$$

$$\theta = 0°$$

$$\text{PF} = 1 \textit{ or } \text{unity}$$

Solved Problem

For the circuit in Fig. 25-1, calculate f_r. Calculate the values of X_L, X_C, Z_T, I_T, V_L, V_C, Q, θ, and PF at f_r. Determine also the bandwidth, BW, and the edge frequencies, f_1 and f_2.

Answer

To solve for f_r, we proceed as follows:

$$f_r = \frac{1}{2\pi\sqrt{LC}}$$

$$= \frac{1}{2\pi\sqrt{100\ \mu\text{H} \times 253.3\ \text{pF}}}$$

$$f_r = 1\ \text{MHz}$$

We can then find X_L and X_C at f_r as follows:

$$X_L = 2\pi f_r L$$

$$= 2 \times 3.14 \times 1\ \text{MHz} \times 100\ \mu\text{H}$$

$$X_L = 628.3\ \Omega$$

$$X_C = \frac{1}{2\pi f_r C}$$

$$= \frac{1}{2 \times 3.14 \times 1\ \text{MHz} \times 253.3\ \text{pF}}$$

$$X_C = 628.3\ \Omega$$

To find Z_T at f_r, we find the net reactance X and combine it phasorially with r_i. We proceed as follows:

$$Z_T = \sqrt{R^2 + X^2}$$

$$= \sqrt{r_i^2 + (X_L - X_C)^2}$$

$$= \sqrt{12.56^2 + (628.3 - 628.3)^2}\ \Omega$$

$$= \sqrt{12.56^2 + 0^2}\ \Omega$$

$$Z_T = 12.56\ \Omega$$

It is important to note that Z_T is at its minimum value for this frequency. As we move above or below f_r, the net reactance is no longer zero, and Z_T increases. Figure 25-2*a* shows how Z_T varies with frequency, and Fig. 25-2*b* shows how I_T varies with frequency.

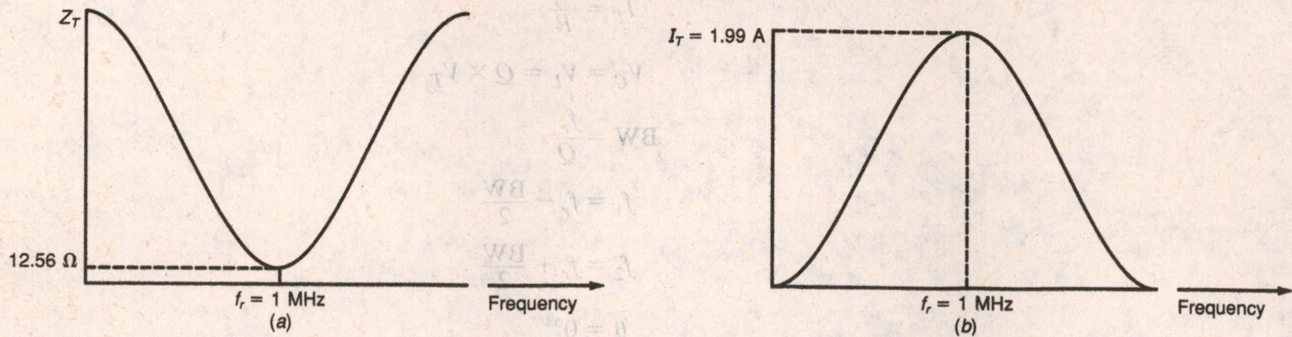

Fig. 25-2 Response curves for series resonant circuits. (*a*) Graph of Z_T versus frequency. (*b*) Graph of I_T versus frequency.

At f_r, the total current I_T is found as shown.

$$I_T = \frac{V_T}{Z_T}$$

$$= \frac{25\ \text{V}}{12.56\ \Omega}$$

$$I_T = 1.99\ \text{A}$$

Since Z_T is minimum at f_r, I_T will be maximum. As we move above or below f_r, I_T decreases from its maximum value. This is a result of the total impedance, Z_T, increasing as we move above or below resonance.

To find the Q of the circuit at f_r, we proceed as follows:

$$Q = \frac{X_L}{r_i}$$

$$= \frac{628.3\ \Omega}{12.56\ \Omega}$$

$$Q = 50$$

Notice that the Q of the coil and the Q of the circuit are the same. There is no for Q because the ohms unit of opposition cancels in the equation.

To find V_L and V_C at f_r, we proceed as shown.

$$V_L = I_T \times X_L$$

$$= 1.99\ \text{A} \times 628.3\ \Omega$$

$$V_L = 1250\ \text{V}$$

$$V_C = I_T \times X_C$$

$$= 1.99\ \text{A} \times 628.3\ \Omega$$

$$V_C = 1250\ \text{V}$$

If we substitute V_T/Z_T for I_T in the equation for V_L, we develop the following relationship. (*Note:* $Z_T = r_i$ at f_r.)

$$V_L = Q \times V_T \text{ (at } f_r\text{)}$$

For Fig. 25-1*a*, we have

$$V_L = V_C = 50 \times 25 \text{ V}$$

$$V_L = 1250 \text{ V}$$

Notice that this answer is the same as the one found using $V_L = I_T \times X_L$.

With the current, I_T, at its maximum value at f_r, the voltage drops across L and C are Q times greater than V_T. The voltages V_C and V_L decrease as we move away from resonance.

The phase angle θ at resonance is found by taking the arctan of the X/R ratio. We have

$$\theta = \arctan \frac{X}{R}$$

$$= \arctan \frac{(X_L - X_C)}{r_i}$$

$$= \arctan \frac{(628.3\ \Omega - 628.3\ \Omega)}{12.56\ \Omega}$$

$$= \arctan \frac{0}{12.56\ \Omega}$$

$$= \arctan 0$$

$$\theta = 0°$$

This phase angle of 0° tells us that the circuit appears purely resistive at f_r, with the total voltage V_T in phase with the total current, I_T.

Since $\theta = 0°$, the PF = 1 because

$$\text{PF} = \cos \theta$$

$$\text{PF} = \cos 0°$$

$$= 1$$

or

$$\text{PF} = \frac{R}{Z}$$

Since $R = r_i$ and $Z = r_i$ at f_r, we have

$$\text{PF} = \frac{r_i}{r_i}$$

$$\text{PF} = 1$$

Frequencies just above and below resonance will produce increased current. For a series resonant circuit, the bandwidth is defined as the range of frequencies where the total current, I_T, is at or above 70.7% of the peak value at resonance. The bandwidth is found as shown.

$$\text{BW} = \frac{f_r}{Q}$$

$$= \frac{1 \text{ MHz}}{50}$$

$$\text{BW} = 20 \text{ kHz}$$

The frequencies in this 20-kHz band centered around f_r will produce a current flow of 1.4 A or more. For series resonance, the edge frequencies are defined as those frequencies for which the total current, I_T, has been reduced to 70.7% of its maximum value. These edge frequencies are found as shown.

$$f_1 = f_r - \frac{BW}{2}$$

$$= 1\text{ MHz} - \frac{20\text{ kHz}}{2}$$

$$f_1 = 990\text{ kHz}$$

$$f_2 = f_r + \frac{BW}{2}$$

$$= 1\text{ MHz} + \frac{20\text{ kHz}}{2}$$

$$f_2 = 1.01\text{ MHz}$$

For a series resonant circuit, the bandwidth BW can be reduced by increasing the Q. This often means going to a higher-quality coil with less r_i. Figure 25-3 shows the response curve of current with the edge frequencies included. These edge frequencies are often referred to as the half-power points.

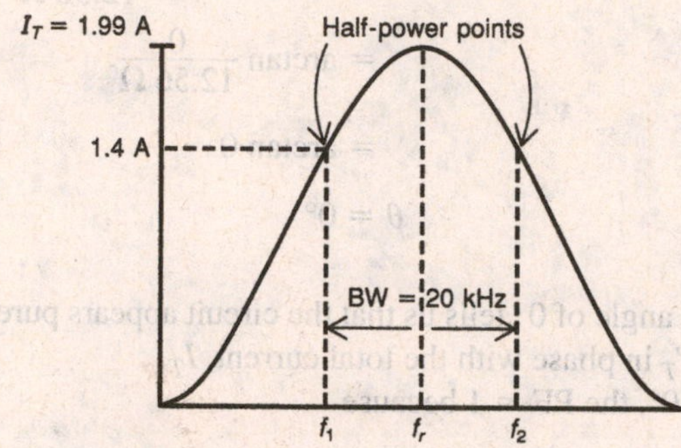

Fig. 25-3 Response curve for a series resonant circuit showing the edge frequencies f_1 and f_2.

PRACTICE PROBLEMS

Sec. 25-1 For Probs. 1–6, refer to Fig. 25-4.

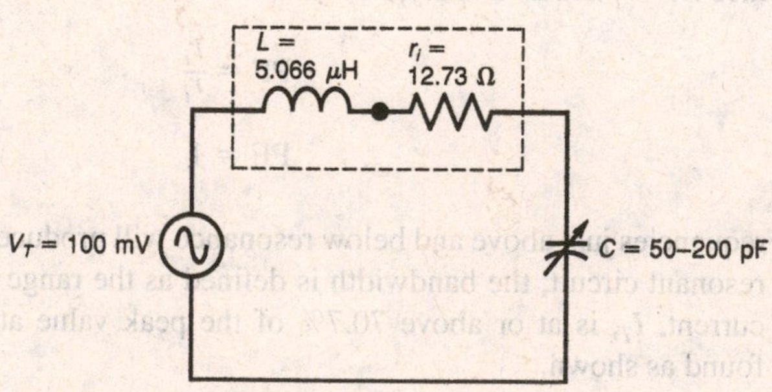

Fig. 25-4

1. With C adjusted to 50 pF, solve for f_r, BW, f_1, and f_2. Also, solve for the following at f_r: Z_T, I_T, X_C, X_L, V_C, V_L, Q, θ, PF, real power, and V_{r_i}.

2. Calculate Z_T, I_T, and V_{r_i} at both f_1 and f_2.
3. At f_1, does the circuit appear inductive, capacitive, or resistive? Calculate θ at f_1.
4. At f_2, does the circuit appear inductive, capacitive, or resistive? Calculate θ at f_2.
5. With C adjusted to 200 pF, solve for f_r, BW, f_1, and f_2. Also solve for the following at f_r: Z_T, I_T, X_C, X_L, V_C, V_L, Q, θ, and V_{r_i}.
6. Compare the bandwidth for the minimum and maximum capacitance settings. Explain the reason for this occurrence.

For Figs. 25-5 through Fig. 25-7, calculate for f_r, BW, f_1, and f_2. Also, solve for the following at f_r: X_C, X_L, Z_T, I_T, V_L, V_C, Q, θ, PF, real power, and V_{r_i}.

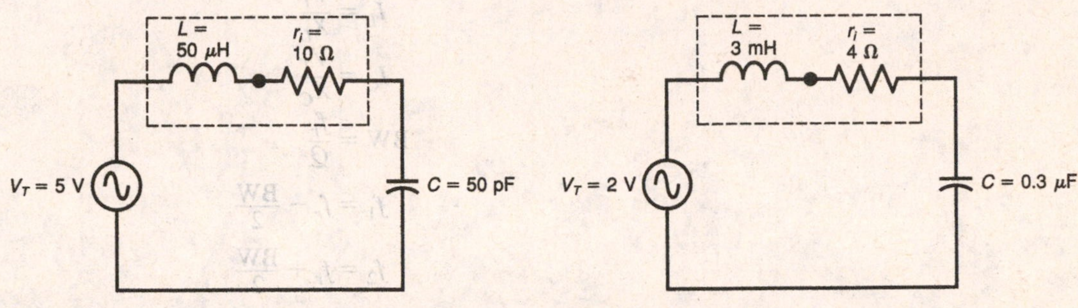

Fig. 25-5

Fig. 25-6

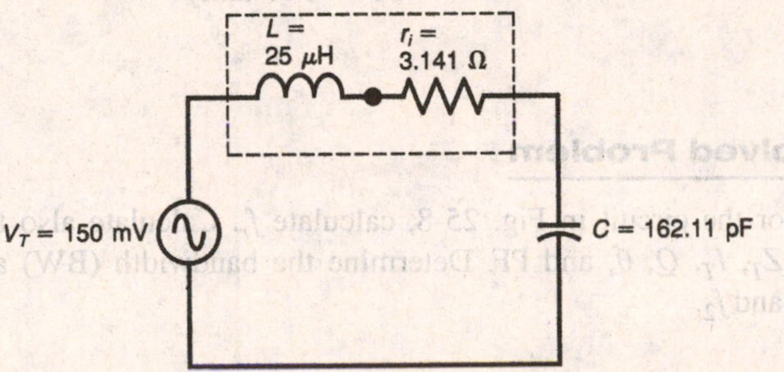

Fig. 25-7

SEC. 25-2 PARALLEL RESONANT CIRCUITS

Figure 25-8 shows a parallel LC circuit. Since the coil does contain some resistance, perfect cancellation of reactive branch currents is not possible, since the inductive branch has slightly more opposition than the capacitive branch at resonance.

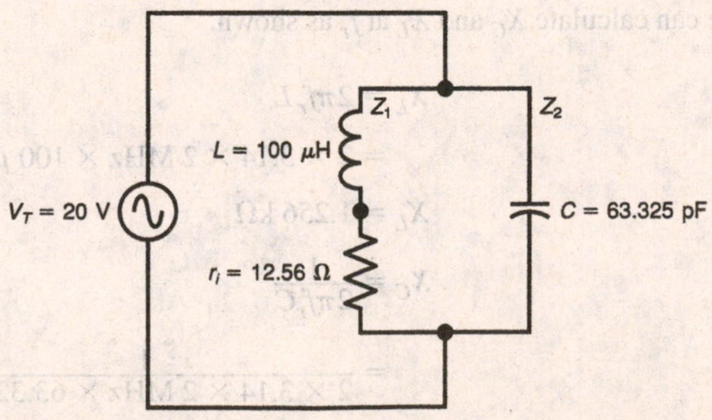

Fig. 25-8 Parallel LC circuit showing the internal resistance of the coil.

For a parallel *RCL* circuit at resonance, we have the following characteristics:

$$f_r = \frac{1}{2\pi\sqrt{LC}}$$

$$Z_T = Q \times X_L \text{ (}Q\text{ of the coil)}$$

$$I_T = \frac{V_T}{Z_T}$$

also

$$I_T = \frac{I_L}{Q}$$

$$I_L = \frac{V_T}{X_L}$$

$$I_C = \frac{V_T}{X_C}$$

$$\text{BW} = \frac{f_r}{Q}$$

$$f_1 = f_r - \frac{\text{BW}}{2}$$

$$f_2 = f_r + \frac{\text{BW}}{2}$$

$$\theta = 0°$$

$$\text{PF} = 1 \textit{ or } \text{unity}$$

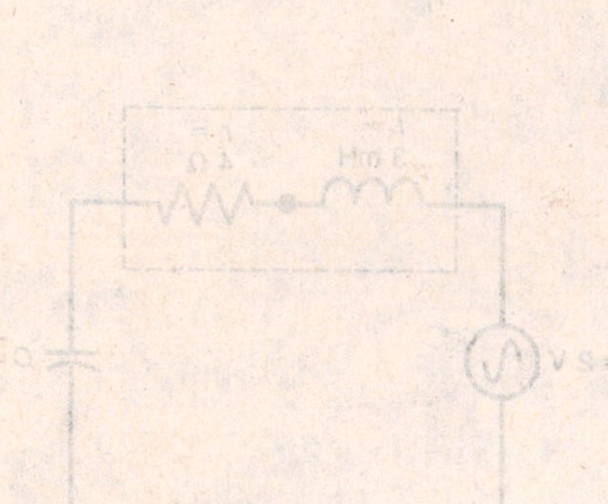

Solved Problem

For the circuit in Fig. 25-8, calculate f_r. Calculate also the values of X_L, X_C, I_C, I_L, Z_T, I_T, Q, θ, and PF. Determine the bandwidth (BW) and the edge frequencies, f_1 and f_2.

Answer

To solve for f_r, we proceed as follows:

$$f_r = \frac{1}{2\pi\sqrt{LC}}$$

$$= \frac{1}{2 \times 3.14\sqrt{100\ \mu\text{H} \times 63.325\ \text{pF}}}$$

$$f_r = 2\ \text{MHz}$$

We can calculate X_C and X_L at f_r as shown.

$$X_L = 2\pi f_r L$$

$$= 2 \times 3.14 \times 2\ \text{MHz} \times 100\ \mu\text{H}$$

$$X_L = 1.256\ \text{k}\Omega$$

$$X_C = \frac{1}{2\pi f_r C}$$

$$= \frac{1}{2 \times 3.14 \times 2\ \text{MHz} \times 63.325\ \text{pF}}$$

$$X_C = 1.256\ \text{k}\Omega$$

We can find I_L and I_C as shown.

$$I_L = \frac{V_T}{X_L}$$

$$= \frac{20 \text{ V}}{1.256 \text{ k}\Omega}$$

$$I_L = 15.92 \text{ mA}$$

$$I_C = \frac{V_T}{X_C}$$

$$= \frac{20 \text{ V}}{1.256 \text{ k}\Omega}$$

$$I_C = 15.92 \text{ mA}$$

The Q of the parallel circuit is equal to the Q of the coil when there is no parallel damping resistance, R_P.

$$Q = \frac{X_L}{r_i}$$

$$= \frac{1.256 \text{ k}\Omega}{12.56 \text{ }\Omega}$$

$$Q = 100$$

To find the impedance, Z_T, at resonance, we use the formula for two parallel-connected impedances. We designate the inductive branch as Z_1 and the capacitive branch as Z_2.

$$Z_T = \frac{Z_1 \times Z_2}{Z_1 + Z_2}$$

$$Z_1 = 12.56 \text{ }\Omega + j1.256 \text{ k}\Omega = 1.256 \text{ k}\Omega \angle{+90^\circ}$$

$$Z_2 = 0 \text{ }\Omega - j1.256 \text{ k}\Omega = 1.256 \text{ k}\Omega \angle{-90^\circ}$$

$$Z_1 + Z_2 = 12.56 \text{ }\Omega - j0 \text{ }\Omega = 12.56 \text{ }\Omega \angle{0^\circ}$$

Multiplying Z_1 and Z_2 in polar form and dividing by $Z_1 + Z_2$ in polar form, we have

$$Z_T = \frac{1.256 \text{ k}\Omega \angle{90^\circ} \times 1.256 \text{ k}\Omega \angle{-90^\circ}}{12.56 \text{ }\Omega \angle{0^\circ}}$$

Notice that this is

$$Z_T = \frac{X_C \times X_L \angle{+90 + (-90)^\circ}}{r_i \angle{0^\circ}}$$

or

$$Z_T = Q \times X_L$$

$$= 100 \times 1256 \text{ }\Omega$$

$$Z_T = 125.6 \text{ k}\Omega$$

We can find I_T as follows:

$$I_T = \frac{V_T}{Z_T}$$

$$= \frac{20 \text{ V}}{125.6 \text{ k}\Omega}$$

$$I_T = 159.22 \text{ }\mu\text{A}$$

At the resonant frequency f_r, Z_T is maximum due to the equal but opposite branch currents canceling. As we move above or below resonance, the net line current increases, causing Z_T to decrease from its maximum value. Since Z_T is maximum at f_r, I_T will be minimum. Figure 25-9 shows the graph of total impedance, Z_T, versus frequency in Fig. 25-9*a*, and the graph of total current, I_T, versus frequency in Fig. 25-9*b*.

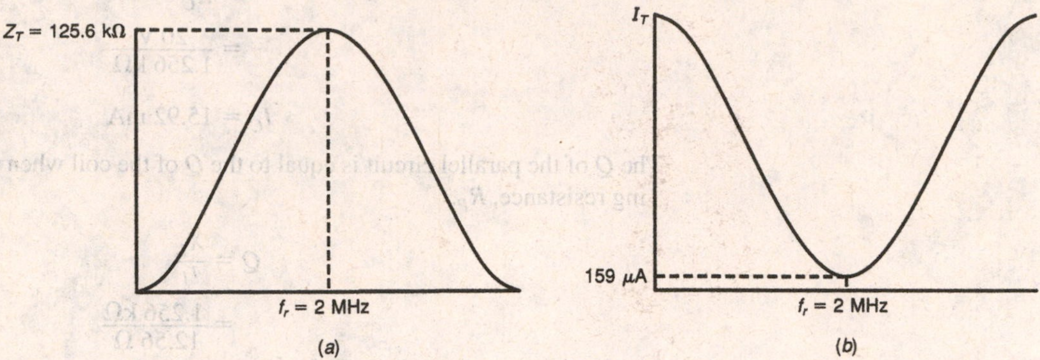

Fig. 25-9 Response curves for parallel resonant circuits. (*a*) Graph of Z_T versus frequency. (*b*) Graph of I_T versus frequency.

In the equation for Z_T, the phase angle θ is nearly 0°. This, in turn, causes the total current I_T to be in phase with V_T. The phase angle θ, therefore, is very close to 0° at resonance for a parallel LC circuit. Since $\theta = 0°$, PF = 1.

To calculate the bandwidth BW, f_1, and f_2, we proceed as follows:

$$\text{BW} = \frac{f_r}{Q}$$

$$= \frac{2\text{ MHz}}{100}$$

$$\text{BW} = 20\text{ kHz}$$

$$f_1 = f_r - \frac{\text{BW}}{2}$$

$$= 2\text{ MHz} - \frac{20\text{ kHz}}{2}$$

$$f_1 = 1.99\text{ MHz}$$

$$f_2 = f_r + \frac{\text{BW}}{2}$$

$$= 2\text{ MHz} + \frac{20\text{ kHz}}{2}$$

$$f_2 = 2.01\text{ MHz}$$

Figure 25-10 shows the response curve of impedance with the edge frequencies included. The bandwidth of a parallel resonant circuit is defined as the range of frequencies where the total impedance, Z_T, is at or above 70.7% of its maximum value at resonance.

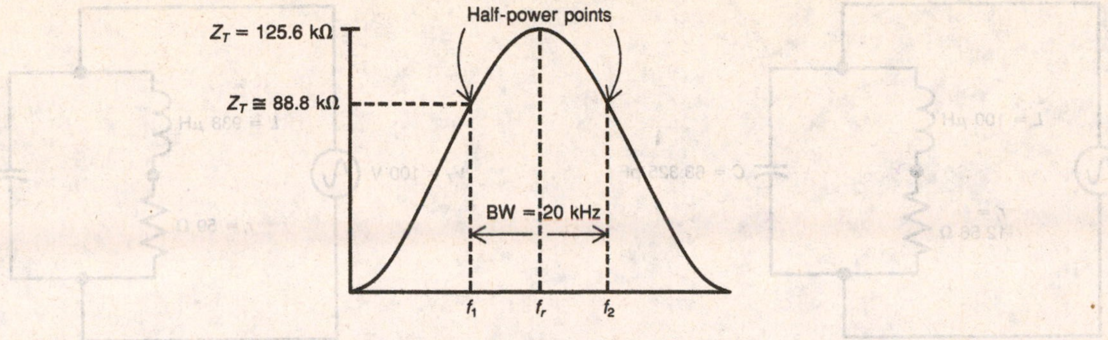

Fig. 25-10 Response curve for a parallel resonant circuit showing the edge frequencies f_1 and f_2.

PRACTICE PROBLEMS

Sec. 25-2 For Probs. 1–5, refer to Fig. 25-11.

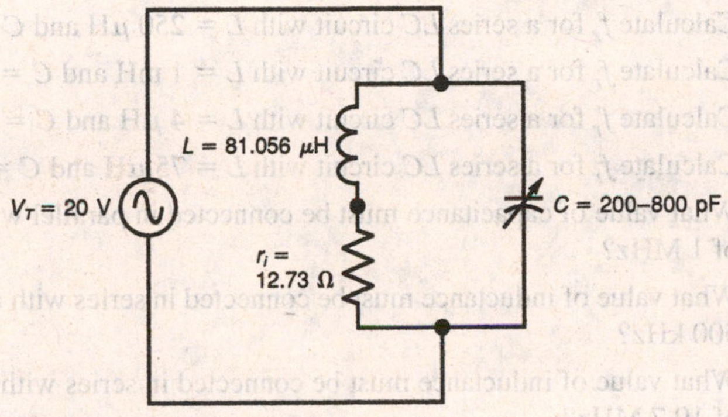

Fig. 25-11

1. With C adjusted to 200 pF, solve for f_r, BW, f_1, and f_2. Also, solve for the following at f_r: X_L, X_C, I_L, I_C, Q, θ, Z_T, I_T, PF, and real power.
2. Calculate Z_T and I_T at both f_1 and f_2.
3. At f_1, does the circuit appear inductive, capacitive, or resistive? Calculate θ at f_1.
4. At f_2, does the circuit appear inductive, capacitive, or resistive? Calculate θ at f_2.
5. With C adjusted to 800 pF, solve for f_r, BW, f_1, and f_2. Also, solve for the following at f_r: X_L, X_C, I_L, I_C, Q, Z_T, I_T, θ, PF, and real power.

For Figs. 25-12 through Fig. 25-15, calculate f_r, BW, f_1, and f_2. Also, solve for the following at f_r: X_L, X_C, I_L, I_C, Q, Z_T, I_T, θ, PF, and real power.

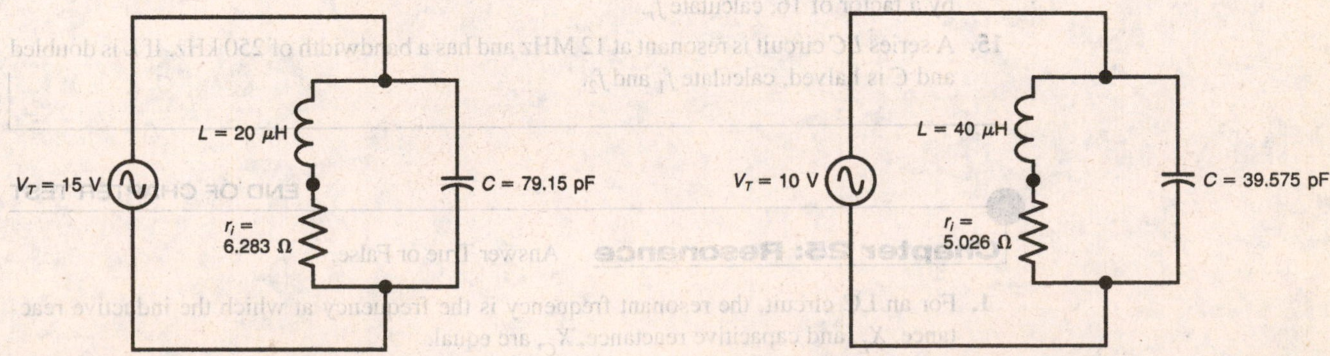

Fig. 25-12

Fig. 25-13

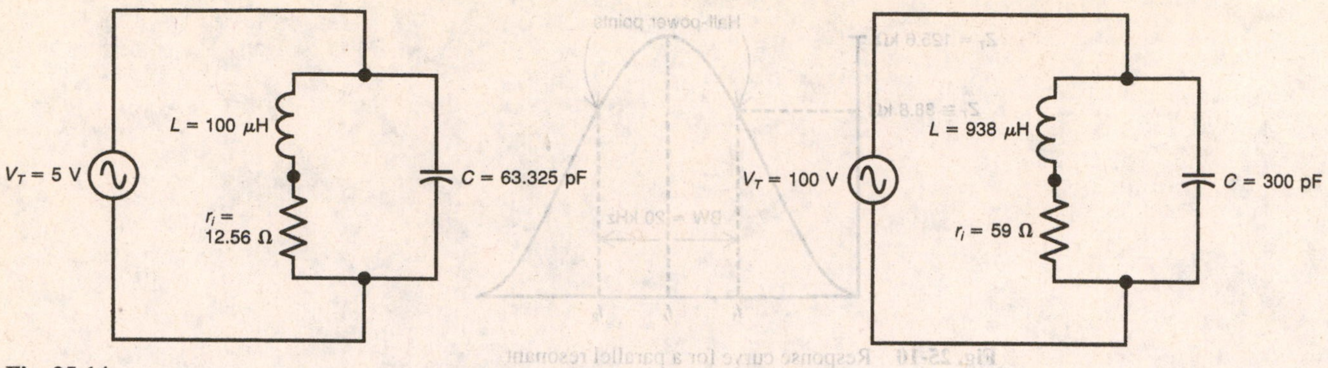

Fig. 25-14

Fig. 25-15

WORD PROBLEMS

Sec. 25-2 Solve the following.

1. Calculate f_r for a series LC circuit with L = 250 μH and C = 5 pF.
2. Calculate f_r for a series LC circuit with L = 1 mH and C = 122.35 pF.
3. Calculate f_r for a series LC circuit with L = 4 μH and C = 13.827 pF.
4. Calculate f_r for a series LC circuit with L = 75 μH and C = 40 pF.
5. What value of capacitance must be connected in parallel with a 33-μH inductor for an f_r of 1 MHz?
6. What value of inductance must be connected in series with a 40-pF capacitance for a f_r of 500 kHz?
7. What value of inductance must be connected in series with a 10-pF capacitance for an f_r of 10.7 MHz?
8. What value of capacitance must be connected in parallel with a 7.5-μH inductor for an f_r of 455 kHz?
9. A response curve for a resonant circuit at 5 MHz has a bandwidth of 100 kHz. If the Q of the circuit is quadrupled, calculate the edge frequencies f_1 and f_2.
10. A response curve for a resonant circuit at 1 MHz has a bandwidth of 25 kHz. If Q is reduced by a factor of 4, calculate f_1 and f_2.
11. A parallel resonant circuit has an X_L value of 5 kΩ, negligible r_i, and a shunt R_P of 150 kΩ. Calculate Q.
12. For a parallel resonant circuit, $I_T = I_L/Q$. Derive this equation.
13. Calculate the lowest and highest values of C needed to tune through the AM broadcast band of 540–1,600 kHz using a coil of 250 μH.
14. A parallel resonant circuit has a resonant frequency of 8 MHz. If the C value is increased by a factor of 16, calculate f_r.
15. A series LC circuit is resonant at 12 MHz and has a bandwidth of 250 kHz. If L is doubled and C is halved, calculate f_1 and f_2.

END OF CHAPTER TEST

Chapter 25: Resonance Answer True or False.

1. For an LC circuit, the resonant frequency is the frequency at which the inductive reactance, X_L, and capacitive reactance, X_C, are equal.
2. For a series resonant circuit, the current, I, is maximum at the resonant frequency.

3. For a parallel resonant circuit, the impedance, Z, is minimum at the resonant frequency.
4. Decreasing either L or C increases the resonant frequency, f_r.
5. In Fig. 25-1, the impedance, Z, at resonance is 12.56 Ω.
6. In Fig. 25-1, the power factor, PF, at resonance is one or unity.
7. In Fig. 25-1, the voltage across either L or C is Q times greater than the total voltage, V_T, at the resonant frequency, f_r.
8. The bandwidth of a resonant circuit increases as the Q increases.
9. For a high Q parallel resonant circuit, the impedance, Z_T, at resonance is calculated as $Q \times X_L$.
10. For either a high Q series or parallel resonant circuit, the phase angle, θ, equals 0° at resonance.

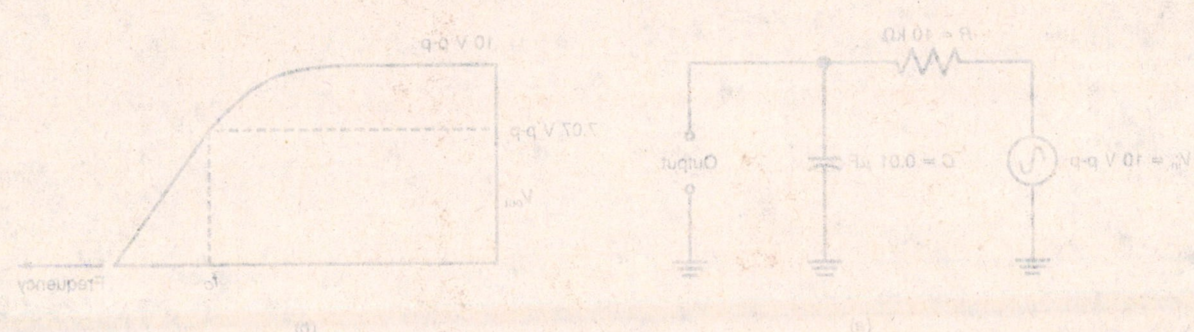

CHAPTER

26

Filters

An electrical filter is a network used to pass a given group of frequencies from its input to its output with little or no attenuation while rejecting or attenuating all others. The frequencies that pass through the filter are said to be in the passband. The frequencies rejected or severely attenuated are said to be in the filter's stop band. The cutoff frequency for a filter is defined as that frequency for which the output has dropped to 70.7% of its maximum value. The cutoff frequency is designated f_C. Inductors and capacitors are used for filtering because their reactance values vary with frequency.

SEC. 26-1 LOW-PASS FILTERS

The basic *RC* circuit in Fig. 26-1*a* can be used as a low-pass filter. The frequency response curve in Fig. 26-1*b* shows how the output voltage from the filter varies with frequency.

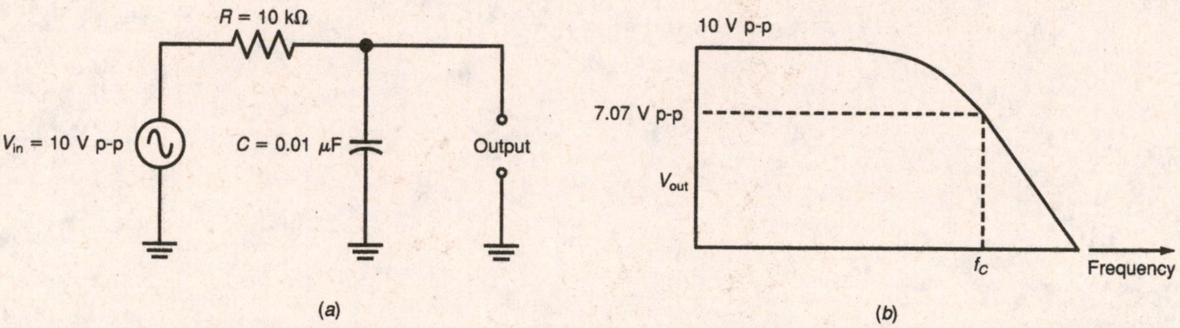

Fig. 26-1 Low-pass filter circuit. (*a*) Series *RC* circuit. Output voltage across *C*. (*b*) Response curve of a low-pass filter.

The cutoff frequency, f_C, for Fig. 26-1*a* can be found using the formula

$$f_C = \frac{1}{2\pi RC}$$

At this frequency the output voltage is down to 70.7% of its maximum value. The value of output voltage for any frequency is found using the formula

$$V_{out} = \frac{X_C}{\sqrt{R^2 + X_C^2}} \times V_{in}$$

To calculate the phase angle, θ, between V_{out} and V_{in} use the formula

$$\theta = \arctan\left(-\frac{R}{X_C}\right)$$

The basic *RL* circuit in Fig. 26-2*a* can also be used as a low-pass filter. The frequency response curve in Fig. 26-2*b* shows how the output voltage varies with frequency.

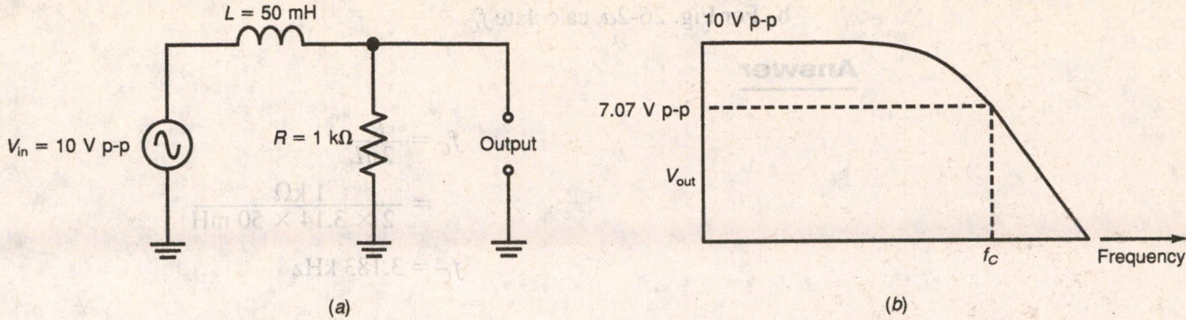

Fig. 26-2 Low-pass filter circuit. (*a*) Series *RL* circuit. Output voltage across *R*. (*b*) Response curve of a low-pass filter.

The cutoff frequency, f_C, for Fig. 26-2*a* can be found using the formula

$$f_C = \frac{R}{2\pi L}$$

At this frequency, the output voltage is down to 70.7% of its maximum value. The value of output voltage for any frequency is found using the formula

$$V_{out} = \frac{R}{\sqrt{R^2 + X_L^2}} \times V_{in}$$

To calculate the phase angle, θ, between V_{out} and V_{in} use the formula

$$\theta = \arctan\left(-\frac{X_L}{R}\right)$$

Low-pass filters can be identified as having capacitors in parallel with the load and/or inductors in series with the load. The use of both a series inductor and a parallel capacitor improves the filtering by providing a sharper cutoff between the stop band and passband.

Capacitors are often connected in parallel with a resistance in order to bypass or shunt AC signals around the resistance above a specified frequency. Figure 26-3 shows *C* connected across resistor R_2.

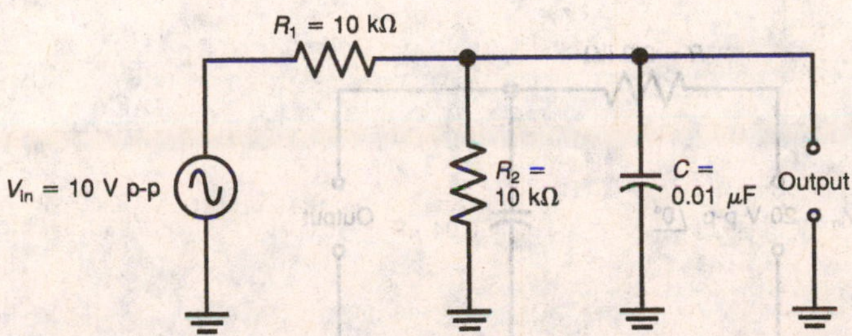

Fig. 26-3 Low-pass filter circuit. Capacitor *C* bypasses R_2 at high frequencies.

To effectively bypass R_2, X_C must be one-tenth of R_2 or less. The result is practically zero AC voltage across R_2 for the frequencies that produce an X_C value of 1 kΩ or less. The circuit is definitely a low-pass filter.

Solved Problems

a. For Fig. 26-1*a*, calculate f_C.

Answer

$$f_C = \frac{1}{2\pi RC}$$

$$= \frac{1}{2 \times 3.14 \times 10\ \text{k}\Omega \times 0.01\ \mu\text{F}}$$

$$f_C = 1.5915\ \text{kHz}$$

b. For Fig. 26-2*a*, calculate f_C.

Answer

$$f_C = \frac{R}{2\pi L}$$

$$= \frac{1\ \text{k}\Omega}{2 \times 3.14 \times 50\ \text{mH}}$$

$$f_C = 3.183\ \text{kHz}$$

c. For Fig. 26-3, calculate the lowest frequency that will be effectively bypassed.

Answer

Since X_C must be one-tenth the value of R_2 for effective bypassing, we find f_{lowest} as follows.

$$f_{\text{lowest}} = \frac{1}{2\pi(R_2/10)C}$$

$$= \frac{1}{2 \times 3.14 \times 1\ \text{k}\Omega \times 0.01\ \mu\text{F}}$$

$$f_{\text{lowest}} = 15.915\ \text{kHz}$$

All frequencies *below* 15.915 kHz will not be effectively bypassed. However, frequencies above 15.915 kHz will be effectively bypassed.

PRACTICE PROBLEMS

Sec. 26-1 Problems 1–10 refer to Fig. 26-4.

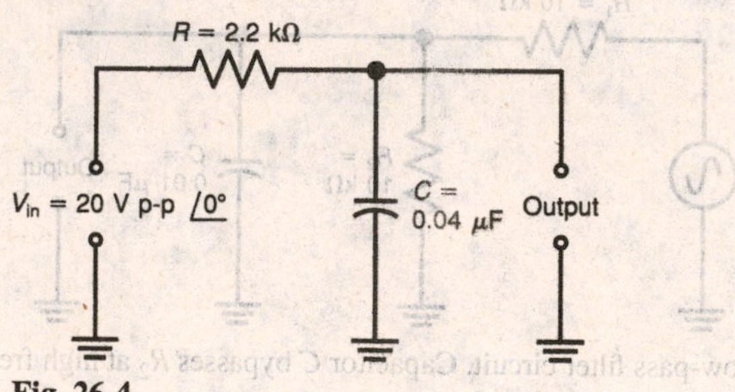

Fig. 26-4

1. Calculate f_C.
2. What is the peak-to-peak value of V_{out} at f_C?
3. Calculate the phase angle θ between V_{out} and V_{in} at f_C.
4. Calculate V_{out} for the frequencies listed: **(a)** 100 Hz, **(b)** 250 Hz, **(c)** 500 Hz, and **(d)** 750 Hz.
5. Calculate the phase angle θ between V_{out} and V_{in} for each frequency listed in Prob. 4.
6. Calculate V_{out} for the frequencies listed: **(a)** 5 kHz, **(b)** 10 kHz, **(c)** 50 kHz, **(d)** 100 kHz, and **(e)** 250 kHz.
7. Calculate the phase angle θ between V_{out} and V_{in} for each frequency listed in Prob. 6.
8. Give a general statement concerning the phase angle θ between V_{in} and V_{out} in both the passband and stop band.

9. Would we have to increase or decrease R to **(a)** lower f_C and **(b)** raise f_C?

10. Would we have to increase or decrease C to **(a)** lower f_C and **(b)** raise f_C?

Problems 11–19 refer to Fig. 26-5.

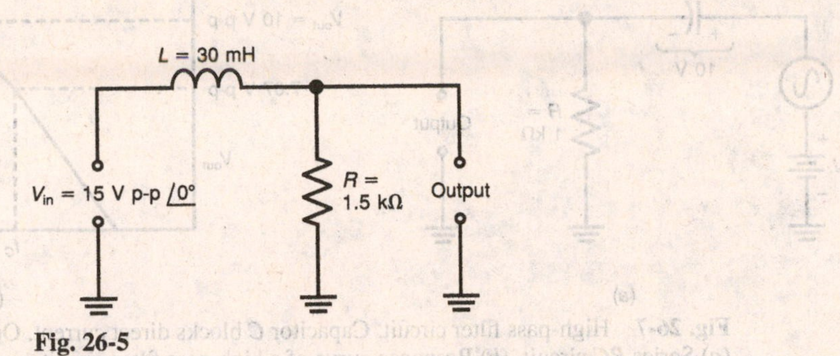

Fig. 26-5

11. Calculate f_C.

12. What is the peak-to-peak output voltage at f_C?

13. Calculate the phase angle θ between V_{out} and V_{in} at f_C.

14. Calculate V_{out} for the frequencies listed: **(a)** 100 Hz, **(b)** 250 Hz, **(c)** 500 Hz, **(d)** 1 kHz, and **(e)** 2 kHz.

15. Calculate V_{out} for the frequencies listed: **(a)** 25 kHz, **(b)** 50 kHz, **(c)** 100 kHz, **(d)** 500 kHz, and **(e)** 1 MHz.

16. Calculate the phase angle θ between V_{out} and V_{in} for each frequency listed in Prob. 14.

17. Calculate the phase angle θ between V_{out} and V_{in} for each frequency listed in Prob. 15.

18. Would we have to increase or decrease R to **(a)** lower f_C and **(b)** raise f_C?

19. Would we have to increase or decrease L to **(a)** lower f_C and **(b)** raise f_C?

Problems 20–22 refer to Fig. 26-6.

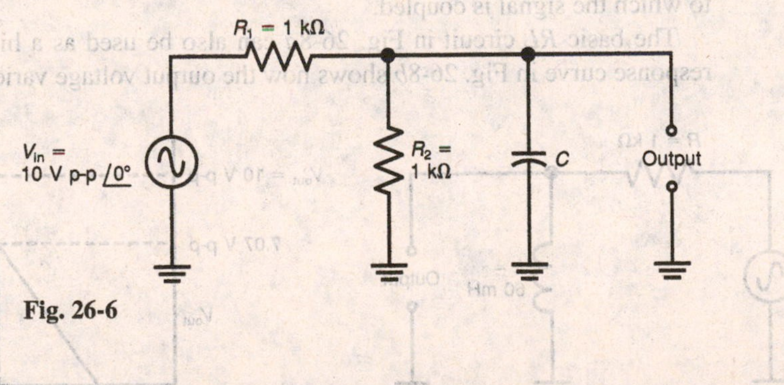

Fig. 26-6

20. Calculate the required values of C necessary to effectively bypass the frequencies listed: **(a)** 5 kHz, **(b)** 40 kHz, and **(c)** 100 kHz.

21. What is the maximum value of output voltage for low frequencies?

22. **(a)** What is the value of output voltage for a frequency of 31.83 kHz if $C = 0.01$ μF? (Use complex numbers.) **(b)** What is the phase angle θ between V_{out} and V_{in} at 31.83 kHz? **(c)** Is the frequency of 31.83 kHz the cutoff frequency for the filter? Why?

SEC. 26-2 HIGH-PASS FILTERS

The basic *RC* circuit in Fig. 26-7*a* can be used as a high-pass filter. The frequency response curve in Fig. 26-7*b* shows how the output voltage from the filter varies with frequency.

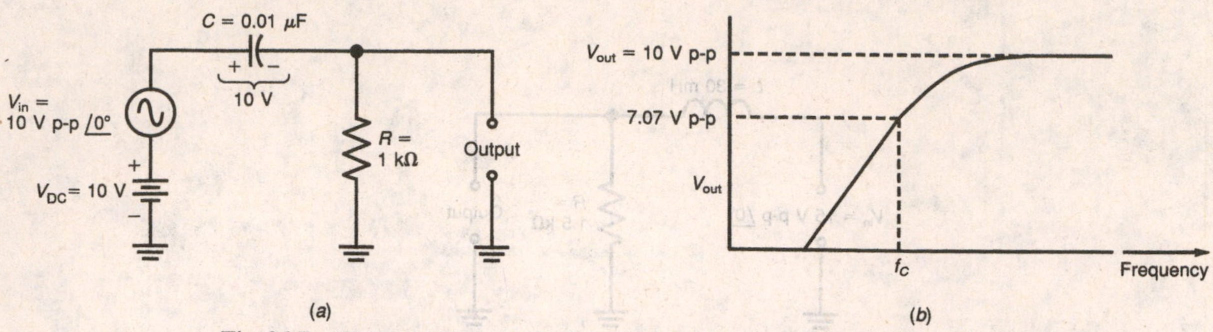

Fig. 26-7 High-pass filter circuit. Capacitor *C* blocks direct current. Output voltage is across *R*. (*a*) Series *RC* circuit. (*b*) Response curve of a high-pass filter circuit.

The cutoff frequency f_C for Fig. 26-7*a* can be found using the formula

$$f_C = \frac{1}{2\pi RC}$$

At this frequency, the output voltage is down to 70.7% of its maximum value. The value of output voltage for any frequency is found using the formula

$$V_{\text{out}} = \frac{R}{\sqrt{R^2 + X_C{}^2}} \times V_{\text{in}}$$

To calculate the phase angle, θ, between V_{out} and V_{in} use the formula

$$\theta = \arctan\left(\frac{X_C}{R}\right)$$

Capacitors are often used to couple the desired AC signals from the output of one circuit to the input of another. The capacitor will also block the DC component of the input waveform from reaching the output. The capacitor will charge to the average value of the input waveform, which is its DC value. To effectively couple signals from one point in a circuit to another, the capacitive reactance, X_C, must be one-tenth or less than the resistance in the circuit to which the signal is coupled.

The basic *RL* circuit in Fig. 26-8*a* can also be used as a high-pass filter. The frequency response curve in Fig. 26-8*b* shows how the output voltage varies with frequency.

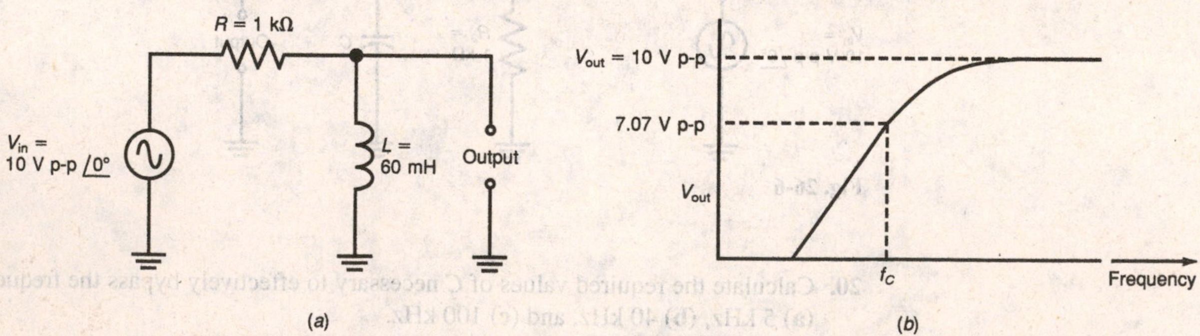

Fig. 26-8 High-pass filter circuit. (*a*) Series *RL* circuit. Output voltage is across *L*. (*b*) Response curve of a high-pass filter circuit.

The cutoff frequency, f_C, for Fig. 26-8*a* can be found using the formula

$$f_C = \frac{R}{2\pi L}$$

At this frequency, the output voltage is down to 70.7% of its maximum value. The value of output voltage for any frequency is found using the formula

$$V_{out} = \frac{X_L}{\sqrt{R^2 + X_L{}^2}} \times V_{in}$$

To calculate the phase angle, θ between V_{out} and V_{in} use the formula

$$\theta = \arctan\left(\frac{R}{X_L}\right)$$

High-pass filters can be identified as having capacitors in series with the load and/or inductors in parallel with the load. The use of both a series capacitor and a parallel inductor improves the filtering by providing a sharper cutoff between the stop band and passband.

Solved Problems

a. For Fig. 26-7, calculate f_C.

Answer

$$f_C = \frac{1}{2\pi RC}$$

$$= \frac{1}{2 \times 3.14 \times 1\ \text{k}\Omega \times 0.01\ \mu\text{F}}$$

$$f_C = 15.915\ \text{kHz}$$

b. For Fig. 26-7, calculate the lowest frequency that will be effectively coupled to the load, R_L.

Answer

Since X_C must be one-tenth R_L for effective coupling, we proceed as shown.

$$f_{lowest} = \frac{1}{2\pi(R_L/10)C}$$

$$= \frac{1}{2 \times 3.14 \times (1\ \text{k}\Omega\ /\ 10) \times 0.01\ \mu\text{F}}$$

$$f_{lowest} = 159.15\ \text{kHz}$$

All frequencies 159.15 kHz and above will be coupled to the load, R_L, with little or no attenuation.

c. For Fig. 26-8, calculate f_C.

Answer

$$f_C = \frac{R}{2\pi L}$$

$$= \frac{1\ \text{k}\Omega}{2 \times 3.14 \times 60\ \text{mH}}$$

$$f_C = 2.652\ \text{kHz}$$

PRACTICE PROBLEMS

Sec. 26-2 Problems 1–11 refer to Fig. 26-9.

1. Calculate f_C.
2. What is the peak-to-peak output voltage at f_C?

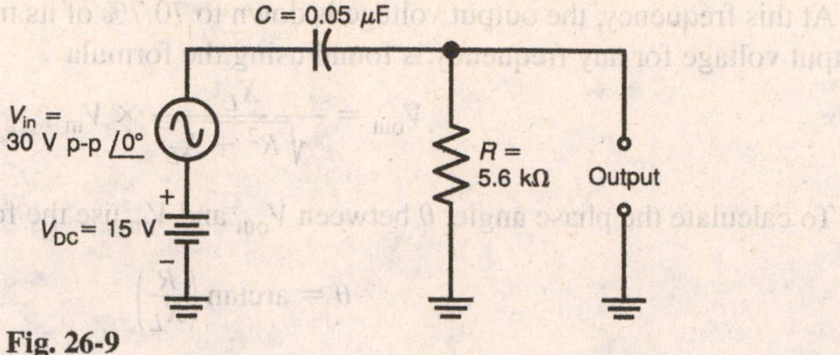

Fig. 26-9

3. Find the following: (a) the DC voltage across R and (b) the DC voltage across C.
4. For the R value of 5.6 kΩ shown, calculate the required values of C to effectively couple the frequencies listed: (a) 5 kHz, (b) 20 kHz, and (c) 100 kHz.
5. What is the maximum value of output voltage for high frequencies?
6. Calculate the phase angle θ between V_{in} and V_{out} at f_C.
7. Calculate V_{out} for the frequencies listed: (a) 5 kHz, (b) 10 kHz, (c) 25 kHz, and (d) 100 kHz.
8. Calculate V_{out} for the frequencies listed: (a) 25 Hz, (b) 50 Hz, and (c) 100 Hz.
9. Give a general statement concerning the phase angle θ between V_{in} and V_{out} in both the passband and stop band. (*Hint:* Find θ between V_{out} and V_{in} for Probs. 7 and 8.)
10. Should R be increased or decreased to (a) lower f_C and (b) raise f_C?
11. Would we have to increase or decrease C to (a) lower f_C and (b) raise f_C?

Problems 12–20 refer to Fig. 26-10.

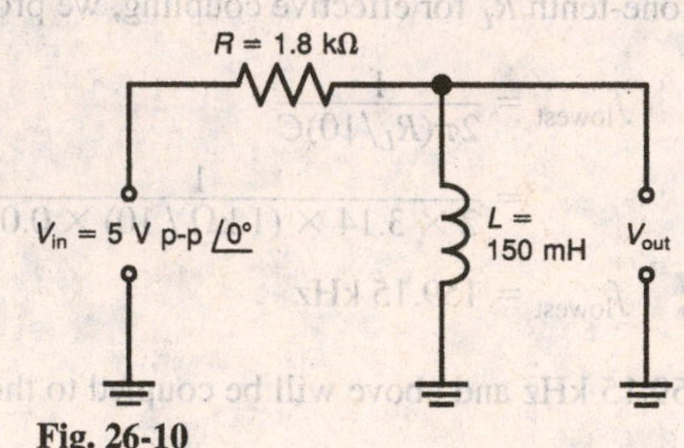

Fig. 26-10

12. Calculate f_C.
13. What is the peak-to-peak output voltage at f_C?
14. What is the maximum value of output voltage for high frequencies?
15. Calculate the phase angle θ between V_{out} and V_{in} at f_C.
16. Calculate V_{out} for the frequencies listed: (a) 10 kHz, (b) 25 kHz, (c) 50 kHz, and (d) 100 kHz.
17. Calculate V_{out} for the frequencies listed: (a) 10 Hz, (b) 50 Hz, (c) 200 Hz, and (d) 1 kHz.
18. Give a general statement concerning the phase angle θ between V_{in} and V_{out} in the passband and stop band.
19. Would we have to increase or decrease R to (a) lower f_C and (b) raise f_C?
20. Should the value of L be increased or decreased in order to (a) lower f_C and (b) raise f_C?

SEC. 26-3 RESONANT FILTERS

Tuned circuits provide a method of filtering a band of radio frequencies because relatively small values of L and C can be used. Resonant filters are called *bandpass* or *bandstop filters*. The width of the band of frequencies passed or attenuated is determined by the Q of a tuned circuit. Bandpass filters can be series resonant circuits in series with the load, or parallel resonant circuits connected in parallel with the load. Bandstop filters can be parallel resonant circuits in series with the load or series resonant circuits in parallel with the load.

PRACTICE PROBLEMS

Sec. 26-3 Identify the filters shown in Figs. 26-11 through 26-21 as low-pass, high-pass, bandpass, or bandstop.

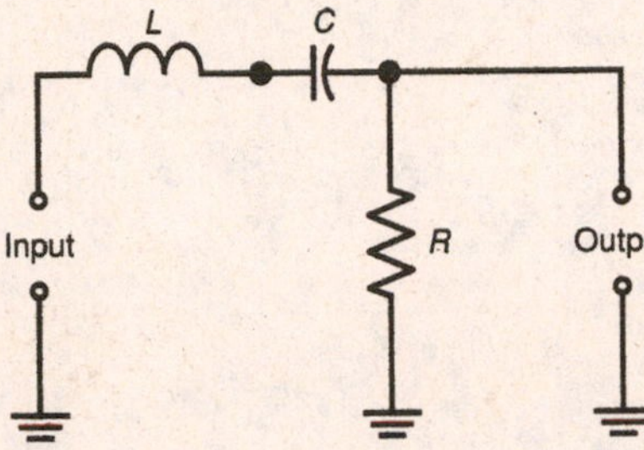

Fig. 26-11

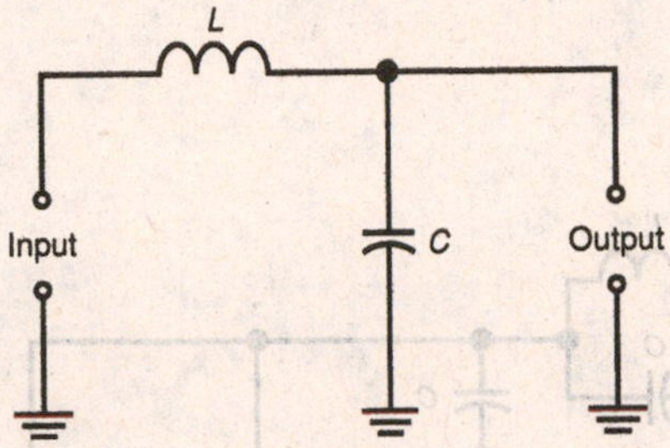

Fig. 26-12

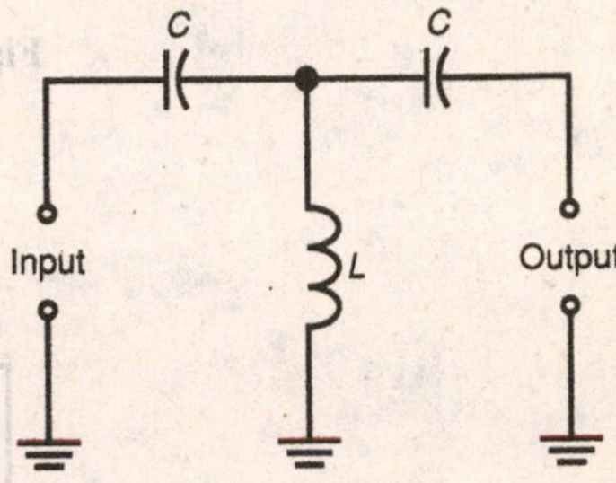

Fig. 26-13

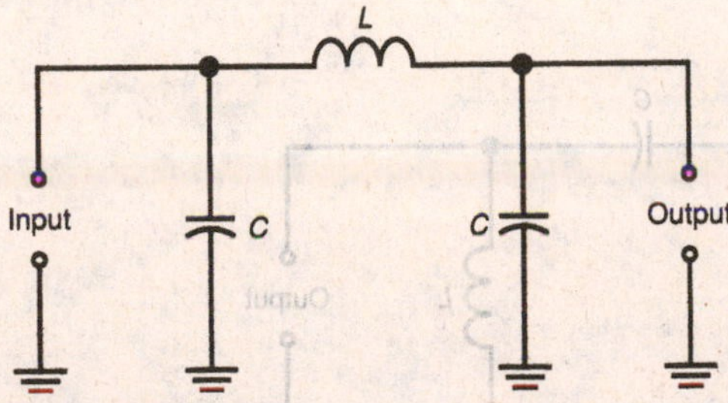

Fig. 26-14

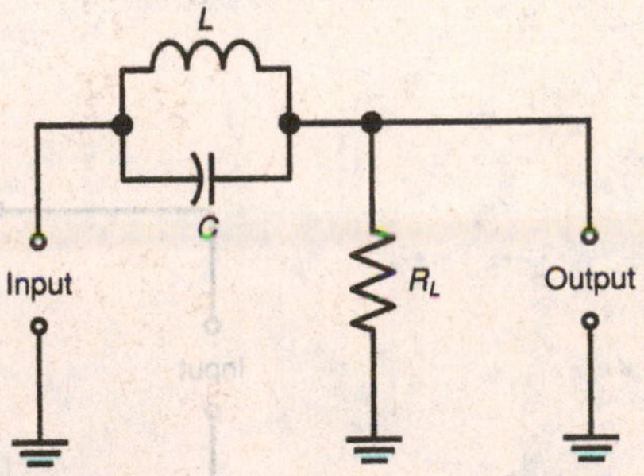

Fig. 26-15

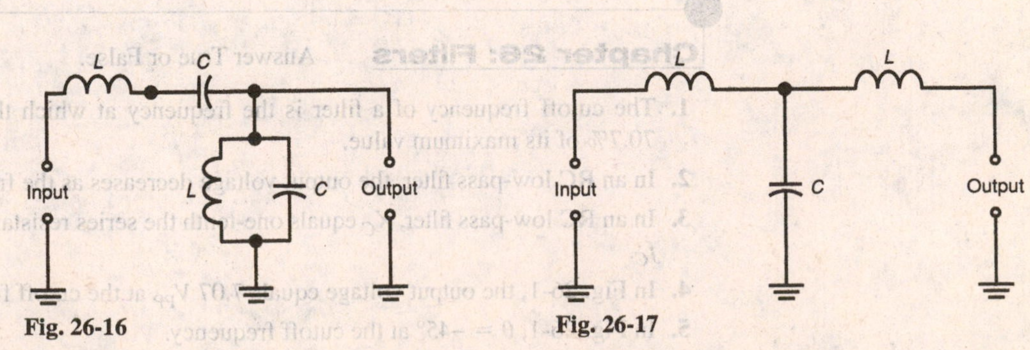

Fig. 26-16

Fig. 26-17

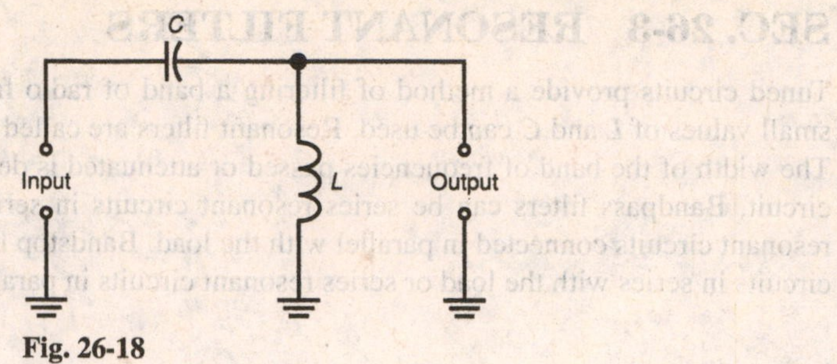

Fig. 26-18

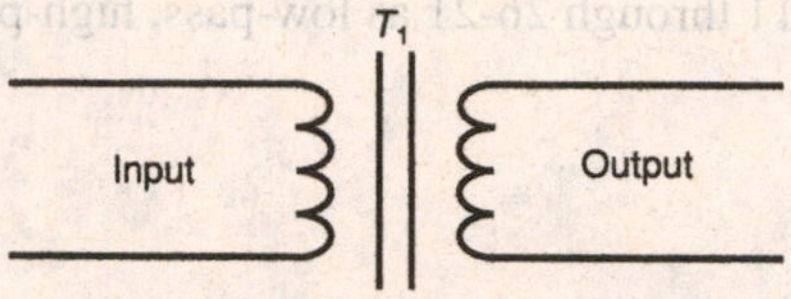

Fig. 26-19

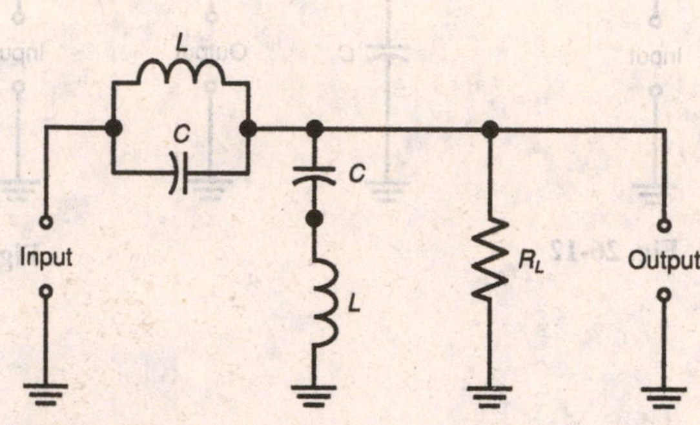

Fig. 26-20

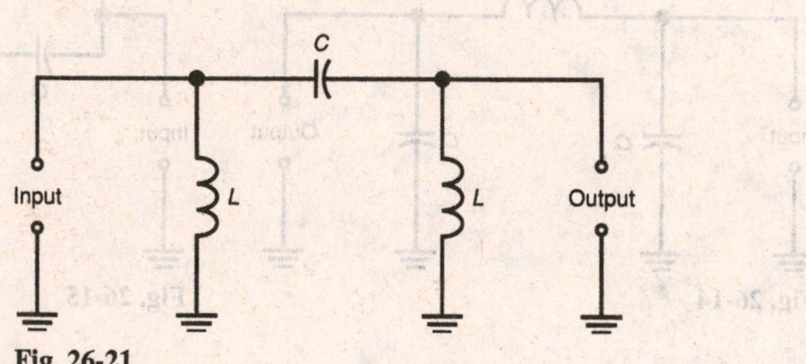

Fig. 26-21

END OF CHAPTER TEST

Chapter 26: Filters

Answer True or False.

1. The cutoff frequency of a filter is the frequency at which the output voltage drops to 70.7% of its maximum value.
2. In an RC low-pass filter, the output voltage decreases as the frequency increases.
3. In an RC low-pass filter, X_C equals one-tenth the series resistance at the cutoff frequency, f_C.
4. In Fig. 26-1, the output voltage equals 7.07 V_{PP} at the cutoff frequency.
5. In Fig. 26-1, $\theta = -45°$ at the cutoff frequency.

6. In an *RL* low-pass filter, the output voltage is taken across the inductor, *L*.
7. In Fig. 26-2, the output voltage is approximately the same as the input voltage for frequencies significantly below the cutoff frequency, f_C.
8. A bypass capacitor must have an X_C value equal to one-tenth or less the resistance across which it is connected for the lowest frequency intended to be bypassed.
9. In an *RC* high-pass filter, the capacitor, *C*, blocks the DC component of the input voltage, V_{in}.
10. In Fig. 26-7, X_C must be one-tenth or less of the series resistance to effectively couple the signal from the input to the output.
11. In Fig. 26-7, the DC voltage across the resistor, *R*, is 10 V.
12. In Fig. 26-8, the output voltage increases as the frequency decreases.
13. The bandwidth of either a bandpass or bandstop filter is determined by the *Q* of the circuit.
14. Bandpass filters consist of series resonant circuits in series with the load and/or parallel resonant circuits in parallel with the load.
15. Bandstop filters consist of series resonant circuits in series with the load and/or parallel resonant circuits in parallel with the load.

ANSWERS

Answers to Odd-Numbered Problems and End of Chapter Tests

Introduction to Powers of 10

Sec. I-1 Practice Problems

1. 5.6×10^{4}
3. 5.0×10^{-2}
5. 5.6×10^{-11}
7. 3.3×10^{5}
9. 1.0×10^{-3}
11. 1.5×10^{-4}
13. 4.7×10^{9}
15. 5.5×10^{-3}
17. 1.0×10^{5}
19. 6.6×10^{-5}
21. 5.0×10^{9}
23. 2.15×10^{2}
25. 6.58×10^{-1}

Sec. I-1 More Practice Problems

1. 26,600,000
3. 0.0551
5. 72.1
7. 0.0000000000136
9. 0.36
11. 0.0000136
13. 756,000,000
15. 68
17. 0.30
19. 0.000000033
21. 4,700
23. 3.3
25. 4,700,000

Sec. I-2 Practice Problems

1. 47.2×10^{3}
3. 650×10^{-3}
5. 1.875×10^{3}
7. 75×10^{-3}
9. 82×10^{-12}
11. 1.68×10^{6}
13. 650×10^{-6}
15. 12.5×10^{3}
17. 470×10^{-12}
19. 330×10^{-9}
21. 68×10^{3}
23. 68×10^{-12}
25. 8.8×10^{-3}

Sec. I-2 More Practice Problems

1. 39 kΩ
3. 42 nV
5. 270 kΩ
7. 330 pF
9. 68 nF
11. 1.5 GHz
13. 150 mW
15. 75 μW
17. 56 kΩ
19. 12 Ω
21. 1.8 MHz
23. 33 μH
25. 1.49 MHz

Sec. I-3 Practice Problems

1. 22,000 μA
3. 4,700 kΩ
5. 1.51 MHz
7. 330,000 nF
9. 0.5 mH
11. 100,000 pF
13. 30 μF
15. 2.5 μF
17. 5,600 kΩ
19. 0.047 GΩ
21. 0.0055 μF
23. 5,600,000 Ω
25. 0.001 nF

Sec. I-4 Practice Problems

1. 1.75×10^{6}
3. 1.25×10^{-3}
5. 7.0×10^{-2}
7. 2.04×10^{-6}
9. 3.5×10^{6}
11. 2.7×10^{3}
13. 4.75×10^{7}
15. 5.0×10^{3}
17. 7.5×10^{5}
19. 1.0×10^{2}

Sec. I-5 Practice Problems

1. 5.0×10^4
3. 7.26×10^7
5. 4.5×10^0
7. 7.5×10^1
9. 1.0×10^4
11. 4×10^{-1}
13. 2.0×10^4
15. 1.0×10^0
17. 6.4×10^2
19. 4.0×10^1

Sec. I-6 Practice Problems

1. 10^{-8}
3. 10^{-1}
5. 10^3
7. 10^{12}
9. 10^6
11. 10^{-15}
13. 10^{-14}
15. 10^{25}

Sec. I-7 Practice Problems

1. 9.0×10^{10}
3. 1.0×10^2
5. 1.6×10^3
7. 1.21×10^{-6}
9. 4.9×10^5
11. 4.0×10^{-10}
13. 2.5×10^{-1}
15. 9.0×10^{10}

Sec. I-8 Practice Problems

1. 6.325×10^2
3. 6.0×10^{-6}
5. 1.581×10^{-1}
7. 2.0×10^{-2}
9. 9.487×10^{-2}
11. 5.916×10^{-3}
13. 2.236×10^4
15. 2.236×10^2

CHAPTER 1 Electricity

Sec. 1-1 Practice Problems

1. (a) electron, (b) proton
3. Like charges repel, unlike charges attract.
5. Dielectric
7. $-Q = 4$ C
9. $+Q = 100$ C
11. $+Q = 0.1$ C
13. $+Q = 3$ C
15. $-Q = 80$ pC

Sec. 1-2 Practice Problems

1. When two unlike charges are in close proximity to each other, they have the ability to do the work of moving an electric charge, such as an electron.
3. Volt (V)
5. Voltage is a measure of the amount of work or energy needed to move a specific amount of charge between two points.
7. $V = 12$ V
9. 120 joules

Sec. 1-3 Practice Problems

1. Current can be defined as the rate of charge flow.
3. I
5. No. Potential difference is necessary to move electric charge.
7. $I = 10$ A
9. $I = 10{,}000$ A or $I = 10$ kA
11. $Q = 1$C
13. $Q = 0.001$ C or $Q = 1$ mC
15. $T = 20$ s

Sec. 1-4 Practice Problems

1. Resistance is the opposition to the flow of current.
3. Ohm (Ω)
5. Conductance is the opposite of resistance.

$$G = \frac{1}{R}$$

7. Siemens
9. (a) $G = 100$ S
   (b) $G = 400$ S
   (c) $G = 2{,}500$ S or 2.5 kS
11. (a) $R = 20\ \Omega$
    (b) $R = 150\ \Omega$
    (c) $R = 7.5\ \Omega$

Sec. 1-5 Practice Problems

1. (a) There must be a source of potential difference or voltage.
   (b) There must be a complete path for current flow.
   (c) Resistance to limit the amount of current flow.
3. Yes.
5. An open circuit is a break in a wire or component that interrupts the path for current flow.
7. The applied voltage is still present even if there is not a complete path for current flow. A voltage source generates its potential difference internally and therefore is not dependant on whether or not there is current in the circuit.
9. The danger of a short circuit is that it can cause excessively high currents to flow and, as a result, can cause wire insulation to burn or melt and in turn start a fire.

CHAPTER 2 Resistor Color Coding

Sec. 2-1 Practice Problems

	Coded Value	Tolerance	Permissible Ohmic Range
Fig. 2-3	100 Ω	±10%	90 Ω–110 Ω
Fig. 2-5	39 Ω	±10%	35.1 Ω–42.9 Ω
Fig. 2-7	0.33 Ω	±10%	0.297 Ω–0.363 Ω
Fig. 2-9	47 kΩ	±5%	44.65 kΩ–49.35 kΩ
Fig. 2-11	91 Ω	±5%	86.45 Ω–95.55 Ω
Fig. 2-13	560 kΩ	±10%	504 kΩ–616 kΩ
Fig. 2-15	24 Ω	±5%	22.8 Ω–25.2 Ω
Fig. 2-17	470 Ω	±10%	423 Ω–517 Ω
Fig. 2-19	3.3 kΩ	±10%	2970 Ω–3630 Ω
Fig. 2-21	1.8 kΩ	±10%	1.62 kΩ–1.98 kΩ
Fig. 2-23	62 Ω	±5%	58.9 Ω–65.1 Ω
Fig. 2-25	2.7 kΩ	±10%	2.43 kΩ–2.97 kΩ
Fig. 2-27	2.2 MΩ	±10%	1.98 MΩ–2.42 MΩ
Fig. 2-29	51 kΩ	±5%	48.45 kΩ–53.55 kΩ
Fig. 2-31	1 kΩ	±5%	950 Ω–1.05 kΩ

Sec. 2-1 More Practice Problems

1. Blue, gray, yellow, gold
3. Brown, black, red, gold
5. Green, blue, black, silver
7. White, brown, yellow, silver
9. Red, red, red, gold
11. Orange, orange, orange no band
13. Brown, red, orange, gold
15. Brown, red, black, silver
17. Gray, red, silver, gold
19. Violet, green, orange, gold
21. Brown, green, green, silver
23. Violet, green, black, gold
25. Yellow, violet, silver, silver

Sec. 2-2 Practice Problems

	Coded Value	Tolerance	±Permissible Ohmic Range
Fig. 2-35	11.5 Ω	±1%	±0.115 Ω
Fig. 2-37	887 Ω	±1%	±8.87 Ω
Fig. 2-39	23.4 kΩ	±0.5%	±117 Ω
Fig. 2-41	301 Ω	±1%	±3.01 Ω
Fig. 2-43	277 kΩ	±0.1%	±277 Ω
Fig. 2-45	94.2 kΩ	±0.1%	±94.2 Ω
Fig. 2-47	750 kΩ	±2%	±15 kΩ
Fig. 2-49	7.77 kΩ	±0.1%	±7.77 Ω
Fig. 2-51	8.98 Ω	±0.5%	±0.0449 Ω
Fig. 2-53	22.6 kΩ	±1%	±226 Ω
Fig. 2-55	110 Ω	±2%	±2.2 Ω
Fig. 2-57	4.53 kΩ	±1%	±45.3 Ω

Sec. 2-2 More Practice Problems

1. Green, blue, red, gold, brown
3. White, gray, gray, red, violet
5. Yellow, yellow, red, brown, blue
7. Blue, blue, green, black, violet
9. Gray, green, blue, silver, green
11. Brown, gray, red, red, green
13. Violet, yellow, brown, silver, violet
15. Orange, green, violet, black, green

Sec. 2-3 Practice Problems

Fig. 2-61. 330 kΩ
Fig. 2-63. 4.7 MΩ
Fig. 2-65. 220 Ω
Fig. 2-67. 39.5 kΩ
Fig. 2-69. 750 kΩ
Fig. 2-71. 100 kΩ
Fig. 2-73. 3.6 Ω
Fig. 2-75. 56 kΩ
Fig. 2-77. 100 kΩ
Fig. 2-79. 47 Ω

CHAPTER 3 Ohm's Law

Sec. 3-1 Practice Problems

Fig. 3-3. $R = 1.5$ kΩ
Fig. 3-5. $R = 1.44$ kΩ, $G = 694.4\ \mu$S
Fig. 3-7. $I = 400\ \mu$A
Fig. 3-9. $R = 68$ kΩ
Fig. 3-11. $R = 2202.6$ Ω
Fig. 3-13. $R = 400$ kΩ
Fig. 3-15. $V = 16.5$ V
Fig. 3-17. $I = 300\ \mu$A
Fig. 3-19. $V = 900$ V
Fig. 3-21. $V = 180$ V
Fig. 3-23. $V = 75$ V
Fig. 3-25. $I = 23.3$ mA
Fig. 3-27. $V = 3.5$ kV, $G = 100\ \mu$S

Sec. 3-1 Word Problems

1. $R = 1.6\ k\Omega$
3. $I = 80\ \mu A$
5. $R = 160\ \Omega$
7. No; $I = 20$ A; fuse will blow, then $I = 0$ A
9. $R = 2.4\ \Omega$
11. $V = 6$ V
13. $R = 100\ \Omega$
15. $V = 212$ V (each lead has a voltage drop of 4 V)
17. $R = 113.3\ \Omega$
19. I is reduced by a factor of 4.

Sec. 3-2 Practice Problems

Fig. 3-29. $P = 80$ mW
Fig. 3-31. $R = 1024\ \Omega$
Fig. 3-33. $I = 1\ \mu A$
Fig. 3-35. $R = 1.25\ k\Omega$
Fig. 3-37. $I = 100\ \mu A$
Fig. 3-39. $P = 1.4$ W, $R = 560\ \Omega$
Fig. 3-41. $I = 1.25$ A, $R = 1.6\ \Omega$
Fig. 3-43. $I = 5.454$ A, $R = 40.3\ \Omega$
Fig. 3-45. $P = 3.52$ kW, $R = 55\ \Omega$
Fig. 3-47. $V = 20$ V, $R = 500\ \Omega$

Sec. 3-2 Word Problems

1. $P = 3.3$ kW
3. $I = 250$ mA
5. $V = 70.7$ V
7. $V = 4.89$ V
9. $R = 144\ \Omega$
11. $P = 225$ mW
13. $I = 57.73$ mA
15. $P = 25$ W

Sec. 3-3 Word Problems

1. \$0.36
3. \$1.44
5. \$55.44
7. \$1.08
9. \$31.68

Sec. 3-4 Word Problems

1. % Efficiency = 67.8%
   Cost = \$5.28
3. % Efficiency = 69%
   Cost = \$7.78
5. 7.17 A
7. % Efficiency = 84.77%
9. $I = 416.67$ mA

CHAPTER 4 Series Circuits

Sec. 4-1 Practice Problems

Fig. 4-3. $R_T = 20\ k\Omega$, $I_T = 3$ mA, $V_1 = 9.9$ V, $V_2 = 14.1$ V, $V_3 = 36$ V, $P_T = 180$ mW, $P_1 = 29.7$ mW, $P_2 = 42.3$ mW, and $P_3 = 108$ mW
Fig. 4-5. $R_T = 70\ k\Omega$, $I_T = 500\ \mu A$, $V_1 = 7.5$ V, $V_2 = 11$ V, $V_3 = 16.5$ V, $P_T = 17.5$ mW, $P_1 = 3.75$ mW, $P_2 = 5.5$ mW, and $P_3 = 8.25$ mW
Fig. 4-7. $R_T = 75\ k\Omega$, $I_T = 160\ \mu A$, $V_1 = 4.32$ V, $V_2 = 5.28$ V, $V_3 = 2.4$ V, $P_T = 1.92$ mW, $P_1 = 691.2\ \mu W$, $P_2 = 844.8\ \mu W$, and $P_3 = 384\ \mu W$
Fig. 4-9. $R_T = 4.5\ k\Omega$, $I_T = 20$ mA, $V_1 = 24$ V, $V_2 = 30$ V, $V_3 = 36$ V, $P_T = 1.8$ W, $P_1 = 480$ mW, $P_2 = 600$ mW, and $P_3 = 720$ mW
Fig. 4-11. $R_T = 2.5\ k\Omega$, $I_T = 4$ mA, $V_1 = 1.56$ V, $V_2 = 3.64$ V, $V_3 = 4.8$ V, $P_T = 40$ mW, $P_1 = 6.24$ mW, $P_2 = 14.56$ mW, and $P_3 = 19.2$ mW

Sec. 4-1 Word Problems

1. $R = 15\ \Omega$
3. $R = 400\ \Omega$
5. $V = 2.5$ V
7. $R = 250\ \Omega$
9. $V = 15$ V
11. $V_{max} = 22.82$ V
13. $R_1 = 500\ \Omega$
    $R_2 = 1\ k\Omega$
    $R_3 = 1.5\ k\Omega$
    $R_4 = 3\ k\Omega$
15. $R_S = 100\ \Omega$

Sec. 4-2 Practice Problems

Fig. 4-13. $R_T = 9\ k\Omega$, $I_T = 2$ mA, $R_2 = 2.2\ k\Omega$, $V_1 = 2.4$ V, $V_2 = 4.4$ V, $P_T = 36$ mW, $P_1 = 4.8$ mW, $P_2 = 8.8$ mW, and $P_3 = 22.4$ mW
Fig. 4-15. $R_T = 60\ k\Omega$, $I_T = 1.5$ mA, $V_1 = 45$ V, $V_3 = 15$ V, $R_1 = 30\ k\Omega$, $P_T = 135$ mW, $P_1 = 67.5$ mW, $P_2 = 45$ mW, and $P_3 = 22.5$ mW
Fig. 4-17. $R_T = 3\ k\Omega$, $I_T = 15$ mA, $V_1 = 10.2$ V, $V_2 = 12.3$ V, $V_3 = 22.5$ V, $P_T = 675$ mW, $P_1 = 153$ mW, $P_2 = 184.5$ mW, and $P_3 = 337.5$ mW
Fig. 4-19. $V_T = 10$ V, $R_T = 200\ \Omega$, $I_T = 50$ mA, $V_2 = 1.65$ V, $V_3 = 2.35$ V, $P_T = 500$ mW, $P_1 = 300$ mW, $P_2 = 82.5$ mW, and $P_3 = 117.5$ mW
Fig. 4-21. $R_T = 80\ k\Omega$, $V_1 = 6$ V, $V_2 = 8.8$ V, $V_4 = 13.2$ V, $V_T = 32$ V, $R_3 = 10\ k\Omega$, $P_T = 12.8$ mW, $P_1 = 2.4$ mW, $P_2 = 3.52$ mW, $P_3 = 1.6$ mW, and $P_4 = 5.28$ mW
Fig. 4-23. $V_T = 100$ V, $I_T = 50$ mA, $R_T = 2\ k\Omega$, $V_1 = 5$ V, $V_2 = 11$ V, $V_3 = 34$ V, $R_4 = 1\ k\Omega$, $P_T = 5$ W, $P_2 = 550$ mW, $P_3 = 1.7$ W, and $P_4 = 2.5$ W
Fig. 4-25. $V_T = 30$ V, $I_T = 2$ mA, $R_T = 15\ k\Omega$, $V_1 = 2$ V, $V_3 = 4.4$ V, $V_4 = 20$ V, $R_4 = 10\ k\Omega$, $P_1 = 4$ mW, $P_2 = 7.2$ mW, $P_3 = 8.8$ mW, and $P_4 = 40$ mW

Fig. 4-27. $V_T = 24$ V, $R_T = 1.2$ kΩ, $V_1 = 4$, $V_2 = 12$ V, $V_3 = 5$ V, $V_4 = 3$ V, $R_2 = 600$ Ω, $R_4 = 150$ Ω, $P_1 = 80$ mW, $P_3 = 100$ mW, and $P_4 = 60$ mW
Fig. 4-29. $V_T = 10$ V, $I_T = 1$ mA, $R_T = 10$ kΩ, $V_1 = 1.2$ V, $V_2 = 1$ V, $V_3 = 5.6$ V, $V_4 = 2.2$ V, $P_T = 10$ mW, $P_1 = 1.2$ mW, $P_2 = 1$ mW, $P_3 = 5.6$ mW, and $P_4 = 2.2$ mW
Fig. 4-31. $V_T = 120$ V, $I_T = 30$ mA, $R_T = 4$ kΩ, $V_2 = 45$ V, $V_3 = 45$ V, $R_2 = 1.5$ kΩ, $P_T = 3.6$ W, $P_1 = 900$ mW, $P_2 = 1.35$ W, and $P_3 = 1.35$ W

Sec. 4-3 Practice Problems

Fig. 4-35. $V_{AG} = 11.1$ V, $V_{BG} = 8.1$ V, $V_{CG} = -9.9$ V
Fig. 4-37. $V_{AG} = -20$ V, $V_{BG} = -15.5$ V, $V_{CG} = 30$ V
Fig. 4-39. $V_{AG} = -44$ V, $V_{BG} = -24$ V, $V_{CG} = 20$ V, $V_{DG} = 56$ V
Fig. 4-41. $V_{AG} = 12$ V, $V_{BG} = 9$ V, $V_{CG} = 3.6$ V, $V_{DG} = -6$ V
Fig. 4-43. $V_{AG} = -20$ V, $V_{BG} = -40$ V, $V_{CG} = -60$ V

Sec. 4-4 Practice Problems

Fig. 4-47. $R_T = 12.5$ kΩ, $I_T = 2$ mA, $V_{R_1} = 20$ V, $V_{R_2} = 5$ V, $P_T = 50$ mW, $P_1 = 40$ mW, $P_2 = 10$ mW, $V_{range} = 0$ to -5 V
Fig. 4-49. $R_T = 4.5$ kΩ, $I_T = 3.33$ mA, $V_{R_1} = 6$ V, $V_{R_2} = 1.66$ V, $V_{R_3} = 7.33$ V, $P_T = 50$ mW, $P_1 = 20$ mW, $P_2 = 5.55$ mW, $P_3 = 24.4$ mW, $V_{range} = -7.33$ V to -9 V
Fig. 4-51. $R_T = 20$ kΩ, $I_T = 1.35$ mA, $V_{R_1} = 9.18$ V, $V_{R_2} = 6.75$ V, $V_{R_3} = 11.07$ V, $P_T = 36.45$ mW, $P_1 = 12.393$ mW, $P_2 = 9.1125$ mW, $P_3 = 14.9445$ mW, $V_{range} = 0$ to 6.75 V
Fig. 4-53. $R_T = 40$ kΩ, $I_T = 400$ μA, $V_{R_1} = 6$ V, $V_{R_2} = 4$ V, $V_{R_3} = 6$ V, $P_T = 6.4$ mW, $P_1 = 2.4$ mW, $P_2 = 1.6$ mW, $P_3 = 2.4$ mW, $V_{range} = -10$ V to -6 V

Sec. 4-4 More Practice Problems

Fig. 4-55. Wiper set to *A*. $R_T = 1$ kΩ, $I_T = 10$ mA, $V_{R_1} = 10$ V, $V_{R_2} = 0$ V, $P_T = 100$ mW, $P_1 = 100$ mW, and $P_2 = 0$ W
Wiper set to *B*. $R_T = 2$ kΩ, $I_T = 5$ mA, $V_{R_1} = 5$ V, $V_{R_2} = 5$ V, $P_T = 50$ mW, $P_1 = 25$ mW, and $P_2 = 25$ mW
Fig. 4-57. Wiper set to *A*. $R_T = 3$ kΩ, $I_T = 8$ mA, $V_{R_1} = 14.4$ V, $V_{R_2} = 0$ V, $V_{R_3} = 9.6$ V, $P_T = 192$ mW, $P_1 = 115.2$ mW, $P_2 = 0$ W, and $P_3 = 76.8$ mW
Wiper set to *B*. $R_T = 4$ kΩ, $I_T = 6$ mA, $V_{R_1} = 10.8$ V, $V_{R_2} = 6$ V, $V_{R_3} = 7.2$ V, $P_T = 144$ mW, $P_1 = 64.8$ mW, $P_2 = 36$ mW, and $P_3 = 43.2$ mW
Fig. 4-59. Wiper set to *A*. $R_T = 8$ kΩ, $I_T = 2.25$ mA, $V_{R_1} = 10.575$ V, $V_{R_3} = 7.425$ V, $V_{R_2} = 0$ V, $P_T = 40.5$ mW, $P_1 = 23.8$ mW, $P_3 = 16.7$ mW, and $P_2 = 0$ W
Wiper set to *B*. $R_T = 9$ kΩ, $I_T = 2$ mA, $V_{R_1} = 9.4$ V, $V_{R_3} = 6.6$ V, $V_{R_2} = 2$ V, $P_T = 36$ mW, $P_1 = 18.8$ mW, $P_3 = 13.2$ mW, and $P_2 = 4$ mW
Fig. 4-61. Wiper set to *A*. $R_T = 2.5$ kΩ, $I_T = 48$ mA, $V_{R_1} = 48$ V, $V_{R_2} = 0$ V, $V_{R_3} = 72$ V, $P_T = 5.76$ W, $P_1 = 2.304$ W, $P_2 = 0$ W, and $P_3 = 3.456$ W
Wiper set to *B*. $R_T = 3$ kΩ, $I_T = 40$ mA, $V_{R_1} = 40$ V, $V_{R_2} = 20$ V, $V_{R_3} = 60$ V, $P_T = 4.8$ W, $P_1 = 1.6$ W, $P_2 = 800$ mW, and $P_3 = 2.4$ W

Sec. 4-5 Practice Problems

Fig. 4-65b. R_3 open
Fig. 4-65c. R_1 shorted
Fig. 4-65d. R_2 open
Fig. 4-65e. R_3 shorted
Fig. 4-65f. R_1 open
Fig. 4-65g. R_2 shorted

CHAPTER 5 Parallel Circuits

Sec. 5-1 Practice Problems

Fig. 5-3. $R_{EQ} = 687.5$ Ω, $I_T = 16$ mA, $I_1 = 11$ mA, $I_2 = 5$ mA, $P_T = 176$ mW, $P_1 = 121$ mW, and $P_2 = 55$ mW
Fig. 5-5. $R_{EQ} = 1.2$ kΩ, $I_T = 22.5$ mA, $I_1 = 15$ mA, $I_2 = 4.5$ mA, $I_3 = 3$ mA, $P_T = 607.5$ mW, $P_1 = 405$ mW, $P_2 = 121.5$ mW, and $P_3 = 81$ mW
Fig. 5-7. $R_{EQ} = 750$ Ω, $I_T = 60$ mA, $I_1 = 30$ mA, $I_2 = 22.5$ mA, $I_3 = 7.5$ mA, $P_T = 2.7$ W, $P_1 = 1.35$ W, $P_2 = 1.0125$ W, and $P_3 = 337.5$ mW
Fig. 5-9. $R_{EQ} = 375$ Ω, $I_T = 40$ mA, $I_1 = 16.6$ mA, $I_2 = 8.3$ mA, $I_3 = 15$ mA, $P_T = 600$ mW, $P_1 = 250$ mW, $P_2 = 125$ mW, and $P_3 = 225$ mW
Fig. 5-11. $R_{EQ} = 2$ kΩ, $I_T = 6$ mA, $I_1 = 3$ mA, $I_2 = 2$ mA, $I_3 = 1$ mA, $P_T = 72$ mW, $P_1 = 36$ mW, $P_2 = 24$ mW, and $P_3 = 12$ mW

Sec. 5-1 Word Problems

1. $I_3 = 2$ mA
3. $R_X = 24$ kΩ
5. $R_X = 6$ kΩ
7. No effect. The V/I ratio remains the same.
9. $R_{EQ} = 46.45$ Ω
11. $R_{EQ} = 135$ Ω
13. 5 kΩ
15. $R_1 = 100$ Ω, $R_2 = 300$ Ω and $R_3 = 150$ Ω
17. Yes—I_T would equal 30.5 A.
19. No—I_T is only 19.5 A.
21. (a) $R = 12$ Ω
(b) $R = 20$ Ω
(c) $R = 15$ Ω
23. $I = 13$A
25. The lights and the microwave will still operate. The dishwasher and waffle maker will not operate.

Sec. 5-2 Practice Problems

Fig. 5-15. $R_{EQ} = 666.6\ \Omega$, $I_1 = 12$ mA, $I_2 = 4$ mA, $R_2 = 3\ k\Omega$, $R_3 = 6\ k\Omega$, $P_T = 216$ mW, $P_1 = 144$ mW, $P_2 = 48$ mW, and $P_3 = 24$ mW
Fig. 5-17. $R_{EQ} = 1.8\ k\Omega$, $I_1 = 1.5$ mA, $I_2 = 3$ mA, $I_3 = 500\ \mu A$, $R_2 = 3\ k\Omega$, $R_3 = 18\ k\Omega$, $P_T = 45$ mW, $P_1 = 13.5$ mW, and $P_2 = 27$ mW
Fig. 5-19. $V_T = 10$ V, $I_T = 62.5$ mA, $I_1 = 50$ mA, $I_3 = 2.5$ mA, $R_2 = 1\ k\Omega$, $R_3 = 4\ k\Omega$, $P_1 = 500$ mW, $P_2 = 100$ mW, and $P_3 = 25$ mW
Fig. 5-21. $V_T = 120$ V, $I_T = 3$ mA, $I_1 = 1$ mA, $I_3 = 1$ mA, $R_2 = 120\ k\Omega$, $R_3 = 120\ k\Omega$, $P_T = 360$ mW, and $P_1 = P_2 = P_3 = 120$ mW
Fig. 5-23. $V_T = 4.95$ V, $I_T = 50$ mA, $I_1 = 16.5$ mA, $I_2 = 11$ mA, $I_3 = 22.5$ mA, $R_2 = 450\ \Omega$, $P_T = 247.5$ mW, $P_1 = 81.675$ mW, and $P_3 = 111.375$ mW
Fig. 5-25. $V_T = 75$ V, $I_T = 300$ mA, $I_1 = 75$ mA, $I_2 = 150$ mA, $R_3 = 1\ k\Omega$, $P_T = 22.5$ W, $P_1 = 5.625$ W, $P_2 = 11.25$ W, and $P_3 = 5.625$ W
Fig. 5-27. $I_T = 50$ mA, $I_1 = 15$ mA, $I_2 = 10$ mA, $I_4 = 10$ mA, $R_3 = 1.2\ k\Omega$, $R_4 = 1.8\ k\Omega$, $P_T = 900$ mW, $P_1 = 270$ mW, $P_2 = 180$ mW, $P_3 = 270$ mW, and $P_4 = 180$ mW
Fig. 5-29. $R_{EQ} = 1.826\ k\Omega$, $I_T = 57.5$ mA, $I_2 = 7.5$ mA, $I_3 = 20$ mA, $I_4 = 25$ mA, $R_1 = 21\ k\Omega$, $R_3 = 5.25\ k\Omega$, $P_T = 6.0375$ W, $P_1 = 525$ mW, $P_2 = 787.5$ mW, $P_3 = 2.1$ W, and $P_4 = 2.625$ W
Fig. 5.31. $R_{EQ} = 200\ \Omega$, $I_T = 100$ mA, $I_1 = 50$ mA, $I_3 = 15$ mA, $I_4 = 5$ mA, $I_5 = 10$ mA, $R_2 = 1\ k\Omega$, $R_3 = 1.33\ k\Omega$, $R_5 = 2\ k\Omega$, $P_T = 2$ W, $P_1 = 1$ W, $P_2 = 400$ mW, $P_3 = 300$ mW, $P_4 = 100$ mW, and $P_5 = 200$ mW

Sec. 5-3 Practice Problems

1. R_2 open
3. R_4 open
5. Fuse F_1 blew due to excessive current drawn by one or more loads. Excessive current melts the fuse element, thereby disconnecting V_T from the resistive loads.
   To locate the defective load, measure the resistance of each branch separately, making sure to disconnect at least one lead from the circuit. When the defective load has been located, replace it with a good one, then replace f_1 and reapply power.
7. 0 V
9. 120 V. An open in one branch will not affect current in other branches.

CHAPTER 6 Series-Parallel Circuits

Sec. 6-1 Practice Problems

Fig. 6-3. $R_T = 10\ k\Omega$, $I_T = 3$ mA, $V_1 = 9.9$ V, $V_2 = 8.4$ V, $V_3 = 8.4$ V, $V_4 = 11.7$ V, $I_2 = 1.5$ mA, and $I_3 = 1.5$ mA
Fig. 6-5. $R_T = 9\ k\Omega$, $I_T = 2$ mA, $V_1 = 9.4$ V, $V_2 = 6.6$ V, $V_3 = 2$ V, $V_4 = 2$ V, $I_3 = 333.3\ \mu A$, and $I_4 = 1.66$ mA
Fig. 6-7. $R_T = 2.5\ k\Omega$, $I_T = 10$ mA, $V_1 = 10$ V, $V_2 = 5$ V, $V_3 = 10$ V, $V_4 = 15$ V, $I_2 = I_3 = 2.5$ mA, and $I_4 = 7.5$ mA
Fig. 6-9. $R_T = 7.5\ k\Omega$, $I_T = 2$ mA, $V_1 = 15$ V, $V_2 = 5$ V, $V_3 = 2.5$ V, $V_4 = 2.5$ V, $V_5 = 7.5$ V, $I_1 = 1.5$ mA, $I_2 = I_5 = 500\ \mu A$, and $I_3 = I_4 = 250\ \mu A$
Fig. 6-11. $R_T = 2\ k\Omega$, $I_T = 25$ mA, $V_1 = 9.75$ V, $V_2 = 10.25$ V, $V_3 = 4.125$ V, $V_4 = 2.75$ V, $V_5 = 3.375$ V, $V_6 = 30$ V, $I_2 = 12.5$ mA, and $I_3 = I_4 = I_5 = 12.5$ mA
Fig. 6-13. $R_T = 1.2\ k\Omega$, $I_T = 30$ mA, $V_1 = 5.4$ V, $V_2 = 22.5$ V, $V_3 = 8.1$ V, $V_4 = 6.15$ V, $V_5 = 5.1$ V, $V_6 = 11.25$ V, $I_2 = 22.5$ mA, and $I_4 = I_5 = I_6 = 7.5$ mA
Fig. 6-15. $R_T = 75\ k\Omega$, $I_T = 2$ mA, $V_1 = 20$ V, $V_2 = 100$ V, $V_3 = 30$ V, $V_4 = 10$ V, $V_5 = 68$ V, $V_6 = 22$ V, $I_2 = 1$ mA, and $I_4 = I_5 = I_6 = 1$ mA
Fig. 6-17. $R_T = 2.5\ k\Omega$, $I_T = 20$ mA, $V_1 = 20$ V, $V_2 = 6$ V, $V_3 = 24$ V, $V_4 = 1.5$ V, $V_5 = .45$ V, $V_6 = 4.05$ V, $V_7 = 135$ mV, $V_8 = 202.5$ mV, $V_9 = 112.5$ mV, $I_2 = 5$ mA, $I_4 = I_6 = 15$ mA, $I_5 = 3.75$ mA, and $I_7 = I_8 = I_9 = 11.25$ mA
Fig. 6-19. $R_T = 5\ k\Omega$, $I_T = 8$ mA, $V_1 = 17.6$ V, $V_2 = 14.4$ V, $V_3 = 14.4$ V, $V_4 = 8$ V, $V_5 = 6$ V, $V_6 = 4.32$ V, $V_7 = 4.32$ V, $V_8 = 4.08$ V, $I_2 = 800\ \mu A$, $I_3 = 1.2$ mA, $I_5 = I_8 = 6$ mA, $I_6 = 3.6$ mA, and $I_7 = 2.4$ mA
Fig. 6-21. $R_T = 4\ k\Omega$, $I_T = 25$ mA, $V_1 = 2.5$ V, $V_2 = 93.75$ V, $V_3 = 3.75$ V, $V_4 = 33.75$ V, $V_5 = 18.75$ V, $V_6 = 41.25$ V, $V_7 = 9.375$ V, $V_8 = 5.125$ V, $V_9 = 4.25$ V, $I_2 = 6.25$ mA, $I_4 = I_6 = 18.75$ mA, $I_5 = 12.5$ mA, and $I_7 = I_8 = I_9 = 6.25$ mA
Fig. 6-23. $R_T = 4.5\ k\Omega$, $I_T = 20$ mA, $V_1 = 66$ V, $V_2 = 16$ V, $V_3 = 2.24$ V, $V_4 = 5.76$ V, $V_5 = 5.76$ V, $V_6 = 8$ V, $V_7 = 24$ V, $I_1 = I_T = 20$ mA, $I_2 = 16$ mA, $I_3 = 8$ mA, $I_4 = 4.8$ mA, $I_5 = 3.2$ mA, $I_6 = 8$ mA, and $I_7 = 4$ mA

Sec. 6-2 Practice Problems

Fig. 6-25. $R_T = 1\ k\Omega$, $I_T = 50$ mA, $V_1 = 10$ V, $V_2 = 25$ V, $V_3 = 15$ V, $V_4 = 6.25$ V, $V_5 = 11.25$ V, $I_2 = 25$ mA, $I_4 = I_5 = I_6 = 25$ mA, and $V_T = 50$ V
Fig. 6-27. $R_T = 1\ k\Omega$, $I_T = 60$ mA, $V_1 = 23.4$ V, $V_2 = 13.2$ V, $V_3 = 23.4$ V, $V_4 = 4.4$ V, $V_6 = 4.4$ V, $I_2 = 40$ mA, and $I_4 = I_5 = I_6 = 20$ mA
Fig. 6-29. $R_T = 50\ k\Omega$, $I_T = 3$ mA, $V_1 = 36$ V, $V_3 = 99$ V, $V_4 = 1.5$ V, $V_5 = 10.2$ V, $V_6 = 3.3$ V, $I_2 = 1.5$ mA, $I_4 = I_5 = I_6 = 1.5$ mA, and $V_T = 150$ V
Fig. 6-31. $R_T = 1.2\ k\Omega$, $I_T = 30$ mA, $V_1 = 9.9$ V, $V_2 = 2.7$ V, $V_3 = 19.8$ V, $V_4 = 19.8$ V, $V_5 = 22.5$ V, $I_2 = 15$ mA, $I_3 = 6$ mA, $I_4 = 9$ mA, $I_5 = 15$ mA, and $R_2 = 180\ \Omega$

Sec. 6-3 Practice Problems

1. $R_X = 12.36\ \Omega$
3. $R_X = 41.66\ \Omega$
5. $R_X = 285{,}400\ \Omega$
7. $R_S = 6{,}843\ \Omega$
9. $\frac{R_1}{R_2} = \frac{1}{100}$
11. (a) $R_{X(max)} = 999.9\ \Omega$
(b) $R_{X(max)} = 99.99\ \Omega$
(c) $R_{X(max)} = 99{,}990\ \Omega$
(d) $R_{X(max)} = 999{,}900\ \Omega$
13. (a) $V_{AD} = 8.18$ V, $V_{BD} = 8.18$ V
(b) $I_T = 11.13$ mA
15. (a) $R_X = 47\ \Omega$
(b) $R_X = 47.4\ \Omega$
(c) $R_X = 47.36\ \Omega$
The ratio arm fraction of $\frac{1}{1}$ in part (a) allows the measurement accuracy to be within $\pm 1\ \Omega$. The ratio arm fraction of $\frac{1}{10}$ in part (b) allows the measurement accuracy to be within $\pm 0.1\ \Omega$. Finally, the ratio arm fraction of $\frac{1}{100}$ in part (c) allows the measurement accuracy to be within $\pm 0.01\ \Omega$. Obviously, the $\frac{1}{100}$ ratio arm fraction provides the most accurate measurement of the unknown resistance, R_X.

CHAPTER 7 Voltage and Current Dividers

Sec. 7-1 Practice Problems

Fig. 7-3.

V_{AG}	Position
7.5 mV	1
750 mV	2
7.5 V	3
75 V	4
150 V	5

Fig. 7-5.

V_{AG}	Position
1.7 V	1
3 V	2
20 V	3
26 V	4
30 V	5

Fig. 7-7.

V_{AG}	Position
50 mV	1
2.5 V	2
5 V	3
25 V	4
50 V	5

Fig. 7-9.

V_{AG}	Position
0.5 V	1
1 V	2
2.5 V	3
5 V	4
10 V	5
25 V	6
50 V	7
100 V	8

Fig. 7-11.

V_{AG}	Position
450 mV	1
11.25 V	2
45 V	3
67.5 V	4
112.5 V	5
135 V	6
168.75 V	7
180 V	8

Sec. 7-2 Practice Problems

Fig. 7-13. $I_1 = 2.5$ A, $I_2 = 0.5$ A
Fig. 7-15. $I_1 = 250$ mA, $I_2 = 1.2$ A, $I_3 = 50$ mA
Fig. 7-17. $I_1 = 2$ A, $I_2 = 1.33$ A, $I_3 = 2.66$ A
Fig. 7-19. $I_1 = 250$ mA, $I_2 = 1.25$ A, $I_3 = 1$ A, $I_4 = 2.5$ A
Fig. 7-21. $I_1 = 60\ \mu$A, $I_2 = 30\ \mu$A, $I_3 = 120\ \mu$A, $I_4 = 240\ \mu$A

Sec. 7-3 Practice Problems

Fig. 7-25. $R_1 = 333\ \Omega$, $R_2 = 450\ \Omega$, $I_T = 50$ mA, $R_{L_A} = 1.25$ kΩ, $R_{L_B} = 1.5$ kΩ, $R_{L_C} = 400\ \Omega$, $P_1 = 300$ mW, $P_2 = 180$ mW, and $P_3 = 30$ mW
Fig. 7-27. $R_1 = 214.28\ \Omega$, $R_2 = 4.5$ kΩ, $I_T = 90$ mA, $R_{L_A} = 3$ kΩ, $R_{L_B} = 750\ \Omega$, $R_{L_C} = 375\ \Omega$, $P_1 = 1.05$ W, $P_2 = 450$ mW, and $P_3 = 750$ mW
Fig. 7-29. $R_1 = 1.8$ kΩ, $R_3 = 1$ kΩ, $R_4 = 4$ kΩ, $I_T = 45$ mA, $R_{L_A} = 4$ kΩ, $R_{L_B} = 7$ kΩ, $R_{L_C} = 333.3\ \Omega$, $R_{L_D} = 500\ \Omega$, $P_1 = 1.125$ W, $P_2 = 600$ mW, $P_3 = 25$ mW, and $P_4 = 100$ mW
Fig. 7-31. $R_1 = 342.85\ \Omega$, $R_2 = 4.8$ kΩ, $R_3 = 1.6$ kΩ, $R_{L_A} = 1.8$ kΩ, $R_{L_B} = 800\ \Omega$, $R_{L_C} = 600\ \Omega$, $P_1 = 105$ mW, $P_2 = 30$ mW, and $P_3 = 90$ mW
Fig. 7-33. $I_T = 25$ mA, $R_1 = 1.5$ kΩ, $R_2 = 2$ kΩ, $R_3 = 333.3\ \Omega$, $R_{L_A} = 1$ kΩ, $R_{L_B} = 1$ kΩ, $R_{L_C} = 1.5$ kΩ, $P_1 = 150$ mW, $P_2 = 50$ mW, $P_3 = 75$ mW, and $P_4 = 500$ mW

Sec. 7-4 Practice Problems

1. L_A open
3. L_B open
5. R_2 open

CHAPTER 8 Direct Current Meters

Sec. 8-1 Practice Problems

Fig. 8-3.

Current Range	Shunt Resistance (R_{sh})	Total Meter Resistance (R_m)
1 mA	111.11 Ω	100 Ω
2 mA	52.63 Ω	50 Ω
5 mA	20.4 Ω	20 Ω
10 mA	10.1 Ω	10 Ω
20 mA	5.025 Ω	5 Ω
50 mA	2.004 Ω	2 Ω

Fig. 8-5.

Current Range	Shunt Resistance (R_{sh})	Total Meter Resistance (R_m)
100 μA	1.25 kΩ	1 kΩ
1 mA	102.04 Ω	100 Ω
2 mA	50.505 Ω	50 Ω
5 mA	20.08 Ω	20 Ω
10 mA	10.02 Ω	10 Ω

Fig. 8-7.

Current Range	Shunt Resistance (R_{sh})	Total Meter Resistance (R_m)
250 μA	555.55 Ω	500 Ω
1 mA	128.2 Ω	125 Ω
5 mA	25.125 Ω	25 Ω
25 mA	5.005 Ω	5 Ω

Sec. 8-1 More Practice Problems

Fig. 8-9. $R_1 = 40\ \Omega$, $R_2 = 8\ \Omega$, and $R_3 = 2\ \Omega$
Fig. 8-11. $R_1 = 900\ \Omega$, $R_2 = 90\ \Omega$, and $R_3 = 10\ \Omega$
Fig. 8-13. $R_1 = 2{,}250\ \Omega$, $R_2 = 225\ \Omega$, $R_3 = 25\ \Omega$
Fig. 8-15. $R_1 = 1{,}225\ \Omega$, $R_2 = 20\ \Omega$, $R_3 = 5\ \Omega$
Fig. 8-17. $R_1 = 950\ \Omega$, $R_2 = 40\ \Omega$, $R_3 = 10\ \Omega$

Sec. 8-2 Practice Problems

Fig. 8-21.

Voltmeter Range	R_{mult}	R_V	Ω/V Rating
2.5 V	2.45 kΩ	2.5 kΩ	1 kΩ/V
10 V	9.95 kΩ	10 kΩ	1 kΩ/V
50 V	49.95 kΩ	50 kΩ	1 kΩ/V
100 V	99.95 kΩ	100 kΩ	1 kΩ/V
250 V	249.95 kΩ	250 kΩ	1 kΩ/V

Fig. 8-23.

Voltmeter Range	R_{mult}	R_V	Ω/V Rating
1 V	95 kΩ	100 kΩ	100 kΩ/V
5 V	495 kΩ	500 kΩ	100 kΩ/V
25 V	2.495 MΩ	2.5 MΩ	100 kΩ/V
100 V	9.995 MΩ	10 MΩ	100 kΩ/V
500 V	49.995 MΩ	50 MΩ	100 kΩ/V

Fig. 8-25.

Voltmeter Range	R_{mult}	R_V	Ω/V Rating
3 V	73 kΩ	75 kΩ	25 kΩ/V
10 V	248 kΩ	250 kΩ	25 kΩ/V
30 V	748 kΩ	750 kΩ	25 kΩ/V
100 V	2.498 MΩ	2.5 MΩ	25 kΩ/V
300 V	7.498 MΩ	7.5 MΩ	25 kΩ/V

Sec. 8-2 More Practice Problems

Fig. 8-27. R_1 = 950 Ω, R_2 = 4 kΩ, R_3 = 5 kΩ, R_4 = 40 kΩ, and R_5 = 50 kΩ

Voltmeter Range	R_V
1 V	1 kΩ
5 V	5 kΩ
10 V	10 kΩ
50 V	50 kΩ
100 V	100 kΩ

Fig. 8-29. R_1 = 48 kΩ, R_2 = 150 kΩ, R_3 = 800 kΩ, R_4 = 4 MΩ, and R_5 = 15 MΩ

Voltmeter Range	R_V
2.5 V	50 kΩ
10 V	200 kΩ
50 V	1 MΩ
250 V	5 MΩ
1,000 V	20 MΩ

Fig. 8-31. R_1 = 122.5 kΩ, R_2 = 375 kΩ, R_3 = 2 MΩ, R_4 = 10 MΩ, and R_5 = 37.5 MΩ

Voltmeter Range	R_V
2.5 V	125 kΩ
10 V	500 kΩ
50 V	2.5 MΩ
250 V	12.5 MΩ
1,000 V	50 MΩ

Fig. 8-33. R_1 = 73 kΩ, R_2 = 175 kΩ, R_3 = 500 kΩ, R_4 = 1.75 MΩ, and R_5 = 5 MΩ

Voltmeter Range	R_V
3 V	75 kΩ
0 V	250 kΩ
30 V	750 kΩ
100 V	2.5 MΩ
300 V	7.5 MΩ

Fig. 8-35. R_1 = 23 kΩ, R_2 = 100 kΩ, R_3 = 500 kΩ, R_4 = 1.875 MΩ, and R_5 = 3.75 MΩ

Voltmeter Range	R_V
1 V	25 kΩ
5 V	125 kΩ
25 V	625 kΩ
100 V	2.5 MΩ
250 V	6.25 MΩ

Sec. 8-3 Practice Problems

1. V_1 =1.55 V, V_2 = 0.909 V, and V_3 = 1.32 V
3. V_1 = 4.39 V, V_2 = 1.99 V, and V_3 = 3.59 V
5. V_1 = 3.83 V, V_2 = 1.82 V, and V_3 = 3.16 V
7. The digital multimeter used in Prob. 3. The paralleling effect did not change total current by a significant amount.

Sec. 8-4 Practice Problems

1. 6 kΩ
3. 5 Ω
5. 3.33 Ω
7. R_S = 40 Ω
9. 1.2 V

CHAPTER 9 Kirchhoff's Laws

Sec. 9-1 Practice Problems

Fig. 9-3. I_2 = 9 A↑, $I_{5\text{-}6}$ = 8 A↑, I_7 = 2A ←, I_{11} = 7 A ←, I_{12} = 5 A ←, I_{13} = 14 A↑, and I_{14} = 21 A ←

Fig. 9-5. I_1 = 18 A →, I_2 = 8 A↑, I_4 = 2 A↑, $I_{5\text{-}6}$ = 8 A↑, I_{10} = 11 A↑, I_{11} = 12 A ←, and I_{13} = 6 A↑

Fig. 9-7. $I_2 = 7$ A↑, $I_3 = 26$ A →, $I_4 = 21$ A↑, $I_7 = 3$ A →, $I_{8\text{-}9} = 8$ A ←, $I_{11} = 10$ A ←, and $I_{14} = 33$ A ←
Fig. 9-9. $I_2 = 4$ A↑, $I_3 = 16$ A →, $I_4 = 13.5$ A↑, $I_{8\text{-}9} = 7$ A↑, $I_{10} = 8$ A↑, $I_{13} = 5$ A↑, and $I_{14} = 20$ A ←
Fig. 9-11. $I_1 = 25$ A →, $I_2 = 12$ A↑, $I_3 = 13$ A →, $I_4 = 11$ A↑, $I_7 = 2$ A →, $I_{11} = 10$ A ←, and $I_{13} = 15$ A↑

Sec. 9-2 Practice Problems

Fig. 9-15. $V_{AG} = +9$ V and $V_{BG} = -7.2$ V
Fig. 9-17. $V_{AG} = 2.25$ V, $V_{BG} = 487.5$ mV, and $V_{CG} = -750$ mV
Fig. 9-19. $V_{AG} = 12.73$ V, $V_{BG} = 13.9$ V, and $V_{CG} = 22$ V
Fig. 9-21. $V_{AG} = 25$ V, $V_{BG} = 15$ V, $V_{CG} = -10$ V, $V_{DG} = 20$ V, and $V_{EG} = 5$ V
Fig. 9-23. **(a)** 5 V, **(b)** −5 V, and **(c)** 0 V

Sec. 9-3 Practice Problems

Fig. 9-27. $V_{R_1} = 1.25$ V, $V_{R_2} = 8.75$ V, $V_{R_3} = 11.25$ V, $I_1 = 1.25$ mA, $I_2 = 8.75$ mA, and $I_3 = 7.5$ mA
Fig. 9-29. $V_{R_1} = 7.278$ V, $V_{R_2} = 22.72$ V, $V_{R_3} = 12.72$ V, $I_1 = 4.04$ mA, $I_2 = 2.77$ mA, and $I_3 = 1.272$ mA
Fig. 9-31. $V_{R_1} = 11.312$ V, $V_{R_2} = 8.688$ V, $V_{R_3} = 1.312$ V, $I_1 = 2.02$ mA, $I_2 = 3.62$ mA, and $I_3 = 1.6$ mA
Fig. 9-33. $V_{R_1} = 58$ V, $V_{R_2} = 42$ V, $V_{R_3} = 33$ V, $I_1 = 14.5$ mA, $I_2 = 42$ mA, and $I_3 = 27.5$ mA
Fig. 9-35. $V_{R_1} = 3.72$ V, $V_{R_2} = 2.28$ V, $V_{R_3} = 5.28$ V, $I_1 = 372$ mA, $I_2 = 152$ mA, and $I_3 = 220$ mA

Sec. 9-4 Practice Problems

Fig. 9-39. $V_{R_1} = 22.4$ V, $V_{R_2} = 82.4$ V, $V_{R_3} = 17.6$ V, $I_1 = 56$ mA, $I_2 = 824$ mA, and $I_3 = 880$ mA
Fig. 9-41. $V_{R_1} = 90$ V, $V_{R_2} = 210$ V, $V_{R_3} = 10$ V, $I_1 = 45$ mA, $I_2 = 35$ mA, and $I_3 = 10$ mA

CHAPTER 10 Network Theorems

Sec. 10-1 Practice Problems

Fig. 10-3. −66 V
Fig. 10-5. 0 V
Fig. 10-7. 1.9V
Fig. 10-9. −8.2 V

Sec. 10-1 More Practice Problems

Fig. 10-11. $I_1 = 9.6$ mA, $I_2 = 2.4$ mA, $I_3 = 7.2$ mA, $V_{R_1} = 9.6$ V, $V_{R_2} = 2.4$ V, and $V_{R_3} = 14.4$ V
Fig. 10-13. $I_1 = -340$ μA, $I_2 = 1.26$ mA, $I_3 = -1.6$ mA, $V_{R_1} = 6.12$ V, $V_{R_2} = 15.12$ V, and $V_{R_3} = 2.88$ V
Fig. 10-15. $I_1 = -2.166$ mA, $I_2 = -6.75$ mA, $I_3 = 4.583$ mA, $V_{R_1} = 3.25$ V, $V_{R_2} = 6.75$ V, and $V_{R_3} = 8.25$ V
Fig. 10-17. $I_1 = -13.125$ mA, $I_2 = -11.875$ mA, $I_3 = -1.25$ mA, $V_{R_1} = 13.125$ V, $V_{R_2} = 11.875$ V, and $V_{R_3} = 1.875$ V
Fig. 10-19. $I_1 = -1.514$ mA, $I_2 = -2.0286$ mA, $I_3 = I_4 = 514.28$ μA, $V_{R_1} = 22.714$ V, $V_{R_2} = 20.286$ V, $V_{R_3} = 4.217$ V, and $V_{R_4} = 3.497$ V

Sec. 10-2 Practice Problems

Fig. 10-23. $I_L = 5.3125$ mA

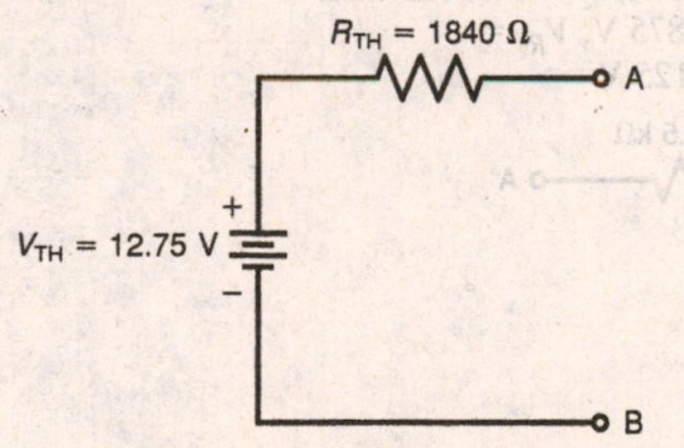

Fig. 10-27. $I_L = 21$ mA

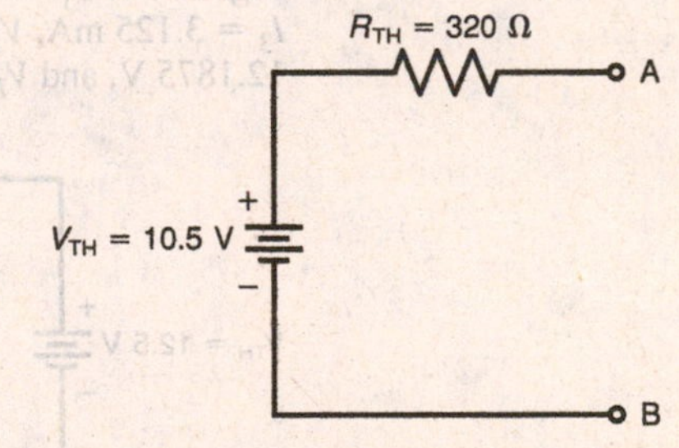

Fig. 10-25. $I_L = 300$ μA

Fig. 10-29. $I_L = 300$ μA

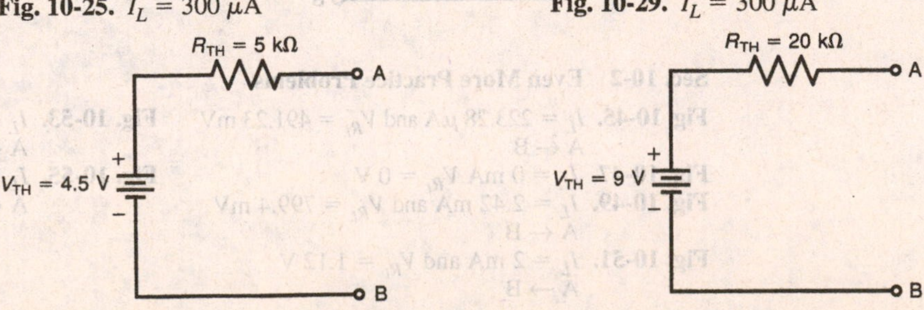

Fig. 10-31. $I_L = 203.5\ \mu A$

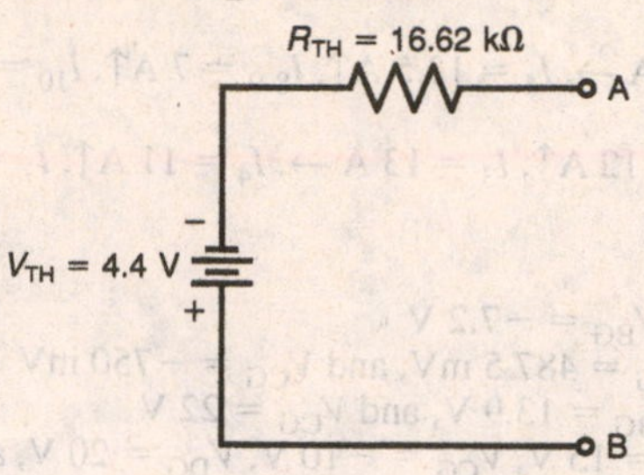

Sec. 10-2 More Practice Problems

Fig. 10-35. $I_1 = 150\ \mu A$, $I_2 = 1.35$ mA, $I_3 = 1.5$ mA, $V_{R_1} = 1.5$ V, $V_{R_2} = 13.5$ V, and $V_{R_3} = 22.5$ V

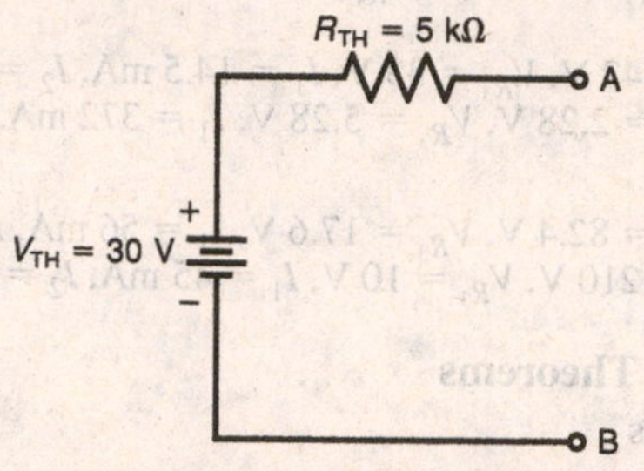

Fig. 10-37. $I_1 = 41.5\ \mu A$, $I_2 = 19.5\ \mu A$, $I_3 = 22\ \mu A$, $V_{R_1} = 4.15$ V, $V_{R_2} = 5.85$ V, and $V_{R_3} = 3.85$ V

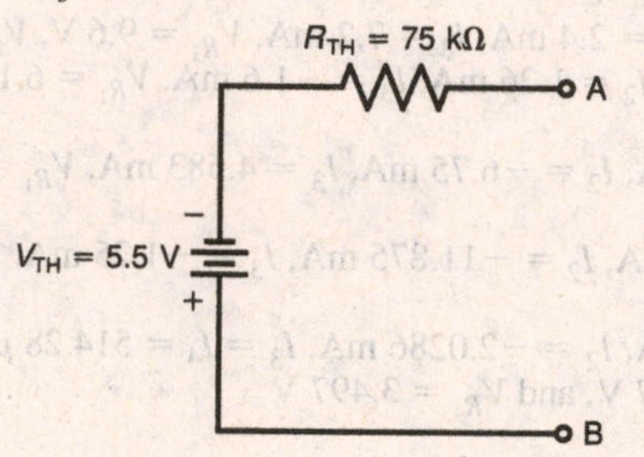

Fig. 10-39. $I_1 = 1.09375$ mA, $I_2 = 2.03125$ mA, $I_3 = 3.125$ mA, $V_{R_1} = 2.1875$ V, $V_{R_2} =$ 12.1875 V, and $V_{R_3} = 7.8125$ V

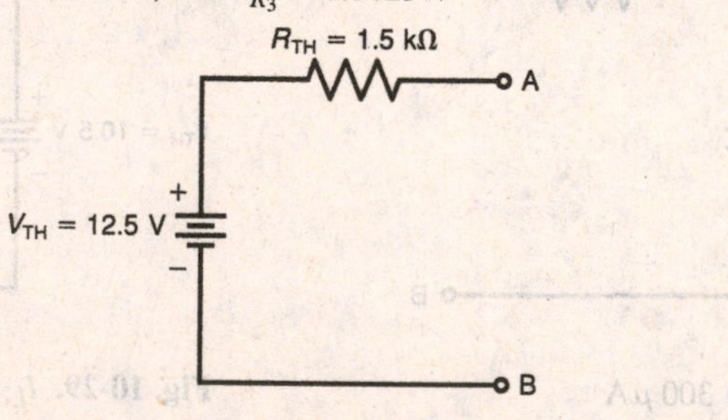

Fig. 10-41. $I_1 = 10.03$ mA, $I_2 = 12.1$ mA, $I_3 = 2.072$ mA, $V_{R_1} = 33.1$ V, $V_{R_2} = 56.9$ V, and $V_{R_3} = 3.1$ V

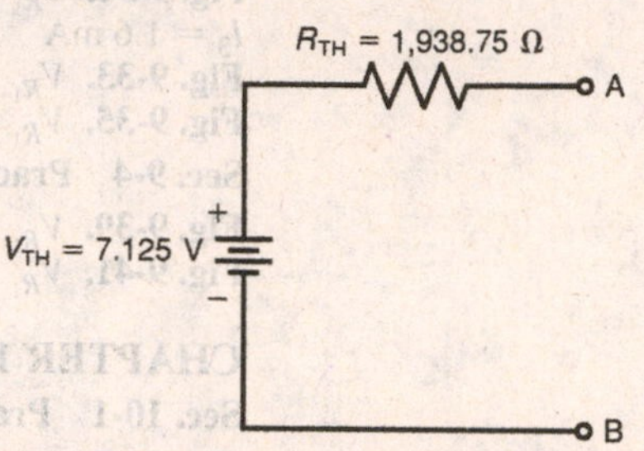

Fig. 10-43. $I_1 = 11.40625$ mA, $I_2 =$ 4.53125 mA, $I_3 = 6.875$ mA, $V_{R_1} = 11.40625$ V, $V_{R_2} = 13.59375$ V, and $V_{R_3} = 8.59375$ V

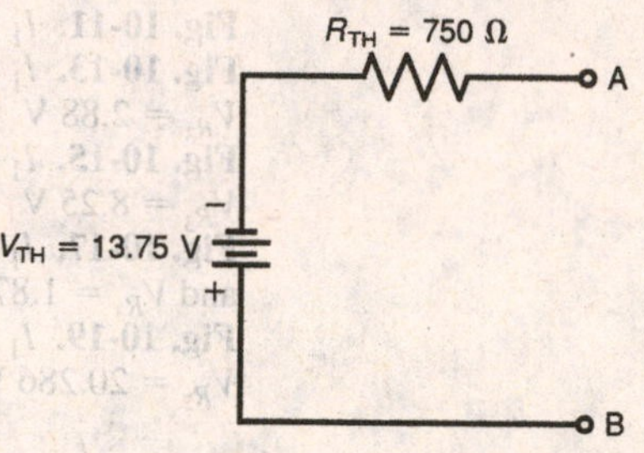

Sec. 10-2 Even More Practice Problems

Fig. 10-45. $I_L = 223.28\ \mu A$ and $V_{R_L} = 491.23$ mV
A ← B

Fig. 10-47. $I_L = 0$ mA $V_{R_L} = 0$ V

Fig. 10-49. $I_L = 2.42$ mA and $V_{R_L} = 799.4$ mV
A ← B

Fig. 10-51. $I_L = 2$ mA and $V_{R_L} = 1.12$ V
A → B

Fig. 10-53. $I_L = 5.99$ mA and $V_{R_L} = 0.719$ V
A ← B

Fig. 10-55. $I_L = 5$ mA and $V_{R_L} = 5$ V
A ← B

Sec. 10-3. Practice Problems

Fig. 10-61. $I_L = 1$ mA

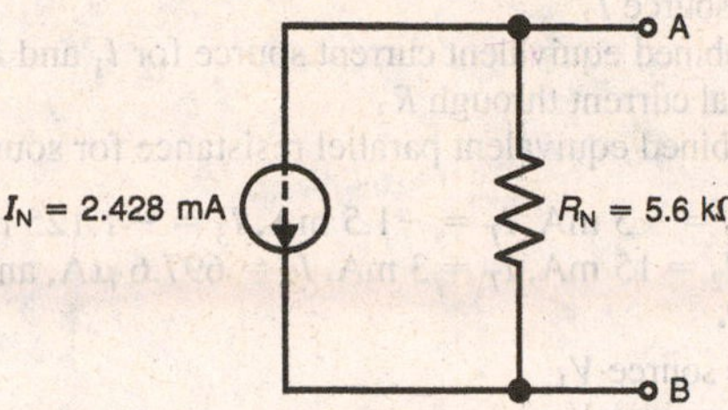

Fig. 10-63. $I_L = 4.5$ mA

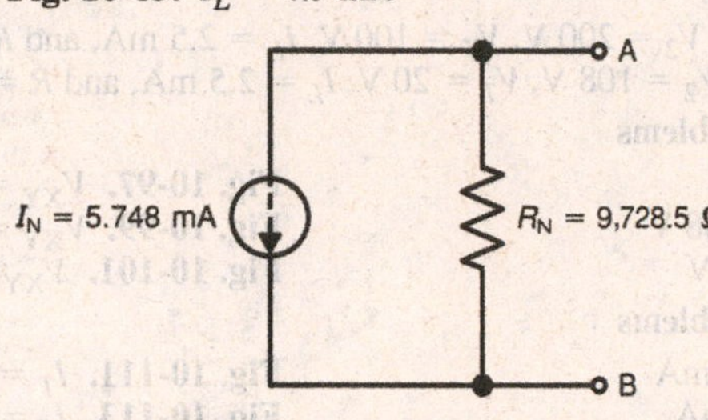

Fig. 10-65. $I_L = 1.06$ mA

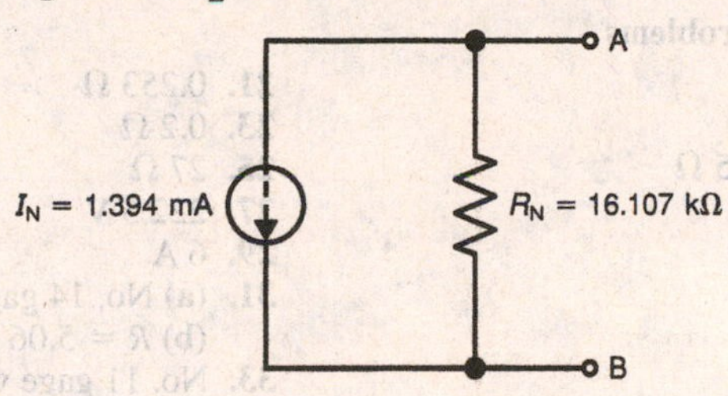

Fig. 10-67. $I_L = 150$ mA

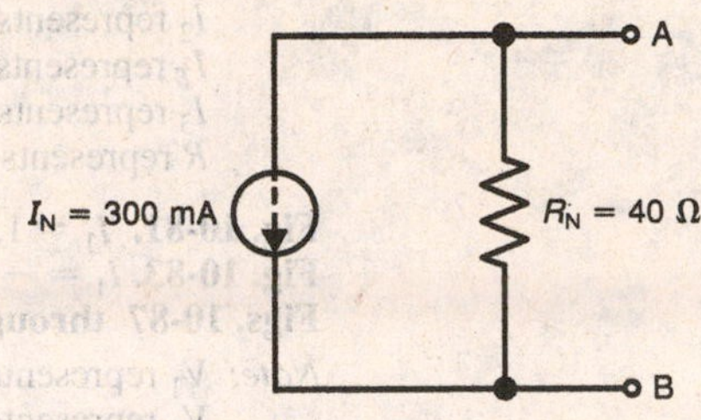

Fig. 10-69. $I_3 = 500\ \mu$A

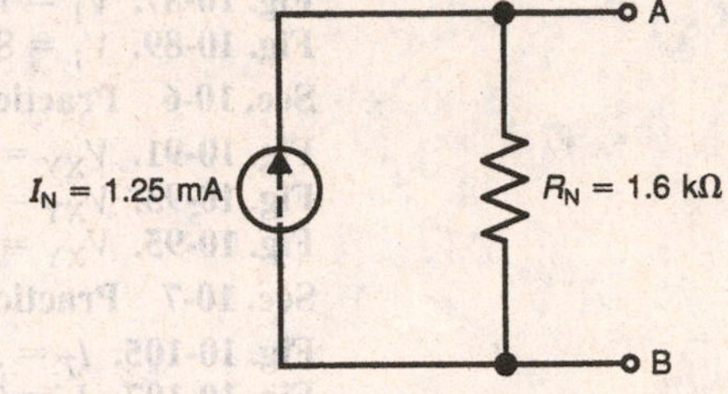

Sec. 10-4. Practice Problems

Fig. 10-73.

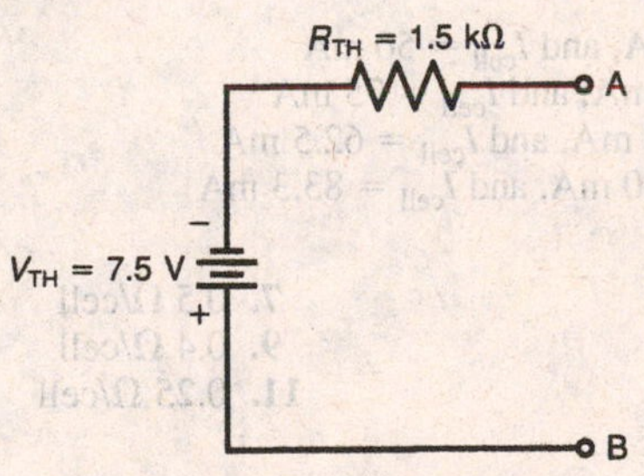

Fig. 10-75.

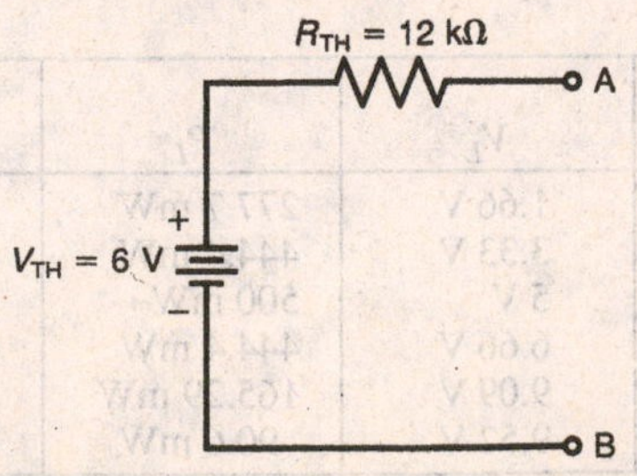

Fig. 10-77.

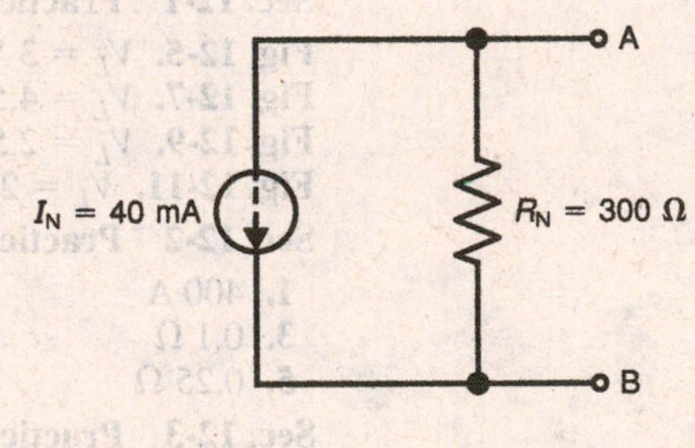

Fig. 10-79.

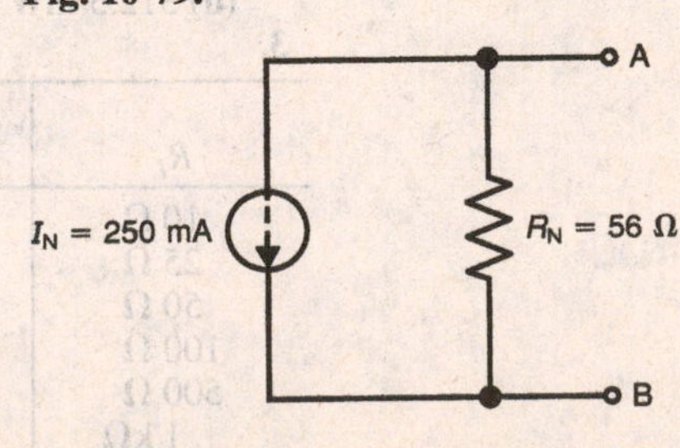

Sec. 10-5 Practice Problems

Note: I_1 represents current source I_1
I_2 represents current source I_2
I_T represents the combined equivalent current source for I_1 and I_2
I_3 represents the actual current through R_3
R represents the combined equivalent parallel resistance for sources I_1 and I_2

Fig. 10-81. $I_1 = 1.5$ mA, $I_2 = -3$ mA, $I_T = -1.5$ mA, $I_3 = -1.125$ mA, and $R = 6$ kΩ
Fig. 10-83. $I_1 = -12$ mA, $I_2 = 15$ mA, $I_T = 3$ mA, $I_3 = 697.6$ μA, and $R = 545.45$ Ω
Figs. 10-87 through 10-89.

Note: V_1 represents voltage source V_1
V_2 represents voltage source V_2
V_T represents the combined equivalent voltage source for V_1 and V_2
I_L represents actual load current
R represents the combined series resistance for sources V_1 and V_2

Fig. 10-87. $V_1 = 100$ V, $V_2 = 200$ V, $V_T = 100$ V, $I_L = 2.5$ mA, and $R = 25$ kΩ
Fig. 10-89. $V_1 = 88$ V, $V_2 = 108$ V, $V_T = 20$ V, $I_L = 2.5$ mA, and $R = 4$ Ω

Sec. 10-6 Practice Problems

Fig. 10-91. $V_{XY} = 5$ V
Fig. 10-93. $V_{XY} = 17.368$ V
Fig. 10-95. $V_{XY} = 6.92$ V
Fig. 10-97. $V_{XY} = -3.85$ V
Fig. 10-99. $V_{XY} = -1.5$ V
Fig. 10-101. $V_{XY} = 8.62$ V

Sec. 10-7 Practice Problems

Fig. 10-105. $I_T = 3.078$ mA
Fig. 10-107. $I_T = 3.75$ mA
Fig. 10-109. $I_T = 5.33$ mA
Fig. 10-111. $I_T = 84.67$ mA
Fig. 10-113. $I_T = 2.7$ mA
Fig. 10-115. $I_T = 6.97$ mA

CHAPTER 11 Conductors and Insulators

Sec. 11-1 Practice Problems

1. 2.53 Ω
3. 24.35 Ω
5. (a) 0.75 Ω, (b) 7.65 Ω
7. 7,898 ft
9. 9,711.7 ft
11. 17.475 Ω
13. 7.57 Ω
15. 53.48 Ω
17. 1,483.68 Ω
19. 49,240.9 ft
21. 0.253 Ω
23. 0.2 Ω
25. 27 Ω
27. 222.5 V
29. 6 A
31. (a) No. 14 gage (approximately), (b) $R = 5.06$ Ω, (c) $R_{strand} = 1048.9$ Ω
33. No. 11 gage wire (calculations do not produce exact answers).
35. $V_{load} = 117$ V AC

CHAPTER 12 Batteries

Sec. 12-1 Practice Problems

Fig. 12-5. $V_L = 3$ V, $I_L = 50$ mA, and $I_{cell} = 50$ mA
Fig. 12-7. $V_L = 4.5$ V, $I_L = 75$ mA, and $I_{cell} = 75$ mA
Fig. 12-9. $V_L = 2.5$ V, $I_L = 125$ mA, and $I_{cell} = 62.5$ mA
Fig. 12-11. $V_L = 2.5$ V, $I_L = 250$ mA, and $I_{cell} = 83.3$ mA

Sec. 12-2 Practice Problems

1. 400 A
3. 0.1 Ω
5. 0.25 Ω
7. 0.5 Ω/cell
9. 0.4 Ω/cell
11. 0.25 Ω/cell

Sec. 12-3 Practice Problems

1. (a) 1 Ω
(b) 5 Ω
(c) 375 mW
(d) 312.5 mW
(e) 83.3%
(f) 1 Ω
(g) 562.5 mW
(h) 50%

3.

R_L	I_L	V_L	P_L	P_T	(% Eff.) P_L/P_T
10 Ω	166.6 mA	1.66 V	277.7 mW	1.66 W	16.6%
25 Ω	133.3 mA	3.33 V	444.4 mW	1.33 W	33.3%
50 Ω	100 mA	5 V	500 mW	1 W	50%
100 Ω	66.6 mA	6.66 V	444.4 mW	666.6 mW	66.6%
500 Ω	18.18 mA	9.09 V	165.29 mW	181.81 mW	90.9%
1 kΩ	9.52 mA	9.52 V	90.6 mW	95.2 mW	95.1%

5. (a) 0.3 Ω
(b) 3.75 V
(c) 11.25 W
(d) 9.375 W
(e) 83.3%
(f) 0.9 W
7. (a) 2.52 V
(b) 56%
9. $R \times 1$ range. This range draws the largest current from the battery. This range, therefore, will lower the terminal voltage more than the higher ranges.

11.

R_L	I_L
0 Ω	1 mA
1 Ω	1 mA
10 Ω	999.9 μA
100 Ω	999 μA
1 kΩ	990 μA
10 kΩ	909 μA

13.

R_L	V_L
0.05 Ω	13.63 V
0.1 Ω	14.28 V
1 Ω	14.92 V
5 Ω	14.98 V
10 Ω	14.99 V
100 Ω	14.999 V
500 Ω	15 V
1 kΩ	15 V
5 kΩ	15 V
10 kΩ	15 V

CHAPTER 13 Magnetism

Sec. 13-1 Practice Problems

1. (a) The symbol for magnetic flux is ϕ.
(b) The symbol for flux density is B.
3. (a) 1 Mx = 1 Magnetic Field Line
(b) 1 Wb = 1×10^8 Mx
5. Relative permeability, μ_r, is the ratio of the flux density, B, in a material such as iron to the flux density, B, in air.
7. 2×10^{-4} Wb
9. 0.25 T
11. 1.5×10^{-5} Wb
13. 15 T
15. 2.5×10^7 Mx
17. 20 G
19. 5×10^5 Mx
21. 1.5 T
23. 1×10^{-2} T
25. 1×10^5 Mx
27. 2×10^5 lines
29. 4×10^{-4} Wb
31. 0.28 T
33. 0.016 T
35. $\mu_r = 480$
37. 10,000 μWb

CHAPTER 14 Electromagnetic

Sec. 14-1 Practice Problems

1. 1.26×10^3 Gb
3. 5,000 A·t
5. 3,000 A·t
7. 40,000 A·t
9. 6.35 A·t
11. 50 A·t
13. 24 mA
15. (a) 1,125 A·t and
(b) 1,417.5 Gb
17. 3.33 A
19. 30 mA

Sec. 14-2 Practice Problems

1. 5.04 Oe
3. 15.75 Oe
5. 11.34 Oe
7. 20,000 A·t/m
9. 55×10^3 A·t/m
11. 1,000 A·t/m

Sec. 14-3 Practice Problems

1. $3.15 \times 10^{-3} \dfrac{\text{T}}{\text{A·t/m}}$
3. 30,000
5. 9.45×10^{-3} T
7. 0.63 T
9. 0.0756 T
11. 1,000

Sec. 14-4 Practice Problems

1. 1.5 μWb
3. 2,500 A·t
5. 1×10^{-2} Wb
7. 2.67×10^{-3} Wb
9. 6×10^6 A·t/Wb

Sec. 14-5 Practice Problems

1. 2 V
3. 18 V
5. 2.4 Wb/s
7. 0 V
9. 5,000 turns
11. Any induced voltage has the polarity that opposes the change causing the induction.

CHAPTER 15 Alternating Voltage and Current

Sec. 15-1 Practice Problems

1. π rad/12 *or* 0.261 rad
3. 85.94°
5. 7.76 rad
7. 83°
9. 3.66 rad
11. 28.64°
13. 4π rad *or* 12.56 rad
15. 2.09 rad
17. 180°
19. 90°
21. 360°
23. **(a)** Amount of flux
(b) Time rate of cutting
(c) The angle at which the conductors cut across the flux
25. 270°

Sec. 15-2 Practice Problems

1. **(a)** 25 V
(b) 43.3 V
(c) 35.35 V
(d) 0 V
(e) −25 V
(f) −46.98 V
(g) −43.3 V
(h) −25 V

3. **(a)** 170 V
(b) 85 V
(c) 0 V
(d) −147.2 V
(e) −170 V
(f) −147.2 V
(g) −120.2 V
(h) 120.2 V

5. 250 V
7. 60 V
9. 16.22 V
11. 70.7 V

Sec. 15-3 Practice Problems

Fig. 15-13. $I_{p\text{-}p}$ = 10 mA, I_p = 5 mA, I_{rms} = 3.535 mA, I_{av} = 3.185 mA, V_p = 5 V, V_{rms} = 3.535 V, V_{av} = 3.185 V, and P_1 = 12.49 mW
Fig. 15-15. $I_{p\text{-}p}$ = 113.12 mA, I_p = 56.56 mA, I_{rms} = 40 mA, I_{av} = 36 mA, $V_{p\text{-}p}$ = 56.56 V, V_{rms} = 20 V, V_{av} = 18 V, and P_1 = 800 mW
Fig. 15-17. V_p = 80 V, V_{rms} = 56.56 V, V_{av} = 50.96 V, and $V_{p\text{-}p}$ = 160 V

Sec. 15-3 Word Problems

1. **(a)** 17.67 V rms; **(b)** 707 mA rms, **(c)** 44 V rms, and **(d)** 7.07 V rms
3. **(a)** 21.21 V peak, **(b)** 314 mV peak, **(c)** 1 kV peak, and **(d)** 169.68 peak
5. **(a)** 1.27 V av, **(b)** 72 μV av, **(c)** 17.49 V av, and **(d)** 27 V av
7. **(a)** 282.8 μA p-p, **(b)** 141.3 V p-p, **(c)** 44 V p-p, and **(d)** 6.28 mA p-p
9. **(a)** 135 V av, **(b)** 63.6 V av, **(c)** 28.66 V av, and **(d)** 56 V av

Sec. 15-4 Word Problems

1. **(a)** 1 ms,
(b) 500 μs,
(c) 100 μs, and
(d) 4 μs
3. **(a)** 75 m and
(b) 0.25 μs
5. **(a)** 1.13 ft and
(b) 1 ms
7. **(a)** 2.5 kHz and
(b) 400 μs
9. **(a)** 4.52 ft and
(b) 250 Hz

Sec. 15-4 Practice Problems

Fig. 15-23. period = 2 μs, f = 500 kHz, V_{av} = 38.22 V, and V @ 60° = 51.96 V, λ = 600 m
Fig. 15-25. period = 400 μs, f = 2.5 kHz, V_{rms} = 8.48 mV, and V @ 240° = −6 mV, λ = 120 km

Sec. 15-5 Practice Problems

1. **(a)** 1.66 μs, **(b)** 2.5 μs, **(c)** 5 μs, **(d)** 8.33 μs, and **(e)** 11.6 μs
3. **(a)** 277.7 μs, **(b)** 1.66 ms, **(c)** 3.33 ms, **(d)** 5.55 ms, and **(e)** 8.3 ms
5. **(a)** 45° and **(b)**

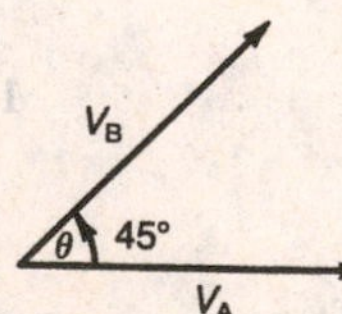

Sec. 15-6 Practice Problems

Fig. 15-31. *prf* = 2.5 kHz, % duty cycle = 50%; and V_{av} = 20 V
Fig. 15-33. *prf* = 10 kHz, % duty cycle = 25%, and V_{av} = −1.25 V
Fig. 15-35. *prf* = 1 kHz, % duty cycle = 30%, and V_{av} = −4 V
Fig. 15-37. *prf* = 100 Hz, % duty cycle = 25%, and V_{av} = −5 V

Sec. 15-6 Word Problems

1. *prf* = 20 kHz,
prt = 50 μs
3. *prf* = 83.3 Hz,
prt = 12 ms
5. *tp* = 30 μs
7. Pulse train

Sec. 15-7 Practice Problems

1. 25 kHz
3. 40 kHz
5. 440 Hz
7. 111.1 m
9. 25 kHz
11. 20 kHz
13. 4 kHz
15. 250 kHz

Sec. 15-8 Practice Problems

Fig. 15-45. $V_{p\text{-}p} = 0.6$ V pp $T = 500$ μs $f = 2$ kHz
Fig. 15-47. $V_{p\text{-}p} = 5$ V pp $T = 80$ μs $f = 12.5$ kHz
Fig. 15-49. $V_{p\text{-}p} = 8$ V pp $T = 680$ μs $f = 1.47$ kHz
Fig. 15-51. $V_{p\text{-}p} = 6.8$ V pp $T = 50$ μs $f = 20$ kHz
Fig. 15-53. $V_{p\text{-}p} = 1.08$ V pp $T = 4$ μs $f = 250$ kHz
Fig. 15-55. $t_p = 1.48$ ms $prt = 2$ ms %Duty Cycle = 74% $prf = 500$ Hz $V_{pk} = +2.5$ V
Fig. 15-57. $t_p = 0.4$ ms $prt = 1$ ms %Duty Cycle = 40% $prf = 1$ kHz $V_{pk} = -1.25$ V
Fig. 15-59. $t_p = 0.3$ ms $prt = 0.5$ ms %Duty Cycle = 60% $prf = 2$ kHz $V_{pk} = -1.65$ V
Fig. 15-61. $t_p = 130$ μs $prt = 300$ μs %Duty Cycle = 43.3% $prf = 3.33$ kHz $V_{pk} = 13$ mV
Fig. 15-63. $t_p = 17.5$ ms $prt = 50$ ms %Duty Cycle = 35% $prf = 20$ Hz $V_{pk} = 0.9$ V

CHAPTER 16 Capacitance

Sec. 16-1 Practice Problems

1. $Q = 2{,}000$ μC
3. $C = 500$ μF
5. $V = 50$ V
7. $Q = 15$ mC
9. $V = 40$ V
11. $C = 4.425$ nF
13. $C = 141.6$ pF
15. Breakdown voltage rating
17. $V = 160$ V

Sec. 16-2 Practice Problems

1. $C_T = 40$ μF
3. $C_T = 0.16$ μF
5. $C_T = 5$ μF
7. $C_T = 4$ nF
9. $C_1 = 53.3$ nF, $C_2 = 160$ nF
11. $V = 20$ V

Sec. 16-3 Practice Problems

Fig. 16-5. $Q = 15$ μC, $V_{C_1} = 30$ V, $V_{C_2} = 10$ V, $C_T = 75$ nF, and $V_{C_3} = 160$ V
Fig. 16-7. $Q = 11.52$ μC, $V_{C_1} = 28.8$ V, $V_{C_2} = 19.2$ V, $V_{C_3} = 192$ V, $C_T = 40$ nF, and $V_T = 288$ V
Fig. 16-9. $Q_T = 18$ μC, $V_{C_1} = 15$ V, $V_{C_2} = 45$ V, $V_{C_3} = 60$ V, $V_{C_4} = 5.625$ V, $V_{C_5} = 16.875$ V, $V_{C_6} = 22.5$ V, and $C_T = 0.15$ μF
Fig. 16-11. $Q_T = 13.5$ μC, $V_{C_1} = 120$ V, $V_{C_2} = 75$ V, $V_{C_3} = 7.5$ V, $V_{C_4} = 7.5$ V, $V_{C_5} = 30$ V, $V_{C_6} = 5$ V, $V_{C_8} = 5$ V, $C_T = 0.1125$ μF, and $V_T = 120$ V

Sec. 16-4 Practice Problems

Fig. 16-17. $C = 5{,}600$ pF $\pm 10\%$
Fig. 16-19. $C = 1.5$ pF ± 0.25 pF
Fig. 16-21. $C = 0.001$ μF, $+100\%$, -0%
Z7 T means that C will vary $+22\%$, -33% over the temperature range of $+10$°C to $+125$°C
Fig. 16-23. $C = 10{,}000$ pF $\pm 5\%$
Fig. 16-25. $C = 0.68$ pF ± 0.1 pF
Fig. 16-27. $C = 8{,}200$ pF $\pm 20\%$
Fig. 16-29. $C = 390$ pF $\pm 10\%$
Z5 U means that C will vary $+22\%$, -56% over the temperature range of $+10$°C to $+85$°C
Fig. 16-31. $C = 12{,}000$ pF $\pm 20\%$
Fig. 16-33. $C = 0.56$ pF ± 0.1 pF
Fig. 16-35. $C = 390$ pF $\pm 20\%$
Fig. 16-37. $C = 270{,}000$ pF $\pm 10\%$
Fig. 16-39. $C = 33{,}000$ pF $\pm 10\%$
Z6 S means that C will vary $\pm 22\%$ over the temperature range of $+10$°C to $+105$°C

CHAPTER 17 Capacitive Reactance

Sec. 17-1 Practice Problems

Fig. 17-3. $X_C = 2$ kΩ, $I = 5$ mA
Fig. 17-5. $C = 7.957$ nF, $I = 10$ mA
Fig. 17-7. $C = 636.6$ pF, $I = 24$ mA
Fig. 17-9. $f = 795.7$ Hz, $I = 2$ mA
Fig. 17-11. $X_C = 500$ kΩ, $C = 636.6$ pF
Fig. 17-13. $V_T = 1$ V, $X_C = 636.6$ Ω
Fig. 17-15. $f = 193$ Hz, $I = 600$ μA
Fig. 17-17. $C = 265.2$ nF, $I = 60$ mA

Sec. 17-1 Word Problems

1. (a) $X_C = 10\ k\Omega$, (b) $X_C = 2\ k\Omega$, and (c) $X_C = 1\ k\Omega$
3. Infinite ohms

Sec. 17-2 Practice Problems

Fig. 17-19. $C_T = 800$ pF, $X_{C_1} = 1.25\ k\Omega$, $X_{C_2} = 5\ k\Omega$, $X_{C_T} = 6.25\ k\Omega$, $I_T = 4$ mA, $V_{C_1} = 5$ V, and $V_{C_2} = 20$ V

Fig. 17-21. $C_T = 25$ nF, $X_{C_1} = 4\ k\Omega$, $X_{C_2} = 1\ k\Omega$, $X_{C_T} = 800\ \Omega$, $I_{C_1} = 6$ mA, $I_{C_2} = 24$ mA, and $I_T = 30$ mA

Fig. 17-23. $C_T = 14.285$ nF, $X_{C_1} = 800\ \Omega$, $X_{C_2} = 400\ \Omega$, $X_{C_3} = 200\ \Omega$, $X_{C_T} = 1.4\ k\Omega$, $I_T = 15$ mA, $V_{C_1} = 12$ V, $V_{C_2} = 6$ V, and $V_{C_3} = 3$ V

Fig. 17-25. $C_T = 0.01\ \mu F$, $X_{C_1} = 1\ k\Omega$, $X_{C_2} = 5\ k\Omega$, $X_{C_3} = 1.25\ k\Omega$, $X_{C_T} = 500\ \Omega$, $I_{C_1} = 20$ mA, $I_{C_2} = 4$ mA, $I_{C_3} = 16$ mA, and $I_T = 40$ mA

Fig. 17-27. $C_2 = 0.05\ \mu F$, $C_T = 0.015\ \mu F$, $X_{C_1} = 1.25\ k\Omega$, $X_{C_2} = 750\ \Omega$, $X_{C_3} = 500\ \Omega$, $X_{C_T} = 2.5\ k\Omega$, $I_T = 80$ mA, $V_{C_1} = 100$ V, and $V_{C_2} = 60$ V

Fig. 17-29. $C_T = 0.01\ \mu F$, $X_{C_1} = 5\ k\Omega$, $X_{C_2} = 10\ k\Omega$, $X_{C_3} = 10\ k\Omega$, $X_{C_T} = 10\ k\Omega$, $I_T = 1$ mA, $V_{C_1} = 5$ V, $V_{C_2} = 5$ V, $V_{C_3} = 5$ V, $I_2 = 500\ \mu A$, and $I_3 = 500\ \mu A$

Sec. 17-3 Word Problems

1. $i_C = 200\ \mu A$
3. $i_C = 800\ \mu A$
5. $i_C = 55$ mA
7. $C = 12.5\ \mu F$
9. 0 A

CHAPTER 18 Capacitive Circuits

Sec. 18-1 Practice Problems

Fig. 18-3. $Z_T = 5\ k\Omega$, $I_T = 3$ mA, $V_R = 9$ V, $V_C = 12$ V, and $\theta = -53.1°$

Fig. 18-5. $Z_T = 25\ k\Omega$, $I_T = 2$ mA, $V_R = 30$ V, $V_C = 40$ V, and $\theta = -53.1°$

Fig. 18-7. $Z_T = 8.94\ k\Omega$, $I_T = 2.23$ mA, $V_R = 17.88$ V, $V_C = 8.94$ V, and $\theta = -26.56°$

Fig. 18-9. $X_C = 1.989\ k\Omega$, $Z_T = 2.226\ k\Omega$, $I_T = 11.22$ mA, $V_R = 11.22$ V, $V_C = 22.33$ V, and $\theta = -63.3°$

Fig. 18-11. $X_C = 677\ \Omega$, $Z_T = 1.063\ k\Omega$, $I_T = 28.2$ mA, $V_R = 23.1$ V, $V_C = 19$ V, and $\theta = -39.58°$

Fig. 18-13. $I_T = 3$ mA, $X_C = 3\ k\Omega$, $V_R = 12$ V, $V_C = 9$ V, $f = 1061$ Hz, and $\theta = -36.9°$

Fig. 18-15. $Z_T = 10\ k\Omega$, $I_T = 750\ \mu A$, $V_C = 6$ V, $V_T = 7.5$ V, $C = 625$ pF, and $\theta = -53.1°$

Fig. 18-17. $Z_T = 4.2\ k\Omega$, $I_T = 14.28$ mA, $V_R = 38.56$ V, $V_C = 45.94$ V, $X_C = 3.217\ k\Omega$, and $C = 989.4$ pF

Sec. 18-2 Practice Problems

Fig. 18-19. $I_R = 200$ mA, $I_C = 200$ mA, $I_T = 282.8$ mA, $Z_{EQ} = 70.7\ \Omega$, and $\theta = 45°$

Fig. 18-21. $I_R = 4.5$ mA, $I_C = 4$ mA, $I_T = 6.02$ mA, $Z_{EQ} = 5.98\ k\Omega$, and $\theta = 41.63°$

Fig. 18-23. $I_R = 90\ \mu A$, $I_C = 60\ \mu A$, $I_T = 108.16\ \mu A$, $Z_{EQ} = 832\ k\Omega$, and $\theta = 33.7°$

Fig. 18-25. $X_C = 1{,}326\ \Omega$, $I_R = 146.34$ mA, $I_C = 90.5$ mA, $I_T = 172$ mA, $Z_{EQ} = 697.4\ \Omega$, and $\theta = 31.73°$

Fig. 18-27. $X_C = 5\ k\Omega$, $I_R = 500\ \mu A$, $I_C = 5$ mA, $I_T = 5.025$ mA, $Z_{EQ} = 4.975\ k\Omega$, and $\theta = 84.29°$

Fig. 18-29. $I_R = 7.937$ A, $V_T = 158.74$ V, $Z_{EQ} = 13.22\ \Omega$, $X_C = 17.63\ \Omega$, $f = 902.3$ Hz, and $\theta = 48.59°$

Fig. 18-31. $V_T = 300$ V, $I_C = 15$ mA, $I_T = 25$ mA, $Z_{EQ} = 12\ k\Omega$, $f = 159.15$ kHz, and $\theta = 36.87°$

Fig. 18-33. $I_T = 5$ mA, $I_C = 2.5$ mA, $I_R = 4.33$ mA, $X_C = 6\ k\Omega$, $R = 3.464\ k\Omega$, and $C = 265.2$ pF

CHAPTER 19 Inductance

Sec. 19-1 Practice Problems

1. $L = 62.5$ mH
3. $V_L = 20$ V
5. $V_L = 3.2$ V
7. $L = 283.5\ \mu H$
9. $L = 1.008$ mH
11. $L = 15.75\ \mu H$
13. $L = 500$ mH
15. $L = 30$ mH

Sec. 19-2 Practice Problems

Fig. 19-5. $V_{S_1} = 600$ V, $V_{S_2} = 24$ V, $I_{S_1} = 600$ mA, $I_{S_2} = 2$ A, $P_{sec\ 1} = 360$ W, $P_{sec\ 2} = 48$ W, $P_{pri} = 408$ W, and $I_p = 3.4$ A

Fig. 19-7. $V_{S_1} = 120$ V, $V_{S_2} = 96$ V, $V_{S_3} = 288$ V, $I_{S_1} = 5$ A, $I_{S_2} = 800$ mA, $I_{S_3} = 1.92$ A, $P_{sec\ 1} = 600$ W, $P_{sec\ 2} = 76.8$ W, $P_{sec\ 3} = 552.96$ W, $P_{pri} = 1{,}229.76$ W, and $I_p = 10.248$ A

Fig. 19-9. $V_{S_1} = 6$ V, $V_{S_2} = 48$ V, $I_{S_1} = 5$ A, $I_{S_2} = 12$ A, $P_{sec\ 1} = 30$ W, $P_{sec\ 2} = 576$ W, $P_{pri} = 606$ W, and $I_p = 5.05$ A

Fig. 19-11. $V_{S_1} = 240$ V, $V_{S_2} = 480$ V, $V_{S_3} = 30$ V, $I_{S_1} = 240$ mA, $I_{S_2} = 320$ mA, $I_{S_3} = 200$ mA, and $R_{L_2} = 1.5\ k\Omega$

Fig. 19-13. $V_{S_1} = 1$ kV, $V_{S_2} = 2$kV, $I_{S_1} = 400$ mA, $I_{S_2} = 100$ mA, $I_{S_3} = 6$ A, $V_{pri} = 125$ V, and $R_{L_3} = 4.167\ \Omega$

Sec. 19-2 Word Problems

1. (a) I_p = 400 mA and
   (b) I_p = 3 A
3. (a) 2.8 A,
   (b) 4.4 A, and
   (c) 3.2 A
5. L_M = 25 mH
7. k = 0.75
9. Tight coupling indicates a high value of k. Loose coupling indicates a low value of k.
11. 1 MHz

Sec. 19-3 Practice Problems

Fig. 19-17. Z_P = 1.6 kΩ
Fig. 19-19. Z_P = 40 Ω
Fig. 19-21. N_P/N_S = 4/1
Fig. 19-23. Z_S = 75 Ω
Fig. 19-25. Z_S = 2.7 kΩ
Fig. 19-27. N_S = 60 T
Fig. 19-29. Z_P = 111.1 Ω
Fig. 19-31. N_S = 250 T

1. Z_{out} = 2,048 Ω
3. 1 W (convert p-p to rms when calculating power)

Sec. 19-4 Practice Problems

1. L_T = 1.5 mH
3. L_T = 11 mH
5. L_1 = 45 μH,
   L_2 = 135 μH, and
   L_3 = 270 μH
7. L_1 = 20 μH and
   L_2 = 6 μH
9. L_T = 100 μH and
   k = 0.166

Sec. 19-5 Practice Problems

Fig. 19-39. L = 4,200 μH ±5%
Fig. 19-41. L = 0.33 μH
Fig. 19-43. L = 22,000 μH ±20%
Fig. 19-45. L = 2.7 μH
Fig. 19-47. L = 5,600 μH ±5%
Fig. 19-49. L = 1,000 μH ±10%
Fig. 19-51. L = 560 μH
Fig. 19-53. L = 12,000 μH ±5%
Fig. 19-55. L = 1.5 μH
Fig. 19-57. L = 100 μH

CHAPTER 20 Inductive Reactance

Sec. 20-1 Practice Problems

Fig. 20-3. X_L = 314.1 Ω and I = 15.91 mA
Fig. 20-5. L = 10 mH and I = 20 mA
Fig. 20-7. L = 1.75 H and I = 45.45 μA
Fig. 20-9. X_L = 1 kΩ and I = 50 mA
Fig. 20-11. X_L = 600 Ω and L = 1.9 H
Fig. 20-13. V_T = 20 V and f = 1,591.5 Hz
Fig. 20-15. X_L = 3,110 Ω and V_T = 15.55 V
Fig. 20-17. L = 200 mH and I = 18.75 mA

Sec. 20-1 Word Problems

1. (a) 1 kΩ, (b) 2 kΩ, and (c) 4 kΩ,
3. 0 Ω

Sec. 20-2 Practice Problems

Fig. 20.19. L_T = 600 mH, X_{L_1} = 2.5 kΩ, X_{L_2} = 7.5 kΩ, X_{L_T} = 10 kΩ, I_T = 6.4 mA, V_{L_1} = 16 V, and V_{L_2} = 48 V
Fig. 20-21. L_T = 24 mH, X_{L_1} = 4 kΩ, X_{L_2} = 6 kΩ, X_{L_T} = 2.4 kΩ, I_1 = 12 mA, I_2 = 8 mA, and I_T = 20 mA
Fig. 20-23. L_T = 150 mH, X_{L_1} = 250 Ω, X_{L_2} = 500 Ω, X_{L_3} = 750 Ω, X_{L_T} = 1.5 kΩ, I_T = 10 mA, V_{L_1} = 2.5 V rms, V_{L_2} = 5 V rms, and V_{L_3} = 7.5 V rms
Fig. 20-25. L_T = 6 mH, X_{L_1} = 2 kΩ, X_{L_2} = 4 kΩ, X_{L_3} = 12 kΩ, X_{L_T} = 1.2 kΩ, I_1 = 12 mA, I_2 = 6 mA, I_3 = 2 mA, and I_T = 20 mA
Fig. 20-27. L_1 = 15 mH, L_T = 100 mH, X_{L_1} = 600 Ω, X_{L_2} = 1 kΩ, X_{L_3} = 2.4 kΩ, X_{L_T} = 4 kΩ, I_T = 4.5 mA, V_{L_1} = 2.7 V, and V_{L_2} = 4.5 V
Fig. 20-29. L_T = 1 mH, X_{L_1} = 800 Ω, X_{L_2} = 16 kΩ, X_{L_3} = 4 kΩ, X_{L_T} = 4 kΩ, V_{L_1} = 4 V, V_{L_2} = 16 V, V_{L_3} = 16 V, I_2 = 1 mA, I_3 = 4 mA, and I_T = 5 mA

Sec. 20.2 Word Problems

1. (a) 5,026.5 Ω and (b) 7,539.8 Ω

CHAPTER 21 Inductive Circuits

Sec. 21-1 Practice Problems

Fig. 21-3. Z_T = 5 kΩ, I_T = 4 mA, V_R = 16 V, V_L = 12 V, and θ = 36.87°
Fig. 21-5. Z_T = 18.028 kΩ, I_T = 832 μA, V_R = 12.48 V, V_L = 8.32 V, and θ = 33.7°
Fig. 21-7. Z_T = 141.42 Ω, I_T = 141.42 mA, V_R = 14.14 V, V_L = 14.14 V, and θ = 45°
Fig. 21-9. X_L = 628.3 Ω, Z_T = 784.65 Ω, I_T = 19.1 mA, V_R = 8.98 V, V_L = 12.01 V, and θ = 53.2°
Fig. 21-11. X_L = 6.283 kΩ, Z_T = 6.396 kΩ, I_T = 15.63 mA, V_R = 18.76 V, V_L = 98.22 V, and θ = 79.1°
Fig. 21-13. I_T = 2 mA, X_L = 4.49 kΩ, V_R = 4.4 V, V_L = 8.98 V, f = 7.146 kHz, θ = 63°
Fig. 21-15. Z_T = 25 kΩ, I_T = 600 μA, V_L = 12 V, V_T = 15 V, L = 212.2 mH, and θ = 53.1°
Fig. 21-17. Z_T = 2 kΩ, I_T = 10 mA, V_R = 10 V, V_L = 17.32 V, X_L = 1.732 kΩ, and L = 27.56 mH

Sec. 21-2 Practice Problems

Fig. 21-19. $I_R = 10$ mA, $I_L = 5$ mA, $I_T = 11.18$ mA, $Z_{EQ} = 894.4\ \Omega$, and $\theta_I = -26.56°$
Fig. 21-21. $I_R = 20$ mA, $I_L = 12.5$ mA, $I_T = 23.58$ mA, $Z_{EQ} = 1.272$ kΩ, and $\theta_I = -32°$
Fig. 21-23. $I_R = 60.6$ mA, $I_L = 35.71$ mA, $I_T = 70.34$ mA, $Z_{EQ} = 284.3\ \Omega$, and $\theta_I = -30.5°$
Fig. 21-25. $I_R = 38.46$ mA, $X_L = 471.23\ \Omega$, $I_L = 31.83$ mA, $I_T = 49.92$ mA, $Z_{EQ} = 300.5\ \Omega$, and $\theta_I = -39.6°$
Fig. 21-27. $I_R = 13.88$ mA, $X_L = 1.256$ kΩ, $I_L = 19.89$ mA, $I_T = 24.26$ mA, $Z_{EQ} = 1.03$ kΩ, and $\theta_I = -55°$
Fig. 21-29. $X_L = 10$ kΩ, $I_R = 4.85$ mA, $I_L = 1.2$ mA, $Z_{EQ} = 2.4$ kΩ, $R = 2.474$ kΩ, and $\theta_I = -13.89°$
Fig. 21-31. $V_T = 20$ V, $I_T = 202$ mA, $Z_{EQ} = 99\ \Omega$, $I_L = 200$ mA, $L = 212.2\ \mu$H, and $\theta_I = -84.29°$
Fig. 21-33. $I_R = 35.77$ mA, $I_L = 17.88$ mA, $V_T = 16.815$ V, $Z_{EQ} = 420.4\ \Omega$, $X_L = 940\ \Omega$, and $L = 12.47$ mH

CHAPTER 22 *RC* and *L/R* Time Constants

Sec. 22-1 Practice Problems

1. $T = 20\ \mu$s
3. 51.88 mA
5. (a) 56.9 V,
   (b) 64.21 V,
   (c) 74.36 V,
   (d) 77.82 V, and
   (e) 87.88 V
7. (a) 60 kV,
   (b) 2,000,000 A/s
9. $L = 40$ mH
11. $L = 60$ mH

Sec. 22-2 Practice Problems

1. 150 μs
3. $i_C = 360.9\ \mu$A
5. (a) 40 V, (b) 20 V, (c) 10.54 V, (d) 2.78 V, and (e) 0.27 V (*or* 0 V)
7. $dv/dt = 266{,}666$ V/s
9. $C = 0.05\ \mu$F
11. $C = 0.2\ \mu$F

Sec. 22-3 Practice Problems

1. (a) $t = 52.7\ \mu$s
   (b) $t = 111.6\ \mu$s
   (c) $t = 346.6\ \mu$s
   (d) $t = 602\ \mu$s
   (e) $t = 1.15$ ms
3. (a) $V_C = 83$ V
   (b) $V_C = 100.3$ V
   (c) $V_C = 113.2$ V
   (d) $V_C = 136.47$ V
   (e) $V_C = 149.3$ V
5. (a) $t = 34.77\ \mu$s
   (b) $t = 133.8\ \mu$s
   (c) $t = 228.74\ \mu$s
   (d) $t = 302.38\ \mu$s
   (e) $t = 591.28\ \mu$s
7. (a) $V_C = 75.31$ V
   (b) $V_C = 86.89$ V
   (c) $V_C = 96.98$ V
   (d) $V_C = 105.39$ V
   (e) $V_C = 115.65$ V
9. $V_R = 400$ V

Sec. 22-4 Practice Problems

These waveforms are included in the Instructor's Solution Manual only.

CHAPTER 23 AC Circuits

Sec. 23-1 Practice Problems

Fig. 23-5. $X_T = 2$ kΩ, $I_T = 10$ mA, $V_L = 10$ V, $V_C = 30$ V, and $\theta = -90°$
Fig. 23-7. $I_L = 25$ mA, $I_C = 12.5$ mA, $I_T = 12.5$ mA, $X_T = 4$ kΩ, and $\theta = -90°$
Fig. 23-9. $X_T = 5$ kΩ, $I_T = 4$ mA, $V_{C_1} = 60$ V, $V_{C_2} = 40$ V, $V_{L_1} = 40$ V, $V_{L_2} = 40$ V, and $\theta = -90°$
Fig. 23-11. $I_{C_1} = 10$ mA, $I_{C_2} = 10$ mA, $I_{L_1} = 10$ mA, $I_{L_2} = 5$ mA, $I_T = 5$ mA, $X_T = 2$ kΩ, and $\theta = 90°$
Fig. 23-13. $V_{L_1} = 80$ V, $X_{C_1} = 25\ \Omega$, $X_{L_2} = 20\ \Omega$, $V_{L_2} = 80$ V, $X_T = 15\ \Omega$, and $\theta = 90°$

Sec. 23-2 Practice Problems

Fig. 23-17. $Z_T = 11.18$ kΩ, $I_T = 1.78$ mA, $V_R = 17.88$ V, $V_C = 17.88$ V, $V_L = 8.94$ V, and $\theta_V = -26.56°$
Fig. 23-19. $Z_T = 21.633$ kΩ, $I_T = 1.85$ mA, $V_R = 33.28$ V, $V_C = 37$ V, $V_L = 14.8$ V, and $\theta_V = -33.7°$
Fig. 23-21. $Z_T = 1.5$ kΩ, $I_T = 60$ mA, $V_R = 72$ V, $V_C = 108$ V, $V_L = 54$ V, and $\theta_V = -36.86°$
Fig. 23-23. $Z_T = 300\ \Omega$, $I_T = 80$ mA, $V_{C_1} = 800$ V, $V_{C_2} = 800$ V, $V_{L_1} = 400$ V, $V_{L_2} = 1.2$ kV, $V_{R_1} = 8$ V, $V_{R_2} = 16$ V, and $\theta = 0°$
Fig. 23-25. $Z_T = 30\ \Omega$, $V_R = 56.56$ V, $X_C = 10\ \Omega$, $V_L = 40$ V, $\theta = 19.47°$, and $R = 28.28\ \Omega$

Sec. 23-3 Practice Problems

Fig. 23-27. $I_L = 900$ mA, $I_C = 600$ mA, $I_R = 1.5$ A, $I_T = 1.529$ A, $Z_{EQ} = 23.53\ \Omega$, and $\theta = -11.3°$
Fig. 23-29. $I_L = 9$ mA, $I_C = 3$ mA, $I_R = 4.5$ mA, $I_T = 7.5$ mA, $Z_{EQ} = 3.6$ kΩ, and $\theta = -53.13°$
Fig. 23-31. $I_L = 20$ mA, $I_C = 30$ mA, $I_R = 300$ mA, $I_T = 300.1$ mA, $Z_{EQ} = 99.95\ \Omega$, and $\theta = 1.9°$
Fig. 23-33. $I_{L_1} = 1$ A, $I_{L_2} = 4$ A, $I_{C_1} = 4$ A, $I_{C_2} = 2$ A, $I_{R_1} = 2$ A, $I_{R_2} = 4$ A, $I_T = 6.083$ A, $Z_{EQ} = 3.288\ \Omega$, and $\theta = 9.46°$
Fig. 23-35. $V_T = 40$ V, $Z_{EQ} = 533.3\ \Omega$, $I_C = 100$ mA, $X_C = 400\ \Omega$, $\theta = 53.13°$, and $R = 888.8\ \Omega$

Sec. 23-4 Practice Problems

1. Real power = 0 W, apparent power = 200 mVA, and PF = 0
3. Real power = 0 W, apparent power = 625 mVA, and PF = 0
5. Real power = 32 mW, apparent power = 35.778 mVA, and PF = 0.894
7. Real power = 113.12 W, apparent power = 120 VA, and PF = 0.942
9. Real power = 200 mW, apparent power = 377.2 mVA, and PF = 0.53

CHAPTER 24 Complex Numbers for AC Circuits

Sec. 24-1 Practice Problems

1. 7.2 A $\angle 33.7°$
3. 15.62 V $\angle -50.2°$
5. 20.61 V $\angle 14°$
7. 25 V $\angle 53.1°$
9. 14.422 kΩ, $\angle 56.3°$
11. 9.397 A + j3.42 A
13. 0 V − j15 V
15. 1.414 kΩ − j1.414 kΩ
17. 17.101 kΩ + j46.984 kΩ
19. 0 V + j120 V

Sec. 24-2 Practice Problems

Fig. 24-7. $Z_T = 509.9\ \Omega \angle -64.44°$, $I_T = 98.05$ mA $\angle +64.44°$, $V_R = 21.57$ V $\angle 64.44°$, $V_C = 66.67$ V $\angle -25.56°$, and $V_L = 21.57$ V $\angle 154.44°$
Fig. 24-9. $Z_T = 5$ kΩ $\angle -53.13°$, $I_T = 5$ mA $\angle +53.13°$, $V_R = 15$ V $\angle +53.13°$, $V_C = 60$ V $\angle -36.87°$, and $V_L = 40$ V $\angle 143.13°$
Fig. 24-11. $Z_T = 70.7\ \Omega$, $\angle -45°$, $I_T = 707$ mA $\angle +45°$, $V_R = 35.36$ V $\angle +45°$, $V_{C_1} = 35.36$ V $\angle -45°$, $V_{C_2} = 70.7$ V $\angle -45°$, and $V_L = 70.7$ V $\angle 135°$
Fig. 24-13. $Z_T = 5.59$ kΩ, $\angle -26.56°$, $Z_T = 5$ kΩ − j2.5 kΩ
Fig. 24-15. $Z_T = 16.766\ \Omega \angle 88.16°$, $Z_T = 0.538\ \Omega + j16.757\ \Omega$
Fig. 24-17. $Z_T = 7.5$ kΩ $\angle -53.13°$, $Z_T = 4.5$ kΩ − j6 kΩ
Fig. 24-19. $Z_T = 282\ \Omega \angle -1.4°$, $Z_T = 281.9\ \Omega - j6.89\ \Omega$
Fig. 24-21. $Z_T = 7.069\ \Omega \angle 22.39°$, $Z_T = 6.536\ \Omega - j2.69\ \Omega$

Sec. 24-3 Practice Problems

Fig. 24-23. $Z_T = 28.77\ \Omega \angle +7.3°$, $I_T = 1.7375$ A $\angle -7.3°$, $I_1 = 1.387$ A $\angle +33.69°$, $I_2 = 400$ mA $\angle -36.87°$, and $I_3 = 790.6$ mA $\angle -71.56°$
Fig. 24-25. $Z_T = 19.64\ \Omega \angle -.9°$, $I_T = 2.036$ A $\angle .9°$, $I_1 = 1.94$ A $\angle -14.03°$, $I_2 = 784.46$ mA $\angle +78.69°$, and $I_3 = 266.6$ mA $\angle -90°$
Fig. 24-27. $Z_T = 4632\ \Omega \angle -27.1°$, $I_T = 5.397$ mA $\angle +27.1°$, $V_{R_1} = 4.03$ V $\angle 8°$, $V_{R_2} = 3.7$ V $\angle 66.8°$, $V_{R_3} = 17.81$ V $\angle 27.1°$, $V_{C_1} = 2.472$ V $\angle -23.2°$, $V_{C_2} = 11.87$ V $\angle -62.9°$, $V_{L_1} = 1.894$ V $\angle 98°$, $V_{X_G} = 4.46$ V $\angle 33.1°$, $I_A = 4.03$ mA $\angle 8°$, and $I_B = 2.06$ mA $\angle 66.8°$

CHAPTER 25 Resonance

Sec. 25-1 Practice Problems

1. $f_r = 10$ MHz, BW = 400 kHz, $f_1 = 9.8$ MHz, $f_2 = 10.2$ MHz. At f_r, we have $Z_T = 12.73\ \Omega$, $I_T = 7.855$ mA, $X_C = X_L = 318.3\ \Omega$, $V_C = V_L = 2.5$ V, $Q = 25$, $\theta = 0°$, PF = 1, real power = 785.5 μW, and $V_{r_i} = 100$ mV
3. Capacitive, $\theta = -45°$
5. $f_r = 5$ MHz, BW = 400 kHz, $f_1 = 4.8$ MHz, $f_2 = 5.2$ MHz. At f_r, we have $Z_T = 12.73\ \Omega$, $I_T = 7.855$ mA, $X_C = X_L = 159.15\ \Omega$, $V_C = V_L = 1.25$ V, $Q = 12.5$, $\theta = 0°$, and V_{r_i} 100 mV

Fig. 25-5. $f_r = 3.183$ MHz, BW = 31.83 kHz, $f_1 = 3.167$ MHz, $f_2 = 3.198$ MHz. At f_r, we have $X_C = X_L = 1$ kΩ, $Z_T = 10\ \Omega$, $I_T = 500$ mA, $V_C = V_L = 500$ V, $Q = 100$, $\theta = 0°$, PF = 1, real power = 2.5 W, and $V_{r_i} = 5$ V
Fig. 25-7. $f_r = 2.5$ MHz, BW = 20 kHz, $f_1 = 2.49$ MHz, $f_2 = 2.51$ MHz. At f_r, we have $X_C = X_L = 392.7\ \Omega$, $Z_T = 3.141\ \Omega$, $I_T = 47.75$ mA, $V_C = V_L = 18.75$ V, $Q = 125$, $\theta = 0°$, PF = 1, real power = 7.163 mW, and $V_{r_i} = 150$ mV

Sec. 25-2 Practice Problems

1. $f_r = 1.25$ MHz, BW = 25 kHz, $f_1 = 1.2375$ MHz, $f_2 = 1.2625$ MHz. At f_r, we have $X_C = X_L = 636.6\ \Omega$, $I_C = I_L = 31.41$ mA, $Q = 50$, $Z_T = 31.83$ kΩ, $I_T = 628.3$ μA, $\theta = 0°$, PF = 1, and real power = 12.56 mW
3. Inductive, $\theta = -45°$
5. $f_r = 625$ kHz, BW = 25 kHz, $f_1 = 612.5$ kHz, $f_2 = 637.5$ kHz. At f_r, we have $X_C = X_L = 318.3\ \Omega$, $I_C = I_L = 62.83$ mA, $Q = 25$, $Z_T = 7.957$ kΩ, $I_T = 2.5$ mA, $\theta = 0°$, PF = 1, and real power = 50.26 mW

Fig. 25-13. f_r = 4 MHz, BW = 20 kHz, f_1 = 3.99 MHz, f_2 = 4.01 MHz. At f_r, we have $X_C = X_L$ = 1005 Ω, $I_C = I_L$ = 9.947 mA, Q = 200, Z_T = 201 kΩ, I_T = 49.75 μA, θ = 0°, PF = 1, and real power = 497.5 μW

Fig. 25-15. f_r = 300 kHz, BW = 10 kHz, f_1 = 295 kHz, f_2 = 305 kHz. At f_r, we have $X_C = X_L$ = 1768 Ω, $I_C = I_L$ = 56.56 mA, Q = 30, Z_T = 53.04 kΩ, I_T = 1.885 mA, θ = 0°, PF = 1, and real power = 188.5 mW

Sec. 25-2 Word Problems

1. f_r = 4.5 MHz
3. f_r = 21.4 MHz
5. C = 767.5 pF
7. L = 22.12 μH
9. f_1 = 4.9875 MHz, f_2 = 5.0125 MHz
11. $Q = R_P/X_L$ = 30
13. C_{lowest} = 347.4 pF, $C_{highest}$ = 39.57 pF
15. f_1 = 11.9375 MHz, f_2 = 12.0625 MHz

CHAPTER 26 Filters

Sec. 26-1 Practice Problems

1. 1.808 kHz
3. −45°
5. (a) −3.19°, (b) −7.9°, (c) −15.47°, and (d) −22.55°
7. (a) −70.1°, (b) −79.7°, (c) −87.93°, (d) −88.96°, and (e) −89.58°
9. (a) Increase, (b) decrease
11. f_C = 7.957 kHz
13. θ = −45°
15. (a) 4.55 V p-p, (b) 2.35 V p-p, (c) 1.189 V p-p, (d) 0.238 V p-p and (e) 0.119 V p-p
17. (a) −72.3°, (b) −80.98°, (c) −85.45°, (d) −89°, and (e) −89.5°
19. (a) Increase, (b) decrease
21. 5 V p-p

Sec. 26-2 Practice Problems

1. 568 Hz
3. (a) 0 V, (b) 15 V
5. 30 V p-p
7. (a) 29.8 V p-p, (b) 29.95 V p-p, (c) 29.99 V, and (d) 29.99 V
9. θ is approximately 0° in passband θ is approximately 90° in stop band
11. (a) Increase, (b) decrease
13. 3.535 V p-p
15. +45°
17. (a) 26.1 mV p-p, (b) 130 mV p-p, (e) 520 mV p-p, and (d) 2.32 V p-p
19. (a) Decrease R, (b) increase R

Sec. 26-3. Practice Problems

Fig. 26-11. Bandpass
Fig. 26-13. High-pass
Fig. 26-15. Bandstop
Fig. 26-17. Low-pass
Fig. 26-19. High-pass
Fig. 26-21. High-pass

ANSWERS TO END OF CHAPTER TESTS

CHAPTER I Introduction to Powers of 10

1. T
2. T
3. T
4. F
5. T
6. T
7. T
8. T
9. T
10. F
11. F
12. T
13. T
14. T
15. T

CHAPTER 1 Electricity

1. T
2. T
3. T
4. F
5. T
6. T
7. F
8. T
9. T
10. T
11. T
12. T
13. T
14. T
15. F
16. F
17. T
18. T
19. T
20. T

CHAPTER 2 Resistor Color Coding

1. F
2. T
3. T
4. F
5. T
6. T
7. T
8. T
9. T
10. T
11. T
12. F
13. F
14. T
15. T

CHAPTER 3 Ohm's Law

1. T
2. T
3. T
4. T
5. T
6. F
7. F
8. T
9. T
10. F

CHAPTER 4 Series Circuits

1. T 5. F 9. T 13. F
2. T 6. T 10. F 14. T
3. T 7. T 11. F 15. T
4. T 8. T 12. T

CHAPTER 5 Parallel Circuits

1. T 5. T 9. F 13. T
2. T 6. T 10. T 14. F
3. F 7. T 11. F 15. F
4. T 8. T 12. T

CHAPTER 6 Series Parallel Circuits

1. F 4. T 7. T 9. T
2. T 5. T 8. F 10. T
3. F 6. F

CHAPTER 7 Voltage and Current Dividers

1. T 4. T 7. T 9. T
2. T 5. F 8. T 10. F
3. T 6. T

CHAPTER 8 Direct Current Meters

1. T 4. T 7. T 9. T
2. T 5. F 8. F 10. T
3. T 6. T

CHAPTER 9 Kirchhoff's Laws

1. T 4. T 7. T 9. T
2. F 5. T 8. F 10. T
3. T 6. T

CHAPTER 10 Network Theorems

1. T 4. F 7. T 9. T
2. T 5. T 8. F 10. F
3. T 6. T

CHAPTER 11 Conductors and Insulators

1. T 4. F 7. T 9. T
2. T 5. T 8. T 10. T
3. T 6. F

CHAPTER 12 Batteries

1. T 4. T 7. F 9. T
2. T 5. T 8. T 10. T
3. T 6. F

CHAPTER 13 Magnetism

1. T 4. F 7. T 9. T
2. T 5. T 8. T 10. T
3. T 6. T

CHAPTER 14 Electromagnetism

1. T 4. T 7. T 9. F
2. T 5. T 8. T 10. T
3. T 6. F

CHAPTER 15 Alternating Voltage and Current

1. T 8. T 15. T 22. T
2. T 9. F 16. F 23. F
3. T 10. T 17. F 24. T
4. T 11. T 18. T 25. T
5. T 12. T 19. T
6. T 13. T 20. T
7. F 14. T 21. T

CHAPTER 16 Capacitance

1. T
2. T
3. T
4. T
5. T
6. T
7. F
8. T
9. T
10. F

CHAPTER 17 Capacitive Reactance

1. T
2. F
3. T
4. T
5. T
6. T
7. T
8. F
9. F
10. T

CHAPTER 18 Capacitive Circuits

1. T
2. F
3. T
4. T
5. T
6. T
7. F
8. T
9. T
10. F

CHAPTER 19 Inductance

1. T
2. T
3. T
4. F
5. F
6. T
7. T
8. T
9. T
10. F
11. T
12. T
13. T
14. T
15. T

CHAPTER 20 Inductive Reactance

1. T
2. T
3. F
4. T
5. T
6. T
7. T
8. T
9. T
10. F

CHAPTER 21 Inductive Circuits

1. F
2. T
3. T
4. F
5. T
6. T
7. T
8. F
9. T
10. T

CHAPTER 22 RC and L/R Time Constants

1. T
2. T
3. T
4. T
5. T
6. F
7. T
8. F
9. T
10. T
11. F
12. T
13. T
14. F
15. T

CHAPTER 23 AC Circuits

1. F
2. T
3. T
4. T
5. T
6. T
7. T
8. F
9. T
10. T

CHAPTER 24 Complex Numbers for AC Circuits

1. T
2. T
3. T
4. F
5. T
6. T
7. F
8. T
9. T
10. T
11. T
12. T
13. T
14. F
15. T

CHAPTER 25 Resonance

1. T
2. T
3. F
4. T
5. T
6. T
7. T
8. F
9. T
10. T

CHAPTER 26 Filters

1. T
2. T
3. F
4. T
5. T
6. F
7. T
8. T
9. T
10. T
11. F
12. F
13. T
14. T
15. F